INTRODUCTION TO NONPARAMETRIC REGRESSION

INTRODUCTION TO NONPARAMETRIC REGRESSION

Kunio Takezawa

National Agricultural Research Center
National Agriculture and Bio-oriented Research Organization
Tsukuba-shi ibaraki-ken, Japan

University of Tsukuba
Graduate School of Life and Environmental Sciences
Tsukuba-shi ibaraki-ken, Japan

WILEY-INTERSCIENCE

A JOHN WILEY & SONS, INC., PUBLICATION

Library of Congress Cataloging-in-Publication Data:

Takezawa, Kunio, 1959–
    Introduction to nonparametric regression / Kunio Takezawa.
      p. cm.
    Includes bibliographical references and index.
    ISBN-13 978-0-471-74583-9 (cloth)
    ISBN-10 0-471-74583-9 (cloth)
      1. Regression analysis—Textbooks. 2. Nonparametric statistics—Textbooks. I. Title.

    QA278.2.T35 2005
    519.5'36—dc22   2005048977

Printed in the United States of America.

10 9 8 7 6 5 4 3 2 1

To all my loves

CONTENTS

PREFACE

See, I know you want it to stay "Pleasant" but there are so many things that are so much better: like Silly ... or Sexy ... or Dangerous ... or Wild ... or Brief ...
(beat)
And every one of those things is in you all the time if you just have the guts to look for them.

— "Pleasantville" (New Line Cinema) (1998) by Gary Ross

The establishment of rigorous theories and clear methodologies are undoubtedly prominent in the development of science. However, attaching too much importance to them can hinder innovative creations or impede developments in flexible thinking. This inherent conflict, associated with prominent scientific theories, is observed in the developments of mathematical theories and calculation techniques for fitting data to a linear curve, in the discovery of a normal distribution, in the construction of an effective theoretical system based on a normal distribution, and so on. People tend to consider that real data should be analyzed using these methodologies, not merely that the mathematical clarity of these theoretical systems should be emphasized. This tendency has hampered the evolution of flexible methods for practical data analysis and leads to rejection of the possibility that we have complex data that cannot be treated with conventional methods, but new methods may break this impasse. Under these circumstances, one field of research has continued to develop new techniques to meet the challenges of the forthcoming era. This field is nonparametric regression, which has developed a variety of mathematically reliable techniques that are not governed by rigid forms such as a linear curve, a flat plate, or a normal distribution.

Currently, nonparametric regression is attracting much attention in the background in various fields: striking progress in computers, advancements in studies on neural networks, enhancement of the demand for data mining, and accumulation of diverse data by enabling the upgrading of measurement and observation techniques. To cope with this situation, many studies have been published, and the number of software programs available for carrying out nonparametric regression on computers is greatly increasing. As a logical consequence, nonparametric regression is no longer geared to a limited number of specialists. It is becoming an indispensable tool for dealing with diverse phenomena concerning nature, human beings, and society. On the other hand, the literature on nonparametric regression tends to divide into two categories: one focusing on rigid discussions on the mathematical side, and the other focusing on manuals for using computer software that includes functionality for nonparametric regression. It is difficult to find literature that treats both aspects. That is, not many reports are designed to meet the expectation that the use of nonparametric regression in practical data analysis based on a clear understanding of the theoretical background will provide analysts with reliable and beneficial results.

In this context, this book aims at helping readers to understand the basic concepts underlying nonparametric regression in order to utilize it practically, even if they are scarcely acquainted with this field and are reading this book on their own initiative with only a fundamental grasp of statistics and the rudiments of linear algebra. Therefore, this book commits itself to providing explanations in plain language by avoiding complex mathematics and excessively generalized expressions in order to allow the value and attractiveness of nonparametric regression to come across intuitively. Furthermore, an effort is made to enable readers to obtain knowledge effectively even if they refer to each chapter independently or read only parts of the book. The author sincerely hopes that this book will familiarize readers in various fields with nonparametric regression, and motivate them to use this methodology. The book's title, *Intoduction to Nonparametric Regression*, reflects this intention.

However, "introduction to nonparametric regression" contains another meaning. This second intention is made a reality by including a number of codes of S-Plus objects, which are utilized to draw graphs in this book and their explanations in "Examples of S-Plus Objects." The use of these objects and the improvement of them permits readers to carry out calculations and graphics for nonparametric regression very easily. These codes were constructed and run using Windows OS. Other OSs will, however, accept them with minor or no modification. It should be noted that the author does not always guarantee desirable results using these codes, and an emphasis on intelligibleness has sacrificed the speed of calculation, the efficient use of memory, and universality to some extent.

Furthermore, readers may also enjoy this book as an introduction to an area of mathematical science. Those who do not regard the practical use of nonparametric regression as a current interest will nevertheless be intrigued by the theoretical system for this field because of their appreciation of the diverse characteristics of regression and modeling; such characteristics are created by the metamorphosis of the two concepts of weighted average and the expansion of the function with one simple intention: smoothing. The author expects that this impression will attract readers' interest to more sophisticated theories.

Nonparametric regression was previously considered a refined but not very useful technique created by smoothing dilettantes. However, it has begun to be recognized as a requisite tool for extracting useful information from data. People no longer consider that in an age of artificial intelligence, nonparametric regression, which is only an insignificant field of statistics, is worthy of little attention. Instead, the fundamental ideas supporting nonparametric regression have become essential for data analysis, regression equation creation, data mining, modeling, and pattern recognition. The effects of the changing times are striking; the term "nonparametric regression" was valid among only a very small segment of the population at a time in the past. The author sincerely hopes that this book will accelerate this important advancement.

Kunio Takezawa

PREFACE TO THE SECOND EDITION

The first edition of this book was published in December 2001. Since then, in Japan, people's awareness of nonparametric regression has been raised further. The number of people who are interested in nonparametric regression in terms of its practical values, a fascination with it as mathematics, and its association with other areas of research has steadily increased. The author believes that this book has played a role in encouraging this tendency.

The goal of this book is to enable readers to understand the basic concepts of nonparametric regression. Therefore, though the state of the art in this field has progressed remarkably even in the latest few years, the basic contents of this book have been negligibly affected by such advanced developments. However, since the author has received many comments as feedback from readers, and has some particular concerns with the first edition, the need for a second edition has become considerable.

Hence, chapters on wavelets and pattern recognition, both of which are particularly necsssary for the practical use of nonparametric regression, have been added to the second edition *(note for readers of English edition: the chapter on wavelets is not included in the English edition)*. Furthermore, the S-Plus objects shown in the book have been modified to operate on S-Plus for Windows 6.0J. Considering the fact that the most serious requests regarding the first edition have been that: (1) readers need a more detailed description of the B-spline, and (2) readers would like to use "R" instead of S-Plus, Appendix A and Appendix B have been added. To accommodate the increased number of pages of the book, the second edition consists of two volumes *(note for readers of English edition: the English edition consists of one volume)*.

The author sincerely hopes that the widespread use of this second edition will enhance the understanding of the real advantages of nonparametric regression even further.

Kunio Takezawa
October 2003

PREFACE TO ENGLISH EDITION

This book is the English edition of the first volume of *Minnanotameno Nonparame-torikku Kaiki, Second Edition* (2003) (Yoshioka Publishing Company, Kyoto, Japan); this original version is written in the Japanese language and consists of two volumes. Problems and Appendix C have been added to the English edition. On the other hand, the chapters on wavelets, neural networks, and the tree-based model have been omitted from the English edition. In addition, the pages entitled "Kyukeijikann (Coffee Break)" between chapters and references written in the Japanese language have been also omitted. In consequence, the English edition consists of one volume. Furthermore, the publisher set up a ftp site related to the materials in this book at the address `ftp://ftp.wiley.com/public/sci_tech_med/`
`nonparametr ic-regression/`. This ftp site includes data sets used in this book.

The original book aims to be the most beginner-friendly text on nonparametric regression in the world and to extend the practical uses of nonparametric regression in a diverse range of fields. The publisher, John Wiley & Sons, Inc., shares this intention and is commended for publishing this English edition. The author sincerely hopes that this book will provide impetus for the understanding and popularization of nonparametric regression among many people worldwide.

Kunio Takezawa

National Agricultural Research Center (NARC), Tsukuba, Japan

ACKNOWLEDGMENTS

I am grateful to Dr. Masahiko Sagae (Department of Information Science, Faculty of Engineering, Gifu University) and his graduate school students, who read the manuscript and helped improve it. I am also very grateful to Mr. Tsutomu Toida (Faculty of Social and Information Studies, Gunma University), Dr. Hirotake Inoue (Kure National College of Technology), Dr. Nobuo Kajikawa (self-achievement adviser), and Mr. Hiroshi Shono (National Research Institute of Far Seas Fisheries) for their useful commnents.

S-Plus has been used for most of the calculations and graphics in this book. Furthermore, to help readers carry out the calculations of nonparametric regression by means of S-Plus, the objects developed for this book are shown as "Examples of S-Plus Objects." Ms. Mika Nakazono (Mathematical Systems Inc., Tokyo, Japan) provided the author with helpful comments on the use of S-Plus objects, and I am particularly grateful to her. I am convinced that the use of S-Plus and its application in nonparametric regression by such talented individuals in Japan will continue to build a sense of confidence in S-Plus and nonparametric regression using S-Plus. Additionally, members of an international s-news mailing list gave the author helpful information on the usage of S-Plus and related issues.

Moreover, many people have become interested in the practical significance of nonparametric regression and have presented their achievements in applying it to various problems. I would like to acknowledge the fact that this circumstance has helped produce this book. Even before nonparametric regression was fully appreciated, thinking people spared no effort in learning its value for practical applications, and developing and extending its usage. Such efforts have placed nonparametric regression research in high gear and have expanded the demand for a book with a central focus on its practical aspects.

Furthermore, some parts of this book have been financially supported by Fundamental Research on a Reproductive Information Base to Develop an Agricultural Production Support System; this is a research project of the Ministry of Agriculture, Forestry and Fisheries of Japan.

K. T.

CHAPTER 1

EXORDIUM

1.1 INTRODUCTION

This chapter describes the practical advantages of nonparametric regression by presenting the fundamental concepts of regression and modeling, and by discussing their desirable forms. The basic concepts of regression are presented first. Then, some popular misconceptions regarding smoothing are examined and critiqued using simple illustrations to indicate that multilateral discussion on mathematics and the accumulation of empirical knowledge are required for smoothing even though smoothing the roughness of data appears to be a simple task. People tend not to accept that discussion on smoothing is valuable because it appears to be easy to grasp it intuitively. This section, however, shows that appropriate smoothing is often crucial in making the most of data. It expresses straightforwardly the desirable shape of regression and the role of nonparametric regression in achieving it, along with providing important knowledge on the handling of data by the use of smoothing. The last section of this chapter defines the field of nonparametric regression more clearly. This chapter is designed to simply present the raison d'être of nonparametric regression and its domain by creating an image of nonparametric regression using a comparison of several methods, rather than by defining nonparametric regression first and then showing the advantages and the usage of it.

Introduction to Nonparametric Regression, Kunio Takezawa.
Copyright © 2006 John Wiley & Sons, Inc.

These observations $\{(\mathbf{X}_i, Y_i)\}$ $(1 \leq i \leq n)$ are typically assumed as data for constructing models. $\mathbf{X}_i$ is a vector (sometimes a scalar) and Y_i is a scalar. The equation below is considered to generate the data.

$$y = m(\mathbf{x}) + \tilde{\epsilon}, \tag{1.1}$$

where $m(\cdot)$ is a nonaccidental function for describing the intrinsic behavior of y. Observing $(\mathbf{x}, y)$ provides $\{(\mathbf{X}_i, Y_i)\}$. $\tilde{\epsilon}$ is error (i.e., model error), which is usually assumed to be distributed randomly around 0.0. Then, the data $\{(\mathbf{X}_i, Y_i)\}$ $(1 \leq i \leq n)$ created using eq(1.1) are depicted as

$$Y_i = m(\mathbf{X}_i) + \tilde{\epsilon}_i, \tag{1.2}$$

where $\{\tilde{\epsilon}_i\}$ are realizations of $\tilde{\epsilon}$. Estimation of the regression equation (regression function) (the term "regression model" is also used) aims at obtaining a useful $m(\cdot)$. This procedure is sometimes simply called "regressing." The resultant regression equation is identified as $\hat{m}(\cdot)$. To emphasize the purpose of prediction, a regression equation can also be a "prediction equation." The function $\hat{m}(\cdot)$ is usually considered effective if it has been derived by extracting as many smooth movements in the data as possible. Therefore, regression is conventionally designed to create $\hat{m}(\cdot)$ which allows the absolute values of $\{\tilde{\epsilon}_i\}$ to take on average small values on the condition that $\hat{m}(\cdot)$ is a smooth function.

The variable $\mathbf{x}$ is called a predictor. It is sometimes identified as an "independent variable." The term may be inappropriate because the word "independent" is misleading; $\mathbf{x}$ could depend on something, and the elements of $\mathbf{x}$ could be dependent on each other. The term "explanatory variable" might also invite a misunderstanding; $\mathbf{x}$ does not always explain y. The other alternatives, "regressor variable" and "regressor," have not yet become common terms.

The variable y is named the target variable (object variable). The term "dependent variable" is also possible, but the word "dependent" may be misleading; y does not always depend completely on $\mathbf{x}$. "Explained variable" presents a similar problem to explanatory variable. "Predictand" could be confused with predictor and predictant. The terms "response variable (response)" and "regressand" have not been widely accepted yet.

The term regression originates from the phenomenon that a repetition of genetic inheritance allows body height and other such factors to become close to the average. However, the term may also carry the implication that regression recovers an original form by eliminating errors. In truth, $\hat{m}(\cdot)$ should not be considered an approximation of the real image of data because $\hat{m}(\cdot)$ often depends on the particular goal even when the same data are utilized. We should regard $\hat{m}(\cdot)$ as a functional relationship that was obtained from a set of data for our purpose. In this respect, "modeling" is the most appropriate expression. The meaning of "model," however, is too broad to describe something like eq(1.1) precisely, and the term regression is deeply rooted in statistics. Therefore, this book uses the term "regression equation" and the result of the formulation of data in a more general form than eq(1.1) is termed a "model." The word "model" is to be used on occasions when a term containing "model" is widely used (e.g., "additive model").

On the other hand, estimation of the probability density function (or simply, the density function) indicates that when $\{\mathbf{X}_i\}$ $(1 \leq i \leq n)$ ($\mathbf{X}_i$ is a datum (a vector),

n is the number of data) are given, analysts estimate a probability density function ($\hat{f}(\cdot)$) which describes the distribution of $\{X_i\}$ appropriately on the basis of the assumption that $\{X_i\}$ are realizations of $f(\mathbf{x})$. When $f(\cdot)$ is considered to be a smooth function, this estimation is carried out to derive a smooth $\hat{f}(\cdot)$. In addition, $\int_{-\infty}^{\infty} \cdots \int_{-\infty}^{\infty} \hat{f}(\mathbf{x})d\mathbf{x} = 1$ should always be at least approximately correct; it must be satisfied rigidly on some occasions. $\hat{f}(\mathbf{x}) \geq 0$ is also a typical condition. While a probability density function is significantly different from a regression equation based on eq(1.2) from a number of points of view, it shares a common property in that they are both results of formulation by extracting inherent characteristics of data for a specific purpose, and both use similar concepts and techniques. These commonalities allow a probability density function to be regarded as a regression equation or a model.

The estimation of the regression equation of eq(1.2) and a probability density function usually requires that the resultant function ($\hat{m}(X_i)$ and $\hat{f}(\mathbf{x})$) be smooth, and much experience justifies this requirement. Regressions using eq(1.1) are categorized into parametric regression and nonparametric regression; the distinction is clarified later in this chapter. Similarly, the methods of obtaining a probability density function are classified into parametric probability density function estimation (parametric density function estimation) and nonparametric probability density function estimation (nonparametric density function estimation). Estimation of the nonparametric probability density function is treated as a category of nonparametric regression in this book because the techniques for estimating nonparametric probability density described here are analogous to those based on eq(1.1).

Smoothing is the term for relatively simple nonparametric regression; when it is apparent that the values of data or the distribution of data are being "ironed out," the procedure is called smoothing. The establishment of the field of smoothing originates with the fact that techniques categorized as smoothing are beneficial data analysis methods when considered in isolation, and the historical circumstance that the evolution of smoothing has become integrated into the development of nonparametric regression, which forms a significant realm within the field of statistical data analysis. Smoothing should not, however, be defined as simple nonparametric regression because parametric regression is also used for smoothing. Hence, smoothing is divided into that by parametric regression and that by nonparametric regression in the strict sense.

The relationships among these terms are drawn as a Venn diagram in figure 1.1, which shows that nonparametric regression is involved in both regression equation estimation and probability density function estimation. As is often the case in the terminology of statistics, the meanings of terms are entangled and the terminology used here is not necessarily universal. Hence, the definitions provided here are to be taken as no more than a rough guide. This is because usually we cannot predict how a method will develop and with what techniques a method will be associated when it is named. Another reason is that those aspects which are important from a mathematical point of view are sometimes far from those which are important for practical applications. That is, the perspectives to be emphasized in the choice of terminology depend significantly on the purpose and context. Naming conventions, therefore, are not always entirely logical.

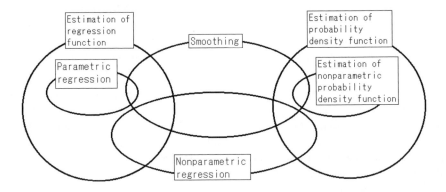

Figure 1.1   Venn diagram to show the relationship of parametric regression and nonparametric regression. This relationship is not necessarily a conclusively established one, and therefore other mappings may be reported in the literature.

1.2 ARE THE MOVING AVERAGE AND FOURIER SERIES SUFFICIENTLY USEFUL?

Figure 1.2 illustrates the monthly average exchange rate of US dollar and yen from February 1987 through May 1999. A long term trend with a superimposed fine oscillation forms the shape; the fine oscillation is the short term variation. The subtraction of fine oscillation (i.e., the short term variation) from the longitudinal data gives the long term variation. The two types of variations reflect different mechanisms in economic change and hence the separation of the data into these two trends allows a clear understanding. It is reasonable that a long term trend should be considered as a gentle curve extracted from data, and the subtraction of this trend from the data gives the short term variation. Such an extraction of long term variation is one role of smoothing.

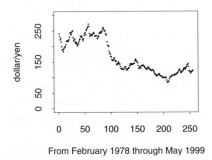

From February 1978 through May 1999

Figure 1.2   Monthly average exchange rate of US dollar and yen from February 1987 through May 1999. Numbers on the x-axis indicate months beginning at February 1987.

The moving average is the most widely used tool for eliciting a smooth trend by removing fine oscillation from the data. Figure 1.3(left) is an example of the result of applying the moving average to the data; the curve is drawn by connecting the values of the moving averages. When data are represented as $\{Y_i\}$ $(1 \leq i \leq n)$ (n is the number of data), the relationship between the line $(\hat{m}(i))$ in figure 1.3(left) and the data is written as

$$Y_i = \hat{m}(i) + \tilde{\epsilon}_i, \tag{1.3}$$

where $\hat{m}(i)$ is a smooth function with a variable (i); it indicates a long term variation. $\tilde{\epsilon}_i$ is random; the average of $\{\tilde{\epsilon}_i\}$ in a certain range of i leads to a value close to zero. Since the solid line in figure 1.3(left) captures the rough trend of the data, this line is a strong candidate for $\hat{m}(i)$. This line, however, includes peculiar behaviors as a result of smoothing the data. Figure 1.3(right), which shows the first 80 data and corresponding $\hat{m}(i)$, makes this problem more apparent. This $\hat{m}(i)$ derived from the moving average is not acceptable as a rough sketch of the variation of exchange rate in this period. A local maximal value is observed where the data imply a local minimum, and vice versa. This $\hat{m}(i)$ cannot be considered as a smooth trend that is extracted from the data. A problem of this kind posed by the moving average is also treated in Chapter 2. Figure 1.3(right) in itself gives a clear confirmation that we should refrain from utilizing the results of the moving average incautiously as a basis for an important conclusion.

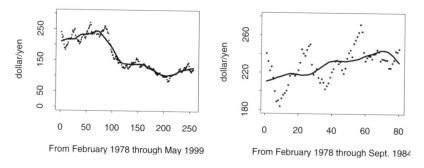

Figure 1.3 Example of the result of moving average using the data in figure 1.2. The result of extraction of the first 80 data from figure 1.3 (left) and corresponding $\hat{m}(i)$ (right).

Therefore, the use of periodical oscillation, which is often employed in the explanation of economic change, is attempted for smoothing. Well-known concepts such as the Kitchen cycle, the Jugler cycle, the Kuznets cycle, and the Kondratieff cycle indicate that economic change is considered as a composition of several waves with different periodicities. This approach may provide a persuasive means of smoothing. The failure of the moving average may be due to the lack of economic knowledge, and therefore the use of the significant economic theory of the existence of periodic phenomena is expected to improve the results. Following this approach, the data are decomposed into sine waves and cosine waves with various periodicities by Fourier transform. The extraction of waves with long periods presents the smooth line in figure 1.4(left). This line may have a problem in the left part. Then, the first 80 data and corresponding $\hat{m}(i)$ were taken out in order to draw figure 1.4(right). However, this line does not describe the behavior of the data appropriately. When data are plotted as shown in this graph, a human being would not draw such a smooth curve to show the characteristics of the data.

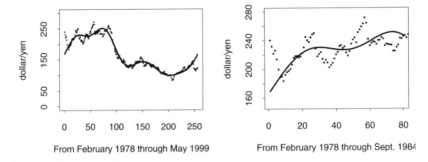

Figure 1.4   A smooth line $(\hat{m}(i))$ that is the sum of longer waves. Waves are obtained by the decomposition into sine waves and cosine waves (left). The result of extraction of the first 80 data from those in figure 1.4(left) and corresponding $\hat{m}(i)$ (right).

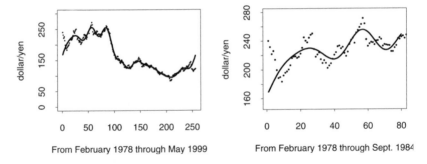

Figure 1.5   A smooth line $(\hat{m}(i))$ that is the sum of longer waves. Waves are obtained by the decomposition into sine waves and cosine waves (the number of waves is greater than that in figure 1.4(right)), (left). The result of extraction of the first 80 data from those in figure 1.4(left) and corresponding $\hat{m}(i)$ (right).

However, the lines in figure 1.4(left) and figure 1.4(right) are constructed using a small number of waves based on the Fourier series; this may cause the odd behavior of the lines. Therefore, a line obtained using the Fourier series and taking finer variations into account is drawn in figure 1.5(left). The first 80 data and corresponding $\hat{m}(i)$ are shown in figure 1.5(right). This line is also far from ideal. These examples show that the potential occurrence of periodicity in economic change is not directly associated with the realization of beneficial smoothing using the Fourier series.

The concepts of the moving average and the Fourier series are easily comprehensible. This seems to be the main reason for their widespread use as a tool for deriving a sketchy trend from the available data. Even the small number of examples shown here, however, prove the insufficiency of these methods as tools for extracting smooth curves from data. The data used here were not intentionally selected to be disadvantageous for these techniques. This proposition can easily be affirmed by comparing the products of these methods with the behavior of the original data when readers encounter smoothed values elsewhere in the literature. It is very easy to identify inappropriate smooth curves that are the outcomes of smoothing of the data; some of the curves do not bear any resemblance to the behavior of the original data.

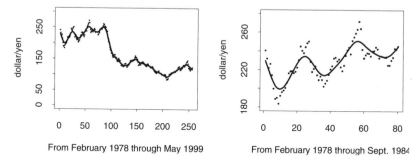

Figure 1.6 Curve obtained using smoothing spline with the data shown in figure 1.2 (left). First 80 data of those in figure 1.6(left) and the corresponding $\hat{m}(i)$ (right).

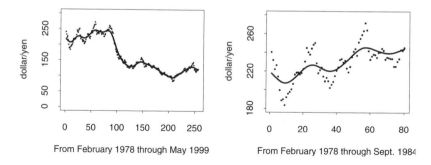

Figure 1.7 Curve obtained using the smoothing spline with the data shown in figure 1.2 (the degree of smoothness is larger than that in figure 1.6(left)), (left). First 80 data of those in figure 1.7(left) and the corresponding $\hat{m}(i)$ (right).

Then, figure 1.6(left) is presented. This figure illustrates the result of calculation using a technique called the smoothing spline (smoothing splines). The smoothing spline is one of the most useful methods among those based on the nonparametric regression methodology. Figure 1.6(right) shows the first 80 data in figure 1.6(left) and the corresponding $\hat{m}(i)$. Furthermore, the curve in figure 1.7(left) is drawn by increasing the smoothness of the curve in figure 1.6(left). Figure 1.7(right) shows the first 80 data and the corresponding $\hat{m}(i)$. This line is quite convincing as a result of applying a slightly greater degree of smoothing. A notable feature of smoothing using the smoothing spline lies in the fact that continuous adjustment of the smoothing is possible. With the data used in the case presented here, a variety of curves with different smoothnesses can be derived. This is one of the advantages of using the smoothing spline over using the moving average or the Fourier series.

Smoothing by the smoothing spline or the like yields far more persuasive curves than those by moving average and Fourier series; "persuasive" here means that the curves have the appearance of those drawn by skillful hands. In a mathematical argument, it may be considered an insufficient justification for the use of the smoothing spline that the ability to draw curves that are similar to those drawn by human beings proves the usefulness of the smoothing spline. However, mathematically strict discussions on smoothing often support intuitive diagnoses by human beings. This fact indicates that the domain in which our intuitive sense operates can be broadened by a good grasp of the association between the intuitive appropriateness of a smoothing curve and its mathematical explanation of smoothing, and by the application of the concepts and techniques developed in simple cases to more complicated problems which defy intuitive understanding, such as problems with a number of predictors.

The smoothing spline is detailed in Chapters 2, 3, and 4. The simple description given here is nothing more than illustration by examples, which indicate that the smoothing spline results in intuitively favored curves and that it is advantageous over the moving average and applications of the Fourier series. These instances, however, provide reasonable grounds for the recommendation of the smoothing spline over the moving average and the Fourier series, if it is available. The smoothing spline is now easily obtainable. Many free software programs are available on the Internet and other places, and in addition, well-developed commercial software programs are also available. The only reason to continue the use of the moving average and the Fourier series, if any, is that these obsolete techniques facilitate comparisons with past findings. If this is not essential, we do not find any compelling reason to carry out smoothing by using the moving average or the Fourier series.

1.3   IS A HISTOGRAM OR NORMAL DISTRIBUTION SUFFICIENTLY POWERFUL?

Figure 1.8(left) shows the daily average temperature (the number of data is 365) in 1986 at an Automated Meteorological Data Aquisition System (AMeDAS) observation station located on Ishigaki Island in Japan (all of the meteorological data in this book are AMeDAS data and have been provided by the Meteorological Agency of the Japanese government). The distribution of these observational data is shown as a histogram. A histogram is almost equivalent to $n \cdot \hat{f}(X)$ in which n is the number of data (here, $n = 365$) and $\hat{f}(X)$ is an estimated probability density function. The binwidth (the width of rectangles constituting a histogram) is 2.06. The height of each bin indicates frequency. For example, the height of the first bin (rectangle) represents the number of data which satisfy $10 \leq X_i \leq 12.06$; the whole data set is written as $\{X_i\}$ ($1 \leq i \leq 365$). The height of the second bin is the number of data which satisfy $12.06 < X_i \leq 14.12$, the height of the third bin is the number of data which satisfy $12.06 < X_i \leq 14.12$, and the height of the next bin is the number of data which satisfy $14.12 < X_i \leq 16.18$; the bin height continues in the same manner. The definition of frequencies here is slightly different from the conventional one; this is in the manner of hist() (a default object in S-Plus). The point of the minimum value of the first bin (the point of $x = 10$ for this example) is called an anchor, while the points between the bins ($x = 10 + 2.06, 10 + 2.06 \cdot 2, 10 + 2.06 \cdot 3, \ldots$ for this example) are termed boundaries (bin edges). The value of 2.06 as a binwidth is calculated by Sturges' rule:

$$h = \frac{X_{max} - X_{min}}{1 + \log_2(n)}, \tag{1.4}$$

where h is the binwidth. X_{max} is the maximum value of $\{X_i\}$ and X_{min} is the minimum value of $\{X_i\}$. n is the number of data (= 365 in this example). $\log_2()$ indicates a base 2 logarithm.

This histogram shows that the data distribution is almost certainly left-skewed. However, it is not certain that this distribution has three modes; in other words, this may not be trimodal. Then, another histogram created using the same data under slightly different conditions is drawn in figure 1.8(right). The height of the first rectangle of this histogram is the number of data which satisfy $9 < X_i \leq 11.06$, and the height of the second one is the number of data which satisfy $11.06 < X_i \leq 13.12$, and so forth. That is, the anchor was moved while other conditions were kept the same. This histogram has only one mode; this means that it is unimodal. Figure 1.8(left) has no sufficient basis for the trimodality. The shift of the anchor with the same binwidth as determined according to Sturges' rule changes the essential characteristics of the histogram.

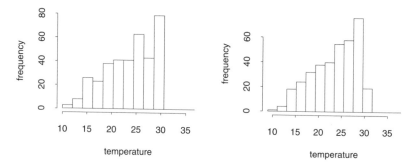

Figure 1.8   Histogram of the daily average temperatures observed in 1986 on Ishigaki Island. The anchor is positioned at $x = 10$, and the binwidth is set at 2.06 (left). The binwidth remains the same as that in figure 1.8(left), and the anchor was moved to $x = 9$ (right).

Therefore, the binwidth is shortened in order to clarify the fine structure of the data. Sturges' rule is based on the binomial distribution and hence it may not be highly applicable to these data. The binwidth is changed to be 1.0, then the conditions for the data placed in the rectangles are $10.0 < X_i \leq 11.0, 11.0 < X_i \leq 12.0, \ldots$. That is, the anchor point is located at $x = 10$ and the binwidth is fixed at $h = 1$. These conditions yield the histogram drawn in figure 1.9(left). This histogram apparently has four modes. Figure 1.9(right) shows the result of the change of the setup: the anchor point is placed at $x = 10.5$ and the binwidth is set as $h = 1$. This one has five modes. Although the number of local maximum points in a data distribution, namely, the number of modes, keenly interests analysts, histograms do not provide clear information. This example implies that an impression given by a histogram may be changed by slight alteration of the means of its construction, although the histogram is a standard means of showing the distribution of data plainly.

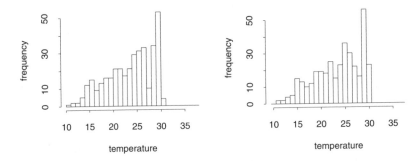

Figure 1.9   Histogram of the daily average temperatures observed in 1986 on Ishigaki Island. The anchor is positioned at $x = 10$, and the binwidth is set at 1 (left). The binwidth remains the same as that in figure 1.9(left), and the anchor was moved to $x = 10.5$ (right).

It may be apparent to an analyst that the histograms have the problem of being too rough because they consist of aligned rectangles that represent the density of data. Points such as $x = 9$ and $x = 10$ are positioned at rounded figures. It is true that such numbers are convenient for analysts, but it is unlikely that rounded figures play a significant role in treating temperature observations. Rather, we should assume that the distribution of the data is intrinsically smooth. Hence, the distribution of the data should be represented using a smooth curve. A simple method for executing this policy is to fit a smooth curve to the data. The most well-known distribution containing this feature is a normal distribution.

The probability density function of the normal distribution is represented as

$$f(x) = \frac{1}{\sqrt{2\pi\sigma^2}} \exp\left(-\frac{(x - x_a)^2}{2\sigma^2}\right), \tag{1.5}$$

where x_a is average, and σ^2 is variance (i.e., σ is standard deviation). x_a is the position of the mode; $f(x_a)$ is the maximum of $f(x)$. $\int_{-\infty}^{\infty} f(x)dx = 1$ is satisfied.

Figure 1.10 is the result of the fitting of a normal distribution to the data shown in figure 1.8 using a maximum likelihood procedure (maximum likelihood method). The resultant parameters are $x_a = 23.568$ and $\sigma = 4.803$. The curve in this figure is based on multiplying $f(x)$ by the number of data ($n = 365$); the integration of the function gives 365. The histogram (figure 1.8(left)) is superimposed. The comparison between the curve of the normal distribution and the histogram conveys the impression that the curve does not represent the distribution of the data appropriately. This is because a normal distribution, which is symmetrical, is not capable of giving an unsymmetrical shape; the distribution of the data is apparently left-skewed. Furthermore, a normal distribution, which is based on the assumption of being unimodal, always leads to unimodal distributions, though it should be possible to study the number of local maximum points.

Analysts are tempted to consider that examination of the shape of the data distribution $\{X_i\}$ ($1 \leq i \leq n$) is not a difficult task. Therefore, insufficient attention is usually paid to the selection of the methods for investigating the distribution of data, in spite of the fact that actual data analysis often requires such procedures. As

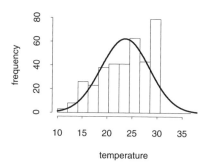

Figure 1.10 Normal distribution obtained using the daily average temperature observed in 1986 on Ishigaki Island. The histogram in figure 1.8(left) is superimposed.

is shown by this example, however, it is not easy to capture the data distribution, in order to clarify its significant characteristics such as the number of local maximal points.

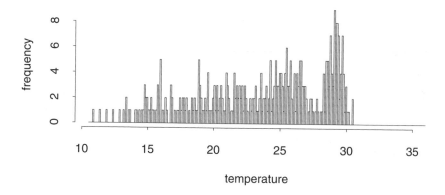

Figure 1.11 Histogram of the daily average temperature observed in 1986 on Ishigaki Island. The binwidth is 0.1. Each frequency represents the number of data with the same value, since the data were taken in units of 0.1 degree.

The data are represented more simply in figure 1.11. The binwidth of this histogram is 0.1. The data were taken in units of 0.1 degree so that two or more data falling in the same bin means that these data have identical values. This histogram supports the intuitive judgment made on the basis of figure 1.10 that the normal distribution should not be fitted because of the left-skewed property of the data distribution, and confirms that this distribution has a relatively complex shape. To make this impression more reliable, this histogram is smoothed using nonparametric regression. Figure 1.12 and figure 1.13 illustrate smoothed histograms of the data in figure 1.11; these smoothings were carried out using the nonparametric regression methodology. The histogram in figure 1.13 possesses a higher degree of smoothness and is smoother than that in figure 1.12. These two histograms highlight the intrinsic characteristics of figure

1.11 by smoothing the roughness appropriately. This method simply assumes that the distribution is smooth; this assumption is of great help in realizing good smoothing. On the other hand, the fitting of the normal distribution is based on more restrictive conditions.

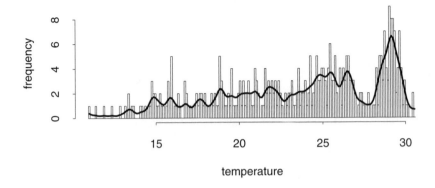

Figure 1.12   Nonparametrically smoothed histogram. Data shown in figure 1.11 are used.

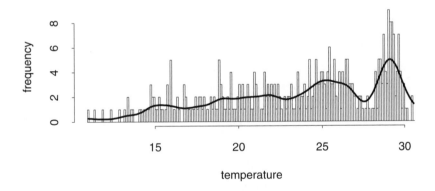

Figure 1.13   Nonparametrically smoothed histogram. The degree of smoothness is higher than that in figure 1.12.

   Daily average temperatures are shown as raw data in figure 1.14. The start point of the horizontal axis ($= 1$) corresponds to January 1st, 1986, and $2, 3, \ldots$ correspond to January 2nd, January 3rd, $\ldots$. The daily average temperature on Ishigaki Island is characterized by stable variation in summer and unstable variation in other seasons. This is the background of the left-skewed distribution. Symmetric distributions such as a normal distribution are obviously not applicable for the approximation of such a distribution. Some may consider that complicated phenomena such as those dealt with in meteorology should result in normally distributed data. This is based on a misconception regarding the central limit theorem. In fact, complicated phenomena give rise to diverse data distributions; even simple phenomena can produce complexly

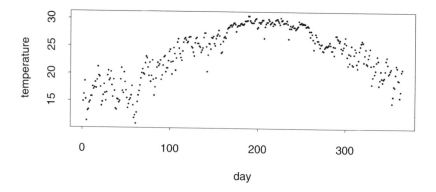

Figure 1.14   Daily average temperatures observed in 1986 on Ishigaki Island.

distributed data. Therefore, these examples clearly indicate that if a simple functional form is assumed to describe the probability density function of data, it can cause important features of the data to be missed.

There is no doubt that theories based on the normal distribution form the foundation of the history of statistics. The simple structure and useful characteristics of the normal distribution are greatly beneficial for practical purposes. However, these facts do not justify the fitting of the normal distribution to data to which a normal distribution cannot give a close approximation. As has been the case for many breakthroughs in science, applications of the breakthroughs to inappropriate subjects may lead to unexpected and erroneous conclusions. If distributions with a small number of parameters, such as a normal distribution, are fitted to unsuitable data, a portion of the information contained in the data may be lost or distorted. That said, it is not certain that a histogram will yield a desirable result simply because there is no assumption of a specific functional form. The shape of the resulting histograms may still be an uncertain one. To prevent such situations from arising, the use of nonparametric-regression-based histograms and probability density functions is an effective approach because they are easy-to-use but powerful tools.

1.4   IS INTERPOLATION SUFFICIENTLY POWERFUL?

Figure 1.15 illustrates the daily average temperatures from January 1st through December 26th in 1992 in the Takada Meteorological Data Aquisition System (AMeDAS) observation station located in the city of Joetsu in Niigata prefecture in Japan. The number of data is 361. The data are represented as $\{X_i, Y_i\}$ $(X_i = i, 1 \leq i \leq 361)$. Such observations are not always made daily; they may be made once every few days. Here, we take the data for temperature measured every 20 days. Hence, the data we have are assumed to be $\{X_j^*, Y_j^*\}$ $(X_j^* = j, Y_j^* = Y_j, j = 20k - 19, 1 \leq k \leq 19)$. If we estimate the almost correct values of $\{X_i, Y_i\}$ using only $\{X_j^*, Y_j^*\}$, the daily average temperature is obtained with a reasonable degree of accuracy. Various types of data obtained through experiments, observations, and censuses generally require such treatment.

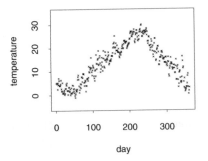

Figure 1.15   Daily average temperatures from January 1st through December 26th in 1992 at Takada observation station (Jyoetsu city, Niigata prefecture, Japan).

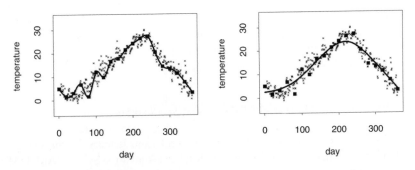

Figure 1.16 $\{X_j^*, Y_j^*\}(X_j^* = j, Y_j^* = Y_j, j = 20k - 19, 1 \le k \le 19)$, which are extracted from the data in figure 1.15, and the interpolated estimates obtained by the cubic natural spline (left). Smoothed estimates are calculated by the smoothing spline using the same data as those in figure 1.16(left), (right).

In this situation, we obtain a smooth curve ($y = \hat{m}(x)$) which interpolates $\{X_j^*, Y_j^*\}$ to estimate the values of $\{Y_i\}$ on $\{X_i\}$. Interpolation here indicates that the equation below is satisfied.

$$Y_j^* = \hat{m}(X_j^*). \tag{1.6}$$

This procedure is based on the concept that the curve described by $y = \hat{m}(x)$ must pass exactly through $\{X_j^*, Y_j^*\}$ because data such as temperature are obtained with high accuracy.

Figure 1.16(left) shows $\{X_j^*, Y_j^*\}$, $\{X_i, Y_i\}$, and a curve given by interpolating $\{X_j^*, Y_j^*\}$ using a natural cubic spline. The smooth curve passes exactly through all of the data. $\{X_i, Y_i\}$ are the unobserved data and hence the estimates at these points given by the curve in figure 1.16(left) indicate the accuracy of the estimation of $\{X_i, Y_i\}$. Then, E_{alias} as defined below is calculated.

$$E_{alias} = \sum_{i=1}^{361}(\hat{m}(X_i) - Y_i)^2. \tag{1.7}$$

As a result, $E_{alias} = 2947.141$ is obtained. A smaller value of E_{alias} means that $\{\hat{m}(X_i)\}$ performs more accurately.

The problem to be discussed here is the advisability of carrying out interpolation when the accuracy of data is considered to be very high. In fact, we must assume that there are considerable errors in such data. This is because however exact the data may be, it is considered to contain errors if it was obtained by discrete observations of a time-series phenomenon like our example, or if the overall appearance is assessed using isolated observations carried out on a spatial phenomenon. The error caused by such factors is called aliasing error.

Figure 1.16(right) is a result of using smoothing instead of interpolation to verify the existence of aliasing error. The smoothing spline is employed for smoothing. The estimates obtained using $y = \hat{m}(x)$ do not pass precisely through $\{X_j^*, Y_j^*\}$, but run close to the data to follow a smoother trajectory than that obtained by interpolation. $E_{alias} = 2583.495$ is obtained for this curve. This value is appreciably smaller than that obtained by interpolation. Thus, smoothing worked better than interpolation for the estimation of $\{Y_i\}$ on $\{X_i\}$.

Such a result is not limited to this example but is a common finding when one analyzes the data obtained by discrete observations. Estimates (estimated values) obtained by smoothing instead of interpolation do not simply have a good appearance; they possess the important advantage of higher accuracy. In fact, estimates for constant pressure lines on weather maps and contour lines on topological maps are results of smoothing. Estimates on such charts do not pass exactly through the data, though the observed data have very high accuracy. The universal significance of smoothing is clear when one considers the fact that most data are obtained through discrete measurements or observations.

However, this conclusion depends on the purpose; the discussion above indicates that smoothing is appropriate for enhancing the overall accuracy of estimates. For example, estimates given on the basis of the curve in figure 1.16(right) lead to a lower value of the maximum daily temperature during this period. Hence, this curve cannot be recommended if the estimates around the peak are the object of one's interest. Estimates by interpolation or estimates by smoothing with special attention given to the behavior around the local maximal and minimal points are more appropriate. One may expect that the selection of a good method will not depend on purpose as long as it is a simple task such as the estimation of the values where data do not exist based on discretely observed data. For practical purposes, however, even for such a simple estimation, a method must be selected with respect to the purpose.

1.5 SHOULD WE USE A DESCRIPTIVE EQUATION?

Regression equations such as a polynomial (polynomial function) and a trigonometric function (Fourier series) are usually used although they do not represent exactly the phenomenon that generates the data. That is to say, when the equation for describing such a phenomenon is not specified, polynomials or trigonometric functions are used to attempt to depict the phenomenon. The resulting regression equations are considered to express the phenomenon approximately. How about cases in which the form of the equation describing a phenomenon is known? One may consider that the use of the known equation as a good expressor of the phenomenon guarantees useful regression, and that since the equation underlying the data is well known, no other forms of equations have to be considered as possibilities for a suitable regression equation. However, this intuitive judgment cannot be sufficiently justified.

To clarify this supposition, simple simulation data generated as shown below are used.

$$X_i = 0.02i \qquad (1 \le i \le 50), \tag{1.8}$$

$$Y_i = -0.5 + X_i - X_i^2 + X_i^3 - X_i^4 + X_i^5 - X_i^6 + X_i^7 + \epsilon_i. \tag{1.9}$$

$\{\epsilon_i\}$ are realizations of $N(0.0, 0.5^2)$ (a normal distribution; the average is 0.0 and the standard deviation is 0.5). In other words, the values of $\{\epsilon_i\}$ are a sample from $N(0.0, 0.5^2)$. The values that the 7th-degree polynomial (eq(1.9) without ϵ_i) gives are drawn in figure 1.17(top left). Simulation data with ϵ_i are plotted in figure 1.17(top right). The polynomial equations are regressed on these 50 data. It is assumed that the analyst knows that the data obey a 7th-degree polynomial. That is, it is supposed that Y_i is obtained by adding a normal distributed error to a 7th-degree polynomial of X_i. Figure 1.17(bottom left) is the result of fitting a 7th-degree polynomial equation to these data by the least squares (least square method). The result of fitting a quadratic equation to the data is illustrated in figure 1.17(bottom right). The value below is calculated in order to compare the performances of the two regression equations.

$$E_{prediction} = \frac{1}{1000} \sum_{j=1}^{1000} \frac{1}{50} \sum_{i=1}^{50} (\hat{m}(X_i) - Y_i^{(j)})^2. \tag{1.10}$$

Here, $\{Y_i^{(j)}\}$ $(1 \le i \le 50, 1 \le j \le 1000)$ are simulation data obtained using eq(1.8) and eq(1.9) with various seeds. A total of 1000 simulation data sets were generated by this process. $\hat{m}(\cdot)$ is the resulting regression equation derived by the least squares using the data in figure 1.17(top right). $E_{prediction}$ performs as an index to show how an obtained regression equation will function with future data. If the value of $E_{prediction}$ is small, $\hat{m}(\cdot)$ is regarded as a reasonably effective equation for prediction. When the regression equation derived using a 7th-degree polynomial

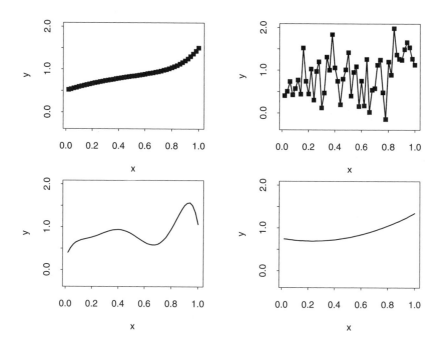

Figure 1.17   Data obtained by removing $\{\epsilon_i\}$ from the simulation data (top left). Simulation data (top right). Estimates obtained by fitting a 7th-degree polynomial to the simulation data (bottom left). Estimates obtained by fitting a quadratic equation to the simulation data (bottom right).

equation is $\hat{m}(\cdot)$, $E_{prediction}$ is 0.2881179. On the other hand, using a quadratic equation as the regression equation gives the value of 0.2559359 for $E_{prediction}$. Although the data were actually generated by a 7th-degree polynomial equation, the quadratic equation is shown to have higher predictability. This example indicates that even if a regression equation is known to depict the phenomenon exactly, the resultant regression equation is not always applicable.

When the form of an equation that describes the generation of a phenomenon is specified, it is natural to assume that a regression using this form will give a persuasive regression equation. In particular, one wants to use an equation or a model that represents a societal or biological phenomenon realistically. Such equations and models are not, however, guaranteed to be effective in prediction or control. Hence, even if the form of the equation representing a phenomenon is given, other forms of equations should also be attempted for regression. In this sense, nonparametric regression is not simply an interim shortcut to be used when the form of the equation underlying a phenomenon is not identified. Even when the form of an equation representing a phenomenon is known, nonparametric regression may lead to results that meet the purpose. On the contrary, a regression equation which is chosen with respect to prediction or control does not always describe the phenomenon realistically. A selected regression equation is no more than a tool that is useful under a certain condition and for a specific purpose.

1.6 PARAMETRIC REGRESSION AND NONPARAMETRIC REGRESSION

Smoothing using the moving average, that using the Fourier series, and that using the smoothing spline are included among the techniques of nonparametric regression. On the other hand, parametric regression makes use of regression equations such as

$$y = ax + b, \tag{1.11}$$

$$y = px^2 + qx + r, \tag{1.12}$$

$$y = a_1\sin(b_1x) + a_2\cos(b_2x), \tag{1.13}$$

$$y = s_1/(1 + s_2\exp(s_3x)), \tag{1.14}$$

where x is a predictor, y is a target variable, and a, b, p, q, r, a_1, b_1, a_2, b_2, s_1, s_2, and s_3 are constants called regression coefficients. A regression coefficient is sometimes more generally termed a parameter. Therefore, parametric regression is the estimation of the values of parameters (regression coefficients) using data at hand. Regression equations in eq(1.11) and eq(1.12) are called linear models because the target variable (y) is a linear function of the regression coefficients in these equations. The derivation of a linear model is called linear regression. In most linear regressions, the solution of simultaneous linear equations provides the values of the regression coefficients. On the other hand, the target variable is not considered a linear function of the regression coefficients in eq(1.13) and eq(1.14). Hence, these equations are addressed as nonlinear models; the derivation of such regression equations is called nonlinear regression. Regression has conventionally been the term to describe fitting such regression equations to the data in order to obtain the regression coefficients, since most conventional regressions employ regression equations of this kind.

This line of thought gives the impression that nonparametric regression seems to be a regression technique that does not derive regression coefficients because parametric regression is a method for giving estimates by calculating regression coefficients (parameters). This is not, however, an appropriate way to differentiate the two concepts. For example, estimates obtained using the moving average are considered to be the results of a regression equation that gives constants in each region, and curves obtained using some terms of a Fourier series are drawn using a regression equation with several sine waves and cosine waves. Furthermore, a smoothing spline uses a regression equation with a specific form (cf. eq(3.213) and eq(4.11)). That is, nonparametric regression is similar to parametric regression because both methodologies utilize a regression equation with a specific form and derive the estimates by estimating the regression coefficients.

Therefore, what distinguishes nonparametric regression from parametric regression is the nonexistence of an orientation toward the reduction of the number of parameters (regression coefficients). Regression with such directionality is parametric regression, and regression without it is nonparametric regression. Parametric regression favors expressions with as small a number of parameters as possible and selects a regression equation with a larger number of parameters only when the use of such a regression equation is indispensable for representing the data well. On the other hand, nonparametric regression is focused chiefly on the goal of deriving beneficial trends from the data and does not give special consideration to the reduction of parameters.

However, this explanation may give rise to an inconsistency: if regression without the intention of the reduction of parameters is called nonparametric regression, the meaning of the prefix "non-", which has a negative sense, remains unclear. That is, if nonparametric regression cannot be associated with regression without parameters, but means regression without the intention of obtaining regression equations with a small number of parameters, the prefix "non-" is unknown. To resolve this issue, we should focus on the term "parameterization." Parameterization indicates the representation of phenomena using as small a number of parameters as possible. Hence, the "non-" of nonparametric regression is a prefix used in the negative sense to mean that regression is exclusively focused on carrying out practical regressions to represent a phenomenon; it has no concern for achieving formulations with simple parameters (i.e., achieving a small number of parameters).

It should be noted that it is sometimes not easy to classify a regression definitively as parametric regression or nonparametric regression because it is not clear whether the intention is to achieve a smaller number of parameters or not, when the regression equation is derived. Additionally, the performance of the resultant regression equation is of greater significance than the intention when the regression equation is derived; clarification of the intention in regression does not affect the value of the regression equation. Therefore, the definition of parametric regression is somewhat ambiguous: it is, in general, regression which uses a regression equation with a smaller number of parameters (i.e., parametric functions). Furthermore, we should take note of the fact that an overparameterized model does not indicate a model with too strong an intention to achieve a smaller number of parameters, but a model with too many parameters, and an underparameterized model means a model with too few parameters.

After all, the use of a certain term to specify a certain regression depends heavily on the context of the discussion and historical circumstances: a choice among the terms nonparametric regression, a model based on fuzzy logic, neural networks, and artificial intelligence depends on the context. There are no clear definitions for classifying these concepts. However, the most crucial issue is the derivation of suitable models or regression equations for a certain purpose, rather than establishing the terminology for such techniques. Therefore, analysts should not place a great deal of emphasis on terminology when they select or develop methods.

The use of the term parameterization as a synonym for the modeling or derivation of regression equations is common in certain disciplines. The background of this usage is that a small number of regression coefficients (parameters) of regression equations usually indicates practical usefulness in the age before computers were well developed. Under these circumstances, parameterization (i.e., the reduction of the number of regression coefficients in regression equations) was a necessary condition for a beneficial regression. Hence, parameterization necessarily resulted in obtaining useful regression equations. However, a large number of parameters is no longer impractical in the computer age, so that it is not surprising that the pursuit of practical benefits has resulted in the adoption of regression equations with many parameters. In today's environment, a small number of parameters is no longer a necessity for a useful regression. Failure to give consideration to this change may result in inappropriate terminology. For example, a regression or a model for describing plant growth and development may be called "parameterization of plant growth and development." This expression may lead to the odd expression, "growth and development are parameterized using nonparametric regression." Avoidance of the use of the term "parameterization" as an equivalent term for the derivation of regression equations and modeling will prevent such instances of confusion.

The same concern is relevant to the estimation of probability density functions. The fitting of a regression equation to data with a small number of parameters, such as with the use of a normal distribution, in order to represent the data distribution using a smooth probability density function, was synonymous with obtaining a beneficial probability density function when computer processing power was more limited. In addition, distributions such as the normal distribution have the advantage of a history of use, through which its characteristics are well known, and a variety of applications have been developed. However, a probability density function with many parameters now causes no inconvenience. Rather, discussions based on an unnatural fitting of data, even though this may be unsuitable for a normal distribution, to a normal distribution can lead to conclusions that do not reflect the features of the data sufficiently. This is a situation of real concern. This consideration emphasizes the importance of capturing the essence of data by estimating a probability density function through the application of nonparametric regression. The tendency commonly observed is that parameterization (i.e., the reduction of the number of parameters) and beneficial fitting are becoming less closely related. Hence, the derivation of a probability density function should not be represented as parameterization.

Times have changed. It can no longer be considered that effective modeling can only be the equivalent of the representation of data with a regression equation or a model with a small number of parameters, when we seek to derive regression equations or probability density functions. Importance should be placed not on the number of parameters but on the construction of effective models in line with our purposes.

CHAPTER 2

SMOOTHING FOR DATA WITH AN EQUISPACED PREDICTOR

2.1 INTRODUCTION

This chapter deals with the smoothing of data in which a predictor and a target variable (object variable) are scalars, and the values of predictors are equispaced. That is, the collected data is $\{(X_i, Y_i)\}$ $(1 \leq i \leq n, \ X_i = i)$ (n is the number of data). The purpose of the methods treated here is that $\{Y_i\}$ is smoothed to obtain $\{\hat{Y}_i\}$ (estimates (estimated values)). The term smoothing means that $\{\hat{Y}_i\}$ becomes a smooth function with i as a predictor. The main goal of this chapter is to provide an understanding of the rudiments of smoothing by discussing the simplest form of smoothing. Nevertheless, this form of data is often used practically to reach significant results and hence the methods discussed in this chapter are beneficial for actual use.

2.2 MOVING AVERAGE AND BINOMIAL FILTER

Moving average is the most typical technique for smoothing one-dimensional equispaced data. In particular, the values of the predictor for most time-series data are equispaced and the target variable is the result of observation or census at each time so that smoothing of the data by moving average to determine a rough trend of the time-series behavior of the target variable is still a common procedure currently. There are many examples that indicate data trends by plotting meteorological data or economic change data on graphs and by drawing the results of smoothing by the moving average on the same graphs. The main reason for the common use of the moving average as a smoothing technique appears to be that the concept of the moving average is simple to understand and thus provides an analyst with a sense of security, which is supported by its long use. However, the moving average faces many problems, as are illustrated in Chapter 1. Therefore, the moving average is formulated in the context of nonparametric regression to examine its characteristics. Let us assume that the width of the range upon taking an average is $(2m + 1)$. That is, the estimate at X_i is the average of $(2m + 1)$ data: $\{Y_j\}$ $((i - m) \leq j \leq (i + m))$. As a result, the resulting estimate $(\hat{Y}_i)$ is

$$\hat{Y}_i = \sum_{j=1-m}^{n+m} w_{ij} Y_j, \tag{2.1}$$

$$w_{ij} = \begin{cases} \dfrac{1}{2m + 1} & \text{if } -m \leq (i - j) \leq m \\ 0 & \text{otherwise.} \end{cases} \tag{2.2}$$

There are several possible ways to derive estimates close to the two ends. The reflection boundary condition is adopted here:

$$Y_0 = Y_1, \quad Y_{-1} = Y_2, \quad \ldots, \quad Y_{-m+1} = Y_m; \tag{2.3}$$

$$Y_{n+1} = Y_n, \quad Y_{n+2} = Y_{n-1}, \quad \ldots, \quad Y_{n+m} = Y_{n-m+1}. \tag{2.4}$$

This procedure is repeated when $m > n$. For example, when $m = 1$, eq(2.1) gives $\hat{Y}_1 = (Y_0 + Y_1 + Y_2)/3$. In light of $Y_0 = Y_1$, substantially, $\hat{Y}_1 = (2Y_1 + Y_2)/3$. Taking this point into account, eq(2.1) should be replaced with the equation below to describe the effects of each datum on an estimate more appropriately.

$$\hat{Y}_i = \sum_{j=1}^{n} w'_{ij} Y_j. \tag{2.5}$$

The region of summation on the right-hand side is different from that of eq(2.1). w'_{ij} is easily derived by the following method. At first, K data sets are prepared: $\{X(k)_i, Y(k)_i\}$ $(1 \leq i \leq n, \ X(k)_i = i, \ Y(k)_k = 1,$ and $Y(k)_i = 0$ for another $Y(k)_i)$ $(1 \leq k \leq K)$. The smoothing procedure described in eq(2.5) with these data sets gives

$$\hat{Y}(k)_i = w'_{ik}. \tag{2.6}$$

That is, $\{\hat{Y}(k)_i\}$, which is the estimate given by the data set, is identical to $\{w'_{ik}\}$. When a smoothing method is known to be described in the form of eq(2.5) but it is not easy to represent $\{w'_{ik}\}$ using a simple equation, $\{w'_{ik}\}$ is obtained by the method above. $\{w'_{ik}\}$ is called weight when the relationship between data and estimates is depicted as eq(2.5). The value of weight signals the degree of importance with which each $\{Y_j\}$ is treated. The wider region of data is used to calculate the moving average as the value of m becomes larger; the degree of the smoothing is strengthened according to the value of m. That is, the estimates become smoother. Therefore, an appropriate value of m is usually selected by drawing estimates using several values of m on a graph.

To realize the behavior of the estimates by the moving average, the daily average air temperature observed in the city of Wakkanai, Hokkaido, Japan, in January 1996 is used. These data were obtained by an Automated Meteorological Data Aquisition System (AMeDAS) observation station (run by the Meteorological Agency of the Japanese government). This example is useful because meteorological data such as temperature data are usually equispaced on a predictor and smoothing by moving average is often carried out. Furthermore, the characteristics of each smoothing method are expected to be revealed by this example, since the temperature in Wakkanai in January exhibits dramatic oscillation. These data and the estimates smoothed by the moving average ($m = 1$) are displayed in figure 2.1(left). The curve connecting the estimates reveals the characteristics of the data almost fully by ironing out the bumps of data to a certain extent. Minor fluctuations, however, remain in the estimates because of inadequate smoothing. The values of weights ($\{w'_{ij}\}$) for this smoothing are indicated in figure 2.1(right). As mentioned before, the weights for the estimates near the two ends take large values for the data on the edge, while those for other estimates are the average of three data: the data at the point where the estimate is derived, and adjacent data to the right and left.

Figure 2.2 and figure 2.3 also illustrate the estimates and weights ($\{w'_{ij}\}$), when $m = 3, 4$ is set out. The increase of the value of m gives smoother estimates. The graphs clearly show that the increase of the value of m accompanies the reduction of the weights ($\{w'_{ij}\}$) which broaden out from the point of estimation. It is noted, however, that a comparison between the estimates and the data shows that these estimates contain unsatisfactory ones as the result of smoothing the data. The most impressive finding is the right-hand side of figure 2.3(left) ($m = 4$). A local minimal value is observed at $i = 21$, and a local maximal value is observed at $i = 25$. The data do not suggest such behavior. Rather, each point appears to be a local maximal point and a local minimal one.

As stated in Chapter 1, this problem is not exclusive to specific data. Estimates obtained by the moving average are often far from the behavior suggested by the data. If the purpose of smoothing is to determine the rough trend, such artificiality may not be a problem on some occasions. However, the estimates drawn on the right-hand side of figure 2.3(left) ($m = 4$) would cause an analyst to be hesitant about using them even as a rough summary of the data trend. Moreover, when the estimates by smoothing are utilized for complex analysis, we must point out the risky nature of obtaining estimates by the moving average. We consider that unlikely conclusions may be derived if we set $i = 25$ to be a local maximal point in figure 2.3(left) ($m = 4$) in order to carry out an analysis with other meteorological data, or to investigate the relationship between the obtained results and the effects of meteorological conditions on living things and social activities.

Therefore, a more suitable smoothing is attempted, to overcome these disadvantages. We postulate that smoothing by the moving average is an unnatural calculation. That is, the average of $(2m + 1)$ data near the point where an estimate is obtained means that the $(2m + 1)$ data are treated equally while the remaining data receive no attention. We assume, however, that the data at the position where the estimation is

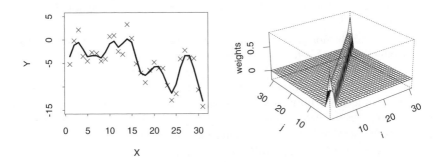

Figure 2.1   Smoothing by moving average ($m = 1$). × indicates data. A curve connects the estimates (left). Weights ($\{w'_{ij}\}$) for obtaining estimates in figure 2.1(left), (right).

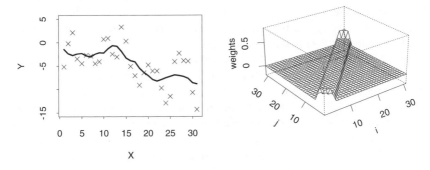

Figure 2.2   Smoothing by moving average ($m = 3$). × indicates data. A curve connects the estimates (left). Weights ($\{w'_{ij}\}$) for obtaining estimates in figure 2.2(left), (right).

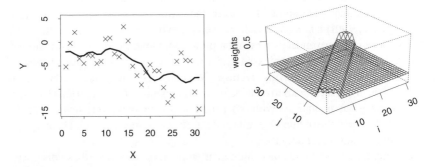

Figure 2.3   Smoothing by moving average ($m = 4$). × indicates data. A curve connects the estimates (left). Weights ($\{w'_{ij}\}$) for obtaining estimates in figure 2.3(left), (right).

carried out serve as a useful reference, and this usefulness declines gradually according to the distance between the data and the estimation point. Therefore, the weights ($\{w'_{ij}\}$) should be reduced gradually as a data point goes far from the estimation point; they should not be like those in figure 2.2(right) and figure 2.3(right).

This standpoint suggests the concept of the binomial filter. While the binomial filter and the moving average have the use of data average close to the estimation point in common, the binomial filter adjusts the values of weights ($\{w'_{ij}\}$) to reduce the values according to the distance between the data point and the estimation point. The binomial filter has the further advantage of intelligibleness and ease of calculation. The estimate by this method is

$$\hat{Y}_i = \sum_{j=1-0.5m}^{n+0.5m} w_{ij} Y_j, \tag{2.7}$$

$$w_{ij} = \begin{cases} \dfrac{{}_mC_{(0.5m+i-j)}}{2^m} & \text{if } 0 \le (0.5m + i - j) \le m \\ 0 & \text{otherwise,} \end{cases} \tag{2.8}$$

where m is a positive even number. ${}_mC_k$ is the number of combinations when k items are chosen from among m ones. It is defined as

$$_mC_k = \frac{m!}{k! \cdot (m - k)!}, \tag{2.9}$$

where "!" indicates factorial. The estimates near the two ends are analogous to those used by the moving average; for example,

$$Y_0 = Y_1, \quad Y_{-1} = Y_2, \quad \dots, \quad Y_{-0.5m+1} = Y_{0.5m}; \tag{2.10}$$

$$Y_{n+1} = Y_n, \quad Y_{n+2} = Y_{n-1}, \quad \dots, \quad Y_{n+0.5m} = Y_{n-0.5m+1}. \tag{2.11}$$

If $0.5m > n$, this procedure is repeated. Some characteristics are shared by the moving average: m is a positive even number and the larger value of m provides smoother estimates. $\{w'_{ij}\}$ for the binomial filter, which has a similar role to that of the moving average, is obtainable.

The origin of the term binomial filter is that the most significant element of eq(2.8) is ${}_mC_{(0.5m+i-j)}$, which is called a binomial factor. When the coefficient of $a^{0.5m-i+j}$ $b^{0.5m+i-j}$, which is given by the expansion of the equation $(a + b)^m$ and which is called a binomial factor, is divided by 2^m, the value equals w_{ij} (not w'_{ij}) under the conditions that m is a positive even number, $0 \le (0.5m - i + j) \le m$, and $0 \le (0.5m + i - j) \le m$. The equation below is of help:

$$\begin{aligned} (a + b)^m &= {}_mC_0 \cdot a^m +_m C_1 \cdot a^{m-1}b +_m C_2 \cdot a^{m-2}b^2 + \cdots \\ &\quad +_m C_k \cdot a^{m-k}b^k + \cdots +_m C_{m-1} \cdot a^1 b^{m-1} +_m C_m \cdot b^m. \end{aligned} \tag{2.12}$$

For instance, when $m = 2$, we obtain

$$(a + b)^2 = (1 \cdot a^2 + 2 \cdot ab + 1 \cdot b^2). \tag{2.13}$$

Accordingly, equations such as these are given:

$$\hat{Y}_2 = \frac{1}{4}(Y_1 + 2Y_2 + Y_3), \tag{2.14}$$

$$\hat{Y}_5 = \frac{1}{4}(Y_4 + 2Y_5 + Y_6). \tag{2.15}$$

The same equations are derived using eq(2.7). When $m = 4$, we have

$$(a + b)^4 = (1 \cdot a^4 + 4 \cdot a^3 b + 6 \cdot a^2 b^2 + 4 \cdot ab^3 + 1 \cdot b^4). \tag{2.16}$$

As a result, the equations below are obtained.

$$\hat{Y}_3 = \frac{1}{16}(Y_1 + 4Y_2 + 6Y_3 + 4Y_4 + Y_5), \tag{2.17}$$

$$\hat{Y}_7 = \frac{1}{16}(Y_5 + 4Y_6 + 6Y_7 + 4Y_8 + Y_9). \tag{2.18}$$

As is well known, binomial factors are calculated using Pascal's triangle:

$$
\begin{array}{ccccccc}
 & & & 1 & & & \\
 & & 1 & & 1 & & \\
 & 1 & & 2 & & 1 & \\
1 & & 3 & & 3 & & 1 \\
1 & & 4 & & 6 & & 4 & & 1 \\
\vdots & \vdots & \vdots & \vdots & \vdots & \vdots & \vdots
\end{array}
$$

First, "1" is written on the first line. The procedure for calculating the value on the next line by the addition of the two adjacent numbers on the above line ("0" is substituted for a vacant place) is repeated. The kth number on the $(m + 1)$th line is identical to the coefficient of $a^{m-k+1}b^{k-1}$ on the right-hand side of eq(2.12). This relationship is proved by the equation

$$(a + b)^m = (a + b)^{m-1}(a + b). \tag{2.19}$$

The comparison of the coefficient of $a^{m-k+1}b^{k-1}$ on both sides indicates that the coefficient of $a^{m-k+1}b^{k-1}$ given by the expansion of the left-hand side coincides with the sum of the coefficient of $a^{m-k}b^{k-1}$ and that of $a^{m-k+1}b^{k-2}$, which are derived from the expansion of $(a + b)^{m-1}$ on the right-hand side. On the other hand, the numbers obtained on each line of Pascal's triangle by dividing by 2^m are

$$
\begin{array}{ccccccc}
 & & & 1 & & & \\
 & & \frac{1}{2} & & \frac{1}{2} & & \\
 & \frac{1}{4} & & \frac{1}{2} & & \frac{1}{4} & \\
\frac{1}{8} & & \frac{3}{8} & & \frac{3}{8} & & \frac{1}{8} \\
\frac{1}{16} & & \frac{1}{4} & & \frac{3}{8} & & \frac{1}{4} & & \frac{1}{16} \\
\vdots & \vdots & \vdots & \vdots & \vdots & \vdots & \vdots
\end{array}
$$

To create the triangle shown above directly, after "1" is written on the first line, the procedure in which the average of the two adjacent numbers on the above line is written down on the next line ("0" is substituted for a vacant place) is repeated. When

m is a positive even number, the numbers on the mth line are the coefficients of the binomial filter (w_{ij} in eq(2.7) (page 27)). In the definition of a binomial filter in eq(2.7), m is defined as a positive even number rather than a positive integer, and thus $0.5m$ appears in eq(2.7). If m is defined as a positive integer, $0.5m$ becomes m; a simpler expression is realized. Nevertheless, the definition of m as a positive even number clarifies its association with the binomial factor and Pascal's triangle. w_{ij} in eq(2.7) can be represented as $w(m)_{ij}$ because w_{ij} is a function of m; it yields

$$w(m+2)_{ij} = \frac{1}{4}w(m)_{i(j-1)} + \frac{1}{2}w(m)_{ij} + \frac{1}{4}w(m)_{i(j+1)}. \tag{2.20}$$

By the same token, $\hat{Y}_i$ given by $w(m)_{ij}$ is described as $\hat{Y}(m)_i$, yielding

$$
\begin{aligned}
\hat{Y}(m+2)_i &= \sum_j w(m+2)_{ij} Y_j \\
&= \sum_j \left(\frac{1}{4}w(m)_{i(j-1)} + \frac{1}{2}w(m)_{ij} + \frac{1}{4}w(m)_{i(j+1)} \right) Y_j \\
&= \frac{1}{4}\hat{Y}(m)_{i-1} + \frac{1}{2}\hat{Y}(m)_i + \frac{1}{4}\hat{Y}(m)_{i+1},
\end{aligned}
\tag{2.21}
$$

where the values of $\{Y_i\}$ if they are not located in the region $(1 \leq i \leq n)$ on the second line are defined by eq(2.10) and eq(2.11), and the summation with "j" on the first and second lines is taken in a wide enough region to calculate the value of $\hat{Y}(m+2)_i$. The equation above allows the computations of $\{\hat{Y}(2)_i\}, \{\hat{Y}(4)_i\}$, $\{\hat{Y}(6)_i\}, \ldots$ sequentially; it is a less costly calculation than that using eq(2.7), although this computational saving is not of significant value with the wide availability of powerful computers today. However, this consideration facilitates the understanding of smoothing by allowing the realization that the larger values of m render a higher degree of smoothing; that is, the gradual increase of the degree of smoothness of the estimates corresponds to the repetition of the smoothing procedure defined by eq(2.21).

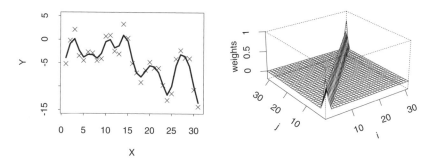

Figure 2.4 Smoothing by binomial filter (left). The weights ($\{w'_{ij}\}$) used to obtain the estimates in figure 2.4(left), (right).

Figure 2.4(left), figure 2.5(left), and figure 2.6(left) illustrate the estimates obtained by binomial filter smoothing. Their comparison with figure 2.1(left) (page 26), figure

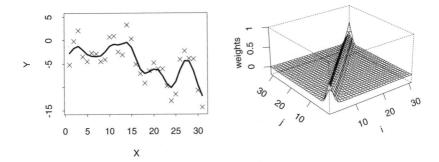

Figure 2.5   Smoothing by binomial filter (left). The weights ($\{w'_{ij}\}$) used to obtain the estimates in figure 2.5(left), (right).

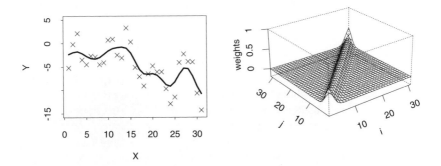

Figure 2.6   Smoothing by binomial filter (left). The weights ($\{w'_{ij}\}$) used for the estimates in figure 2.6(left), (right).

2.2(left) (page 26), and figure 2.3(left) (page 26) reveals considerable differences. The binomial filter provides more accurate estimates. For example, the behavior of the estimates in the right-hand half of figure 2.5(left) is more realistic than that in figure 2.3(left). The former is reminiscent of a smooth curve that a human being would superimpose on a data plotted graph. The background of such estimates consists of the weights ($\{w_{ik}^*\}$ calculated using eq(2.6)), which are displayed in figure 2.4(right), figure 2.5(right), and figure 2.6(right). The weight of the datum near an estimation point takes a larger value, and it declines as the datum becomes very distant from the estimation point, with the exception of the estimation points close to the two ends. The appearance of the weights implies that the binomial filter outperforms the moving average. This is because these contours of weights — which are based on the concept that when an estimate at a point is derived, great importance is attached to data near these points and the importance declines with the distance from the estimation point — are consistent with the commonsense understanding that data obtained under similar conditions take on more importance and the data obtained under less similar conditions take on less importance. This intuitive judgment is

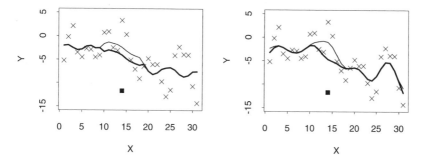

Figure 2.7   Result of moving average when the value of 15 is subtracted from Y_{14}; ■ indicates the moved data, and the thick line shows the estimate ($m = 4$) (left). Result of binomial filter when the value of 15 is subtracted from Y_{14}; ■ indicates the moved data, and the thick line shows the estimate ($m = 6$) (right).

ascertained by observing the changes in estimates of the moving average and the binomial filter when one of the data is moved. Figure 2.7(left) compares the estimates calculated using the data which were obtained by subtracting 15 from Y_{14} in the original data, and those calculated using the original ones. As is assumed on the basis of the weights ($\{w_{ik}'\}$), the constant influence of the movement of Y_{14} extends to the estimates, which are positioned 4 away from the data, but does not extend to the estimates positioned more than 4 away. On the other hand, figure 2.7(right) shows the result of using the binomial filter with the similar procedure. The influence of the movement of Y_{14} declines as the estimates become distant from X_{14}. Thus, the weights ($\{w_{ik}'\}$) bear great importance for smoothing partly because they represent the response to the movements of data.

The appropriateness of the response of the estimates to the changes in data can be taken into account when we investigate whether a common method can be applied to data sets with similar values. For instance, let us suppose that observations of the daily average air temperature in January in another year give the same measurements as those of 1996, except that the value of Y_{14} is 15 degrees less than Y_{14} of 1996. Thus, figure 2.7(left) and figure 2.7(right) show the estimates obtained by these procedures. Figure 2.7(left) and figure 2.7(right) can be viewed as the graphs illustrating the problem of which estimates should be used to compare the two data sets. Since one of the goals of smoothing is to extract a rough trend from a data set, the comparison of the rough trend from a data set with that from another data set should lead to an appropriate analysis. To achieve this goal, it is vital that the comparison of estimates with those obtained by the data set with one data moved should be carried out appropriately. With this approach, we can confirm the practical significance of the values of weights.

2.3   HAT MATRIX

The estimates obtained using eq(2.5) (page 24) are calculated by the weighted summation of the adjacent values. This form of estimation is not confined to binomial filters and the like. For example, the relationship between the values of the target variable of the data and the corresponding estimates is represented in this form when multiple regression (multiple linear regression) is carried out by the least squares (least square method). The relationship between the values of predictors and the corresponding estimates, as described by eq(2.5), means that the estimates can be derived by a linear estimator. When the sum of the weights equals unity, the procedure is called weighted average. Note that w'_{ij} in eq(2.5) is usually replaced with h_{ij} in order to represent an estimator that allows the derivation of the estimates by a linear estimator:

$$\hat{Y}_i = \sum_{j=1}^{n} h_{ij} Y_j. \tag{2.22}$$

The matrix ($\mathbf{H}$) with h_{ij} as an ij element is termed a hat matrix. The term hat matrix is derived from the fact that an estimate is often represented with a " hat " ($\hat{}$) on a letter and hence the matrix for deriving estimates has come to be called a hat matrix. The hat matrix for smoothing, in particular, is occasionally called a smoothing matrix (smoother matrix). The hat matrix plays a significant role in regression in general, for deriving estimates using a linear estimator. In fact, many statistics concerning such regressions are calculated with a hat matrix. The examples shown above demonstrate unequivocally that the values of the elements of a hat matrix should be intelligible ones to obtain desirable estimates. This tendency is not limited to smoothing. The decency in a hat matrix is a point which we should take into consideration in evaluating the appropriateness of a linear estimator. Regression methods other than linear estimators sometimes use concepts that are analogous to the hat matrix to examine the appropriateness of regression equations (regression functions) and models. Then, the hat matrix for fitting polynomials (polynomial functions) to the Wakkanai data by the least squares is derived in order to clarify the meaning of the hat matrix. This is because the hat matrix for the fitting is obtainable because the fitting of polynomial by the least squares is a type of regression by a linear estimator. Moreover, it is beneficial to understand the characteristics of the fitting of polynomials by the least squares, since this method is often utilized for practical purposes. It should be noted that though $X_i = i$ has been presumed so far, the arguments in this section do not require this assumption.

The data have the form of $\{(X_i, Y_i)\}$ $(1 \le i \le n)$ (X_i is a predictor, and Y_i is a target variable; if a predictor is equispaced, $X_i = i$). The fitting of the qth polynomial to the data by the least squares is realized by minimizing the values below to obtain $\{a_0, a_1, \ldots, a_q\}$.

$$E_{squares} = \sum_{i=1}^{n}\left(a_0 + \sum_{j=1}^{q} a_j X_i^j - Y_i\right)^2 = \sum_{i=1}^{n}\left(\sum_{j=0}^{q} a_j X_i^j - Y_i\right)^2. \qquad (2.23)$$

The sum of the squared difference between data and estimates, such as $E_{squares}$, is called the residual sum of squares. To minimize $E_{squares}$, the matrix and vectors below are defined.

$$\mathbf{X} = \begin{pmatrix} 1 & X_1 & X_1^2 & \cdots & X_1^q \\ 1 & X_2 & X_2^2 & \cdots & X_2^q \\ \vdots & \vdots & \vdots & \ddots & \vdots \\ 1 & X_n & X_n^2 & \cdots & X_n^q \end{pmatrix}, \qquad (2.24)$$

$$\mathbf{y} = \begin{pmatrix} Y_1 \\ \vdots \\ Y_n \end{pmatrix}, \qquad (2.25)$$

$$\mathbf{a} = \begin{pmatrix} a_0 \\ \vdots \\ a_q \end{pmatrix}. \qquad (2.26)$$

This $\mathbf{X}$ is a form of design matrix. A design matrix also plays an important role in multiple regression using the least squares. Model matrix is another name for the design matrix. Then, to minimize $E_{squares}$, as defined in eq(2.23), it is differentiated with respect to each of $\{a_0, a_1, \ldots, a_q\}$ to yield the $(q+1)$ equations:

$$\sum_{i=1}^{n}\sum_{j=0}^{q} a_j X_i^{j+k} = \sum_{i=1}^{n} X_i^k Y_i \qquad (0 \le k \le q). \qquad (2.27)$$

Eq(2.27) can be rewritten as shown below using eq(2.24), eq(2.25), and eq(2.26).

$$\mathbf{X}^t \mathbf{X} \mathbf{a} = \mathbf{X}^t \mathbf{y}. \qquad (2.28)$$

The superscript t indicates the transpose of a matrix. Eq(2.28) is called a normal equation. The equivalence between eq(2.27) and eq(2.28) is ensured by representing the matrices and vectors in eq(2.28) as their elements. First, the left-hand side of eq(2.28) becomes

$$\mathbf{X}^t \mathbf{X} \mathbf{a} = \begin{pmatrix} 1 & 1 & \cdots & 1 \\ X_1 & X_2 & \cdots & X_n \\ X_1^2 & X_2^2 & \cdots & X_n^2 \\ \vdots & \vdots & \ddots & \vdots \\ X_1^q & X_2^q & \cdots & X_n^q \end{pmatrix} \begin{pmatrix} 1 & X_1 & X_1^2 & \cdots & X_1^q \\ 1 & X_2 & X_2^2 & \cdots & X_2^q \\ \vdots & \vdots & \vdots & \ddots & \vdots \\ 1 & X_n & X_n^2 & \cdots & X_n^q \end{pmatrix} \begin{pmatrix} a_0 \\ a_1 \\ a_2 \\ \vdots \\ a_q \end{pmatrix}$$

$$
= \begin{pmatrix}
n & \sum_{i=1}^{n} X_i & \cdots & \sum_{i=1}^{n} X_i^{q} \\
\sum_{i=1}^{n} X_i & \sum_{i=1}^{n} X_i^{2} & \cdots & \sum_{i=1}^{n} X_i^{q+1} \\
\sum_{i=1}^{n} X_i^{2} & \sum_{i=1}^{n} X_i^{3} & \cdots & \sum_{i=1}^{n} X_i^{q+2} \\
\vdots & \vdots & \ddots & \vdots \\
\sum_{i=1}^{n} X_i^{q} & \sum_{i=1}^{n} X_i^{q+1} & \cdots & \sum_{i=1}^{n} X_i^{2q}
\end{pmatrix}
\begin{pmatrix} a_0 \\ a_1 \\ \vdots \\ a_q \end{pmatrix}
$$

$$
= \begin{pmatrix}
\sum_{i=1}^{n} \sum_{j=0}^{q} a_j X_i^{j} \\
\sum_{i=1}^{n} \sum_{j=0}^{q} a_j X_i^{j+1} \\
\sum_{i=1}^{n} \sum_{j=0}^{q} a_j X_i^{j+2} \\
\vdots \\
\sum_{i=1}^{n} \sum_{j=0}^{q} a_j X_i^{j+q}
\end{pmatrix},
\tag{2.29}
$$

and the right-hand side becomes

$$
\mathbf{X}^t \mathbf{y} = \begin{pmatrix}
1 & 1 & \cdots & 1 \\
X_1 & X_2 & \cdots & X_n \\
X_1^{2} & X_2^{2} & \cdots & X_n^{2} \\
\vdots & \vdots & \ddots & \vdots \\
X_1^{q} & X_2^{q} & \cdots & X_n^{q}
\end{pmatrix}
\begin{pmatrix} Y_1 \\ Y_2 \\ \vdots \\ Y_n \end{pmatrix}
= \begin{pmatrix}
\sum_{i=1}^{n} Y_i \\
\sum_{i=1}^{n} X_i Y_i \\
\sum_{i=1}^{n} X_i^{2} Y_i \\
\vdots \\
\sum_{i=1}^{n} X_i^{q} Y_i
\end{pmatrix}.
\tag{2.30}
$$

Eq(2.28) provides estimates of the regression coefficients (**a**):

$$
\hat{\mathbf{a}} = (\mathbf{X}^t \mathbf{X})^{-1} \mathbf{X}^t \mathbf{y}.
\tag{2.31}
$$

The uniqueness of $\hat{\mathbf{a}}$ which minimizes $E_{squares}$ is proved as follows.

$\hat{\mathbf{a}}^*$ is assumed; the number of elements of $\hat{\mathbf{a}}^*$ is the same as that of $\hat{\mathbf{a}}$. Then, $\Delta E_{squares}$ is obtained as shown below.

$$
\begin{aligned}
\Delta E_{squares} &= \| \mathbf{y} - \mathbf{X}\hat{\mathbf{a}}^* \|^2 - \| \mathbf{y} - \mathbf{X}\hat{\mathbf{a}} \|^2 \\
&= (\mathbf{y} - \mathbf{X}\hat{\mathbf{a}}^*)^t (\mathbf{y} - \mathbf{X}\hat{\mathbf{a}}^*) - (\mathbf{y} - \mathbf{X}\hat{\mathbf{a}})^t (\mathbf{y} - \mathbf{X}\hat{\mathbf{a}}) \\
&= -\hat{\mathbf{a}}^{*t}\mathbf{X}^t\mathbf{y} - \mathbf{y}^t\mathbf{X}\hat{\mathbf{a}}^* + \hat{\mathbf{a}}^{*t}\mathbf{X}^t\mathbf{X}\hat{\mathbf{a}}^* + \hat{\mathbf{a}}^t\mathbf{X}^t\mathbf{y} + \mathbf{y}^t\mathbf{X}\hat{\mathbf{a}} - \hat{\mathbf{a}}^t\mathbf{X}^t\mathbf{X}\hat{\mathbf{a}} \\
&= -\hat{\mathbf{a}}^{*t}\mathbf{X}^t\mathbf{X}\hat{\mathbf{a}} - \mathbf{y}^t\mathbf{X}\hat{\mathbf{a}}^* + \hat{\mathbf{a}}^{*t}\mathbf{X}^t\mathbf{X}\hat{\mathbf{a}}^* + \hat{\mathbf{a}}^t\mathbf{X}^t\mathbf{X}\hat{\mathbf{a}} + \mathbf{y}^t\mathbf{X}\hat{\mathbf{a}} \\
&\quad - \hat{\mathbf{a}}^t\mathbf{X}^t\mathbf{X}\hat{\mathbf{a}} \\
&= -\hat{\mathbf{a}}^{*t}\mathbf{X}^t\mathbf{X}\hat{\mathbf{a}} - \mathbf{y}^t\mathbf{X}\hat{\mathbf{a}}^* + \hat{\mathbf{a}}^{*t}\mathbf{X}^t\mathbf{X}\hat{\mathbf{a}}^* + \mathbf{y}^t\mathbf{X}\hat{\mathbf{a}},
\end{aligned}
\tag{2.32}
$$

where $\| \ \|$ indicates the vector length. Eq(2.31) is used to obtain equality between the fourth line and fifth line. Eq(2.31) gives

$$
\begin{aligned}
(\mathbf{X}^t\mathbf{y} - \mathbf{X}^t\mathbf{X}\hat{\mathbf{a}})^t(\hat{\mathbf{a}} - \hat{\mathbf{a}}^*) &= \mathbf{y}^t\mathbf{X}\hat{\mathbf{a}} - \mathbf{y}^t\mathbf{X}\hat{\mathbf{a}}^* - \hat{\mathbf{a}}^t\mathbf{X}^t\mathbf{X}\hat{\mathbf{a}} + \hat{\mathbf{a}}^t\mathbf{X}^t\mathbf{X}\hat{\mathbf{a}}^* \\
&= 0.
\end{aligned}
\tag{2.33}
$$

Substitution of this result into eq(2.32) yields

$$
\begin{aligned}
\Delta E_{squares} &= -\hat{\mathbf{a}}^{*t}\mathbf{X}^t\mathbf{X}\hat{\mathbf{a}} + \hat{\mathbf{a}}^{*t}\mathbf{X}^t\mathbf{X}\hat{\mathbf{a}}^* + \hat{\mathbf{a}}^t\mathbf{X}^t\mathbf{X}\hat{\mathbf{a}} - \hat{\mathbf{a}}^t\mathbf{X}^t\mathbf{X}\hat{\mathbf{a}}^* \\
&= \| \mathbf{X}(\hat{\mathbf{a}} - \hat{\mathbf{a}}^*) \|^2 \\
&\geq 0.
\end{aligned}
\tag{2.34}
$$

Since equality is satisfied when $\hat{a} = \hat{a}^*$, it is proved that $\hat{a}$ as obtained by eq(2.31) is the unique solution to the minimization of $E_{squares}$. This $\hat{a}$ renders $\{\hat{Y}_i\}$ $(1 \leq i \leq n)$:

$$\hat{y} = X\hat{a} = X(X^tX)^{-1}X^ty, \qquad (2.35)$$

where $\hat{y}$ is

$$\hat{y} = \begin{pmatrix} \hat{Y}_1 \\ \vdots \\ \hat{Y}_n \end{pmatrix}. \qquad (2.36)$$

Then, eq(2.35) is rewritten as

$$\hat{y} = Hy, \qquad (2.37)$$

where H is

$$H = X(X^tX)^{-1}X^t. \qquad (2.38)$$

This H ($n \times n$ matrix) is the hat matrix for the fitting of polynomials by the least squares. That is, the procedure for the fitting of polynomials by the least squares means that a vector consisting of the data is multiplied by a hat matrix, and this is accompanied by the optimization of regression coefficients to obtain a regression equation.

It is noted that the hat matrix defined by eq(2.38) is a function of the values of the predictor of data $(\{X_i\})$ and is independent of the value of the target variable of the data $(\{Y_i\})$. In the case of meteorological data, the values of the predictor of the daily average air temperature in January, for example, are fixed in every year: $X_i = 1, 2, \ldots, 31$. Hence, the result of fitting of the same degree polynomial equation to the data from different years is identical to the multiplication of a certain hat matrix. It seems that the matrix describing the relationship between the target variable of data and the estimates may be a function of the value of the target variable of the data, because the values of both the predictor and the target variable are taken into account for the optimization. However, this is not the case. This point is noted not simply with regard to the fitting of polynomials by the least squares, but with regard to the general regressions for estimation using a linear estimator, such as in the fitting of a multiple regression equation by the least squares.

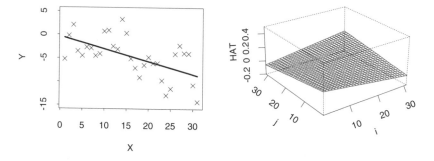

Figure 2.8   The fitting of a linear equation (left). The values of the elements of the hat matrix corresponding to figure 2.8(left), (right).

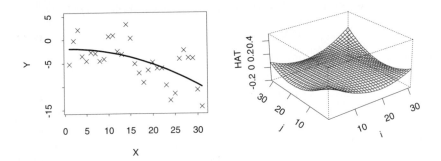

Figure 2.9   The fitting of a quadratic equation (left). The values of the elements of the hat matrix corresponding to figure 2.9(left), (right).

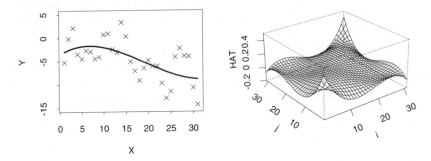

Figure 2.10   The fitting of a cubic equation (left). The values of the elements of the hat matrix corresponding to figure 2.10(left), (right).

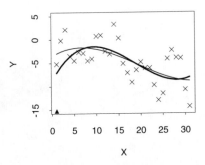

Figure 2.11   The result of the fitting of the cubic equation to the data in which the value of Y_1 (▲) is calculated by the substitution of 10 from the original value (thick line).

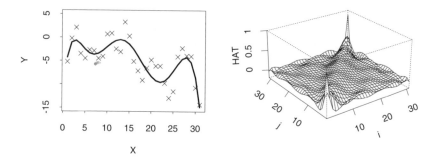

Figure 2.12   The fitting of the 7th-degree polynomial (left). The values of the elements of the hat matrix correspond to figure 2.12(left), (right).

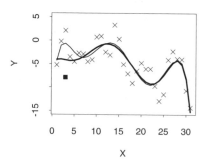

Figure 2.13   The result of fitting the 7th-degree polynomial to the data in which the value of Y_3 (■) is calculated by the substitution of 10 from the original value (thick line).

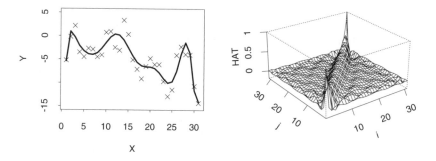

Figure 2.14   The fitting of the 11th-degree polynomial (left). The values of the elements of the hat matrix corresponding to figure 2.14(left), (right).

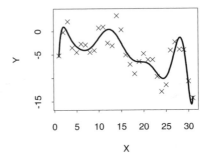

Figure 2.15   The result of estimation for the values of predictor ($\{1, 1.1, 1.2, \ldots, 31\}$) obtained using the same regression equation as for figure 2.14(left).

Figure 2.8 and figure 2.10 demonstrate the results of the fitting of polynomials to the Wakkanai data and the values of the corresponding hat matrix. Estimates obtained by applying the linear equation, the quadratic equation, and the cubic equation, shown in figure 2.8(left), figure 2.9(left), and figure 2.10(left), respectively, illustrate a too rough sketch; the degree of smoothing is apparently too large. In addition, many values of the elements of the hat matrix are negative, while some values distant from the estimation points are large. These findings indicate that the movement of a value in a data set brings about an unexpected change of the estimates. In fact, the estimates that result from the fitting of a cubic equation to the data in which the value of Y_1 is calculated by subtracting 10 from the original value (figure 2.11) indicate that more than a few estimates are displaced in a positive direction from estimates obtained by the original data. This is due to the fact that some elements of the hat matrix take negative values, and their absolute values are not small. This phenomenon is highly significant because the values of the elements of the hat matrix are determined by the values of the predictor of the data; the hat matrix demonstrated in figure 2.10(right) remains the same when a data set from a different year or different location is applied. We assume that the new data are the same as the Wakkanai data except that the value of Y_1 is the result of subtracting 10 from the Wakkanai data. The comparison between the two data sets subjected to the fittings of cubic equations is equivalent to the comparison of the two sets of estimates in figure 2.11. The changes of the estimates by the movement of Y_1 alone imply that the comparison of the two data sets using the results of the fitting of cubic equations is undesirable.

On the other hand, fitting a 7th-degree polynomial gives a local minimal point around $i = 22$ (figure 2.12(left)); however, this is impossible considering the behavior of the data. These estimates exemplify the well-known characteristic that the fitting of high-degree polynomials results in an unimaginable behavior of the estimate curves, because of the unfavorable nature of polynomials. The reason for this is clarified by deriving the hat matrix (figure 2.12(right)). That is, examining the regression using polynomials from the viewpoint of the hat matrix provides a reason for the odd behavior of the estimates obtained using the high-degree polynomials: the weights for the data given by high-degree polynomials oscillate widely. Incidentally, the direction of the changes of the estimates obtained by fitting the 7th-degree polynomial, using the data in which the value of Y_3 is obtained by subtracting 15 from the original value of Y_3 (figure 2.13), depends on the position of the estimation point. This tendency reflects the ripple of the values of the elements of the hat matrix (figure 2.12(right)) and clearly implies that such a regression can lead to unfavorable results. In particular, the responses of the estimates to the data positioned close to the two sides are strange. The contour of the hat matrix suggests this problem.

Furthermore, figure 2.14(left) demonstrates the result of fitting the 11th-degree polynomial. The estimates here are similar to those in figure 2.6(left) (page 30), which are estimates obtained using a binomial filter ($m = 10$). The values of the elements of the hat matrix (figure 2.14(right)) are also analogous to those obtained using the binomial filter ($m = 10$) (figure 2.6(right) (page 30)). Nevertheless, we hesitate to adopt the result of fitting the 11th-degree polynomial because the purpose of the fitting of polynomials is not only estimation at $\{X_i\}$ but also estimation between $\{X_i\}$. Figure 2.15 illustrates the issue. The curve based on estimates at $\{1, 1.1, 1.2, 1.3, 1.4, \ldots, 31\}$ has an odd-looking local minimal point near the right-hand end. The values of the elements of the hat matrix (figure 2.14(right)) do not imply this behavior. Therefore, though the hat matrix provides important statistics for identifying the relationship between data and estimates, it does not cover all the characteristics of regression.

The discussion so far has made it clear that the bases for the preference for using polynomials as regression equations for fitting to data are that they are familiar and that approximation using polynomials reminds us of the Taylor expansion and the Weierstrass polynomial approximation theorem; these reasons are no longer sufficient. We have no good reason to employ polynomials, in particular, high-degree polynomials. Polynomials are simply a form of regression which are representable using a hat matrix, and they do not lead to results of noteworthy quality compared with regression with another hat matrix. Rather, they often give questionable estimates. As a result, we do not recommend their usage: if the fittings of low-degree polynomials are poor, higher-degree polynomials are attempted; this procedure is familiar to most of us. It is advisable to employ a hat matrix that renders natural smoothing. This is one reason justifying the practical value of smoothing based on nonparametric regression.

Next, four important characteristics of a hat matrix corresponding to the fitting of polynomials by the least squares are introduced. Though the fitting of polynomials by the least squares has limited practical use, its mathematical characteristics play a significant role in understanding regressions for the calculation of estimates using linear estimators. We do not assume that $\{X_i\}$ is equispaced (i.e., $\{X_i\} = i$) in deriving these four characteristics. Hence, these characteristics hold if $\{X_i\}$ is not equispaced.

The first characteristic is that a hat matrix satisfies the equation below, regardless of the value of k.

$$\sum_{j=1}^{n} [\mathbf{H}]_{kj} = 1. \tag{2.39}$$

This relationship is confirmed when we consider the data of $Y_i \equiv 1$ ($1 \leq i \leq n$); apparently the trivial estimates are $\hat{Y}_i \equiv 1$ ($1 \leq i \leq n$), regardless of the degree of polynomials. Therefore, if $\mathbf{H}$ is a symmetric matrix, we have

$$\sum_{j=1}^{n}[\mathbf{H}]_{jk} = 1. \tag{2.40}$$

The second characteristic is

$$\mathbf{H} \cdot \mathbf{H} = \mathbf{H}. \tag{2.41}$$

That is, the repetition of the multiplication of a hat matrix results in the identical matrix. A matrix with this property is called an idempotent matrix. This relationship is proved by assuming the following situation: the estimates obtained by fitting a polynomial by the least squares are used as data to fit the same degree polynomial by the least squares. The resulting estimate is $\hat{\mathbf{y}}$, so that $\mathbf{H} \cdot \mathbf{Hy} = \mathbf{H}\hat{\mathbf{y}} = \hat{\mathbf{y}} = \mathbf{Hy}$ is obtained. Incidentally, a calculation based on eq(2.38) provides the same result:

$$
\begin{aligned}
\mathbf{H} \cdot \mathbf{H} &= \mathbf{X}(\mathbf{X}^t\mathbf{X})^{-1}\mathbf{X}^t \cdot \mathbf{X}(\mathbf{X}^t\mathbf{X})^{-1}\mathbf{X}^t = \mathbf{X}(\mathbf{X}^t\mathbf{X})^{-1} \cdot \mathbf{X}^t\mathbf{X}(\mathbf{X}^t\mathbf{X})^{-1} \cdot \mathbf{X}^t \\
&= \mathbf{X}(\mathbf{X}^t\mathbf{X})^{-1}\mathbf{X}^t = \mathbf{H}. \tag{2.42}
\end{aligned}
$$

A symmetric and idempotent matrix is termed a projection matrix; it is not peculiar to the hat matrix for the fitting of polynomials. However, an idempotent and not necessarily symmetric hat matrix is occasionally defined as a projection matrix, and hence a symmetric and idempotent hat matrix is more accurately defined as an orthogonal projection matrix; it should also be noted that the term "projection" has other meanings in the field of statistics in general. Furthermore, another point of terminology is that the only orthogonal projection matrix is termed a hat matrix; and other matrices for smoothing are called smoothing matrices (page 129 of [2]). The details of the projection matrix are provided in Chapter 6 of [2], for example.

The clarity of the meaning of orthogonal projection is promoted by the equation

$$
\begin{aligned}
\hat{\mathbf{y}}^t(\mathbf{y} - \hat{\mathbf{y}}) &= (\mathbf{Hy})^t(\mathbf{y} - \mathbf{Hy}) \\
&= \mathbf{y}^t\mathbf{H}^t(\mathbf{y} - \mathbf{Hy}) \\
&= \mathbf{y}^t(\mathbf{Hy} - \mathbf{HHy}) \\
&= \mathbf{y}^t(\mathbf{Hy} - \mathbf{Hy}) \\
&= 0. \tag{2.43}
\end{aligned}
$$

That is, $\hat{\mathbf{y}}$ is orthogonal to $(\mathbf{y} - \hat{\mathbf{y}})$ (the vector that consists of residuals). If $\mathbf{H}$ is an idempotent and symmetric matrix, it is an orthogonal projection matrix in this sense. This geometric relationship is illustrated in figure 2.16(left). In short, $\hat{\mathbf{y}}$ is the result of the orthogonal projection of $\mathbf{y}$ to the plane given by $\mathbf{H}$.

If a hat matrix is idempotent, the result of the repetition of the orthogonal projection is the same as that of a single orthogonal projection. As a result, if we suppose the average of $\{Y_i\}$ to be $\bar{Y}$, we have

$$
\begin{aligned}
\sum_{i=1}^{n}(Y_i - \bar{Y})^2 &= \sum_{i=1}^{n}(Y_i - \bar{Y} + \hat{Y}_i - \hat{Y}_i)^2 \\
&= \sum_{i=1}^{n}(\hat{Y}_i - \bar{Y})^2 + \sum_{i=1}^{n}(Y_i - \hat{Y}_i)^2 + 2\sum_{i=1}^{n}(\hat{Y}_i - \bar{Y})(Y_i - \hat{Y}_i)
\end{aligned}
$$

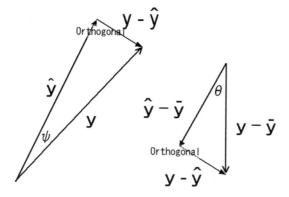

Figure 2.16   Relationship among $\mathbf{y}$, $\hat{\mathbf{y}}$, and $(\mathbf{y} - \hat{\mathbf{y}})$($\psi$ is the angle between $\mathbf{y}$ and $\hat{\mathbf{y}}$) (left). Relationship among $(\mathbf{y} - \bar{\mathbf{y}})$, $(\hat{\mathbf{y}} - \bar{\mathbf{y}})$, and $(\mathbf{y} - \hat{\mathbf{y}})$($\theta$ is the angle between $(\mathbf{y} - \bar{\mathbf{y}})$ and $(\hat{\mathbf{y}} - \bar{\mathbf{y}})$) (right).

$$= \sum_{i=1}^{n} (\hat{Y}_i - \bar{Y})^2 + \sum_{i=1}^{n} (Y_i - \hat{Y}_i)^2 - 2\bar{Y} \sum_{i=1}^{n} (Y_i - \hat{Y}_i). \quad (2.44)$$

Equality between the second line and the third line is obtained by the use of eq(2.43).

Here, let us prove that as long as a regression equation contains a constant term, $\sum_{i=1}^{n}(Y_i - \hat{Y}_i)$ equals 0. Assume that a regression equation is $g(x) + a_0$; $g(x)$ is a regression equation that does not contain a constant term, and a_0 is a constant.

$$\sum_{i=1}^{n} (Y_i - g(X_i) - a_0)^2. \quad (2.45)$$

Differentiation of the equation above with respect to a_0 to be 0 gives

$$\sum_{i=1}^{n} (Y_i - g(X_i) - a_0) = 0 \quad (2.46)$$

(Q.E.D.). Then, the substitution of eq(2.44) into eq(2.46) gives

$$\sum_{i=1}^{n} (Y_i - \bar{Y})^2 = \sum_{i=1}^{n} (\hat{Y}_i - \bar{Y})^2 + \sum_{i=1}^{n} (Y_i - \hat{Y}_i)^2. \quad (2.47)$$

The three terms of this equation represent different variabilities. $\sum_{i=1}^{n}(Y_i - \bar{Y})^2$ is the total variability of the target variable. $\sum_{i=1}^{n}(\hat{Y}_i - \bar{Y})^2$ is referred to as the variability explained by the regression equation; eq(2.40) shows that the average of $\hat{Y}_i$ is equal to that of $\bar{Y}$. $\sum_{i=1}^{n}(Y_i - \hat{Y}_i)^2$ corresponds to the variability that remains unexplained by the regression equation. Figure 2.16(right) depicts the relationships among the three vectors: $(\mathbf{y} - \bar{\mathbf{y}})$, $(\hat{\mathbf{y}} - \bar{\mathbf{y}})$, and $(\mathbf{y} - \hat{\mathbf{y}})$. We define θ as the angle between $(\mathbf{y} - \bar{\mathbf{y}})$ and $(\hat{\mathbf{y}} - \bar{\mathbf{y}})$. In addition, the division of the two sides of eq(2.47) by $\sum_{i=1}^{n}(Y_i - \bar{Y})^2$ provides

$$1 = \frac{\sum_{i=1}^{n}(\hat{Y}_i - \bar{Y})^2}{\sum_{i=1}^{n}(Y_i - \bar{Y})^2} + \frac{\sum_{i=1}^{n}(Y_i - \hat{Y}_i)^2}{\sum_{i=1}^{n}(Y_i - \bar{Y})^2}. \tag{2.48}$$

This $\frac{\sum_{i=1}^{n}(\hat{Y}_i - \bar{Y})^2}{\sum_{i=1}^{n}(Y_i - \bar{Y})^2}$ indicates the ratio of the variability expressed by the regression equation obtained by the least squares to that of the target variable of data. This ratio is referred to as the coefficient of determination and is ordinarily termed R^2. R^2 corresponds to the squared ratio of the length of $(\hat{\mathbf{y}} - \bar{\mathbf{y}})$ to that of $(\mathbf{y} - \bar{\mathbf{y}})$. Figure 2.16 reveals that R^2 is the squared cosine of the angle between the vector $(\mathbf{y} - \bar{\mathbf{y}})$ and the vector $(\hat{\mathbf{y}} - \bar{\mathbf{y}})$. It follows that if we call the angle θ, the definition of the inner product leads to

$$\cos(\theta) = \frac{\sum_{i=1}^{n}(\hat{Y}_i - \bar{Y})(Y_i - \bar{Y})}{\sqrt{\sum_{i=1}^{n}(\hat{Y}_i - \bar{Y})^2}\sqrt{\sum_{i=1}^{n}(Y_i - \bar{Y})^2}}. \tag{2.49}$$

This $\cos(\theta)$ is termed the coefficient of correlation between $\{\hat{Y}_i\}$ and $\{Y_i\}$. That is, R^2 is the squared coefficient of correlation between $\{\hat{Y}_i\}$ and $\{Y_i\}$.

The third important characteristic of the hat matrix corresponding to the fitting of polynomials by the least squares is that the diagonal elements of the hat matrix ($[\mathbf{H}]_{kk}$) satisfy

$$\frac{1}{n} \leq [\mathbf{H}]_{kk} \leq 1. \tag{2.50}$$

Since regression using polynomials satisfies eq(2.41) (page 40), and the corresponding hat matrix is symmetric, we discover the relationship

$$\begin{aligned}
[\mathbf{H}]_{kk} &= \sum_{j=1}^{n}[\mathbf{H}]_{kj}[\mathbf{H}]_{jk} \\
&= \sum_{j=1}^{n}[\mathbf{H}]_{kj}[\mathbf{H}]_{kj} \\
&\geq ([\mathbf{H}]_{kk})^2.
\end{aligned} \tag{2.51}$$

Thus, we obtain

$$0 \leq [\mathbf{H}]_{kk} \leq 1. \tag{2.52}$$

Furthermore, assuming $0 \leq [\mathbf{H}]_{kk} < \frac{1}{n}$, we obtain $0 \leq \hat{Y}_k < \frac{1}{n}$ if $Y_k = 1$ and $Y_i = 0$ $(i \neq k)$. Considering eq(2.39) (page 39), estimates other than $\hat{Y}_k$ derived to minimize the residual sum of squares are found to be $\hat{Y}_i = \frac{1 - \hat{Y}_k}{n - 1}$ $(i \neq k)$. Thus, the residual sum of squares becomes

$$\begin{aligned}
\left(\frac{1 - \hat{Y}_k}{n - 1}\right)^2 (n - 1) + (1 - \hat{Y}_k)^2 &= \frac{n}{n - 1}(1 - \hat{Y}_k)^2 \\
&> \frac{n}{n - 1}\left(1 - \frac{1}{n}\right)^2 \\
&= \frac{n - 1}{n}.
\end{aligned} \tag{2.53}$$

On the other hand, if the estimates ($\hat{Y}_i = \frac{1}{n}$ for all i) are adopted, the residual sum of squares is $\frac{n-1}{n}$, which is smaller than those obtained using the estimates adopted previously. That is, the estimates ($\hat{Y}_i = \frac{1}{n}$ for all i) are desirable in terms of the residual sum of squares, and these estimates can be estimates for any degree polynomials. It is concluded that the estimates ($\hat{Y}_i = \frac{1}{n}$ $(1 \le i \le n)$) $(1 \le i \le n)$ outperform those that satisfy $\hat{Y}_k < \frac{1}{n}$ and $\hat{Y}_i = \frac{1-\hat{Y}_k}{n-1}$ $(i \ne k)$. This is the contradiction that results from the assumption that the fitting of polynomials by the least squares can provide the relationship $[\mathbf{H}]_{kk} < \frac{1}{n}$. Therefore, by *reductio ad absurdum*, it is proved that $[\mathbf{H}]_{kk} \ge \frac{1}{n}$. This conclusion and eq(2.52) provide eq(2.50) (Q.E.D.).

The fourth characteristic is that the trace (the sum of the diagonal elements) of a hat matrix for the fitting of a polynomial by the least squares equals the sum of unity and the degree of the polynomial; the sum is identical to the number of regression coefficients. This relationship is given by

$$
\begin{aligned}
\text{trace}(\mathbf{X}(\mathbf{X}^t\mathbf{X})^{-1}\mathbf{X}^t) &= \sum_{i=1}^{n}\sum_{j=1}^{q+1}\sum_{k=1}^{q+1}[\mathbf{X}]_{ij}[(\mathbf{X}^t\mathbf{X})^{-1}]_{jk}[\mathbf{X}^t]_{ki} \\
&= \sum_{j=1}^{q+1}\sum_{k=1}^{q+1}[(\mathbf{X}^t\mathbf{X})^{-1}]_{jk}\sum_{i=1}^{n}[\mathbf{X}^t]_{ki}[\mathbf{X}]_{ij} \\
&= \sum_{j=1}^{q+1}\sum_{k=1}^{q+1}[(\mathbf{X}^t\mathbf{X})^{-1}]_{jk}[(\mathbf{X}^t\mathbf{X})]_{kj} \\
&= \text{trace}(\mathbf{I}) \\
&= q+1,
\end{aligned}
\tag{2.54}
$$

where $[(\mathbf{X}^t\mathbf{X})^{-1}]_{jk}$ is a jk-element of the matrix $(\mathbf{X}^t\mathbf{X})^{-1}$, and $\mathbf{I}$ is an identity matrix (the size of $\mathbf{I}$ is $(q+1) \times (q+1)$).

2.4 LOCAL LINEAR REGRESSION

The previous section confirmed that the fitting of polynomials by the least squares is not a favorable method for smoothing. This is due to the fact that only one polynomial is fitted to the entire region in which the data are located. The fitting of only one equation to a large number of data with complex behavior occasionally yields undesirable results: the equation fits well to the data in specific regions but does not in other regions, or the shape of the regression curves does not correspond to the data because the curves unduly reflect the features of parametric functions. This consideration leads us to the concept of partitioning the data into several regions for fitting curves to data in each region and connecting them so as to gain appropriate estimates. This concept is called the piecewise fitting of regression equations.

Among the methods that use this approach, smoothing by a spline function (or simply, spline) is most popular. The simplest method is the piecewise fitting of linear equations on the condition that the estimates take continuous values at the boundaries of regions. This function is called a first degree spline function or linear spline function. Figure 2.17(left) demonstrates the result of the fitting using the Wakkanai

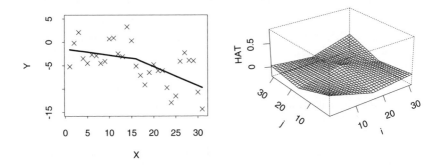

Figure 2.17   Fitting of a linear spline (broken lines). The internal knot is located at point $i = 16$ (left). Values of elements of the hat matrix that corresponds to figure 2.17(left), (right).

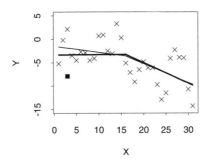

Figure 2.18   Estimates by the fitting of linear splines (broken lines) when the value of Y_3, represented by ■, is calculated by subtracting 10 from the original Y_3. The internal knot is located at point $i = 16$.

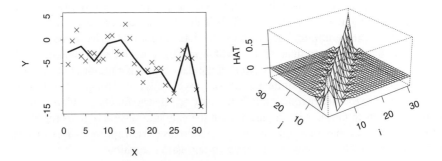

Figure 2.19   Fitting of linear spline (broken lines). The internal knots are placed at points $i = 4, 7, 11, 15, 19, 23, 27$ (left). Values of elements of the hat matrix that corresponds to figure 2.17(left), (right).

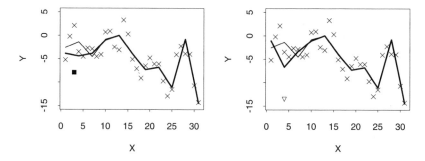

Figure 2.20 The estimates obtained by the fitting of linear splines (thick broken line) when the value of Y_3, represented as ■, is calculated by subtracting 10 from the original Y_3. The estimates derived from the original data are superimposed (thin broken line). The internal knots are placed at points $i = 4, 7, 11, 15, 19, 23, 27$ (left). Estimates obtained by the fitting of linear splines (thick broken line) when the value of Y_4 is calculated by subtracting 10 from the original Y_4; the data is represented as $\triangledown$. The estimates derived from the original data are superimposed (thin broken line). The internal knots are placed at points $i = 4, 7, 11, 15, 19, 23, 27$ (right).

data: one linear equation is fitted to the data located in the region between $i = 1$ and $i = 16$, and another linear equation is fitted to the data located in the region between $i = 16$ and $i = 31$, under the condition that the estimates by the two regression equations coincide at point $i = 16$. The points located at the two ends of each piecewise regression equation are called knots (breakpoints). In particular, the knots that are not located at the endpoints of the region where data are located are termed internal knots (internal breakpoints). The hat matrix for smoothing using this method is calculated on the basis of eq(2.6) (page 24), since this smoothing is considered to be that by using a linear estimator. Figure 2.17(right) depicts the values of elements of the hat matrix for calculating the estimates in figure 2.17(left). Furthermore, figure 2.18 illustrates the estimates (thick line) when the value of Y_3 is obtained by subtracting 10 from the original Y_3. The responses of estimates to the movement of data are similar to those inferred from the hat matrix. That is, the influence of the movement of the data increases in the region to the left of the moved data and declines in the region to the right of the moved data; the direction of the shift of the estimates around the internal knot is opposite to that of the movement of the data. The hat matrix and the resulting relationship between data and estimates are not very unnatural, so that this estimation summarizes the data very well.

Then, the number of internal knots is increased to seven: $i = 4, 7, 11, 15, 19, 23, 27$. Figure 2.19(left) demonstrates the estimates, which reveal the trend of the data well. This impression is reinforced by the fact that the hat matrix in figure 2.19(right) is analogous to that in figure 2.6(right) (page 30), which is obtained using a binomial filter ($m = 10$). Nevertheless, this hat matrix contains negative values with considerably large absolute values located near diagonal elements. The data are replaced with those in which Y_3 is obtained by subtracting 10 from the original Y_3 to investigate the effect of negative values in the hat matrix. The estimates are shown in figure 2.20(left). Some estimates are shifted in the opposite direction of the movement of the data. This slight shift may be negligible. However, the estimates obtained using the data in which the value of Y_4 is calculated by subtracting 10 from the original Y_4 (figure 2.20(right)) convey the impression that the responses of the movement of the data are somewhat unnatural.

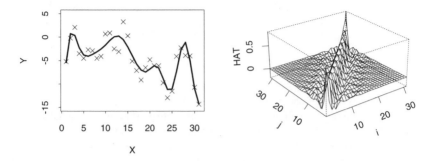

Figure 2.21   Fitting of cubic splines. The internal knots are located at $i = 4, 7, 11, 15, 19, 23, 27$ (left). Values of the elements of the hat matrix that corresponds figure 2.21(left), (right).

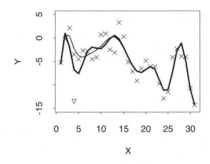

Figure 2.22   Estimates obtained by fitting cubic splines using the data in which the value of Y_4 (∇) is derived by substacting 10 from the original Y_4 (thick line). The estimates derived from the original data are superimposed (thin line). The internal knots are placed at $i = 4, 7, 11, 15, 19, 23, 27$.

Let us compare these results obtained by a linear spline function with those obtained by a cubic spline function. The cubic spline function consists of piecewise cubic equations; the estimates, the first derivatives, and the second derivatives of adjacent cubic equations coincide at the internal knots. Figure 2.21(left) shows the estimates obtained by the least squares using the cubic spline function with seven knots at $i = 4, 7, 11, 15, 19, 23, 27$ without boundary conditions at the two ends. The corresponding hat matrix is demonstrated in figure 2.21(right). The estimates in figure 2.21(left) change more gradually than those in figure 2.19(left) and hence the estimates obtained by the cubic spline function appear to outperform those obtained by the linear spline function. However, the values of the elements of the hat matrix in figure 2.21(right) vary in a complex manner compared to those in figure 2.19(right).

In consequence, the shifts of estimates brought about by the movement of Y_4 become unsatisfactory (figure 2.22). Hence, we should note that the comparative merits and demerits of the results of the cubic spline function and those of the linear spline function depend on the user's purpose. Though the cubic spline function is preferable when the smoothness of estimates is desired, it is at a disadvantage because movements of the data can cause surprising shifts of the estimates.

Thus, when spline functions are fitted, the degree of the spline function, the number of knots and their positions (the knots are not required to be equispaced), and the boundary conditions should be chosen with comprehensive consideration of the discrepancies between data and estimates, the smoothness of estimates, and the responses of estimates to movements of data in order to apply the most suitable regression method. It is difficult to develop a universal method to automate this process.

In conclusion, many new problems arise when attempting piecewise fittings in order to avoid the difficulties caused by the fitting of one equation to data located over the entire region. Two reasons are pointed out: the problems concerning the number and the positions of knots have to be solved for piecewise fittings because the data, although not having any intrinsic divisions, must be divided into several regions, and the existence of knots renders the appearance of the hat matrix unnatural. These problems cannot be overcome, regardless of the function that is employed as the piecewise function. If complex functions (with many regression coefficients) are adopted for a piecewise fitting, an analyst may face difficulties both by fitting complex functions and by using piecewise functions. Therefore, local regression is developed to retain the advantages of piecewise fitting while avoiding the laborious process of selecting knots. Here, local linear regression is adopted, since it is well known that this is a very simple but eminently practical method.

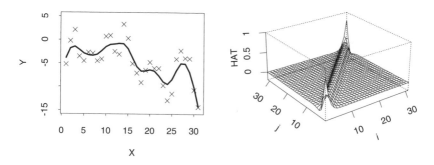

Figure 2.23   Smoothing by local linear regression ($h = 1.5$) (left). Values of the elements of the hat matrix that corresponds to figure 2.23(left), (right).

Ordinal simple regression minimizes the value:

$$E_{linear} = \sum_{i=1}^{n}(aX_i + b - Y_i)^2, \qquad (2.55)$$

where $\{(X_i, Y_i)\}$ ($1 \leq i \leq n$) is the data, X_i is equispaced, the value of the predictor is calculated by $X_i = i$, and a and b are regression coefficients. Ordinal simple regression usually fits one equation to the entire region where data are located. That

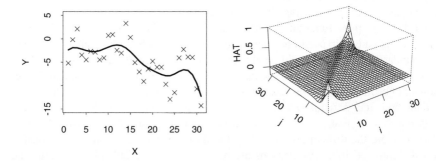

Figure 2.24   Smoothing by local linear regression ($h = 2.5$) (left). Values of the elements of the hat matrix that corresponds to figure 2.24(left), (right).

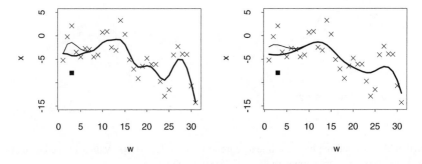

Figure 2.25   Estimates obtained by local linear regression using the data in which the value of Y_3 (■) is calculated by subtracting 10 from the original Y_3 (thick line, $h = 1.5$) (left). Estimates obtained by local linear regression using the data in which the value of Y_3 (■) is calculated by subtracting 10 from the original Y_3 (thick line, $h = 2.5$) (right).

is, one value is used for a, and another value for b. On the other hand, the fitting of the linear spline divides the data existing in one region into several regions. The values of a and b are fixed in each region but different regions require different coefficients; this is piecewise fitting. On the other hand, local linear regression provides the values of a and b, which depend on the position of the estimation point. This method is called local fitting. Since the local linear regression fits linear equations locally, a and b are functions of the values of the predictor. Hence, these regression coefficients (a and b) are represented as $\{a_j\}$ and $\{b_j\}$ ($1 \le j \le n$), respectively; X_j is the position where the estimate is calculated. Then, the value below is minimized to derive the regression coefficients.

$$E(X_j) = \sum_{i=1}^{n} w_{ij}(a_j X_i + b_j - Y_i)^2. \tag{2.56}$$

Thus, when a_j and b_j obtained by minimizing $E(X_j)$ are defined as $\hat{a}_j$ and $\hat{b}_j$, the estimate $(\hat{Y}_j)$ is written as

$$\hat{Y}_j = \hat{a}_j X_j + \hat{b}_j. \tag{2.57}$$

w_{ij}, which is termed a kernel function, gives the weights used to determine how important the data $((X_i, Y_i))$ are to obtain $\hat{a}_j$ and $\hat{b}_j$. Here, w_{ij} can be described as weight because it is an element of a matrix. However, a better term is kernel function, since a similar calculation is carried out for estimation at the position where data do not exist (cf. eq(3.126)). In addition, the term weight function sometimes indicates a function that weighs data so as to improve the behavior of estimates close to the two ends of a region where data exist and to derive regression equations that reflect the variation of the variance of error depending on a position, although weight function means a kernel function on some occasions. On the other hand, the term weight is used to describe the values of kernel functions and weight functions. The value of a kernel function takes a larger value when X_i is close to X_j (i.e., the value of i is near the value of j), and it takes a smaller value but not a negative one when the distance between X_i and X_j becomes greater. Hence, a simple kernel function is the Gaussian function:

$$w_{ij} = \exp\left(-\frac{1}{2}\left(\frac{X_i - X_j}{h}\right)^2\right). \tag{2.58}$$

h is a positive constant for determining the smoothness of estimates, called bandwidth. The extent of smoothness is increased by the enhancement of the values of h because a larger h means that a wider range of data are taken into account to estimate a_j and b_j.

Figure 2.23(left) is an example of smoothing by a kernel function ($h = 1.5$) using the Wakkanai data. The corresponding values of the elements of the hat matrix are demonstrated in figure 2.23(right). (For the method of calculating a hat matrix, refer to eq(3.132) (page 143).) On the other hand, figure 2.24(left) displays a hat matrix when $h = 2.5$ in eq(2.58). The corresponding values of the hat matrix are shown in figure 2.24(right). Both of the hat matrices take smooth downhill curves from the position of data approaching 0, except at the two ends. Negative values rarely exist. Comparison of figure 2.24(right) with the hat matrices of spline functions (figure 2.17(right) (page 44), figure 2.19(right) (page 44), figure 2.21(right) (page 46)) shows the superiority of that in figure 2.24(right). This fact results in the natural responses of estimates to the movements of data, as is evident in figure 2.25(left) and figure 2.25(right). However, some elements of the hat matrix of a local linear regression for the estimates near the two ends are negative and the responses of the estimates to the movements of data reflect this. The regression equations (eq(2.57)) used for calculating each estimate are

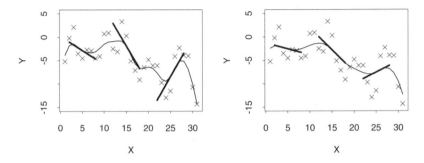

Figure 2.26   Linear equation fitted locally at $i = 5, 15, 25$ in local linear regression ($h = 1.5$) (left). Linear equation fitted locally at $i = 5, 15, 25$ in local linear regression ($h = 2.5$) (right).

illustrated to allow intuitive understanding of the concept of local regression. Figure 2.26(left) and figure 2.26(right) display the regression equations for calculating the estimates at $i = 5, 15$, and 25 when $h = 1.5$ and $h = 2.5$, respectively. The relationship between the curves connecting the estimates and the linear curves is reminiscent of a tangent line to a curve. The concept of derivative is motivated by the idea that a linear equation is derived to approximate a curve in the neighborhood of a specific point, and a linear equation analogous to a tangent line is obtained by narrowing the region in the neighborhood of a specific point. There seems to be such a prehistory in the development of differential calculus. That is, the linear curve used in calculating the estimate at each point in the application of local linear regression is the curve that was obtained by tracing the development of differential calculus and adopting the early concept of estimating a linear equation in order to approximate a finite region of a curve, and in addition by rendering the method more sophisticated by introducing the concept of the kernel function. In this sense, the concept of local linear regression is based on the history of curve analysis. The practical demand for a means of smoothing rough data is closely associated with the establishment of differential calculus, which is an important development in the history of mathematics.

Next, an example of the regression coefficients is demonstrated. Figure 2.27(left) shows $\{\hat{a}_j\}$ and figure 2.27(right) shows $\{\hat{b}_j\}$ ($h = 1.5$ and $h = 2.5$). The variation of the regression coefficients with $h = 2.5$ is less marked than that with $h = 1.5$. As a result, the smoothness of the curve connecting the estimates is enhanced. That is, the connection between regression coefficients (i.e., estimates) for neighboring data is strengthened by broadening the region in the neighborhood of a specific point on a curve.

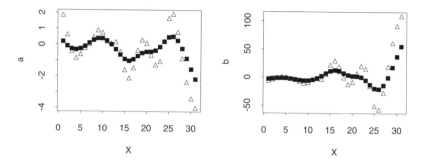

Figure 2.27   Values of regression coefficients of a linear regression obtained at each point in local linear regression ($\{\hat{a}_j\}$) (left). Values of regression coefficients of a linear regression ($\{\hat{b}_j\}$) (right). $\triangle$ indicates the values when $h = 1.5$, and $\blacksquare$ indicates those when $h = 2.5$.

2.5   SMOOTHING SPLINE

The concept of regression by the least squares develops into penalized regression, which aims at regression that renders the residual sum of squares small and satisfies various conditions to a certain extent; this is realized by adding terms to equations such as eq(2.55) (page 47) to represent the conditions which the estimates or the regression coefficients should satisfy. It is possible that smoothing based on this concept is achieved by minimizing the sum of the two terms: one is for reducing the residual sum of squares, the other is for requiring the smooth behavior of estimates. For instance, we minimize

$$E_{ss-equi} = \sum_{i=1}^{n}(\hat{Y}_i - Y_i)^2 + \lambda \sum_{i=2}^{n-1}(\hat{Y}_{i-1} - 2\hat{Y}_i + \hat{Y}_{i+1})^2. \qquad (2.59)$$

The first term on the right-hand side reflects the expectation that the values of the data should be close to those of the estimates. This aim is common to that of the use of ordinary least squares. The second term on the right-hand side imposes the condition that the estimates should vary smoothly; this term is called the roughness penalty. λ is a positive constant called a smoothing parameter which plays the role of adjusting the degree of smoothness. A large value of λ results in highly smooth estimates. This is analogous to enhancing the smoothness of estimates by increasing the bandwidth (h) in the local linear regression (eq(2.58) (page 49)). As a consequence, the two terms on the right-hand side of this equation conflict in terms of estimation in most cases. That is, an attempt to reduce the value of the first term prompts the increase of the roughness of the estimates; as a consequence, the value of the second term (where λ is omitted) is augmented. As a result, the positive constant λ determines the trade-off.

This method is called the smoothing spline (smoothing splines). In addition to the smoothing spline, the trade-off between fidelity to the data and smoothness of the estimates is an important concept in smoothing in general, and in most areas of regression. As long as smoothing is our purpose, the estimates should be smooth. However, if the discrepancy between the estimates and the data is too wide, the estimates cannot offer good value. Hence, the balance between the two sides should be taken into account in smoothing. The same is true for the binomial filter. That is, though it is desirable that the estimates in figure 2.4(left) (page 29) ($m = 2$) are close to the data, large roughness of the estimates is not acceptable. On the other hand, though it is favorable that the estimates in figure 2.6(left) (page 30) have small roughness, the considerable discrepancy between the estimates and data raises a question. These remarks also apply to polynomial regression (figure 2.8(left) (page 35) and figure 2.12(left) (page 37)), the fitting of linear splines (figure 2.17(left) (page 44) and figure 2.19(left) (page 44)), and local linear regression (figure 2.23(left) (page 47) and figure 2.24(left) (page 48)). Thus a fundamental concept in smoothing is the trade-off between the two requirements: estimates should be close to data, and they should be smooth. In addition, this concept plays a significant role in nonparametric regression in general. One of the advantages of the smoothing spline is that it reveals this concept explicitly. We can gain some insight on how these two requirements are satisfied by calculating the first term and the second term (λ is omitted) on the right-hand side of eq(2.59). Such a straightforward calculation demonstrates the reliability of the use of the smoothing spline and is the basis for the development of various nonparametric regression methods and other diverse theories on nonparametric regressions based on the smoothing spline.

To minimize eq(2.59), eq(2.59) is differentiated with respect to each $\{\hat{Y}_i\}$ to be 0. Thus, the simultaneous equation, which consists of n equations, is obtained:

$$
\begin{aligned}
&(\hat{Y}_1 - Y_1) + \lambda(\hat{Y}_1 - 2\hat{Y}_2 + \hat{Y}_3) = 0, \\
&(\hat{Y}_2 - Y_2) + \lambda(-2\hat{Y}_1 + 5\hat{Y}_2 - 4\hat{Y}_3 + \hat{Y}_4) = 0, \\
&(\hat{Y}_i - Y_i) + \lambda(\hat{Y}_{i-2} - 4\hat{Y}_{i-1} + 6\hat{Y}_i - 4\hat{Y}_{i+1} + \hat{Y}_{i+2}) = 0, \quad (3 \leq i \leq n-2) \\
&(\hat{Y}_{n-1} - Y_{n-1}) + \lambda(\hat{Y}_{n-3} - 4\hat{Y}_{n-2} + 5\hat{Y}_{n-1} - 2\hat{Y}_n) = 0, \\
&(\hat{Y}_n - Y_n) + \lambda(\hat{Y}_{n-2} - 2\hat{Y}_{n-1} + \hat{Y}_n) = 0.
\end{aligned}
\tag{2.60}
$$

These equations are rewritten as

$$
\hat{\mathbf{y}} + \lambda \mathbf{S}\hat{\mathbf{y}} = \mathbf{y},
\tag{2.61}
$$

where $\mathbf{S}$ is a matrix:

$$
\begin{pmatrix}
1 & -2 & 1 & 0 & \cdots & \cdots & \cdots & \cdots & \cdots & \cdots & \cdots & \cdots & 0 \\
-2 & 5 & -4 & 1 & 0 & \cdots & \cdots & \cdots & \cdots & \cdots & \cdots & \cdots & 0 \\
1 & -4 & 6 & -4 & 1 & 0 & \cdots & \cdots & \cdots & \cdots & \cdots & \cdots & 0 \\
0 & 1 & -4 & 6 & -4 & 1 & 0 & \cdots & \cdots & \cdots & \cdots & \cdots & 0 \\
\vdots & \vdots & \vdots & \vdots & \vdots & \vdots & \vdots & \vdots & \vdots & \vdots & \vdots & \vdots & \vdots \\
\vdots & \vdots & \vdots & \vdots & \vdots & \vdots & \vdots & \vdots & \vdots & \vdots & \vdots & \vdots & \vdots \\
0 & \cdots & \cdots & \cdots & \cdots & 0 & 1 & -4 & 6 & -4 & 1 & 0 \\
0 & \cdots & \cdots & \cdots & \cdots & \cdots & 0 & 1 & -4 & 6 & -4 & 1 \\
0 & \cdots & \cdots & \cdots & \cdots & \cdots & \cdots & 0 & 1 & -4 & 5 & -2 \\
0 & \cdots & \cdots & \cdots & \cdots & \cdots & \cdots & \cdots & 0 & 1 & -2 & 1
\end{pmatrix}
\tag{2.62}
$$

Eq(2.61) is transformed into

$$
\hat{\mathbf{y}} = (\mathbf{I} + \lambda \mathbf{S})^{-1}\mathbf{y},
\tag{2.63}
$$

where $\mathbf{I}$ is the identity matrix, and $()^{-1}$ is the inverse matrix. The comparison between this equation and eq(2.22) (page 32) indicates that $(\mathbf{I}+\lambda\mathbf{S})^{-1}$ functions as $\mathbf{H}$. Though eq(2.59) does not look like a regression that results in a hat matrix at first glance, the relationship between the data and estimates can be represented using a hat matrix.

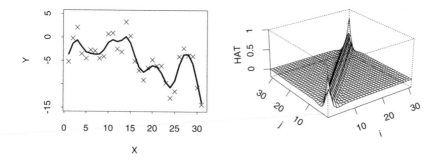

Figure 2.28 Smoothing by the smoothing spline ($\lambda = 1$) (left). Values of the elements of the hat matrix that corresponds to figure 2.28(left), (right).

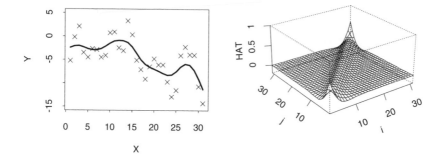

Figure 2.29 Smoothing by the smoothing spline ($\lambda = 10$) (left). Values of the elements of the hat matrix that corresponds to figure 2.28(left), (right).

Figure 2.28(left) is an example of smoothing by using the smoothing spline: the Wakkanai data are used and $\lambda = 1$ (eq(2.61)). The corresponding hat matrix is displayed in figure 2.28(right). The change of λ to $\lambda = 10$ yields figure 2.29(left). The corresponding hat matrix is illustrated in figure 2.29(right). Slight ripples are observed in the values of the hat matrices, and negative values which are slightly distant from the diagonal elements are noted; they do not affect the estimates significantly. As a result, the responses of the estimates to the movements of data are appropriate (figure 2.30(left), figure 2.30(right)).

On the other hand, the values of the squared residuals $((\hat{Y}_i - Y_i)^2)$ of the estimates in figure 2.28(left) and figure 2.29(left) are demonstrated in figure 2.31(left). A value of the squared residual obtained with $\lambda = 10$ is larger than that obtained with $\lambda = 1$ in most points. On the other hand, figure 2.31(right) shows the roughness of each estimate at various positions $((\hat{Y}_{i-1} - 2\hat{Y}_i + \hat{Y}_{i+1})^2)$ ($\lambda = 1, 10$). Most of the values are small when $\lambda = 10$. That is, since greater importance is placed on the fidelity of

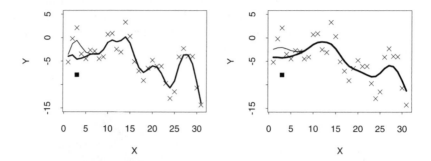

Figure 2.30   Estimates (thick line) by smoothing spline ($\lambda = 1$); the value of Y_3 (■) is obtained by subtracting 10 from the original Y_3 (left). Estimates (thick line) by smoothing spline ($\lambda = 10$); the value of Y_3 (■) is obtained by subtracting 10 from the original Y_3 (right).

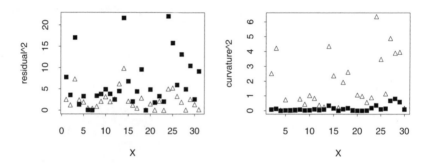

Figure 2.31   Values of squared residual obtained by the smoothing spline at each i (($\hat{Y}_i - Y_i)^2$) (left). Values for representing the roughness obtained by the smoothing spline at each i (($\hat{Y}_{i-1} - 2\hat{Y}_i + \hat{Y}_{i+1})^2$) (right). $\triangle$ represents the values for $\lambda = 1$ and ■ are the values for $\lambda = 10$.

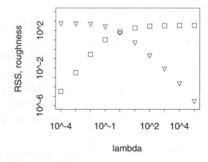

Figure 2.32   Responses of the residual sum of squares ($\sum_{i=1}^{n}(\hat{Y}_i - Y_i)^2$) and the roughness ($\sum_{i=2}^{n-1}(\hat{Y}_{i-1} - 2\hat{Y}_i + \hat{Y}_{i+1})^2$) to the variation of the value of λ. $\square$ represents values of the residual sum of squares. $\triangledown$ represents values of the roughness.

the estimates to the data than on the smoothness of the estimates when $\lambda = 1$, the values in figure 2.31(left) are mostly small, while those in figure 2.31(right) are large; the opposite result occurs when $\lambda = 10$.

Next, figure 2.32 demonstrates the values of the two terms on the right-hand side of eq(2.59) when $\lambda = 10^{-4}, 10^{-3}, \ldots, 10^{5}$; the values of the second term are obtained by omitting λ (i.e., $\sum_{i=2}^{n-1} (\hat{Y}_{i-1} - 2\hat{Y}_i + \hat{Y}_{i+1})^2$). The increase of the value of λ gradually shifts the emphasis from the fidelity of estimates to data to the reduction of the roughness of the estimates. The value of λ is varied exponentially in these graphs because it is usually the case that the value of the smoothing parameter is altered exponentially when the appearance of the estimates obtained by the smoothing spline with various smoothing parameter values is examined, or the value of the smoothing parameter is optimized. Graphs like those shown in figure 2.31(left), figure 2.31(right), and figure 2.32 are possible for local linear regression; such graphs enable us to determine the effects of the bandwidth, which can be regarded as a kind of smoothing parameter, on the squared residuals and the roughness of estimates.

The policy of the smoothing spline toward obtaining estimates aims at reducing the values of both the residual sum of squares and the roughness of estimates; as a result, compromised estimates are derived. This policy is considered to be a mathematical expression in order to imitate a human being who draws a smooth line passing through the data. This viewpoint is justified by the fact that in the graphs which show both the estimates and the data, it often appears that the estimates are derived by a human being. On the other hand, the concept underlying local linear regression is that even a complexly behaved curve is expected to be approximately linear. The assumption of approximate local linearity has the effect of reducing the roughness of the estimates ($(\hat{Y}_{i-1} - 2\hat{Y}_i + \hat{Y}_{i+1})^2$). Therefore, though local linear regression and the smoothing spline appear to be distinct, they hold in common the fact that they result from the goal that estimates which have the appearance of being drawn by a human being observing the data should be derived by mathematically rigid forms and procedures; therefore, their estimates are close on most occasions.

2.6 ANALYSIS ON EIGENVALUE OF HAT MATRIX

A hat matrix for smoothing usually satisfies

$$\mathbf{H} = \mathbf{H}^t, \tag{2.64}$$

$$\sum_{j=1}^{n} [\mathbf{H}]_{ij} = 1 \quad (1 \le i \le n). \tag{2.65}$$

The characteristics of eq(2.64) are described: $\mathbf{H}$ is a symmetric matrix. A symmetric hat matrix means that the degree of influence of Y_i on $\hat{Y}_j$ is the same as that of Y_j on $\hat{Y}_i$. For instance, the hat matrix for fitting polynomials by the least squares (eq(2.38) (page 35)) and that for fitting smooth equispaced data ($\{X_i\}$) using the smoothing spline (eq(2.61) (page 52)) are apparently symmetric matrices. As for the moving average and the binomial filter, their hat matrices are symmetric when a reflective boundary condition is adopted. Furthermore, eq(2.65) shows that an estimate located at one of the data positions is a weighted average; most of the smoothing methods satisfy this equation. In the case of fitting the polynomials by the least squares, eq(2.39) (page 39) has the same form as eq(2.65). The moving average and the binomial filter evidently satisfy this condition (eq(2.65)). Smoothing data, which are expressed by $Y_i = 1$ ($1 \leq i \leq n$), using local linear regression or the smoothing spline, produces the following estimates: $\hat{Y}_i = 1$ ($1 \leq i \leq n$), regardless of the degree of smoothing, and thus eq(2.65) is proved.

Additionally, the relationships discussed in this section assume only eq(2.64) and eq(2.65), and hence they are satisfied even if $\{X_i\}$ is not equispaced.

First, if $\mathbf{H}$ is a symmetric matrix, the transformation called diagonalization, which will be familiar to most readers, is realized:

$$\mathbf{H} = \mathbf{UAU}^t, \qquad (2.66)$$

where $\mathbf{U}$ is an orthogonal matrix, and $\mathbf{A}$ is a diagonal matrix (the values of all the elements are 0, except for the diagonal elements).

$$\mathbf{A} = \begin{pmatrix} \alpha_1 & 0 & 0 & \cdots & 0 \\ 0 & \alpha_2 & 0 & \cdots & 0 \\ 0 & 0 & \alpha_3 & \cdots & 0 \\ \vdots & \vdots & \vdots & \ddots & \vdots \\ 0 & 0 & 0 & \cdots & \alpha_n \end{pmatrix}. \qquad (2.67)$$

These $\{\alpha_i\}$ ($1 \leq i \leq n$) are the eigenvalues (characteristic values) of $\mathbf{H}$. The eigenvalues are solutions of an eigenequation (characteristic equation).

$$|\, \mathbf{H} - \alpha\mathbf{I} \,| = 0, \qquad (2.68)$$

where $\mathbf{I}$ is an identity matrix, and $|\ \ |$ means a determinant. The solutions of this equation ($\{\alpha_i\}$) are all real numbers. When n equations below for $\mathbf{u}$ (a column vector consisting of n elements) are created using $\{\alpha_i\}$, let a vector satisfy the ith equation and have unity length (the unit vector will be called $\mathbf{u}_i$).

$$\mathbf{Hu}_i = \alpha_i \mathbf{u}_i \quad (1 \leq i \leq n). \qquad (2.69)$$

$\{\mathbf{u}_i\}$ is an eigenvector. The matrix (with size of $n \times n$) that is given by aligning $\mathbf{u}_i$ is $\mathbf{U}$ (eq(2.66)). $\mathbf{U}$ is a regular matrix (i.e., an invertible matrix); it satisfies

$$\mathbf{U}^t = \mathbf{U}^{-1}. \qquad (2.70)$$

This characteristic is equivalent to $\mathbf{U}$ being an orthogonal matrix. Thus, we have

$$\begin{aligned} \mathbf{H} &= \mathbf{UAU}^t \\ &= \sum_{i=1}^{n} \alpha_i \mathbf{u}_i \cdot \mathbf{u}_i^t. \end{aligned} \qquad (2.71)$$

The decomposition of a symmetric matrix to $\sum_{i=1}^{n} \alpha_i \mathbf{u}_i \cdot \mathbf{u}_i^t$, as is shown in the equation above, is called spectral resolution (spectral decomposition). $\{\alpha_i\}$ is a spectrum of $\mathbf{H}$. Therefore, $\mathbf{u}_i \cdot \mathbf{u}_i^t$ is a projection matrix (or more precisely, an orthogonal projection matrix). Hence, the vector given by a projection matrix ($\{\mathbf{u}_i \cdot \mathbf{u}_i^t \mathbf{y}\}$ in our framework) is termed the projection of $\mathbf{y}$.

Additionally, the eigenvalue of an idempotent matrix is 0 or 1. This is proved as follows.

The multiplication of $\mathbf{H}$ with eq(2.69) yields

$$\mathbf{H}^2 \mathbf{u}_i = \mathbf{H}(\mathbf{H}\mathbf{u}_i) = \mathbf{H}(\alpha_i \mathbf{u}_i) = \alpha_i(\mathbf{H}\mathbf{u}_i) = \alpha_i^2 \mathbf{u}_i. \tag{2.72}$$

Furthermore, the idempotent characteristic of $\mathbf{H}$ leads to

$$\mathbf{H}^2 \mathbf{u}_i = \mathbf{H}\mathbf{u}_i = \alpha_i \mathbf{u}_i. \tag{2.73}$$

Combining eq(2.72) and eq(2.73) leads to the relationship

$$\alpha_i^2 = \alpha_i. \tag{2.74}$$

Therefore, if $\mathbf{H}$ is idempotent, its eigenvalue is either 0 or 1 (Q.E.D.).

On the other hand, the proof of the theorem — that if $\mathbf{H}$ is an idempotent matrix and at the same time is a symmetric matrix, its eigenvalues are 0 or 1 — is as follows.

The estimates calculated by the multiplication of a hat matrix (symmetric matrix) with data are multiplied by a hat matrix, and the results are multiplied by a hat matrix. Repetition of this procedure k times using eq(2.71) renders the following estimate:

$$\mathbf{H}^k \mathbf{y} = \mathbf{U}\mathbf{A}^k \mathbf{U}^t \mathbf{y}. \tag{2.75}$$

Since $\mathbf{H}$ is idempotent, we have the relationship

$$\mathbf{A}^k = \mathbf{A}. \tag{2.76}$$

Because k is an arbitrary positive integer, α_i always take the value of either 0 or 1 (Q.E.D.).

Eq(2.71) also provides

$$
\begin{aligned}
\text{trace}(\mathbf{H}) &= \text{trace}(\mathbf{U}\mathbf{A}\mathbf{U}^t) \\
&= \sum_{i=1}^{n}\sum_{j=1}^{n}\sum_{k=1}^{n}[\mathbf{U}]_{ij}[\mathbf{A}]_{jk}[\mathbf{U}]_{ik} \\
&= \sum_{i=1}^{n}\sum_{j=1}^{n}\sum_{k=1}^{n}[\mathbf{U}^{-1}]_{ki}[\mathbf{U}]_{ij}[\mathbf{A}]_{jk} \\
&= \sum_{j=1}^{n}\sum_{k=1}^{n}[\mathbf{I}]_{kj}[\mathbf{A}]_{jk} \\
&= \sum_{k=1}^{n}\alpha_k.
\end{aligned}
\tag{2.77}
$$

When polynomials are fitted by the least squares, eq(2.54) (page 43) shows that the value obtained by eq(2.77) is $(q+1)$ (q represents the degree of the polynomial). Then, $(q+1)$ values from $\{\alpha_i\}$ ($1 \leq i \leq n$) are 1 and $(n-q-1)$ values are 0.

On the other hand, the repetition of the multiplication of some values of $\mathbf{H}$ with $\mathbf{y}$ produces gradually smoother estimates. In this case, $\mathbf{H}$ is not idempotent. The multiplication of $\mathbf{H}$ many times converges to certain values on most occasions. This means that $k \to \infty$ in eq(2.75) leads to $\mathbf{H}^k$ in which the values of the elements are finite. That is to say, the absolute value of $\{\alpha_i\}$ is not more than 1, and one or more values of $\{\alpha_i\}$ take the value of 1.

In addition, eq(2.71) gives

$$\mathbf{U}^t\hat{\mathbf{y}} = \mathbf{A}\mathbf{U}^t\mathbf{y}. \tag{2.78}$$

We define

$$\mathbf{y}^* = \mathbf{U}^t\mathbf{y}, \tag{2.79}$$

$$\hat{\mathbf{y}}^* = \mathbf{U}^t\hat{\mathbf{y}}. \tag{2.80}$$

Let $\hat{Y}_i^*$ be the ith element of $\hat{\mathbf{y}}^*$, Y_i^* be the ith element of $\mathbf{y}^*$, and u_{ij} be the jth element of $\mathbf{u}_i$. As a result, we obtain

$$Y_i^* = \sum_{j=1}^{n} u_{ij} Y_j = \ < \mathbf{u}_i \cdot \mathbf{y} >, \tag{2.81}$$

$$\hat{Y}_i^* = \sum_{j=1}^{n} u_{ij} \hat{Y}_j = \ ' < \mathbf{u}_i \cdot \hat{\mathbf{y}} >, \tag{2.82}$$

where $< \mathbf{u}_i \cdot \mathbf{y} > \ = \mathbf{u}_i^t \cdot \mathbf{y}$. That is, $< \cdot >$ indicates an inner product between two vectors. Hence, the coordinate of $\hat{\mathbf{y}}$ in an orthogonal coordinate system which is constructed by using bases ($\{\mathbf{u}_i\}$ ($1 \leq i \leq n$)) is $\hat{\mathbf{y}}^*$, and that of $\mathbf{y}$ in the same system is $\mathbf{y}^*$. Then we have

$$\mathbf{y} = \sum_{i=1}^{n} Y_i^* \mathbf{u}_i = \sum_{i=1}^{n} < \mathbf{u}_i \cdot \mathbf{y} > \mathbf{u}_i, \tag{2.83}$$

$$\hat{\mathbf{y}} = \sum_{i=1}^{n} \hat{Y}_i^* \mathbf{u}_i = \sum_{i=1}^{n} < \mathbf{u}_i \cdot \hat{\mathbf{y}} > \mathbf{u}_i. \tag{2.84}$$

Furthermore, eq(2.78) can be rewritten as

$$\hat{\mathbf{y}}^* = \mathbf{A}\mathbf{y}^*, \tag{2.85}$$

and this equation can be transformed into

$$\hat{Y}_i^* = \alpha_i Y_i^* \quad (1 \leq i \leq n). \tag{2.86}$$

The substitution of eq(2.86) into eq(2.84) produces the following:

$$\hat{\mathbf{y}} = \sum_{i=1}^{n} \alpha_i Y_i^* \mathbf{u}_i = \sum_{i=1}^{n} \alpha_i < \mathbf{u}_i \cdot \mathbf{y} > \mathbf{u}_i. \tag{2.87}$$

This equation implies that smoothing is a simple procedure involving the multiplication of each Y_i^* by a specific constant in a coordinate system which is constructed by using vectors ($\{\mathbf{u}_i\}$). In other words, the estimate ($\hat{\mathbf{y}}$) is represented as the sum of the vectors, each of which has been obtained by multiplication between α_i and the vector ($\{< \mathbf{u}_i \cdot \mathbf{y} > \mathbf{u}_i\}$) given by the orthogonal projection of $\mathbf{y}$ into the space spanned by $\{\mathbf{u}_i\}$. Figure 2.33 demonstrates this relationship when data can be represented using two eigenvalues of the hat matrix. If some values of $\{\alpha_i\}$ are negative, assuming that α_k is negative, the direction of $\alpha_k < \mathbf{u}_k \cdot \mathbf{y} > \mathbf{u}_k$ is opposite to that of $< \mathbf{u}_k \cdot \mathbf{y} > \mathbf{u}_k$. That is, the direction of the vector given by the orthogonal projection of the data into the space spanned by $\mathbf{u}_k$ is opposite to that of the vector given by the orthogonal projection of the estimate into the same space. If this phenomenon has a considerable effect, the resulting estimates are ones that we cannot regard as results of smoothing the data. This is possibly responsible for the inexplicable relationship between the data and the estimates. Additionally, the estimates calculated by the

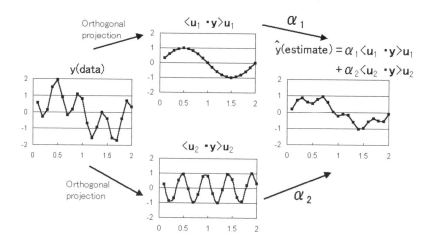

Figure 2.33   The implication of smoothing with respect to the eigenvalues of a hat matrix and the eigenvectors of a hat matrix.

additional multiplication of $\hat{\mathbf{y}}$ by a hat matrix are defined as $\hat{\mathbf{y}}_{(2)}$:

$$
\begin{aligned}
\hat{\mathbf{y}}_{(2)} &= \sum_{i=1}^{n} \alpha_i < \mathbf{u}_i \cdot \hat{\mathbf{y}} > \mathbf{u}_i \\
&= \sum_{i=1}^{n} \alpha_i < \mathbf{u}_i \cdot \left(\sum_{j=1}^{n} \alpha_j < \mathbf{u}_j \cdot \mathbf{y} > \mathbf{u}_j \right) > \mathbf{u}_i \\
&= \sum_{i=1}^{n} \sum_{j=1}^{n} \alpha_i \alpha_j < \mathbf{u}_i \cdot \mathbf{u}_j >< \mathbf{u}_j \cdot \mathbf{y} > \mathbf{u}_i \\
&= \sum_{i=1}^{n} \alpha_i^2 < \mathbf{u}_i \cdot \mathbf{y} > \mathbf{u}_i,
\end{aligned}
\tag{2.88}
$$

where the equation below is utilized to derive the equality between the third and fourth lines of eq(2.88).

$$< \mathbf{u}_i \cdot \mathbf{u}_j > = \begin{cases} 1 & \text{if } i = j \\ 0 & \text{if } i \neq j. \end{cases} \tag{2.89}$$

Hence, we define the estimates obtained by k multiplications of $\hat{\mathbf{y}}$ by $\mathbf{H}$ as $\hat{\mathbf{y}}_{(k)}$:

$$\hat{\mathbf{y}}_{(k)} = \sum_{i=1}^{n} \alpha_i^k < \mathbf{u}_i \cdot \mathbf{y} > \mathbf{u}_i. \tag{2.90}$$

Therefore, if one of $\{\alpha_i\}$ is 1, and the absolute values of the remaining are less than 1, the repetition of the procedure in which $\hat{Y}_i^*$ given by multiplying Y_i^* by α_i is renamed as Y_i^* produces estimates ($\{\hat{Y}_i^*\}$): one of $\{\hat{Y}_i^*\}$ takes the same value as that given by the orthogonal projection of the data, and all of the remaining are 0. This implies that the repetition of smoothing by $\mathbf{H}$, which gives this sort of $\{\alpha_i\}$, renders estimates that are the result of multiplication of the eigenvector by the inner product between an eigenvector (the length is 1), which corresponds to the eigenvalue taking the value of 1, and the data. Similarly, if two of $\{\alpha_i\}$ are 1, and the absolute values of the remaining α_i are less than 1, the estimates given by repeated smoothing by $\mathbf{H}$ are represented as a linear combination of the two eigenvectors corresponding to the two eigenvalues ($= 1$). The coefficients of this linear combination are inner products between the data and their eigenvectors.

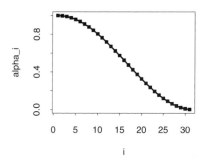

Figure 2.34 Eigenvalues of a hat matrix corresponding to a binomial filter ($n = 31, m = 2$). These values are aligned in descending order ($\{\alpha_1, \alpha_2, \ldots, \alpha_{31}\}$).

Figure 2.34 displays the eigenvalues ($\{\alpha_1, \alpha_2, \ldots, \alpha_{31}\}$) aligned in descending order; these eigenvalues are derived from the hat matrix (a symmetric matrix, figure 2.4(right) (page 29)) corresponding to a binomial filter (eq(2.7) (page 27)) ($n = 31, m = 2$). α_1 is the only eigenvalue that takes the value of 1.0, and the others are smaller than 1.0. The eigenvectors corresponding to $\alpha_1, \alpha_2, \ldots, \alpha_6$, which are extracted from the eigenvectors corresponding to these eigenvalues, are demonstrated in figure 2.35 in the same order of eigenvalues, and the eigenvectors corresponding to $\alpha_{26}, \alpha_{27}, \ldots, \alpha_{31}$ are shown in figure 2.36 in the same order of eigenvalues. As an eigenvalue becomes smaller, the corresponding eigenvector exhibits more rapid vibration. This phenomenon indicates that the estimates smoothed by the binomial filter ($m = 2$) are the result of the damping of the rapid vibration to a greater extent.

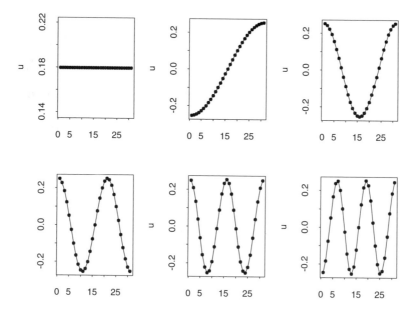

Figure 2.35 Eigenvectors corresponding to $\alpha_1, \alpha_2, \ldots, \alpha_6$, which are extracted from the eigenvalues in figure 2.34. On the top row, eigenvectors corresponding to $\alpha_1, \alpha_2, \alpha_3$ are shown from left to right. On the bottom row, eigenvectors corresponding to $\alpha_4, \alpha_5, \alpha_6$ are shown from left to right.

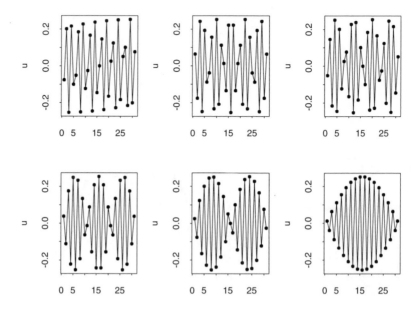

Figure 2.36   Eigenvectors corresponding to $\alpha_{26}, \alpha_{27}, \ldots, \alpha_{31}$, which are extracted from the eigenvalues in figure 2.34. On the top row, eigenvectors corresponding to $\alpha_{26}, \alpha_{27}, \alpha_{28}$ are shown from left to right. On the bottom row, eigenvectors corresponding to $\alpha_{29}, \alpha_{30}, \alpha_{31}$ are shown from left to right.

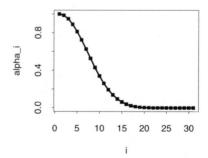

Figure 2.37   Eigenvalues of a hat matrix corresponding to a binomial filter ($n = 31, m = 10$). These values are aligned in descending order ($\{\alpha_1, \alpha_2, \ldots, \alpha_{31}\}$).

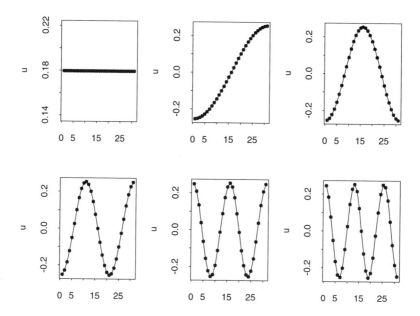

Figure 2.38 Eigenvectors corresponding to $\alpha_1, \alpha_2, \ldots, \alpha_6$, which are extracted from the eigenvalues in figure 2.37. On the top row, eigenvectors corresponding to $\alpha_1, \alpha_2, \alpha_3$ are shown from left to right. On the bottom row, eigenvectors corresponding to $\alpha_4, \alpha_5, \alpha_6$ are shown from left to right.

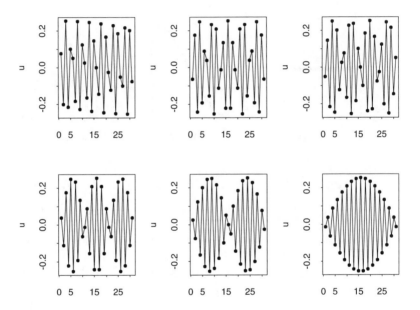

Figure 2.39   Eigenvectors corresponding to $\alpha_{26}, \alpha_{27}, \ldots, \alpha_{31}$, which are extracted from the eigenvalues in figure 2.37. On the top row, eigenvectors corresponding to $\alpha_{26}, \alpha_{27}, \alpha_{28}$ are shown from left to right. On the bottom row, eigenvectors corresponding to $\alpha_{29}, \alpha_{30}, \alpha_{31}$ are shown from left to right.

Figure 2.37 illustrates the eigenvalues ($\{\alpha_1, \alpha_2, \ldots, \alpha_{31}\}$) aligned in descending order; these eigenvalues are derived from the hat matrix (a symmetric matrix, figure 2.4(right) (page 30)) corresponding to a binomial filter (eq(2.6) (page 30)) ($n = 31$, $m = 10$). The eigenvectors corresponding to $\alpha_1, \alpha_2, \ldots, \alpha_6$ selected from eigenvectors corresponding to all eigenvalues are illustrated in figure 2.38, and eigenvectors corresponding to $\alpha_{26}, \alpha_{27}, \ldots, \alpha_{31}$ are illustrated in figure 2.39.

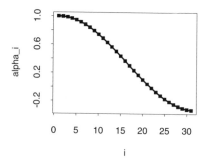

Figure 2.40   Eigenvalues of the hat matrix of the moving average ($n = 31$, $m = 1$) are shown in descending order ($\{\alpha_1, \alpha_2, \ldots, \alpha_{31}\}$).

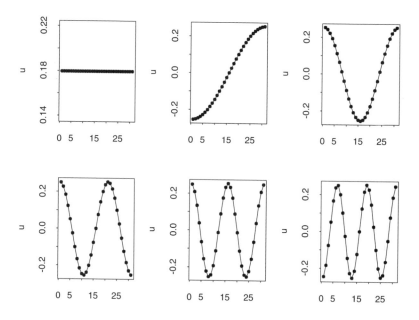

Figure 2.41   Eigenvectors corresponding to $\alpha_1, \alpha_2, \ldots, \alpha_6$ as extracted from figure 2.40. Eigenvectors corresponding to $\alpha_1, \alpha_2, \alpha_3$ are shown on the top row from left to right. Eigenvectors corresponding to $\alpha_4, \alpha_5, \alpha_6$ are shown on the bottom row from left to right.

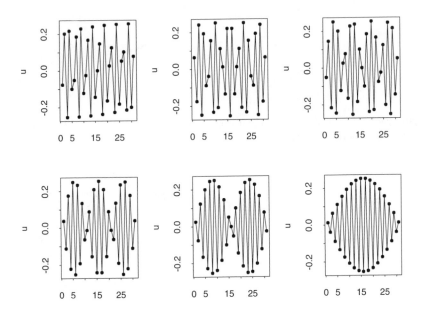

Figure 2.42 Eigenvectors corresponding to $\alpha_{26}, \alpha_{27}, \ldots, \alpha_{31}$ as extracted from figure 2.40. Eigenvectors corresponding to $\alpha_{26}, \alpha_{27}, \alpha_{28}$ are shown on the top row from left to right. Eigenvectors corresponding to $\alpha_{29}, \alpha_{30}, \alpha_{31}$ are shown on the bottom row from left to right.

Next, figure 2.40 displays the eigenvalues of the hat matrix (symmetric matrix figure 2.1(right) (page 26), $n = 31, m = 1$) of the moving average (eq(2.1) (page 24)) in descending order. The value of α_1 is uniquely 1, and the remaining eigenvalues are less than 1. Figure 2.41 illustrates the eigenvectors corresponding to $\alpha_1, \alpha_2, \ldots, \alpha_6$, and figure 2.42 shows those corresponding to $\alpha_{26}, \alpha_{27}, \ldots, \alpha_{31}$. Here again, the elements of the eigenvector corresponding to the eigenvalue of 1 are constant, and the oscillations of the other eigenvectors become more marked in the ascending order.

However, unlike the binomial filter, some eigenvalues are negative ($\alpha_{22}, \alpha_{23},$ $\ldots, \alpha_{31}$). Eq(2.85) indicates that this phenomenon results in the situation that the projection of data into the space spanned by the eigenvector corresponding to a negative eigenvalue ($< \mathbf{u}_k \cdot \mathbf{y} > \mathbf{u}_k$) takes the opposite direction of the projection of the estimates obtained by the moving average into the same space ($\alpha_k < \mathbf{u}_k \cdot \mathbf{y} > \mathbf{u}_k$). As a result, if the data is identical to one of the eigenvectors in figure 2.42, for example, the signs of all the estimates given by the moving average ($m = 1$) will be opposite to the original data; such estimates are not an acceptable result of smoothing. It should be noted here that even if all of the eigenvalues of a hat matrix are positive or 0, this does not ensure that the resulting estimates will not exhibit undesirable behavior. For instance, though the method of maintaining longer wavelengths than a certain threshold and eliminating waves shorter than this wavelength threshold after a Fourier transform is considered to be a smoothing method in which all of the eigenvalues of the hat matrix must take 1 or 0, it usually does not render desirable estimates (cf. Chapter 1). This problem involved in the Fourier transform is closely related to the phenomenon known as the Gibbs effect in signal analysis.

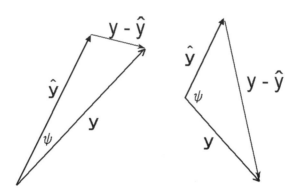

Figure 2.43   Relationships among $\mathbf{y}$, $\hat{\mathbf{y}}$, and $(\mathbf{y} - \hat{\mathbf{y}})$ ($0 \leq \psi \leq \frac{\pi}{2}$) (left). Relationships among $\mathbf{y}$, $\hat{\mathbf{y}}$, and $(\mathbf{y} - \hat{\mathbf{y}})$ ($\frac{\pi}{2} < \psi \leq \pi$) (right).

If a hat matrix is a symmetric matrix, all of the eigenvalues being positive or 0 is equivalent to a hat matrix being a quasi-definite matrix (positive semidefinite matrix, nonnegative definite matrix). Furthermore, $\mathbf{H}$ being a quasi-definite matrix that is equivalent to the equation below is satisfied for any $\mathbf{b}$ (a column vector with n real elements).

$$\mathbf{b}^t \mathbf{H} \mathbf{b} \geq 0. \tag{2.91}$$

The meaning of this equation is clarified by replacing **b** with **y**.

$$
\begin{aligned}
\mathbf{y}^t \mathbf{H} \mathbf{y} &= \mathbf{y}^t \hat{\mathbf{y}} \\
&= \| \mathbf{y} \| \cdot \| \hat{\mathbf{y}} \| \cos(\psi) \\
&\geq 0,
\end{aligned}
\tag{2.92}
$$

where $\| \ \|$ is the vector length. ψ is an angle between $\mathbf{y}$ and $\hat{\mathbf{y}}$. If a hat matrix is a symmetric matrix, $0 \leq \psi \leq \frac{\pi}{2}$ is always satisfied. Hence, the relationship between $\mathbf{y}$ and $\hat{\mathbf{y}}$ is always that depicted in figure 2.43(left) while that in figure 2.43(right) is impossible.

Please note that the eigenvectors corresponding to a moving average ($m = 1$) are identical to those that correspond to a binomial filter ($m = 2$). This is explained by assuming a hat matrix of the former as $\mathbf{H}_1$ and that of the latter as $\mathbf{H}_2$; the eigenvector of $\mathbf{H}_1$ is the same as that of $\mathbf{H}_2$ because $\mathbf{H}_2 = 0.75 \cdot \mathbf{H}_1 + 0.25 \cdot \mathbf{I}$.

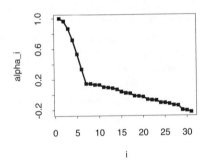

Figure 2.44   Eigenvalues of a hat matrix obtained by a moving average ($n = 31, m = 4$). Values are aligned in descending order ($\{\alpha_1, \alpha_2, \ldots, \alpha_{31}\}$).

A moving average ($m = 4$) (figure 2.44, figure 2.45, figure 2.46) provides negative eigenvalues ($\alpha_{18}, \alpha_{19}, \ldots, \alpha_{31}$). If projections obtained by applying matrices, which spectral resolution of a hat matrix yields to data, are positioned in spaces that are spanned by eigenvectors corresponding to these negative eigenvalues, the sign of these projections will be changed by this moving average. Thus, the moving average may cause unnatural behavior of the estimates.

Next, the results of a similar calculation using a smoothing spline (eq(2.59) (page 51)) are shown. When $n = 31$ and $\lambda = 100$, the eigenvalues can be aligned in descending order ($\{\alpha_1, \alpha_2, \ldots, \alpha_{31}\}$) as illustrated in figure 2.47. The eigenvectors corresponding to $\alpha_1, \alpha_2, \ldots, \alpha_6$ and those corresponding to $\alpha_{26}, \alpha_{27}, \ldots, \alpha_{31}$ are shown in figure 2.48 and figure 2.49, respectively. The values of α_1 and α_2 are 1, and the values of the other eigenvalues are also positive but less than 1. The projections into the space that is spanned by the eigenvectors corresponding to α_1 and α_2 constitute linear curves. Assume that $\{Y_i\}$ in eq(2.61) (page 52) takes the form $Y_i = a_0 + a_1 \cdot i$ ($1 \leq i \leq n$) (a_0 and a_1 are constants), and the corresponding $\mathbf{y}$ is defined as $\mathbf{y}_{(linear)}$.

$$
\mathbf{H} \mathbf{y}_{(linear)} = \mathbf{y}_{(linear)}.
\tag{2.93}
$$

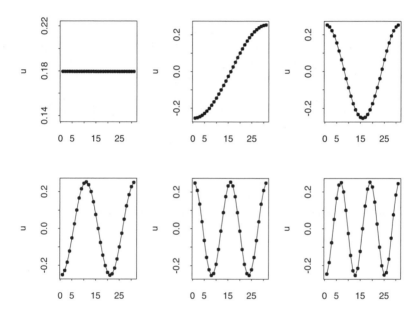

Figure 2.45   Eigenvectors corresponding to $\alpha_1, \alpha_2, \ldots, \alpha_6$ extracted from figure 2.44. Eigenvectors corresponding to $\alpha_1, \alpha_2, \alpha_3$ are displayed on the top row from left to right. Eigenvectors corresponding to $\alpha_4, \alpha_5, \alpha_6$ are displayed on the bottom row from left to right.

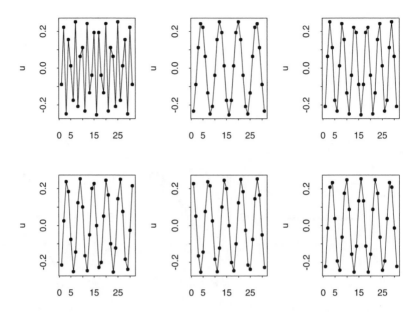

Figure 2.46   Eigenvectors corresponding to $\alpha_{26}, \alpha_{27}, \ldots, \alpha_{31}$ extracted from figure 2.44. Eigenvectors corresponding to $\alpha_{26}, \alpha_{27}, \alpha_{28}$ are displayed on the top row from left to right. The eigenvectors corresponding to $\alpha_{29}, \alpha_{30}, \alpha_{31}$ are displayed on the bottom row from left to right.

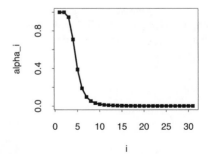

Figure 2.47   Eigenvalues of the smoothing spline ($n = 31$, $\lambda = 100$) are illustrated in descending order ($\{\alpha_1, \alpha_2, \ldots, \alpha_{31}\}$).

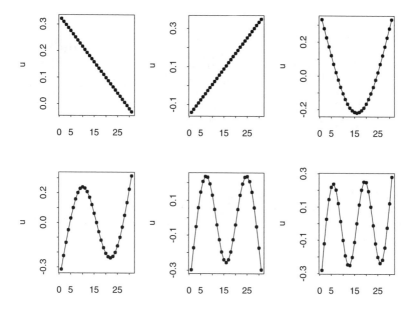

Figure 2.48   Eigenvectors corresponding to $\alpha_1, \alpha_2, \ldots, \alpha_6$ extracted from figure 2.47. The eigenvectors corresponding to $\alpha_1, \alpha_2, \alpha_3$ are displayed on the top row from left to right. The eigenvectors corresponding to $\alpha_4, \alpha_5, \alpha_6$ are displayed on the bottom row from left to right.

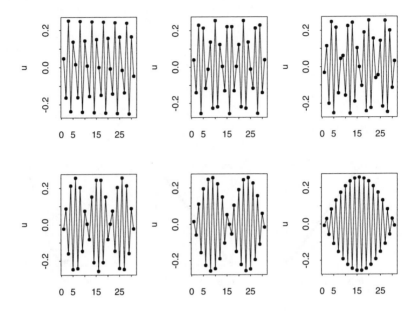

Figure 2.49   Eigenvectors corresponding to $\alpha_{26}, \alpha_{27}, \ldots, \alpha_{31}$ extracted from figure 2.47. Eigenvectors corresponding to $\alpha_{26}, \alpha_{27}, \alpha_{28}$ are displayed on the top row from left to right. Eigenvectors corresponding to $\alpha_{29}, \alpha_{30}, \alpha_{31}$ are displayed on the bottom row from left to right.

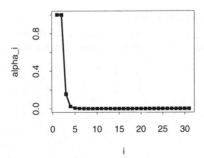

Figure 2.50   Eigenvalues of the smoothing spline ($n = 31$, $\lambda = 10000$) are illustrated in descending order ($\{\alpha_1, \alpha_2, \ldots, \alpha_{31}\}$).

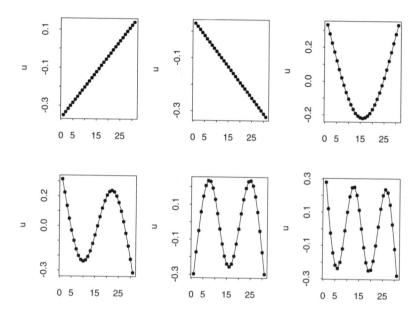

Figure 2.51   Eigenvectors corresponding to $\alpha_1, \alpha_2, \ldots, \alpha_6$ extracted from figure 2.50. The eigenvectors corresponding to $\alpha_1, \alpha_2, \alpha_3$ are displayed on the top row from left to right. The eigenvectors corresponding to $\alpha_4, \alpha_5, \alpha_6$ are displayed on the bottom row from left to right.

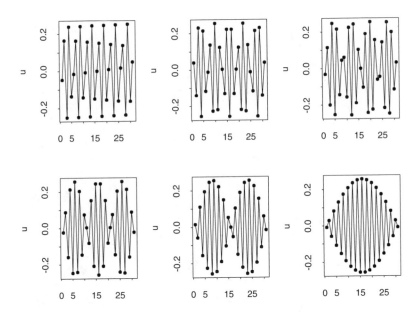

Figure 2.52   Eigenvectors corresponding to $\alpha_{26}, \alpha_{27}, \ldots, \alpha_{31}$ extracted from figure 2.50. The eigenvectors corresponding to $\alpha_{26}, \alpha_{27}, \alpha_{28}$ are displayed on the top row from left to right. The eigenvectors corresponding to $\alpha_{29}, \alpha_{30}, \alpha_{31}$ are displayed on the bottom row from left to right.

This equation means that one of the solutions of eq(2.69) (page 56) is $\alpha = 1$ with **u** in which the ith element is represented as a linear equation of i. Each of the eigenvectors is orthogonal, and hence there are two eigenvectors in which the ith element is represented as a linear equation of i. That is, this eigenequation has a multiple root ($\alpha_1 = \alpha_2 = 1$) and the corresponding eigenvectors are described as vectors in which the ith element is represented as a linear equation of i. This implies that smoothing by the smoothing spline maintains the projections into the space spanned by eigenvectors represented as a linear curve as it is, and reduces the lengths of the projections on the space spanned by other eigenvectors (oscillated ones). Furthermore, all of the eigenvalues of the smoothing spline are no less than 0 and no more than 1; more desirable estimates than those obtained by the moving average are expected.

The eigenvalues corresponding to the smoothing spline ($\lambda = 10000$) are displayed in descending order in figure 2.50. The eigenvectors corresponding to $\alpha_1, \alpha_2, \ldots, \alpha_6$ and those corresponding to $\alpha_{26}, \alpha_{27}, \ldots, \alpha_{31}$ are shown in figure 2.51 and figure 2.52, respectively. As expected, the values of α_1 and α_2 remain 1. The other eigenvalues are, however, closer to 0 than those in figure 2.47. This supports the notion that the increase of the value of λ results in estimates that are close to a linear curve. Furthermore, the eigenvectors in figure 2.48 and figure 2.49 are similar in appearance to those in figure 2.51 and figure 2.52. This similarity is not accidental. Since **S** (eq(2.63) (page 52)) is a symmetric matrix, we have $\mathbf{S} = \mathbf{U}_s \mathbf{A}_s \mathbf{U}_s^t$ ($\mathbf{U}_s$ is an orthogonal matrix and $\mathbf{A}_s$ is a diagonal matrix). Then, utilizing the fact that a hat matrix is represented as $\mathbf{H} = (\mathbf{I} + \lambda \mathbf{S})^{-1}$, we obtain

$$
\begin{aligned}
\mathbf{H} &= (\mathbf{I} + \lambda \mathbf{S})^{-1} \\
&= (\mathbf{I} + \lambda \mathbf{U}_s \mathbf{A}_s \mathbf{U}_s^t)^{-1} \\
&= (\mathbf{U}_s (\mathbf{I} + \lambda \mathbf{A}_s) \mathbf{U}_s^t)^{-1} \\
&= \mathbf{U}_s (\mathbf{I} + \lambda \mathbf{A}_s)^{-1} \mathbf{U}_s^t.
\end{aligned}
\tag{2.94}
$$

H is diagonalized by $\mathbf{U}_s$ because $(\mathbf{I} + \lambda \mathbf{A}_s)^{-1}$ is a diagonal matrix. Furthermore, since $\mathbf{A}_s$ does not depend upon λ, $\mathbf{U}_s$, which is used for the diagonalization of $\mathbf{A}_s$, does not depend upon λ either. Hence, while **H** depends upon λ, $\mathbf{U}_s$, which diagonalizes **H**, does not depend upon λ. Therefore, while the change of the value of the smoothing parameter alternates the eigenvalues of **H**, the eigenvectors of **H** remain the same.

Furthermore, the same calculation is carried out for polynomial regressions by the least squares: a linear equation is employed as a polynomial ($q = 1$ in eq(2.23) (page 33)). When $n = 31$, $\{\alpha_1, \alpha_2, \ldots, \alpha_{31}\}$ are shown in figure 2.53. The eigenvectors corresponding to $\alpha_1, \alpha_2, \ldots, \alpha_6$ and those corresponding to $\alpha_{26}, \alpha_{27}, \ldots, \alpha_{31}$ are illustrated in figure 2.54 and figure 2.55, respectively. Though the linear regression and the smoothing spline have a commonality in that both α_1 and α_2 are 1, all the other eigenvalues corresponding to the linear regression are 0. Eq(2.74) (page 57) and eq(2.54) (page 43) explain this.

Furthermore, as is evident in figure 2.54 (left and center graphs on the top row), the eigenvectors corresponding to α_1 and α_2 can be represented as a linear equation; a consideration similar to that for the smoothing spline elucidates this. That is, when the values of the target variable are represented as a linear regression, replacing **y**

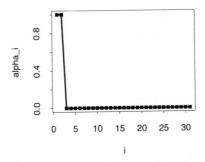

Figure 2.53   Eigenvalues of simple regression are aligned in descending order.

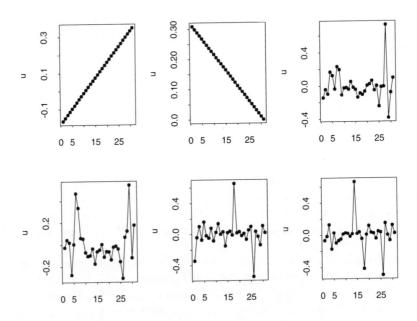

Figure 2.54   Eigenvectors corresponding to $\alpha_1, \alpha_2, \ldots, \alpha_6$ extracted from figure 2.53. The eigenvectors corresponding to $\alpha_1, \alpha_2, \alpha_3$ are displayed on the top row from left to right. The eigenvectors corresponding to $\alpha_4, \alpha_5, \alpha_6$ are displayed on the bottom row from left to right.

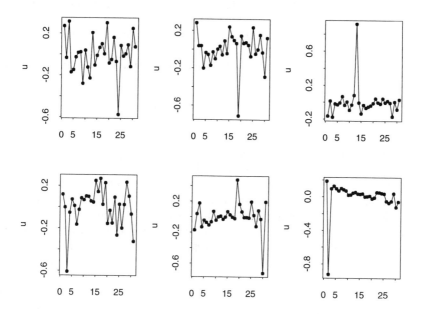

Figure 2.55   Eigenvectors corresponding to $\alpha_{26}, \alpha_{27}, \ldots, \alpha_{31}$ extracted from figure 2.53. The eigenvectors corresponding to $\alpha_{26}, \alpha_{27}, \alpha_{28}$ are displayed on the top row from left to right. The eigenvectors corresponding to $\alpha_{29}, \alpha_{30}, \alpha_{31}$ are displayed on the bottom row from left to right.

with $\mathbf{y}_{(linear)}$ allows $\mathbf{y}_{(linear)}$ to be treated as estimates by fitting a linear equation to $\mathbf{y}_{(linear)}$; the equation below, which is analogous to eq(2.93), is derived.

$$\mathbf{H}\mathbf{y}_{(linear)} = \mathbf{y}_{(linear)}. \tag{2.95}$$

This equation implies that a hat matrix corresponding to the fitting of a linear equation by the least squares provides eigenvalues that take the value of 1, and the corresponding eigenvectors are represented as a linear equation of the predictor. Since all of the vectors specified by $\mathbf{y}_{(linear)}$ are represented as a linear combination of two vectors, there are two eigenvectors corresponding to 1. The two eigenvectors are shown in figure 2.54 (left and center on the top row). However, all other eigenvectors are significantly different from those of the smoothing spline. This result invalidates an intuitive judgment that when the value of λ in the smoothing spline increases to the utmost limit, it arrives at simple regression.

Therefore, in the fitting of a linear equation by the least squares, eq(2.87) becomes

$$\hat{\mathbf{y}} = \sum_{i=1}^{2} Y_i^* \mathbf{u}_i = \sum_{i=1}^{2} < \mathbf{u}_i \cdot \mathbf{y} > \mathbf{u}_i. \tag{2.96}$$

$\{< \mathbf{u}_i \cdot \mathbf{y} > \mathbf{u}_i\}$ $(i = 1, 2)$ represents an orthogonal projection of $\mathbf{y}$ into the space that is spanned by $\mathbf{u}_i$. That is, the procedure of the transformation of $\mathbf{y}$ into $\hat{\mathbf{y}}$ using eq(2.96) is equivalent to the orthogonal projections of $\mathbf{y}$ into the space (the Φ_1 space) that is spanned by $\mathbf{u}_1$ and $\mathbf{u}_2$. All vectors ($\tilde{\mathbf{y}}$) in the Φ_1 space, which are represented as a linear combination of $\mathbf{u}_1$ and $\mathbf{u}_2$, are equivalent to all vectors (consisting of n elements) that are specified as linear equations. Hence, the finding that the orthogonal projection of $\mathbf{y}$ into the Φ_1 space results in $\hat{\mathbf{y}}$ indicates that $\hat{\mathbf{y}}$ is the result of the minimization of $\| \tilde{\mathbf{y}} - \mathbf{y} \|$ among vectors ($\tilde{\mathbf{y}}$), which are represented precisely as a linear equation.

The relationships among these vectors are illustrated in figure 2.56. This diagram shows that the orthogonal projections of $\mathbf{y}$ into the Φ_1 space (actually, the hyperplane) gives $\hat{\mathbf{y}}$. It also implies that the projection of $\mathbf{y}$ into the space that is orthogonal to both $\mathbf{u}_1$ and $\mathbf{u}_2$ is considered to be errors, and the subtraction of the errors from $\mathbf{y}$ results in $\hat{\mathbf{y}}$. Figure 2.16(left) (page 41) depicts this relationship in the Φ_2 space. Figure 2.56 also shows that if $\mathbf{y}$ is in the Φ_1 space, that is, if the data precisely follow a linear equation, $\hat{\mathbf{y}}$, which is given by the regression of a linear equation by the least squares, is identical to $\mathbf{y}$. This reflects the fact that the hat matrix corresponding to the fitting of a linear equation by the least squares is idempotent. Furthermore, a hat matrix ($\mathbf{H}$) in this situation satisfies

$$\begin{aligned} \text{trace}(\mathbf{H}) &= \sum_{i=1}^{2} \| \mathbf{u}_i \|^2 \\ &= \text{rank}(\mathbf{H}) \\ &= 2. \end{aligned} \tag{2.97}$$

Eq(2.96) and eq(2.97) are generalized into the geometric characteristics of the least squares, which gives a symmetric and idempotent hat matrix.

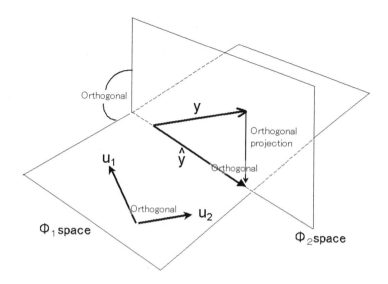

Figure 2.56 Geometric implication of obtaining estimates ($\hat{y}$) by fitting a linear equation by the least squares to data (**y**).

2.7 EXAMPLES OF S-PLUS OBJECT

(A) Smoothing by the moving average
Object: `move1()`

```
function(yy, mm)
{
#   (1)
    nd <- length(yy)
#   (2)
    yyr <- yy[(nd):(nd - mm + 1.)]
    yyl <- yy[mm:1.]
    y2 <- c(yyl, yy, yyr)
#   (3)
    ey <- rep(0, length = nd)
#   (4)
    for(ii in 1.:nd) {
        ey <- mean(y2[ii:(ii + 2. * mm)])
    }
#   (5)
    return(ey)
}
```

(1) The number of elements of yy (the number of data) is defined as nd.

(2) The variable m in eq(2.1) (page 24) is named mm. The vectors of y2 are derived using the boundary conditions expressed in eq(2.3) (page 24) and eq(2.4).
(3) The variable ey is set out for storing estimates.
(4) The estimates are calculated by eq(2.1), and they are saved in ey.
(5) The variable ey is an output argument.

An example of an object to use move1() (construction of figure 2.1(left) (page 26))

```
function()
{
#   (1)
    mm <- 1.
#   (2)
    xx <- seq(from = 1., by = 1., length = 31.)
#   (3)
    yy <- scan("c:\\datasets\\wak2.csv")
#   (4)
    ey <- move1(yy, mm)
#   (5)
    par(mfrow = c(1., 1.), mai = c(1.5, 1.5, 0.5, 0.5),
      oma = c(5., 5., 5., 5.))
#   (6)
    plot(xx, ey, type = "n", ylim = c(-15., 5.), xlab = "X",
      ylab = "Y")
    lines(xx, ey, lwd = 4.)
#   (7)
    points(xx, yy, pch = 4.)
}
```

(1) The variable m in eq(2.1) (page 24) is defined as mm.
(2) The variable xx is a vector for the predictor. In this example, since xx is a vector $(1, 2, \ldots, 31)$, seq() is used to generate sequential numbers starting at 1 and increasing by 1; the number of elements of the resulting vector is 31.
(3) "c:\\datasets\\wak2.csv" indicates a file named "wak2.csv" located in the "datasets" folder in the c-drive (a hard disk, floppy disk, magnetic optical disk, and so forth). The file wak2.csv is written in text format (ASCII format); 31 data are stored in the file:

-5.2
-0.2
2.1
-3.5
-4.5
$\vdots$

scan() is carried out to retrieve this data, which is named yy.
(4) Smoothing by the moving average is performed by move1(), and the resulting estimates are stored as ey.

(5) Space for drawing a graph is set up. Refer to the S-Plus manuals for the details of par().

(6) A graph is drawn to show the relationship between xx and ey; the association between the two variables is represented by a broken line.

(7) "×" is added to indicate the relationship between xx and yy on the graph created in (6).

(B) Weights of the moving average
Object: moveh1()

```
function(nd, mm2)
{
#   (1)
    iden <- diag(nd)
#   (2)
    ww <- apply(iden, 2., move1, mm = mm2)
#   (3)
    return(ww)
}
```

(1) An identity matrix (with size of nd × nd) is defined as iden.

(2) The object move1() described in (A) is used to calculate a matrix (hat matrix) to represent weights (w'_{ik} in eq(2.6) (page 24)), and this matrix is called ww. The estimates are calculated using apply() from the data, which is one of the column vectors in iden. A matrix consisting of the estimates as its column vectors is defined as ww. The value of 2 is assigned as the value for the second argument of apply(); it indicates that the second suffix of the matrix iden is fixed in each calculation. All the results obtained by increasing the value of the suffix by one continuously are outputted. Since the first argument is a matrix, each column vector of the matrix is processed by applying move1(). If 1 is assigned as the value for the second argument, each row vector is processed. Since the assignment "mm = mm2" is specified as the fourth argument, mm2 is used as mm of move1().

(3) The object ww is outputted.

An example of an object to use moveh1() (construction of figure 2.1(right) (page 26))

```
function()
{
#   (1)
    nd <- 31.
#   (2)
    mm <- 1.
#   (3)
    ww <- moveh1(nd, mm)
#   (4)
    par(mfrow = c(1, 1), mai = c(1.5, 1.5, 1.5, 1.5),
        oma = c(5, 5, 5, 5))
#   (5)
```

```
      persp(ww, zlim = c(-0.1, 0.7), xlab = "i", ylab = "j",
        zlab = "weights", lab = c(3., 3., 3.))
#   (6)
      invisible()
}
```

(1) The number of data (nd) is fixed at 31.
(2) The value of m in eq(2.1) (page 24) is given as mm.
(3) The weights (hat matrix) are calculated using moveh1(), and the matrix is called ww.
(4) Space for drawing a graph is set up.
(5) A perspective plot is drawn to illustrate the values of ww.
(6) If the object ends with persp(), various values concerning the perspective plot are outputted. To prevent this phenomenon, invisible() is added.

(C) Smoothing by the binomial filter

```
function(yy, mm)
{
#   (1)
      nd <- length(yy)
#   (2)
      mm2 <- mm * 0.5
#   (3)
      yyw1 <- yy
      yyw2 <- yy
      rlim <- mm2
      yyr <- NULL
      count <- 0.
      while(rlim > nd) {
          yyw1 <- rev(yyw1)
          yyr <- c(yyr, yyw1)
          rlim <- rlim - nd
          count <- count + 1.
      }
      switch(count %% 2. + 1.,
          yyr <- c(yyr, yy[nd:(nd - rlim + 1.)]),
          yyr <- c(yyr, yy[1.:rlim]))
      llim <- mm2
      yyl <- NULL
      while(llim > nd) {
          yyw2 <- rev(yyw2)
          yyl <- c(yyw2, yyl)
          llim <- llim - nd
      }
      switch(count %% 2. + 1.,
          yyl <- c(yy[llim:1.], yyl),
          yyl <- c(yy[(nd - llim + 1.):nd], yyl))
```

```
    y2 <- matrix(c(yyl, yy, yyr), ncol = 1.)
#  (4)
    ww <- matrix(0., ncol = nd + mm, nrow = nd + mm)
#  (5)
    imat <- row(ww)
    jmat <- col(ww)
#  (6)
    check <- 0. <= (mm2 + imat - jmat) & (mm2 + imat - jmat) <= mm
#  (7)
    ww[check] <- exp(lgamma(mm + 1.) -
      lgamma(mm2 + imat[check] - jmat[check] + 1.) -
      lgamma(mm2 - imat[check] + jmat[check] + 1.) -
      mm * logb(2.))
#  (8)
    ey <- ww %*% y2
    ey <- as.vector(ey[(mm2 + 1.):(nd + mm2)])
#  (9)
    return(ey)
}
```

(1) The number of elements of yy (the number of data) is specified as nd.

(2) The variable $0.5m$ in eq(2.7) (page 27) is named mm2.

(3) The variable y2, which is a column vector (a matrix consisting of one column vector), is generated using the boundary conditions in eq(2.10) (page 27) and eq(2.11). "count %% 2" gives a remainder when count is divided by 2. Hence, "switch(count %% 2 + 1, yyr <- c(yyr, yy[nd:(nd - rlim + 1)]), yyr <- c(yyr, yy[1:rlim]))" indicates that if the remainder when count is divided by 2 is 0, "yyr <- c(yyr, yy[nd:(nd - rlim + 1)])" is carried out, and if the remainder is 1, "yyr <- c(yyr, yy[1:rlim])" is executed. If "y2 <- c(yyl, yy, yyr)" is employed, y2 becomes a simple vector; it is not specified as either a row vector or a column vector. However, here, y2 is not a simple vector but a column vector because it is multiplied by ww.

(4) A matrix for storing weights (with size of (nd+mm) × (nd+mm)) is named ww, and 0 is inputted to all of its elements.

(5) The sizes of imat and jmat are the same as those of ww. The values of the elements of each column of imat are the row number, and the values of the elements of each row of jmat are the column number. That is, imat is

$$
\begin{pmatrix}
1 & 1 & 1 & 1 & \cdots \\
2 & 2 & 2 & 2 & \cdots \\
3 & 3 & 3 & 3 & \cdots \\
4 & 4 & 4 & 4 & \cdots \\
\vdots & \vdots & \vdots & \vdots & \ddots
\end{pmatrix},
\tag{2.98}
$$

and jmat is

$$\begin{pmatrix} 1 & 2 & 3 & 4 & \cdots \\ 1 & 2 & 3 & 4 & \cdots \\ 1 & 2 & 3 & 4 & \cdots \\ 1 & 2 & 3 & 4 & \cdots \\ \vdots & \vdots & \vdots & \vdots & \ddots \end{pmatrix}. \qquad (2.99)$$

(6) A variable check (a logical object) is set up; the size is the same as that of ww. Each element takes F if the value of weight is 0, and T if it is not 0.

(7) If gamma() is employed to obtain the value of the gamma function (Euler's gamma function) (when the gamma function is named $\Gamma()$, it is proved that $\Gamma(p) = (p-1)!$ where p is a positive integer) and calculate eq(2.8) (page 27), the assignment of a large value as an argument for gamma() leads to the result Inf (indicating infinity). Then, lgamma() (an object for calculating the logarithm of the value which the gamma function gives) is utilized for the calculation.

(8) Estimates are derived by using a binomial filter and take a vector form.

(9) The object ey is outputted.

An example of an object to use binom1() (construction of figure 2.4 (left) (page 29))

```
function()
{
#    (1)
     mm <- 2.
#    (2)
     xx <- seq(from = 1., by = 1., length = 31.)
#    (3)
     yy <- scan("c:\\datasets\\wak2.csv")
#    (4)
     ey <- binom1(yy, mm)
#    (5)
     par(mfrow = c(1, 1), mai = c(1.5, 1.5, 0.5, 0.5),
       oma = c(5, 5, 5, 5))
#    (6)
     plot(xx, ey, type = "n", ylim = c(-15., 5.),
       xlab = "X", ylab = "Y")
     lines(xx, ey, lwd = 4.)
#    (7)
     points(xx, yy, pch = 4.)
}
```

(1) The variable m in eq(2.7) (page 27) is specified as mm.

(2) A vector consisting of the values of the predictor is named xx.

(3) Data is retrieved from "c:\\datasets\\wak2.csv" for storage in yy.

(4) Smoothing by the binomial filter is carried out by applying binom1() to obtain estimates (ey).

(5) Space for drawing a graph is set up.

(6) A graph is drawn to show the relationship between xx and ey. Their association is illustrated by a broken line.

(7) The graph drawn in (6) is superimposed with "×" to show the relationship between xx and yy.

(D) Fitting of polynomials (construction of figure 2.8(left) (page 35), a method for polynomial equation using lm())

```
function()
{
#   (1)
     nd <- 31.
     xx <- seq(from = 1., by = 1., length = nd)
     yy <- scan("c:\\datasets\\wak2.csv")
     data1 <- data.frame(x = xx, y = yy)
#   (2)
     fit.lm <- lm(y ~ poly(x, degree = 1.), data = data1)
#   (3)
     ey <- fitted.values(fit.lm)
#   (4)
     par(mfrow = c(1., 1.), mai = c(1.5, 1.5, 0.5, 0.5),
       oma = c(5., 5., 5., 5.))
     plot(xx, ey, type = "n", ylim = c(-15., 5.), xlab = "X",
       ylab = "Y")
     lines(xx, ey, lwd = 4.)
     points(xx, yy, pch = 4.)
}
```

(1) Data for lm() are stored collectively in data1 (a data frame); xx is named x and yy is named y in data1.

(2) Using data1 (a data frame), a regression equation y = (a linear function of x) is fitted by the least squares. If "degree = 1" is replaced by "degree = 2", fitting of a quadratic equation is realized, and if it is replaced by "degree = 3", a cubic equation is fitted.

(3) The polynomial equation estimates (a vector with nd elements) given by the values of the predictor, which are the same as those of data to derive this polynomial equation, are stored in ey.

(4) The relationships between data and estimates are drawn in a graph.

The object above employs lm(y ~ poly(x, degree = 1), data = data1), in which the degree of the polynomial is designated, to fit a polynomial. The degree of the polynomial is specified directly by "degree = 1". When ndim (an object for saving the degree of a polynomial) is used to give the degree of the polynomial, (2) is replaced with the object below. The object substitute() is explained in Chapter 6.

```
     ndim <- 1
     formula.name <- substitute(y ~ poly(x, degree = degree1),
```

```
    list(degree1 = ndim))
    fit.lm <- lm(formula.name, data = data1)
```

`poly()` is an object for calculation using orthogonal polynomials and hence reliable results are derived even if the degree of a polynomial is large. However, the estimation of estimates using the regression equations given by `poly()` at the points where no data exists requires `predict.gam()` (cf. an object on page 502). Hence, the use of `predict.gam()` increases the amount of calculation required. Analysts should therefore select from among the use of `poly()`, utilization of the design matrix, the employment of `lsfit()`, and so forth, in light of their purpose.

Without `poly()`, an object for the use of `lm()` after generating a design matrix (eq(2.24) (page 33)) will be similar to the following:

```
function()
{
#   (1)
    nd <- 31.
    xx <- seq(from = 1., by = 1., length = nd)
    yy <- scan("c:\\datasets\\wak2.csv")
    ndim <- 1.
#   (2)
    xm <- matrix(rep(0, length = nd * ndim), nrow = nd)
    for(i in 1.:ndim) {
        xm[, i] <- xx^i
    }
#   (3)
    data1 <- data.frame(x = xm, y = yy)
    fit.lm <- lm(y ~ ., data = data1)
    ey <- fitted.values(fit.lm)
#   (4)
    par(mfrow = c(1., 1.), mai = c(1.5, 1.5, 0.5, 0.5),
      oma = c(5., 5., 5., 5.))
    plot(xx, ey, type = "n", ylim = c(-15., 5.), xlab = "X",
      ylab = "Y")
    lines(xx, ey, lwd = 2.)
    points(xx, yy, pch = 4.)
}
```

(1) Data are given, and the degree of the polynomial is specified as ndim.
(2) A design matrix (xm) (eq(2.24) (page 33) (the result of omitting a column vector $(1, 1, \ldots, 1)^t$ from $\mathbf{X}$)) is derived.
(3) Multiple regression is carried out using xm (predictor values of data) and yy (target variable values of data).
(4) The relationships between data and estimates are drawn in the graph.

The object above calculates xm in (2). This part can be replaced with the three lines below. Since this method does not use for(i in 1:ndim), the calculation is less expensive.

```
xx <- 1:nd
power <- rep(1:ndim, rep(nd, ndim))
xm <- matrix(xx^power, nrow = nd)
```

Another method is possible to derive xm:

```
powerf <- function(jj, x1)
    {
        pw <- x1^jj
        return(pw)
    }
xm <- apply(matrix(c(1:ndim), nrow = 1), 2, powerf, x1 = xx)
```

After defining powerf() (a universal power function), a design matrix is obtained using this power function.

Furthermore, when lsfit() is utilized, (3) is replaced with (3)':

```
#  (3)'
    fit.lin <- lsfit(xm, yy)
    ey <- yy - fit.lin$residuals
```

To calculate the hat matrices accompanied by the fitting of polynomials by the least squares (figure 2.8(right) (page 35), figure 2.9(right) (page 36), figure 2.10(right) (page 36), figure 2.12(right) (page 37), figure 2.14(right) (page 37)), either eq(2.6) (page 24) or eq(2.38) (page 35) is utilized. In this case, a hat matrix is represented as a simple equation (eq(2.38) (page 35)). Hence, the use of this equation seems a more sophisticated approach than that of eq(2.6). In numerical calculation, however, the use of mathematically refined formations does not necessarily result in beneficial outcomes. With respect to this example, when the number of data is large and/or the degree of a polynomial is high, the calculation of the inverse matrix in eq(2.38) (page 35) becomes ill-conditioned and hence the calculation fails or considerable errors arise. When S-Plus is used, the calculation of hat matrices in figure 2.12(right) and figure 2.14(right) using eq(2.38) leads to an ill-conditioned inverse matrix; the object will stop. That is, the calculation of the inverse matrix is rejected. Hence, the use of eq(2.6) is preferable. This is because regression by applying lm() rather than the derivation of the inverse matrix leads to accurate numerical calculation. In particular, the use of poly() in lm() facilitates the calculation for high-degree polynomials. However, the direct use of eq(2.38) to obtain the values represented in figure 2.8(right), figure 2.9(right), and figure 2.10(right) is advisable to ensure that the results obtained using eq(2.38) coincide with those obtained using eq(2.6).

(E) Fitting of a polynomial, and calculation of estimates that are placed at the positions where no data exist (construction of figure 2.15(left) (page 38). Use of results from lm())

```
function()
{
#  (1)
```

```
      nd <- 31.
      xx <- seq(from = 1., by = 1., length = nd)
      ll <- list(r1 = 0.)
      mm <- scan("c:\\datasets\\wak2.csv", ll, sep = ",")
#   (2)
      yy <- mm$r1
#   (3)
      assign("data1", data.frame(x = xx, y = yy), frame = 1.)
      ex <- seq(from = 1., by = 0.1, length = nd * 10. - 9.)
#   (4)
      data2 <- data.frame(x = ex)
      fit.lm <- lm(y ~ poly(x, degree = 11.), data = data1)
      ey <- predict.gam(fit.lm, newdata = data2)
#   (5)
      par(mfrow = c(1., 1.), mai = c(1.5, 1.5, 0.5, 0.5),
        oma = c(5., 5., 5., 5.))
      plot(ex, ey, type = "n", ylim = c(-16., 4.), xlab = "X",
        ylab = "Y")
      lines(ex, ey, lwd = 4.)
      points(xx, yy, pch = 4.)
}
```

(1) The number of data (nd), the values of the predictor (xx), and the values of target variables (yy) are given.

(2) Data (xx and yy) are stored collectively in a data frame. The data frame named data1 is placed in an expression frame (top-level frame, frame 1) (designated "frame = 1") (cf. Chapter 4).

(3) The positions at which the estimates are derived are inputted as a vector (ex). ex is stored in a data frame (data2). On this occasion, the name for ex in data2 must coincide with the name for the predictor (x) in data1.

(4) Polynomial regression (the 11th-degree polynomial is used in this example) is carried out by the object lm(). The result is saved in fit.lm.

Then, estimates located at the positions which data2 designates are calculated by applying fit.lm (a vector) to give ey (a vector).

(5) The relationships among data and estimates are drawn.

(F) Fitting of the spline function (construction of figure 2.19(left). A method of deriving the spline function represented by B-spline (B-splines) by applying lm())

```
function()
{
#   (1)
      nd <- 31.
      xx <- seq(from = 1., by = 1., length = nd)
      yy <- scan("c:\\datasets\\wak2.csv")
#   (2)
      data1 <- data.frame(x = xx, y = yy)
```

```
#   (3)
    fit.lm <- lm(y ~ bs(x, knots = c(4., 7., 10., 13., 16.,
      19., 22., 25., 28.), degree = 1.), data = data1)
#   (4)
    ey <- fitted.values(fit.lm)
#   (5)
    par(mfrow = c(1., 1.), mai = c(1.5, 1.5, 0.5, 0.5),
      oma = c(5., 5., 5., 5.))
    plot(xx, ey, type = "n", ylim = c(-15., 5.),
      xlab = "X", ylab = "Y")
    lines(xx, ey, lwd = 4.)
    points(xx, yy, pch = 4.)
}
```

(1) The number of data (nd), the values of the predictor (xx), and the values of target variables (yy) are provided.

(2) Data for using lm() are stored collectively in a data frame (data1). In the data frame, xx is named x, and yy is named y.

(3) Internal knots are placed at i=(4, 7, 10, 13, 16, 19, 22, 25, 28). The spline function represented by B-spline is fitted by the least squares; the result is saved in fit.lm.

   On this occasion, "degree = 1" designates the fitting of the 1st-degree spline (broken line). The replacement of "degree = 1" by "degree = 3" specifies the fitting of the cubic spline function. A spline function in a finite region is fitted without boundary conditions by lm() and bs(). B-spline is often employed for the fitting of spline functions; this is because estimates with high accuracy are obtained by the use of B-spline.

(4) Estimates corresponding to xx are derived using fit.lm, and are stored in ey.

(5) The relationships among data and estimates are drawn.

(G) Local linear regression for equispaced predictor
Object: lline()

```
function(yy, hh)
{
#   (1)
    llin <- function(ex1, xdata, ydata, band)
    {
#   (2)
        wts <- exp((-0.5 * (ex1 - xdata)^2.)/band^2.)
#   (3)
        data1 <- data.frame(x = xdata, y = ydata, www = wts)
#   (4)
        fit.lm <- lm(y ~ x, data = data1, weights = www)
#   (5)
        est <- fit.lm$coef[1.] + fit.lm$coef[2.] * ex1
#   (6)
        return(est)
```

```
      }
#   (7)
      nd <- length(yy)
      xx <- seq(from = 1., by = 1., length = nd)
#   (8)
      xxmat <- matrix(xx, ncol = 1.)
#   (9)
      ey <- apply(xxmat, 1., llin, xdata = xx, ydata = yy,
        band = hh)
      ey <- as.vector(ey)
#   (10)
      return(ey)
}
```

(1) A function (llin()) for performing smoothing by local linear regression is defined. The position for deriving an estimate is given as ex1 (this is a scalar not a vector). The objects xdata and ydata are vectors which store the predictor values of the data and the target variables of data, respectively. band is a bandwidth (h in eq(2.58) (page 49)). Note that ex1 does not necessarily coincide with any of the elements of xdata. Furthermore, this function is useful even if xdata is not equispaced.

(2) Weights of data are calculated using a kernel function given by eq(2.58), and then they are named wts.

(3) The objects xdata, ydata, and wts are stored in a data frame named data1. xdata, ydata, and wts are named x, y, and www in the data frame, respectively.

(4) Simple regression by applying lm() is carried out using data1. The results are saved as fit.lm.

(5) The object fit.lm\$coef[1] indicates $\hat{b}_j$ in eq(2.57) (page 49), and fit.lm\$coef [2] indicates $\hat{a}_j$. Estimates at ex1 are calculated by eq(2.57); the result is stored as est.

(6) The object est is the output of llin().

(7) The number of data is specified as nd, and the predictor values of data are specified as xx.

(8) Since the object xx has a vector form, xx is transformed into a column vector xxmat to be used in apply().

(9) The estimates are calculated by llin() at the points of xxmat. Since the second argument of apply() is fixed at 1, ey (a vector with nd elements), in which each element is an estimate, is obtained by the application of llin() to each element of the column vector (xxmat). On this occasion, xx, yy, and hh are used as xdata, ydata, and band, respectively. Thus, the resulting ey is transformed into a vector.

(10) The object ey is outputted.

An example of an object to use lline() (construction of figure 2.23(left) (page 47))

```
function()
{
#   (1)
```

```
    yy <- scan("c:\\datasets\\wak2.csv")
    nd <- length(yy)
    xx <- seq(from = 1., by = 1., length = nd)
#   (2)
    hh <- 1.5
#   (3)
    ey <- lline(yy, hh)
#   (4)
    par(mfrow = c(1, 1), mai = c(1.5, 1.5, 0.5, 0.5),
      oma = c(5, 5, 5, 5))
    plot(xx, ey, type = "n", ylim = c(-15., 5.), xlab = "X",
      ylab = "Y")
    lines(xx, ey, lwd = 4.)
    points(xx, yy, pch = 4.)
}
```

(1) The retrieved data is named yy, and values are assigned to nd and xx.
(2) The value of hh (h in eq(2.58) (page 49)) is fixed at 1.5.
(3) Local linear regression for the equispaced predictor is carried out by applying lline(yy, hh). The estimates are identified as ey.
(4) The relationships among data and estimates are drawn in a graph.

(H) Smoothing spline for an equispaced predictor
Object: smspe()

```
function(yy, lambda)
{
#   (1)
    nd <- length(yy)
#   (2)
    ss <- c(1, -2, 1, rep(0, nd - 3))
    ss <- rbind(ss, c(-2, 5, -4, 1, rep(0, length = nd - 4)))
    for(ii in 1:(nd - 4)) {
        ss <- rbind(ss, c(rep(0, ii - 1), 1, -4, 6, -4, 1,
        rep(0, nd - ii - 4)))
    }
    ss <- rbind(ss, c(rep(0, length = nd - 4), 1, -4, 5, -2))
    ss <- rbind(ss, c(rep(0, length = nd - 3), 1, -2, 1))
#   (3)
    ssi <- diag(nd) + lambda * ss
#   (4)
    ey <- solve(ssi, yy)
    ey <- as.vector(ey)
#   (5)
    return(ey)
}
```

(1) The number of elements of yy is fixed to be equal to the number of data (nd).

(2) $\mathbf{S}$ (with size of nd×nd) in eq(2.62) (page 52) is named ss. Vectors are connected by rbind() to generate ss here. When constructing a vector or a matrix, the procedure of expanding a vector or a matrix gradually using c(), rbind(), or cbind() is effective in clarifying the content of an object. However, since additional memory (RAM) is consumed each time a vector or a matrix (ss in this example) is renewed, this procedure requires large memory. When the size of a vector or a matrix to be constructed is known beforehand, a vector or a matrix with the desired size filled with 0 or other values should be prepared first. Then, the values of the elements are provided when necessary. This approach enables the efficient use of memory. With this respect, part of (2) should be replaced with (2)' below.

```
#   (2)'
ss <- matrix(rep(0., length = nd * nd), ncol = nd)
ss[1., 1.:3.] <- c(1., -2., 1.)
ss[2., 1.:4.] <- c(-2., 5., -4., 1.)
for(ii in 3.:(nd - 2.)) {
    ss[ii, (ii - 2.):(ii + 2.)] <- c(1., -4., 6., -4., 1.)
}
ss[(nd - 1.), (nd - 3.):nd] <- c(1., -4., 5., -2.)
ss[nd, (nd - 2.):nd] <- c(1., -2., 1.)
```

(3) $(\mathbf{I} + \lambda\mathbf{S})$ in eq(2.63) (page 52) is calculated to be ssi.
(4) $\hat{y}$ in eq(2.63) (page 52) is computed and the result is named ey (a vector).
(5) The object ey is outputted.

An example of an object to use smspe() (construction of figure 2.28(left) (page 53))

```
function()
{
#   (1)
    yy <- scan("c:\\datasets\\wak2.csv")
#   (2)
    lambda <- 1.
#   (3)
    ey <- smspe(yy, lambda)
#   (4)
    par(mfrow = c(1, 1), mai = c(1.5, 1.5, 0.5, 0.5),
      oma = c(5, 5, 5, 5))
    xx <- seq(from = 1., by = 1., length = length(yy))
    plot(xx, ey, type = "n", ylim = c(-15., 5.),
      xlab = "X", ylab = "Y")
    lines(xx, ey, lwd = 4.)
    points(xx, yy, pch = 4.)
}
```

(1) Data are retrieved and named yy.
(2) The value of lambda (λ in eq(2.63) (page 52)) is fixed at 1.

(3) Smoothing by spline smoothing for equispaced data is performed using smspe();
the estimates are named ey.
(4) The relationships among data and estimates are drawn in a graph.

```
function()
{
#   (1)
    nd <- 31
    mm <- 4
#   (2)
    iden <- diag(nd)
#   (3)
    ww <- apply(iden, 2, move1, mm = mm)
#   (4)
    eigen1 <- eigen(ww, symmetric = T)
    value1 <- eigen1$values
#   (5)
    par(mfrow = c(1, 1), mai = c(1.5, 1.5, 0.5, 0.5),
        oma = c(5, 5, 5, 5))
    xx <- seq(from = 1, to = 31, by = 1)
    plot(xx, value1, type = "n", xlab = "i", ylab = "alpha_i")
    lines(xx, value1, lwd = 4)
    points(xx, value1, pch = 15, cex = 0.8)
}
```

(1) The number of data is stored as nd. m in eq(2.1) (page 24) is named mm.
(2) An identical identity matrix (with size of nd $\times$ nd) is created as iden.
(3) Weights (w'_{ik} in eq(2.6) (page 24)) are calculated using move1() shown in (A);
the result is stored as ww.
(4) The eigenvalues and eigenvectors of ww are derived by applying eigen(): the
result is saved as eigen1. The eigenvalues are extracted as eigen1$values for stor-
age as value1. The resulting eigenvalues are aligned in descending order. Note that
"vector1 <- eigen1$vectors" provides eigenvectors (vector1); vector1[, i]
is an eigenvector corresponding to the i-th eigenvalue.
(5) The values of the eigenvalues are drawn in a graph in descending order.

REFERENCES

1. T. Hastie, R. Tibshirani, and J. H. Friedman (2001). *The Elements of Statistical Learning: Data Mining, Inference, and Prediction*, Springer-Verlag.

2. D.C. Lay (1996). *Linear Algebra and Its Applications*, 2nd edition, Addison Wesley Longman.

Problems

2.1   Consider the data shown in Chapter 1: the monthly average exchange rate of US dollar to yen from February 1987 through May 1999.
(a) Obtain the numerical data of the exchange rate from the WWW site of the publisher of this book or any other places on the Internet.
(b) Edit the data to create the csv (comma separated value format) file such as

1,	240.37
2,	231.86
3,	221.86
4,	226.18
5,	214.11
$\vdots$	$\vdots$

This csv file will be named "dol.csv". A text editor or a spreadsheet software is useful for these purposes.
(c) Create an S-Plus or R object for retrieving this data. `scan()` and `read.table()` can be used.
(d) Using S-Plus or R, plot the data to construct a graph similar to that in figure 1.2 (page 4).
(e) Change the size and horizontal-to-vertical ratio of the graph produced in **(d)**. Alter the plot characters in the graph. Add the title such as "Exchange Rate" and locate a text such as "monthly average exchange rate of US dollar to yen from February 1987 through May 1999" at an approximate position in the graph.

2.2   The reflection boundary condition is a typical boundary condition for smoothing by a moving average or a binomial filter. Other boundary conditions, however, are also available. The periodic boundary condition is also a typical one. This boundary condition replaces eq(2.3) (page 24) and eq(2.4) by

$$Y_0 = Y_n, \quad Y_{-1} = Y_{n-1}, \quad \ldots, \quad Y_{-m+1} = Y_{n-m+1}; \qquad (2.100)$$

$$Y_{n+1} = Y_1, \quad Y_{n+2} = Y_2, \quad \ldots, \quad Y_{n+m} = Y_m. \qquad (2.101)$$

(a) Construct an S-Plus or R object for smoothing by a moving average based on this boundary condition.
(b) Create an S-Plus or R object for smoothing by a binomial filter based on this boundary condition.
(c) Using the objects created in **(a)** and **(b)**, smooth the 20 data below. m can take various values.
2.46, -0.59, 1.14, -0.94, 0.62, -0.63, -0.43, 2.30, 1.29, 0.25,
1.92, -0.17, 0.22, -2.13, -3.03, -1.29, -3.24, 1.04, -0.64, 1.85
(d) Using the objects created in **(a)** and **(b)**, smooth the 20 data below.
-0.63, -0.43, 2.30, 1.29, 0.25, 1.92, -0.17, 0.22, -2.13, -3.03,
-1.29, -3.24, 1.04, -0.64, 1.85, 2.46, -0.59, 1.14, -0.94, 0.62
    The data presented in **(d)** is produced by the periodic shift of the data shown in **(c)**. Confirm that this periodic shift does not affect the essential behavior of the estimates.
(e) Show examples of data in which the periodic boundary condition can be used.

2.3   Gain an understanding of the fitting of polynomials by the least squares using the objects presented in (D) of this chapter. The 30 data below (the daily average air temperature observed at the Aoshima observation point in the city of Miyazaki, Miyazaki prefecture, Japan, in November 1997, obtained by an Automated Meteorological Data Aquisition System (AMeDAS)) are useful for this purpose, though any other available data can be used.

9.6, 12.8, 14.6, 15.6, 15.5, 15.1, 15.6, 13.8, 13.9, 16.1,
17.3, 18, 19.9, 20, 19.9, 18.2, 15.8, 11.2, 9.6, 15.8,
16.7, 17.5, 13.7, 15.7, 20.6, 21.2, 16.7, 16, 20.7, 17.6.

(a) Fit a 5th-degree polynomial equation to this data. Construct a graph to compare the estimates obtained by this regression equation with the data. Confirm that the estimates are not sufficient as the summary of the data. Attempt the use of other degree polynomial equations for the fitting.

(b) Replace the first data (9.6) with the value of 15.0 and compare the estimates given by this data with those given by the original data. Attempt other alterations of the data and observe the responses of the estimates.

(c) Try to fit 20th-degree polynomials using

```
fit.lm <- lm(y ~ I(x^20) + I(x^19) + I(x^18) + I(x^17) +
I(x^16) + I(x^15) + I(x^14) + I(x^13) + I(x^12) + I(x^11) +
I(x^10) + I(x^9) + I(x^8) + I(x^7) + I(x^6) + I(x^5) +
I(x^4) + I(x^3) + I(x^2) + I(x), data = data1)
```

Find the error message such as "Problem in lm.fit.qr(unset(x), unset(y)): computed fit is singular, rank 19". Replace the line above with

```
fit.lm <- lm(y ~ poly(x, degree = 20), data = data1)
```

Then, verify the success at this time.

(d) Using the same data, utilize the estimates supplied by fitting a 4th-degree polynomial by the least squares as values of a target variable (the values of a predictor remain the same) to carry out a 4th-degree polynomial regression by the least squares. Confirm that the resultant estimates are identical to the values of the target variable.

(e) Utilize the estimates supplied by fitting a 2nd-degree polynomial by the least squares as values of a target variable (the values of a predictor remain the same) to carry out a 4th-degree polynomial by the least squares. Confirm that the resultant estimates are identical to the values of the target variable.

(f) Utilize the estimates supplied by fitting a 4th-degree polynomial by the least squares as values of a target variable (the values of a predictor remain the same) to carry out a 2nd-degree polynomial by the least squares. Confirm that the resultant estimates are not identical to the values of the target variable.

(g) By fitting a 4th-degree polynomial by the least squares, check eq(2.43) (page 40).

2.4   Using the objects shown in (E) of this chapter, gain a better understanding of the fitting of polynomials by the least squares. Use the 30 data displayed in Problem **2.3**.

(a) Create an S-Plus or R object for fitting a polynomial by the least squares, calculating the estimates at 1, 1.1, 1.2, 1.3, 1.4, ..., 29.9, 30, instead of 1, 2, 3, 4, 5, ..., 29, 30, and constructing a graph to compare the data and the estimates.

(b) Using the object provided in **(a)**, fit an 18th-degree polynomial to the data and calculate the estimates. Check that some estimates are unacceptably away from the data.

2.5   Using the 30 data displayed in Problem **2.3**, consider the values of elements of a hat matrix given by fitting a polynomial equation by the least squares.

(a) Produce a couple of S-Plus or R objects for computing the values of elements of a hat matrix given by fitting a polynomial by the least squares. Some procedures are possible: (1) eq(2.38) (page 35) is used as is. (2) Simultaneous equations are solved to avoid calculating the inverse matrix, though eq(2.38) is adopted basically. (3) The values of elements of a hat matrix are computed by observing the responses of the estimates to the movements of data. Compare the results obtained using these objects in terms of computational cost.

(b) Using 4th-degree and 5th-degree polynomials, confirm eq(2.54) (page 43).

(c) When the distribution of the values of a predictor are ill-conditioned (e.g., two or more values of the predictor are similar), the values of elements of a hat matrix depend on the algorithm for computing them. Consider the best object that yields accurate values of elements of a hat matrix even if the values of the predictor are ill-conditioned.

2.6   Alter the object presented in **(F)** of this chapter to acquire intuitive understanding of fitting a spline by the least squares.

(a) Adopt $\{1, 10.3, 16.5, 22.8, 31\}$ as the positions of knots of a linear spline instead of $\{1, 4, 7, 10, 13, 16, 19, 22, 25, 28, 31\}$. This is realized by replacing a part of (3) in **(F)** with

```
#   (3)'
    fit.lm <- lm(y ~ bs(x, knots = c(10.3, 16.5, 22.8.),
       degree = 1.), data = data1)
```

Note that the assignment of knots = requires the positions of internal knots; knots at the endpoints of the region (in this example, $\{1, 31\}$) are not included. In addition, the internal knots are not required to be positioned at data points.

(b) Using the object created in **(a)**, observe the responses of the estimates given by the movements of data and describe the findings obtained. Construct a graph of the values of elements of a hat matrix in a manner similar to those of figure 2.17(left) (page 44) and figure 2.19(left) (page 44).

(c) Modify the object created in **(a)** to realize extrapolation. For example, the replacement of (4) of the object presented in **(E)** by (4)' as below gives estimates from -5 through 40.

```
#   (4)'
    xx2 <- seq(from = -5, to = 40, by = 1)
    data2 <- data.frame(x = xx2)
    ey <- predict.gam(fit.lm, newdata = data2)
```

(d) Revise the object created in **(a)** to be able to use cubic splines, for example, the replacement of (3) of the objects presented in **(F)** as below.

```
#   (3)''
```

```
fit.lm <- lm(y ~ bs(x, knots = c(10.3, 16.5,
    22.8.), degree = 3.), data = data1)
```

(e) Using the object created in **(d)**, observe the responses of the estimate given by the movements of data and describe the findings obtained. Create a graph of the values of elements of a hat matrix in a manner similar to those of figure 2.17(left) (page 44) and figure 2.19(left) (page 44).

2.7   Consider the objects shown in **(G)** to obtain the basic concepts of local linear regression.

(a) Create an S-Plus or R object for producing a graph to show how the estimates given by local linear regression are derived; the graph should be similar to that in figure 2.26 (page 50) (hint: `fit.lm$coef[1.]` in the object of `lline()` is the constant term of the linear regression function and `fit.lm$coef[2.]` is the gradient).

(b) Using the data presented in Problem **2.3**, operate the object constructed in **(a)** with various bandwidths.

(c) Smooth the data presented in Problem **2.3** using `lline()` with various bandwidths.

(d) Alter some values of the data presented in Problem **2.3** and smooth them using `lline()`; observe the effects of such alteration. On the basis of these results, calculate the elements of a hat matrix. Then, construct a graph similar to those in figure 2.23 (page 47) and figure 2.24 (page 48). Determine whether the resultant hat matrix is a symmetric matrix.

2.8   Derive eq(2.60) (page 52) from eq(2.59) (page 51) (Advice: though such calculation appears simple, errors can be generated).

2.9   Utilize the objects shown in **(H)** to obtain the basic concepts of the smoothing spline.

(a) Smooth the data presented in Problem **2.3** using the various smoothing parameters.

(b) Alter some values of the data presented in Problem **2.3** and observe the effects of such alteration. On the basis of these results, calculate and illustrate the values of elements of a hat matrix of the smoothing spline. Confirm that the resultant hat matrix is symmetric.

(c) Compute the values of elements of a hat matrix by directly using eq(2.62) (page 52) and eq(2.63) (page 52); $(\mathbf{I} + \lambda \mathbf{S})^{-1}$ in eq(2.63) is a hat matrix. Compare these values with those obtained in **(b)**.

2.10   Verify that the hat matrix given by a polynomial equation by the least squares (Problem **2.5**) is idempotent and that the eigenvalues are either 0 or 1. In addition, confirm eq(2.77) (page 57) numerically using the hat matrix.

2.11   Consider the representation of the hat matrix given by a binomial filter in the form of eq(2.71) (page 56). Use the reflection boundary condition.

(a) Represent the hat matrix given by a binomial filter in the form of eq(2.71); the values of a predictor are $\{1, 2, 3, \ldots, 30\}$, for example. The object shown in **(I)** of this chapter will be useful.

(b) Illustrate the values of eigenvalues and eigenvectors to construct figures such as figure 2.34 (page 60), figure 2.35 (page 61), and figure 2.36 (page 62).

2.12   Consider the representation of the hat matrix given by the smoothing spline (Problem **2.9**) in the form of eq(2.71) (page 56).
(a) Represent the hat matrix given by the smoothing spline in the form of eq(2.71).
(b) Illustrate the values of eigenvalues and eigenvectors, and construct figures such as figure 2.47 (page 70), figure 2.48 (page 71), and figure 2.49 (page 72).

2.13   Consider the representation of the hat matrix given by fitting a polynomial equation by the least squares (Problem **2.5**) in the form of eq(2.71) (page 56).
(a) Represent the hat matrix given by fitting a polynomial equation by the least squares (Problem **2.5**) in the form of eq(2.71) (page 56).
(b) Compute the values of eigenvalues and eigenvectors, and construct figures such as figure 2.53 (page 76), figure 2.54 (page 76), and figure 2.55 (page 77).
(c) Replace the values of a predictor ($\{1, 2, 3, \ldots, 30\}$) by $\{1, 2.0001, 3, \ldots, 30\}$ and observe changes in the graphs; some graphs are markedly altered by this small perturbation, while others remain almost the same. Discuss the background of this phenomenon.

2.14   Prove eq(2.97) (page 78) and extend it to general degree polynomials.

2.15   Consider the diagonal elements of a hat matrix of a simple regression.
(a) Assume that predictor values are $\{X_1, X_2, \ldots, X_n\}$ (they are not necessarily equispaced) and that the values of a target variable are $\{Y_1, Y_2, \ldots, Y_n\}$. The regression equation ($y = a_0 + a_1 x$; a_0 and a_1 are the regression coefficients) are derived by minimizing

$$E_{simple} = \sum_{i=1}^{n} (a_0 + a_1 X_i - Y_i)^2. \tag{2.102}$$

Prove that the resultant regression coefficients are written as

$$\hat{a}_0 = \bar{Y} - \hat{a}_1 \bar{X}, \tag{2.103}$$

$$\hat{a}_1 = \frac{S_{xy}}{S_{xx}}, \tag{2.104}$$

where

$$\bar{X} = \frac{1}{n} \sum_{i=1}^{n} X_i, \tag{2.105}$$

$$\bar{Y} = \frac{1}{n} \sum_{i=1}^{n} Y_i, \tag{2.106}$$

$$S_{xx} = \sum_{i=1}^{n} (X_i - \bar{X})^2, \tag{2.107}$$

$$S_{xy} = \sum_{i=1}^{n} (X_i - \bar{X})(Y_i - \bar{Y}) = \sum_{i=1}^{n} (X_i - \bar{X})Y_i = \sum_{i=1}^{n} X_i(Y_i - \bar{Y}). \tag{2.108}$$

(b) Assume that among $\{Y_1, Y_2, \ldots, Y_n\}$, only Y_k is replaced by $Y_k + \Delta Y_k$ and $\{Y_1, Y_2, \ldots, Y_{k-1}, Y_{k+1}, \ldots, Y_n\}$ remains the same; $\{X_1, X_2, \ldots, X_n\}$ also remains the same. The estimate corresponding to $Y_k + \Delta Y_k$ is defined as $\hat{Y}_k + \Delta \hat{Y}_k$.

Derive the two equations below.

$$\hat{Y}_k = \bar{Y} + \frac{(X_k - \bar{X})S_{xy}}{S_{xx}}, \tag{2.109}$$

$$\hat{Y}_k + \Delta\hat{Y}_k = \bar{Y} + \frac{\Delta Y_k}{n} + \frac{(X_k - \bar{X})S_{xy} + (X_k - \bar{X})^2 \Delta Y_k}{S_{xx}}, \tag{2.110}$$

where S_{xx} and S_{xy} are defined in eq(2.107) and eq(2.108), respectively.
(c) Using the result of (b), obtain

$$\Delta\hat{Y}_k = \frac{\Delta Y_k}{n} + \frac{(X_k - \bar{X})^2 \Delta Y_k}{S_{xx}}. \tag{2.111}$$

(d) On the basis of (c), explain

$$[\mathbf{H}]_{kk} = \frac{1}{n} + \frac{(X_k - \bar{X})^2}{S_{xx}}, \tag{2.112}$$

where $\mathbf{H}$ is a hat matrix.
(e) Using eq(2.112), confirm that $\sum_{i=1}^{n}[\mathbf{H}]_{ii} = 2$.
(f) On the basis of the equation below and eq(2.38) (page 35), derive eq(2.112).

$$(\mathbf{X}^t\mathbf{X})^{-1} = \frac{1}{n\sum_{i=1}^{n} X_i^2 - (\sum_{i=1}^{n} X_i)^2} \begin{pmatrix} \sum_{i=1}^{n} X_i^2 & -\sum_{i=1}^{n} X_i \\ -\sum_{i=1}^{n} X_i & n \end{pmatrix}. \tag{2.113}$$

2.16   Consider the use of orthonormal polynomials for fitting of polynomials by the least squares.
(a) The design matrix for fitting of polynomials by the least squares is written as eq(2.24) (page 33). However, the matrix below can also be used as a design matrix for the same purpose.

$$\mathbf{X} = \begin{pmatrix} r_0(X_1) & r_1(X_1) & r_2(X_1) & \cdots & r_q(X_1) \\ r_0(X_2) & r_1(X_2) & r_2(X_2) & \cdots & r_q(X_2) \\ \vdots & \vdots & \vdots & \ddots & \vdots \\ r_0(X_n) & r_1(X_n) & r_2(X_n) & \cdots & r_q(X_n) \end{pmatrix}, \tag{2.114}$$

where $\{r_1(\cdot), r_2(\cdot), \ldots, r_q(\cdot)\}$ are orthonormal polynomials; $r_j(\cdot)$ is a jth-degree polynomial. Since the polynomials are orthonormal, the equation below holds.

$$\sum_{k=1}^{n} r_i(X_k)r_j(X_k) = \begin{cases} 1 & \text{if } i = j \\ 0 & \text{if } i \neq j. \end{cases} \tag{2.115}$$

Furthermore, $r_0(\cdot) \equiv \frac{1}{\sqrt{n}}$. The object poly(), which is included in S-Plus by default, creates orthonormal polynomials. When $\{X_i\} = \{1, 2, 3, \ldots, 10\}$ is set and a cubic equation is employed ($q = 3$), verify eq(2.115). Note that the output of poly() does not contain the first column of the matrix defined in eq(2.114); the first column must be combined.

(b) When a qth-degree polynomial is fitted by the least squares, estimates corresponding to $\{Y_i\}(1 \leq i \leq n)$ are defined as $\{\hat{Y}_i\}(1 \leq i \leq n)$. Then, $\hat{Y}_i$ is represented as

$$\hat{Y}_i = \sum_{j=0}^{q} a_j X_i^j = \sum_{j=0}^{q} b_j r_j(X_i), \tag{2.116}$$

where $\{a_j\}$ are coefficients of the polynomial equation, and $\{b_j\}$ are those of the orthonormal polynomial. Then, derive

$$b_j = \sum_{i=1}^{n} Y_i r_j(X_i). \tag{2.117}$$

In addition, obtain

$$\hat{Y}_i = \sum_{j=0}^{q} \left(r_l(X_i) \left(\sum_{k=1}^{n} Y_k r_j(X_k) \right) \right). \tag{2.118}$$

Explain the meaning of eq(2.118) in the context of eq(2.87) (page 58).
(c) Assume that $\{X_i\} = \{1, 2, 3, \ldots, 10\}$ and $\{Y_i\} = \{11.9, 12.2, 12.5, 13.3, 13.5,$ $15.0, 16.4, 18.1, 21.0, 24.4\}$. When a cubic equation ($q = 3$) is employed, compute $\{\hat{Y}_i\}$ using eq(2.118). Check that the resultant $\{\hat{Y}_i\}$ are identical to those obtained using an ordinary method (i.e., using eq(2.24) and eq(2.31) (page 34)).
(d) Set $\{X_i\} = \{1, 2, 3, \ldots, 10\}$, and confirm that the corresponding hat matrix (eq(2.38) (page 35)) is not affected by the choice between eq(2.24) and eq(2.114); use a cubic equation ($q = 3$) for example.

## CHAPTER 3

NONPARAMETRIC REGRESSION FOR ONE-DIMENSIONAL PREDICTOR

3.1 INTRODUCTION

Chapter 2 mainly concerned problems with a one-dimensional equispaced predictor. That is, when data is represented as $\{(X_i, Y_i)\}$ $(1 \leq i \leq n)$ and $X_i - X_{i-1} = $ positive constant $(2 \leq i \leq n)$ is satisfied, the goal is to obtain the estimates (estimated values) $(\{\hat{Y}_i\}$ $(1 \leq i \leq n))$. Some of the basic concepts of nonparametric regression were introduced and examined by discussing the methods of addressing this problem using smoothing methods based on nonparametric regression. As a result, it was proved that nonparametric regression realizes better smoothing than does the fitting of parametric functions such as a polynomial (polynomial function).

In this chapter, we expand those methods to deal with data with a non-equispaced predictor. This expansion enables estimation between points where predictor values of data exist regardless of whether the values of a predictor are equispaced or not. By this generalization, smoothing based on nonparametric regression becomes more sophisticated practically and theoretically. For this purpose, regression for a one-dimensional predictor is formulated as: "If data has the form $\{(X_i, Y_i)\}$ $(1 \leq i \leq n)$ (on many occasions, $X_1 < X_2 < \cdots < X_n$), practical regression equations (regression functions) $(m(x))$ are estimated. $m(x)$ is assumed to satisfy

$$Y_i = m(X_i) + \epsilon_i, \qquad (3.1)$$

where ϵ_i is error contained in data. Error is often assumed to be distributed around 0.0 at random. $m(x)$ is a function that represents the intrinsic behavior of data and does not depend upon luck."

Parametric regression makes use of a specific parametric function as $m(x)$ to derive the coefficients of the function. On the other hand, nonparametric regression assumes not that $m(x)$ is a parametric function, but that $m(x)$ is described by a number of parameters to obtain a more practical $m(x)$. In line with this methodology, a variety of methods of deriving $m(x)$ are possible. In analogy with other regressions, the selection of methods relies on the purpose of utilizing $\hat{m}(x)$ as well as the characteristics of data.

In this Chapter, we assume that $\hat{m}(x)$ is a smooth function of x. This is consistent with the policy in Chapter 2: $\{\hat{Y}_i\}$ varies smoothly with respect to i. Although smoothing methods based on nonparametric regression for $\hat{m}(x)$, which contains non-continuous and/or nondifferential points, have been developed so far, such methods are beyond the scope of this book. The assumption that $m(x)$ is a smooth function of x is appropriate for a great diversity of data obtained by experiments, observations, and censuses, and understanding of the methods based on this assumption is requisite to that of other nonparametric regressions.

In this chapter, regression is explained comprehensively, and typical nonparametric regression methods of treating data in which the predictor is not necessarily equispaced are described. These methods yield smooth $\hat{m}(x)$ using data with one predictor. The techniques dealt with in chapter 2 are reduced to weighted averages. Although at first glance, these seem to be smoothing methods based on viewpoints different from that of the weighted average, simple calculation clarifies that the essence of these methods is the weighted average. The same is true of the methods for data with non-equispaced predictors. This chapter centers on techniques based on weighted average and presents several methods of obtaining $\hat{m}(x)$ via smoothing based on nonparametric regression.

First, the Nadaraya-Watson estimator is introduced. It is a typical kernel estimator. Kernel estimators derive estimates by directly using weighted averages. Therefore, understanding the kernel estimator leads to overall understanding of smoothing using the moving average. Furthermore, it is of high practical value. Next, techniques that are essentially varieties of weighted averaging, though at first glance they do not seem to be so, are introduced. They are local polynomial regression, smoothing spline (smoothing splines), LOESS, and supersmoother. Local polynomial regression and smoothing spline are based on those described in Chapter 2 and are generalized further. Additionally, a method called LOWESS (LOcally WEighted Scatterplot Smoother), which deviates from the weighted average, is introduced; this technique is equipped with an enhancing practical value because it is robust to outliers. As was mentioned in Chapter 2, optimization of the degree of smoothing plays an importrant role in realizing effective smoothing, and hence, methods of adjusting the degree of smoothing for each smoothing technique are discussed in this chapter.

3.2 TRADE-OFF BETWEEN BIAS AND VARIANCE

Two of the basic concepts in discussing the appropriateness of regression equations or models, including those based on nonparametric regression, are bias and variance. They are briefly discussed here. More mathematical details can be found in books on the basics of regression analysis and other literature. If necessary, one of $\{(X_i, Y_i)\}$ $(1 \leq i \leq n)$, that is, (X_i, Y_i), is called datum, and $\{(X_i, Y_i)\}$ $(1 \leq i \leq n)$ as a whole is called a data set. This data set is one sample.

When a regression equation is created using data, there are two aspects of the propriety of the regression equation. One is bias. If the procedure that yields a regression equation every time a data set is obtained is repeated, each data set provides a regression equation for deriving estimates. If the number of data set is large, the distribution formed as a result of such estimates is regarded as a sampling distribution. Bias is an index of how accurately the mean of a sampling distribution (i.e., the mean of obtained estimates) represents the intrinsic behavior of data. The mean of the estimates derived by regression equations supplied by a number of data sets should be close to the values obtained using $m(x)$, the function that precisely describes the intrinsic behavior of data. Then, assume that there are an infinite number of data sets and calculate the mean of estimates given by the data sets. The resulting difference between the mean and $m(x)$ is called bias. The amount of bias of an estimate is a significant index when discussing the characteristics of regression equations. The other important index is variance. Variance represents the possible discrepancy between an estimate given by a regression equation provided by one data set and the mean of estimates obtained from an infinite number of data sets. The mean of estimates provided by regression equations with an infinite number of data sets may be useful (i.e., it has only a small amount of bias). However, if an estimate obtained using a regression equation derived from one data set is far from the mean of estimates that were each created by a regression equation obtained using a respective data set (the number of data sets is infinite), the estimate with a small amout of bias cannot be sufficiently justified, since we usually have only one data set. The discrepancy between the two values is represented by the variance of a sampling distribution of estimates. This variance is termed the variance of an estimate and functions as an index when discussing the characteristics of regression equations, along with bias.

Bias $(Bias[\hat{m}(x)])$ of an estimate given by a regression equation $(\hat{m}(x))$ is defined as

$$Bias[\hat{m}(x)] = E[\hat{m}(x)] - m(x), \tag{3.2}$$

where $m(x)$ is a function that represents the intrinsic behavior of data and does not depend on luck. $E[\cdot]$ indicates the expectation of the value placed in [ ]. "Expectation" is the mean of values that were each derived from data sets; it is assumed that there are an infinite number of data sets obtained under the same conditions. In other words, $E[\cdot]$ indicates population mean. Therefore, $E[\hat{m}(x)]$ is a mean of estimates that were each derived from a regression equation based on one data set, when it is assumed that there are an infinite number of data sets. That is, eq(3.2) becomes

$$Bias[\hat{m}(x)] = \lim_{K \to \infty} \left(\sum_{k=1}^{K} \frac{\hat{m}(x, D_k)}{K} \right) - m(x), \tag{3.3}$$

where $\hat{m}(x, D_k)$ is an estimate provided by a regression equation based on data set D_k $(1 \leq k \leq K)$. All data sets ($\{D_k\}$) were obtained in the same manner. Note that the expression "bias is large" usually means that the absolute value of $Bias[\hat{m}(x)]$ is large, rather than $Bias[\hat{m}(x)]$ itself.

Additionally, variance ($Var[\hat{m}(x)]$) of the estimate given by the regression equation ($\hat{m}(x)$) is defined as

$$Var[\hat{m}(x)] = E[(\hat{m}(x) - E[\hat{m}(x)])^2]. \tag{3.4}$$

That is to say, this value is obtained by the following procedure. The squared difference between the mean of estimates ($E[\hat{m}(x)]$) that were each derived from a regression equation based on a data set (the number of data sets is infinite) and each estimate is calculated many times and the mean of the values is derived. This procedure is described in a similar manner to that of eq(3.3) as

$$Var[\hat{m}(x)] = \lim_{K' \to \infty} \lim_{K \to \infty} \left(\frac{\sum_{k'=1}^{K'} \left(\hat{m}(x, D_{k'}) - \sum_{k=1}^{K} \frac{\hat{m}(x, D_k)}{K} \right)^2}{K'} \right). \tag{3.5}$$

However, the chief concern about deriving a regression equation is the discrepancy between an estimate calculated using a regression equation and the value given by $m(x)$. This discrepancy is indicated by Mean Squared Error (MSE). MSE is defined as

$$MSE[\hat{m}(x)] = E[(\hat{m}(x) - m(x))^2]. \tag{3.6}$$

In the same form as eq(3.3), eq(3.6) is rewritten as

$$MSE[\hat{m}(x)] = \lim_{K \to \infty} \left(\frac{\sum_{k=1}^{K} (m(x) - \hat{m}(x, D_k))^2}{K} \right). \tag{3.7}$$

There is a simple relationship among $MSE[\hat{m}(x)]$, $Bias[\hat{m}(x)]$, and $Var[\hat{m}(x)]$:

$$MSE[\hat{m}(x)] = (Bias[\hat{m}(x)])^2 + Var[\hat{m}(x)]. \tag{3.8}$$

This relationship is ascertained by the calculation

$$
\begin{aligned}
(Bias[\hat{m}(x)])^2 &= (E[\hat{m}(x)] - m(x))^2 \\
&= (E[\hat{m}(x)])^2 - 2m(x)E[\hat{m}(x)] + m(x)^2, \tag{3.9} \\
Var[\hat{m}(x)] &= E[(\hat{m}(x) - E[\hat{m}(x)])^2] \\
&= E[\hat{m}(x)^2] + (E[\hat{m}(x)])^2 - 2(E[\hat{m}(x)])^2 \\
&= E[\hat{m}(x)^2] - (E[\hat{m}(x)])^2, \tag{3.10} \\
MSE[\hat{m}(x)] &= E[(\hat{m}(x) - m(x))^2] \\
&= E[\hat{m}(x)^2] - 2m(x)E[\hat{m}(x)] + m(x)^2, \tag{3.11}
\end{aligned}
$$

where $E[m(x)] = m(x)$ is utilized. Since $m(x)$ does not depend on luck, $m(x)$ remains the same for every D_k. That is, the relationship below is satisfied.

$$
\begin{aligned}
E[m(x)] &= \lim_{K \to \infty} \left(\sum_{k=1}^{K} \frac{m(x)}{K} \right) \\
&= m(x).
\end{aligned}
\tag{3.12}
$$

In many regressions, if the fidelity of estimates given by regression equations to data is emphasized, variance becomes large. On the other hand, if the smoothness of those estimates is given higher priority, the (absolute) value of bias is increased. That is, there is a trade-off between variance and bias. Thus, this conflict in the characteristics of bias and variance is called the "trade-off of bias versus variance (bias-variance trade-off)." To achieve the primary goal of obtaining a regression equation with small MSE, both bias and variance should have reasonable values that make MSE small. Hence, the concept of the trade-off of bias versus variance should always be taken into account in attempts to derive a beneficial regression equation.

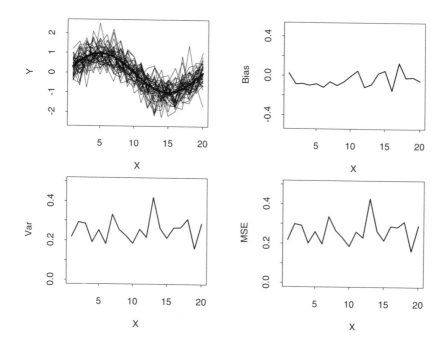

Figure 3.1 The data generated by eq(3.13) is used directly as estimates. Comparison between $m(x)$ and estimates (top left). Bias ($Bias[\hat{Y}_i]$) (top right). Variance ($Var[\hat{Y}_i]$) (bottom left). Mean squared error ($MSE[\hat{Y}_i]$) (bottom right).

To illustrate the meaning of the "trade-off of bias versus variance (bias-variance trade-off)," let us show the results of a simple numerical simulation. The predictor used here is $X_i = i$ ($1 \leq i \leq 20$), and the target variable (object variable) is

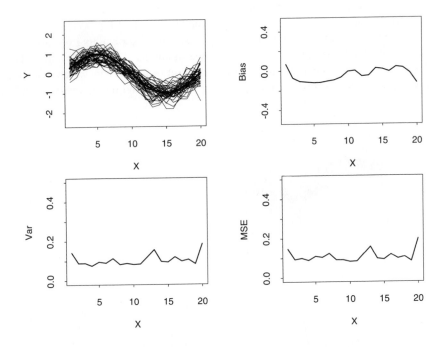

Figure 3.2   The data generated by eq(3.13) is used to derive estimates employing a binomial filter ($m = 2$). Comparison between $m(x)$ and estimates (top left). Bias ($Bias[\hat{Y}_i]$) (top right). Variance ($Var[\hat{Y}_i]$) (bottom left). Mean squared error ($MSE[\hat{Y}_i]$) (bottom right).

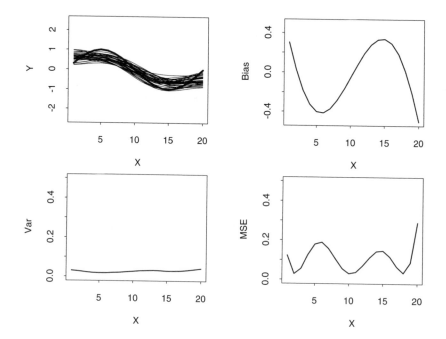

Figure 3.3   The data generated by eq(3.13) is used to derive estimates employing a binomial filter ($m = 40$). Comparison between $m(x)$ and estimates (top left). Bias ($Bias[\hat{Y}_i]$) (top right). Variance ($Var[\hat{Y}_i]$) (bottom left). Mean squared error ($MSE[\hat{Y}_i]$) (bottom right).

represented as

$$Y_i = \sin(0.1\pi X_i) + \epsilon_i, \tag{3.13}$$

where $\{\epsilon_i\}$ is the realization of $N(0.0, 0.5^2)$ (a normal distribution with mean (0.0), and standard deviation (0.5)). A total of 40 data sets are created for smoothing using a binomial filter (Chapter 2 (page 27)). Bias $(Bias[\hat{Y}_i])$, variance $(Var[\hat{Y}_i])$, and mean squared error $(MSE[\hat{Y}_i])$ are obtained. To approximate eq(3.3) (page 105) and eq(3.5) (page 106) using a finite number of data, $Var[\hat{Y}_i]$ and $Bias[\hat{Y}_i]$ are defined as

$$Bias[\hat{Y}_i] = \sum_{k=1}^{K} \frac{\hat{Y}_i(D_k)}{K} - Y_i^+, \tag{3.14}$$

$$Var[\hat{Y}_i] = \frac{\sum_{k'=1}^{K'} \left(\hat{Y}_i(D_{k'}) - \sum_{k=1}^{K} \frac{\hat{Y}_i(D_k)}{K} \right)^2}{K' - 1}, \tag{3.15}$$

where $\{\hat{Y}_i(D_k)\}$ is the estimate derived using the kth data set $(1 \leq k \leq K)$, and Y_i^+ is $m(X_i)$ $(= \sin(0.1\pi X_i))$. Furthermore, since the number of data sets is 40 (not infinite), the denominator of eq(3.4) for computing variance is $(K' - 1)$ rather than K'. Then, $((Bias[\hat{Y}_i])^2 + Var[\hat{Y}_i])$ is used as an approximation of $MSE[\hat{Y}_i]$. In this example, $K = K' = 40$ and $Var[\hat{Y}_i]$, $Bias[\hat{Y}_i]$, and $MSE[\hat{Y}_i]$ are computed. Figure 3.1 displays the result when data as they are (i.e., raw data) are used as estimates. This graph shows that $Var[\hat{Y}_i]$ values are scattered near the variance of the errors contained in data ($\{\epsilon_i\}$) (in this example, the variance is 0.25 because the standard deviation of $\{\epsilon_i\}$ is 0.5), and the absolute value of $Bias[\hat{Y}_i]$ is small. The reason is that when we have an infinite number of data sets, $Var[\hat{Y}_i]$ is 0.25 for every $\hat{Y}_i$, and $Bias[\hat{Y}_i]$ is 0.0 for every $\hat{Y}_i$.

Next, figure 3.2 illustrates the estimates smoothed using a binomial filter $(m = 2)$. Comparison with figure 3.1 indicates that the values of $Var[\hat{Y}_i]$ are smaller and absolute values of $Bias[\hat{Y}_i]$ are larger. Furthermore, the values of $Bias[\hat{Y}_i]$ tend to take negative values around local maxima and positive values around local minima. The values of $MSE[\hat{Y}_i]$ are substantially less than those in figure 3.1. These results indicate that the estimates provided by smoothing are superior to data themselves in term of MSE.

In addition, the result of increasing the degree of smoothing by setting $m = 40$ is shown in figure 3.3. The values of $Var[\hat{Y}_i]$ are further reduced. On the other hand, the absolute values of $Bias[\hat{Y}_i]$ are increased, and there is a more notable tendency for the bias to take negative values around the positions where $m(X_i)$ is locally maximum, and positive values where $m(X_i)$ is locally minimum. The values of $MSE[\hat{Y}_i]$ are inhomogeneous compared to figure 3.1 and are large near two ends and around the positions where $m(X_i)$ takes local maximum or local minimum values. Comparison of figure 3.1, figure 3.2, and figure 3.3 indicates that increasing the degree of smoothness makes variance decline as compensation for increasing (absolute values of) bias. It is preferred that the MSE takes a small value as a result of small values of both variance and squared bias. However, since the responses of these two values to the degree of smoothness are contradictory, the two values cannot be reduced simultaneously. This is the trade-off of bias versus variance.

If the value of $Bias[\hat{m}(x)]$ is 0 at every x, MSE is identical to variance. Hence, MSE is decreased only by reducing the value of variance. For example, in polynomial regression using the least squares (least square method) (cf. Chapter 2 (page 33)), when $m(x)$ is represented as $m(x) = a_0 + \sum_{j=1}^{q} a_j x^j$ (q is a positive integer), fitting of the qth-degree polynomial leads to regression coefficients that have the following characteristics:

$$\mathbf{a} = E[\hat{\mathbf{a}}], \tag{3.16}$$

where $\mathbf{a}$ and $\hat{\mathbf{a}}$ are defined as

$$\mathbf{a} = \begin{pmatrix} a_0 \\ a_1 \\ \vdots \\ a_q \end{pmatrix}, \tag{3.17}$$

$$\hat{\mathbf{a}} = \begin{pmatrix} \hat{a}_0 \\ \hat{a}_1 \\ \vdots \\ \hat{a}_q \end{pmatrix}. \tag{3.18}$$

Eq(3.16) indicates that $\hat{\mathbf{a}}$ given by the least squares is an unbiased estimator of $\mathbf{a}$. Eq(3.16) is proved by

$$\begin{aligned} E[\hat{\mathbf{a}}] &= E\left[(\mathbf{X}^t\mathbf{X})^{-1}\mathbf{X}^t(\mathbf{m} + \boldsymbol{\epsilon})\right] \\ &= (\mathbf{X}^t\mathbf{X})^{-1}\mathbf{X}^t\mathbf{m} \\ &= (\mathbf{X}^t\mathbf{X})^{-1}\mathbf{X}^t\mathbf{X}\mathbf{a} \\ &= \mathbf{a}, \end{aligned} \tag{3.19}$$

where the superscript t means the transpose of a matrix.

Furthermore, $\mathbf{m}$, $\hat{\mathbf{m}}$, and $\boldsymbol{\epsilon}$ are

$$\begin{aligned} \mathbf{m} &= \begin{pmatrix} m(X_1) \\ m(X_2) \\ \vdots \\ m(X_n) \end{pmatrix} \\ &= \begin{pmatrix} 1 & X_1 & X_1^2 & \cdots & X_1^q \\ 1 & X_2 & X_2^2 & \cdots & X_2^q \\ \vdots & \vdots & \vdots & \ddots & \vdots \\ 1 & X_n & X_n^2 & \cdots & X_n^q \end{pmatrix} \begin{pmatrix} a_0 \\ a_1 \\ a_2 \\ \vdots \\ a_q \end{pmatrix}, \end{aligned} \tag{3.20}$$

$$\hat{\mathbf{m}} = \begin{pmatrix} \hat{m}(X_1) \\ \hat{m}(X_2) \\ \vdots \\ \hat{m}(X_n) \end{pmatrix}, \tag{3.21}$$

$$
\epsilon = \begin{pmatrix} \epsilon_1 \\ \epsilon_2 \\ \vdots \\ \epsilon_n \end{pmatrix}. \tag{3.22}
$$

Moreover, eq(3.19) is derived by

$$
E\left[[(\mathbf{X}^t\mathbf{X})^{-1}\mathbf{X}^t\epsilon]_i \right] = \sum_{j=1}^{n} [(\mathbf{X}^t\mathbf{X})^{-1}\mathbf{X}^t]_{ij} E[\epsilon_j] = 0 \quad (0 \le i \le q). \tag{3.23}
$$

Hence, $E[\hat{m}(x)]$ is represented as

$$
\begin{aligned}
E[\hat{m}(x)] &= E\left[\hat{a}_0 + \sum_{j=1}^{q} \hat{a}_j x^j \right] \\
&= a_0 + \sum_{j=1}^{q} a_j x^j \\
&= m(x). \tag{3.24}
\end{aligned}
$$

That is, $\hat{m}(x)$ is also an unbiased estimator.

Note that to make $\hat{a}$ and $\hat{m}(x)$ unbiased estimators, $m(x)$ must be represented exactly as a polynomial along with $E[\epsilon_i] = 0$ $(1 \le i \le n)$. For real data, however, $m(x)$ is rarely represented as a specific-degree polynomial in the strict sense of the term. If $m(x)$ does not obey a specific-degree polynomial precisely, bias arises because a regression equation whose form is different from that of $m(x)$ must be employed. Hence, when a polynomial regression is carried out using data obtained from an actual phenomenon, a possibility of satisfying the characteristic of $Bias[\hat{m}(x)] = 0$, if any, is small. Furthermore, if a regression equation that satisfies $Bias[\hat{m}(x)] = 0$ enhances the value of $Var[\hat{Y}_i]$, analysts may be wise to use a regression equation that does not result in $Bias[\hat{m}(x)] = 0$. On the other hand, nonparametric regression is not expected to have a form of an equation that describes a phenomenon precisely because nonparametric regression uses multipurpose regression equations with a number of regression coefficients. Then, although nonparametric regression is not aimed at realizing the possibility of $Bias[\hat{m}(x)] = 0$, it is aimed at deriving a better regression equation than that obtained with a small number of regression coefficients, in terms of MSE.

Figure 3.4 shows the results of fitting a quadratic equation using the least squares to the same data as before; bias, variance, and MSE are also shown. Since the data in this example were obtained using eq(3.13) (page 110), $m(x)$ does not precisely obey a quadratic equation. Hence, considerable bias arises. This is the main reason behind the large value of MSE. Thus, since $m(x)$ of real-world data often does not obey polynomials, the effect of satisfying eq(3.16) (page 111) is not very large when fitting polynomials by the least squares.

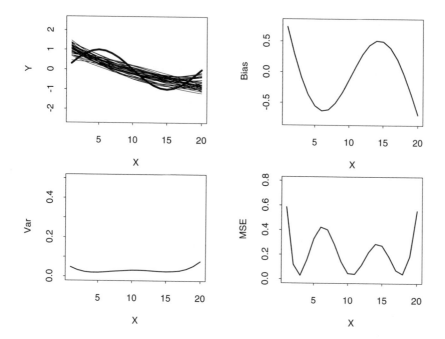

Figure 3.4 Fitting of a quadratic equation using the least squares to the data given by eq(3.13). Comparison with estimates using $m(x)$ (top left). Bias ($Bias[\hat{Y}_i]$) (top right). Variance ($Var[\hat{Y}_i]$) (bottom left). Mean squared error ($MSE[\hat{Y}_i]$) (bottom right).

3.3 INDEX TO SELECT BENEFICIAL REGRESSION EQUATIONS

Analysts should use and assess a regression equation, obtained using data, in actual situations to elucidate how useful the regression equation is. A regression equation may be of great use in one situation, but not in another situation. Hence, the usefulness of the resultant regression equations is evaluated only in a real setting. However, since diverse regression equations are created from one data set, it is laborious to use all the equations practically and to discuss their relative merits in choosing the best one. When regression equations are used for control or prediction, it is difficult to test various regression equations and investigate the results on many occasions.

Therefore, analysts must examine practical values of resulting regression equations before they are used in practice. $MSE[\hat{m}(x)]$ is a possible index for this purpose. However, $MSE[\hat{m}(x)]$ is an index for a specific value of x. The average value of $MSE[\hat{m}(x)]$ in the region where the value of a predictor exists plays a more important role. Then, $MISE[\hat{m}(x)]$ (Mean Integrated Squared Error) is identified as

$$
\begin{aligned}
MISE[\hat{m}(x), f(x)] &= \int_{-\infty}^{\infty} MSE[\hat{m}(x)]f(x)dx \\
&= \int_{-\infty}^{\infty} E[(m(x) - \hat{m}(x))^2]f(x)dx \\
&= E\left[\int_{-\infty}^{\infty} (m(x) - \hat{m}(x))^2 f(x)dx\right], \quad (3.25)
\end{aligned}
$$

where $f(x)$ is a probability density function (density function) that represents the distribution of the value of a predictor (x). The equality between the second and third lines in eq(3.25) is proved by eq(3.7) (page 106). The term "Mean Integrated Squared Error (MISE)" is represented by the equation on the third line, and hence, the equation on the second line might be called the "Integrated Mean Squared Error (IMSE)." Both values are, however, customarily called Mean Integrated Squared Error (MISE). Furthermore, it should be noted that $MISE[\hat{m}(x), f(x)]$ depends on $f(x)$. This indicates that $MISE[\hat{m}(x), f(x)]$ is useful as an index of the benefits of a regression equation only when the distribution of the value of the predictor of data is close to that of the value of a predictor in practical applications of the regression equation. Otherwise, appropriate adjustments are required to cope with individual situations.

The simplest $f(x)$ is that for fixed predictor values of data. That is, each $\{X_i\}$ of data ($\{(X_i, Y_i)\}$ ($1 \le i \le n$)) is known to take a specific value. For example, when Y_i represents the daily average temperature of the ith day, X_i takes a certain fixed value such as i. In such a case, $f(x)$ is represented as

$$
f(x) = \frac{\sum_{i=1}^{n} \delta(x - X_i)}{n}, \quad (3.26)
$$

where $\delta(\cdot)$ is a delta function. Let us assume that the result of substituting the above equation into eq(3.25) is $MISE'[\hat{m}(x)]$. Then, we have the average mean squared error:

$$
MISE'[\hat{m}(x)] = \frac{E[\sum_{i=1}^{n} (m(X_i) - \hat{m}(X_i))^2]}{n}. \quad (3.27)
$$

$MISE'[\hat{m}(x)]$ contains $m(x)$, which is an unknown function; this function ($m(x)$) cannot be estimated from data. In addition, the accurate calculation of expectations is not possible. Hence, an approximate calculation of $MISE'[\hat{m}(x)]$ by some method is required. For this purpose, Average Predictive Squared Error (APSE) is defined as follows (this value is sometimes called Prediction Mean Squared Error (PMSE)):

$$APSE[\hat{m}(x)] = \frac{E[\sum_{i=1}^{n}(Y_i^* - \hat{m}(X_i))^2]}{n}, \tag{3.28}$$

where $\{Y_i^*\}$ ($1 \leq i \leq n$) are predictor values of data ($\{(X_i, Y_i^*)\}$) obtained by an experiment in which predictor values remain the same as $\{(X_i, Y_i)\}$, which is used to estimate $\hat{m}(x)$. Using the notation in eq(3.3) (page 105), $\hat{m}(x)$ is given by D_1, and the predictor values of D_2 are $\{Y_i^*\}$. Then, Y_i^* is written as

$$Y_i^* = m(X_i) + \epsilon_i^*, \tag{3.29}$$

where ϵ_i^* is independent of ϵ_i in eq(3.1) (page 104), but its statistical property remains the same. The substitution of eq(3.29) into eq(3.28) yields

$$\begin{aligned} APSE[\hat{m}(x)] &= \frac{E[\sum_{i=1}^{n}(m(X_i) + \epsilon_i^* - \hat{m}(X_i))^2]}{n} \\ &= MISE'[\hat{m}(x)] + 2E[\epsilon_i^*(m(X_i) - \hat{m}(X_i))] + E[(\epsilon_i^*)^2] \\ &= MISE'[\hat{m}(x)] + \sigma^2, \end{aligned} \tag{3.30}$$

where $\sigma^2 (= E[(\epsilon_i^*)^2])$ is the variance of ϵ_i^*, which equals that of ϵ_i. The equality between the second line and the third line is derived using the relationships

$$E\left[\sum_{i=1}^{n} m(X_i) \cdot \epsilon_i^*\right] = 0, \tag{3.31}$$

$$E\left[\sum_{i=1}^{n} \hat{m}(X_i) \cdot \epsilon_i^*\right] = 0. \tag{3.32}$$

Eq(3.31) is derived under the condition that $E[\epsilon_i^*] = 0$ is satisfied, and $m(X_i)$ does not depend on luck. Eq(3.32) is due to the independence of ϵ_i, which is the error of the data for $\hat{m}(X_i)$, on ϵ_i^*, and $E[\epsilon_i^*] = 0$.

Eq(3.30) shows that the derivation of $m(x)$, which minimizes $MISE'[\hat{m}(x)]$, is equivalent to the minimization of $APSE[\hat{m}(x)]$. Hence, to estimate $APSE[\hat{m}(x)]$, Cross-Validation (CV) is defined as

$$CV[\hat{m}(x)] = \frac{\sum_{k=1}^{n}(Y_k - \hat{m}^{-k}(X_k))^2}{n}, \tag{3.33}$$

where $\hat{m}^{-k}(x)$ is a regression equation derived from the data: (X_k, Y_k) is omitted from $\{(X_i, Y_i)\}$. That is, the use of a regression equation obtained from data with the kth data omitted to estimate the kth value of the predictor is comparable to the assumption of the kth data as pseudodata in the future. This procedure enables analysts to estimate prediction errors of estimates for future data. To emphasize that this method omits one data instead of more than one data, this technique is called leave-one-out cross-validation, the leave-one-out method, or Ordinary Cross-Validation (OCV).

Thus, we have

$$E[CV[\hat{m}(x)]] \approx APSE[\hat{m}(x)], \tag{3.34}$$

where "$\approx$" indicates that the two sides are approximately equal. This relationship is obtained from

$$
\begin{aligned}
E[(Y_k - \hat{m}^{-k}(X_k))^2] &= E[(Y_k - m(X_k) + m(X_k) - \hat{m}^{-k}(X_k))^2] \\
&= E[(Y_k - m(X_k))^2] + E[(m(X_k) - \hat{m}^{-k}(X_k))^2] \\
&= \sigma^2 + E[(m(X_k) - \hat{m}^{-k}(X_k))^2] \\
&\approx \sigma^2 + E[(m(X_k) - \hat{m}(X_k))^2] \\
&= \sigma^2 + MSE[\hat{m}(X_k)]. \tag{3.35}
\end{aligned}
$$

The equality between the first line and the second line is based on

$$
\begin{aligned}
E[(Y_k - m(X_k))(m(X_k) - \hat{m}^{-k}(X_k))] &= E[\epsilon_k \cdot (m(X_k) - \hat{m}^{-k}(X_k))] \\
&= 0. \tag{3.36}
\end{aligned}
$$

Eq(3.36) is derived by the fact that $(m(X_k) - \hat{m}^{-k}(X_k))$ is independent of ϵ_k because (X_k, Y_k) is not used for obtaining $\hat{m}^{-k}(X_k)$.

Eq(3.33) looks like the Residual Sum of Squares (RSS) divided by n.

$$RSS[\hat{m}(x)] = \sum_{i=1}^{n}(Y_i - \hat{m}(X_i))^2. \tag{3.37}$$

This $\hat{m}(X_i)$ is the estimate corresponding to X_i; this regression equation is obtained using all data. Although $CV[\hat{m}(x)]$ has a similar form to $RSS[\hat{m}(x)]$, there is a sharp discrepancy between the two values. $RSS[\hat{m}(x)]$ represents the extent to which past data can be explained by a regression equation. This is a retrospective view, so to speak. On the other hand, $CV[\hat{m}(x)]$ indicates the degree to which future data can be predicted; this prediction is performed using quasi-future data. This is based on a prospective view. That is, the two views are grounded on different aspects. If analysts only need to explain past data, a regression equation that provides a small value of $RSS[\hat{m}(x)]$ is desirable. However, since their main purpose is usually the prediction of future events by a regression equation, $CV[\hat{m}(x)]$ plays an important role. Additionally, it corresponds to an approximate estimation of $APSE[\hat{m}(x)]$ (eq(3.34)). Furthermore, it leads to estimation of $MISE'[\hat{m}(x)]$ (eq(3.30) (page 115)). Furthermore, since discussion so far does not assume that regression to gain $\hat{m}(x)$ is based on a specific form, $CV[\hat{m}(x)]$ is of use in comparing regression equations of various forms. In this context, the statistics offers versatility.

In particular, $CV[\hat{m}(x)]$ is easily calculable if an estimate is represented as

$$\hat{m}(X_i) = \sum_{j=1}^{n}[\mathbf{H}]_{ij}Y_j, \tag{3.38}$$

where $\mathbf{H}$ (a matrix, the size is $n \times n$) is a hat matrix. $X_i = i$ is usually assumed in the discussion in Chapter 2. However, even if a predictor of data does not have such a limitation, an equation with the same form as that of eq(3.38) takes on significant meaning. This is because eq(3.38) holds when regression is carried out to derive estimates using a linear estimator. Regression for deriving estimates using linear estimators indicates a slightly more general concept than regression using the weighted average. That is, a regression equation ($\hat{m}(x)$) is written as

$$\hat{m}(x) = \sum_{j=1}^{n} W_j(x) Y_j. \tag{3.39}$$

$\hat{m}(x)$ is represented as a linear combination of $\{Y_j\}$. $\{W_j(x)\}$ is termed an equivalent kernel. Since eq(3.38) is obtained when $x = X_i$, regression represented as eq(3.39) yields eq(3.38). In addition, it is a regression by weighted averages when $W_j(x)$ satisfies

$$\sum_{j=1}^{n} W_j(x) = 1. \tag{3.40}$$

If regression is described as eq(3.38), eq(3.33) is represented using diagonal elements ($\{[\mathbf{H}]_{ii}\}$):

$$CV[\hat{m}(x)] = \sum_{i=1}^{n} \frac{(Y_i - \hat{m}(X_i))^2}{n \cdot (1 - [\mathbf{H}]_{ii})^2}. \tag{3.41}$$

This equation yields $CV[\hat{m}(x)]$ by a less time-consuming calculation than eq(3.33) as it is. Even though the amout of calculation is small, $CV[\hat{m}(x)]$ given by eq(3.41) is not an approximate value of $CV[\hat{m}(x)]$; the two values are exactly the same.

Diagonal elements of $\mathbf{H}$ ($\{[\mathbf{H}]_{ii}\}$) appear in eq(3.41). $[\mathbf{H}]_{ii}$ is termed leverage (leverage value) or influence. Since it indicates the amount of influence of Y_i on $\hat{Y}_i$, it plays an important role in regressions in which a linear estimator provides estimates.

Eq(3.41) is proved as follows.

A data set ($\{(X_1, Y_1), (X_2, Y_2), \ldots, (X_{k-1}, Y_{k-1}), (X_{k+1}, Y_{k+1}), \ldots, (X_n, Y_n)\}$) is obtained by omitting (X_k, Y_k) from the original data set ($\{(X_1, Y_1), (X_2, Y_2), \ldots, (X_n, Y_n)\}$). The regression equation derived from the least squares using this data set is named $\hat{m}^{-k}(x)$, and the residual sum of squares is called RSS^{-k}. Then, D^+ is defined as a data set ($\{(X_1, Y_1), (X_2, Y_2), \ldots, (X_{k-1}, Y_{k-1}), (X_k, \hat{m}^{-k}(X_k)), (X_{k+1}, Y_{k+1}), \ldots, (X_n, Y_n)\}$) that is obtained by adding a data $(X_k, \hat{m}^{-k}(X_k))$ to the data set ($\{(X_1, Y_1), (X_2, Y_2), \ldots, (X_{k-1}, Y_{k-1}), (X_{k+1}, Y_{k+1}), \ldots, (X_n, Y_n)\}$). The resulting equation from the least squares using the same regression equation as above is defined as $\hat{m}^+(x)$. The residual sum of squares for this regression is identified as RSS^+. Furthermore, each regression equation is unique.

Here we assume $\hat{m}^{-k}(X_k) \neq \hat{m}^+(X_k)$. This implies that $\hat{m}^{-k}(x)$ is not the same function as $\hat{m}^+(x)$. Then, there are three cases of the magnitude relation between RSS^+ and RSS^{-k}. Let us examine each of them.

(1) If $RSS^+ > RSS^{-k}$, the use of $\hat{m}^{-k}(x)$ instead of $\hat{m}^+(x)$ provides a smaller value of the residual sum of squares for D^+. This contradicts the fact that $\hat{m}^+(x)$ is obtained from the least squares using D^+.

(2) If $RSS^+ < RSS^{-k}$, $\hat{m}^+(x)$ instead of $\hat{m}^{-k}(x)$ yields a smaller value of the residual sum of squares for D^{-k}. This contradicts the fact that $\hat{m}^{-k}(x)$ is derived from the least squares using D^{-k}.

(3) If $RSS^+ = RSS^{-k}$, $\hat{m}^+(x)$ provides a smaller value of the residual sum of squares for D^{-k} than $\hat{m}^{-k}(x)$ because of the assumption that $\hat{m}^{-k}(X_k) \neq \hat{m}^+(X_k)$. This contradicts the fact that $\hat{m}^{-k}(x)$ is obtained from the least squares using D^{-k}.

Cases (1), (2), and (3) prove that $\hat{m}^{-k}(X_k) = \hat{m}^+(X_k)$. That is, we obtain the relationship

$$\hat{m}^{-k}(X_k) = \hat{m}^+(X_k) \tag{3.42}$$

$$= \sum_{i=1, i \neq k}^{n} [\mathbf{H}]_{ki} Y_i + [\mathbf{H}]_{kk} \hat{m}^{-k}(X_k). \tag{3.43}$$

Therefore, we have

$$\hat{m}^{-k}(X_k) = \frac{\sum_{i=1, i \neq k}^{n} [\mathbf{H}]_{ki} Y_i}{1 - [\mathbf{H}]_{kk}}. \tag{3.44}$$

Substitution of this equation into eq(3.33) leads to

$$
\begin{aligned}
CV[\hat{m}(x)] &= \frac{1}{n} \sum_{k=1}^{n} \left(Y_k - \frac{\sum_{i=1, i \neq k}^{n} [\mathbf{H}]_{ki} Y_i}{1 - [\mathbf{H}]_{kk}} \right)^2 \\
&= \frac{1}{n} \sum_{k=1}^{n} \left(\frac{Y_k (1 - [\mathbf{H}]_{kk}) - \sum_{i=1, i \neq k}^{n} [\mathbf{H}]_{ki} Y_i}{1 - [\mathbf{H}]_{kk}} \right)^2 \\
&= \frac{1}{n} \sum_{k=1}^{n} \left(\frac{Y_k - \sum_{i=1}^{n} [\mathbf{H}]_{ki} Y_i}{1 - [\mathbf{H}]_{kk}} \right)^2 \\
&= \sum_{k=1}^{n} \frac{(Y_k - \hat{m}(X_k))^2}{n \cdot (1 - [\mathbf{H}]_{kk})^2}.
\end{aligned}
\tag{3.45}
$$

This is the same equation as eq(3.41) (page 117) (Q.E.D.).

This proof is based on the weak assumption that the regression equations obtained from the least squares are expressed using a hat matrix and that the regression equation is unique. Therefore, eq(3.41) and eq(3.44) are useful for all regressions that satisfy these conditions. For the smoothing spline, eq(3.41) and eq(3.44) are derived because a similar proof is possible if the residual sum of squares is replaced by the residual sum of squares with a roughness penalty (e.g., eq(3.209) (page 170)).

Although cross-validation is an effective statistic, some defects were pointed out. Consequently, Generalized Cross-Validation (GCV) was developed. "Generalized" in the term "generalized cross-validation" does not mean that this statistic has a wide scope of application, but that it is a result of the enhancement of the feature of cross-validation. Generalized cross-validation is defined as

$$GCV = \frac{\sum_{i=1}^{n} (Y_i - \hat{m}(X_i))^2}{n \cdot \left(1 - \frac{\sum_{i=1}^{n} [\mathbf{H}]_{ii}}{n} \right)^2}, \tag{3.46}$$

where $\sum_{i=1}^{n} [\mathbf{H}]_{ii} (= \text{trace}(\mathbf{H}))$ plays a significant role in this equation. This value is called degrees of freedom.

However, it should be noted that there are several definitions of degrees of freedom in regression for estimation using a linear estimator; analysts must pay attention to which definition is used on each occasion. There are four main definitions:

$$\nu_1 = \text{trace}(\mathbf{H})$$

$$\nu_2 = \text{trace}(\mathbf{H}^t\mathbf{H}) = \sum_{i=1}^{n}\sum_{j=1}^{n}([\mathbf{H}]_{ij})^2$$

$$\nu_3 = \frac{\text{trace}((\mathbf{H}^t\mathbf{H})^2)}{\text{trace}(\mathbf{H}^t\mathbf{H})}$$

$$\nu_4 = \text{trace}(2\mathbf{H} - \mathbf{H}^t\mathbf{H}). \tag{3.47}$$

In S-Plus, ν_2 is called an equivalent number of freedom, and ν_3 is called a look-up degree of freedom. Furthermore, ν_4 takes on significant meaning because $Var[Y_i - \hat{m}(X_i)]$ is

$$\sum_{i=1}^{n} Var[Y_i - \hat{m}(X_i)]$$

$$= \sum_{i=1}^{n} E[(Y_i - \hat{m}(X_i) - E[Y_i - \hat{m}(X_i)])^2]$$

$$= \sum_{i=1}^{n} E[(Y_i - \hat{m}(X_i) - m(X_i) + E[\hat{m}(X_i)])^2]$$

$$= \sum_{i=1}^{n} E[(Y_i - m(X_i))^2] - \sum_{i=1}^{n} 2E[(Y_i - m(X_i))(\hat{m}(X_i) - E[\hat{m}(X_i)])]$$

$$+ \sum_{i=1}^{n} E[(\hat{m}(X_i) - E[\hat{m}(X_i)])^2]$$

$$= \sigma^2(n - 2\nu_1 + \nu_2)$$

$$= \sigma^2(n - \nu_4), \tag{3.48}$$

where $E[(Y_i - m(X_i))(\hat{m}(X_i) - E[\hat{m}(X_i)])]$ is the covariance between Y_i and $\hat{m}(X_i)$; it is sometimes represented as $Cov[Y_i, \hat{m}(X_i)]$. Since $E[(\hat{m}(X_i) - E[\hat{m}(X_i)])^2]$ is a variance of $\hat{m}(X_i)$, it is identical to $Var[\hat{m}(X_i)]$.

The equation from the fourth and fifth lines of eq(3.48) to the sixth line utilizes the relationship of $\sum_{i=1}^{n} E[(Y_i - m(X_i)) (\hat{m}(X_i) - E[\hat{m}(X_i)])] = \nu_1\sigma^2$, and $\sum_{i=1}^{n} E[(\hat{m}(X_i) - E[\hat{m}(X_i)])^2] = \nu_2\sigma^2$. $\sum_{i=1}^{n} E[(Y_i - m(X_i))(\hat{m}(X_i) - E[\hat{m}(X_i)])] = \nu_1\sigma^2$ is the result of the following calculation:

$$E[(Y_i - m(X_i))(\hat{m}(X_i) - E[\hat{m}(X_i)])]$$

$$= E\left[\epsilon_i\left(\sum_{j=1}^{n}[\mathbf{H}]_{ij}(m(X_j) + \epsilon_j) - \sum_{j=1}^{n}[\mathbf{H}]_{ij}m(X_j)\right)\right]$$

$$= E\left[\epsilon_i\left(\sum_{j=1}^{n}[\mathbf{H}]_{ij}\epsilon_j\right)\right]$$

$$= E\left[\sum_{j=1}^{n}[\mathbf{H}]_{ij}\epsilon_i\epsilon_j\right]$$

$$
\begin{aligned}
&= \sum_{j=1}^{n} [\mathbf{H}]_{ij} \delta_{ij} \sigma^2 \\
&= [\mathbf{H}]_{ii} \sigma^2.
\end{aligned} \tag{3.49}
$$

Therefore, we have

$$
\sum_{i=1}^{n} E[(Y_i - m(X_i))(\hat{m}(X_i) - E[\hat{m}(X_i)])] = \nu_1 \sigma^2, \tag{3.50}
$$

where the equation on the fourth line and the fifth line in eq(3.49) makes use of

$$
E[\epsilon_i \epsilon_j] = \sigma^2 \cdot \delta_{ij}. \tag{3.51}
$$

δ_{ij} is defined as

$$
\delta_{ij} = \begin{cases} 1 & \text{if } i = j \\ 0 & \text{if } i \neq j. \end{cases} \tag{3.52}
$$

Next, the relationship of $\sum_{i=1}^{n} E[(\hat{m}(X_i) - E[\hat{m}(X_i)])^2] = \nu_2 \sigma^2 \; (= \sum_{i=1}^{n} Var[\hat{m}(X_i)])$ is obtained by

$$
\begin{aligned}
&E[(\hat{m}(X_i) - E[\hat{m}(X_i)])^2] \\
&= E\left[\left(\sum_{j=1}^{n} [\mathbf{H}]_{ij} (m(X_j) + \epsilon_j) - \sum_{j=1}^{n} [\mathbf{H}]_{ij} m(X_j) \right)^2 \right] \\
&= E\left[\sum_{j=1}^{n} \left([\mathbf{H}]_{ij} \epsilon_j \right) \sum_{k=1}^{n} \left([\mathbf{H}]_{ik} \epsilon_k \right) \right] \\
&= E\left[\sum_{j=1}^{n} \sum_{k=1}^{n} [\mathbf{H}]_{ij} \epsilon_j [\mathbf{H}]_{ik} \epsilon_k \right] \\
&= \sum_{j=1}^{n} \sum_{k=1}^{n} [\mathbf{H}]_{ij} [\mathbf{H}]_{ik} \delta_{jk} \sigma^2 \\
&= \sigma^2 \cdot \sum_{j=1}^{n} ([\mathbf{H}]_{ij})^2.
\end{aligned} \tag{3.53}
$$

Therefore, we obtain

$$
\begin{aligned}
\sum_{i=1}^{n} E[(\hat{m}(X_i) - E[\hat{m}(X_i)])^2] &= \sum_{i=1}^{n} \sum_{j=1}^{n} ([\mathbf{H}]_{ij})^2 \sigma^2 \\
&= \nu_2 \sigma^2.
\end{aligned} \tag{3.54}
$$

Thus, eq(3.48) is proved.

Since CV and GCV usually are similar for typical data, analysts can use either of them. However, these two values are substantially different on some occasions. Such a difference often indicates that either the data or the regression method or both have a problem; analysts should cope with the situation carefully. Furthermore, while eq(3.41) (page 117) requires the calculation of each value of $\{[\mathbf{H}]_{ii}\}$, eq(3.46) (page 118) requires nothing more than the trace (the sum of diagonal elements) of $\mathbf{H}$. Therefore, numerical manipulations sometimes make the amount of calculation required to obtain GCV less than that in the use of CV.

On the other hand, Akaike's Information Criterion (AIC) is another widely applicable statistics. If errors obey a normal distribution, AIC for a regression using a linear estimator is defined as

$$AIC = n \cdot \log \left(\sum_{i=1}^{n} \frac{(\hat{Y}_i - Y_i)^2}{n} \right) + 2 \sum_{i=1}^{n} [\mathbf{H}]_{ii}. \tag{3.55}$$

This equation can be rewritten as

$$\exp \left(\frac{AIC}{n} \right) = \left(\sum_{i=1}^{n} \frac{(\hat{Y}_i - Y_i)^2}{n} \right) \cdot \exp \left(\frac{2 \cdot \sum_{i=1}^{n} [\mathbf{H}]_{ii}}{n} \right)$$

$$= \frac{\sum_{i=1}^{n} \frac{(\hat{Y}_i - Y_i)^2}{n}}{\exp \left(\frac{-\sum_{i=1}^{n} [\mathbf{H}]_{ii}}{n} \right)^2}. \tag{3.56}$$

Therefore, if the value of n is large enough, we have

$$\exp \left(\frac{AIC}{n} \right) \approx \frac{\sum_{i=1}^{n} \frac{(\hat{Y}_i - Y_i)^2}{n}}{\left(1 - \frac{\sum_{i=1}^{n} [\mathbf{H}]_{ii}}{n} \right)^2}. \tag{3.57}$$

This equation leads to the relationship

$$\exp \left(\frac{AIC}{n} \right) \approx GCV. \tag{3.58}$$

Although GCV and AIC are derived from different contexts, these two statistics are equivalent if the number of data is infinite; this reinforces the reliability of both. However, this approximate expression often does not hold for actual data. Hence, several statistics are recommended to be calculated for model selection, such as tuning of the smoothing parameter.

AIC and its pertinent topics are summarized as follows.

If the data $(\{(X_i, Y_i)\} \ (1 \leq i \leq n))$ is generated by the equation $(y = m(x) + \epsilon)$ (where ϵ is realization of $N(0.0, \sigma^2)$ (normal distribution; mean is 0.0, and variance is σ^2)), the probability that independently obtained n numbers of ϵ take the values of $(\epsilon_1, \epsilon_2, \ldots, \epsilon_n)$ is represented as a probability density function:

$$p(\epsilon_1, \epsilon_2, \ldots, \epsilon_n) = \left(\frac{1}{\sqrt{2\pi\sigma^2}} \right)^n \exp \left(-\frac{1}{2\sigma^2} \sum_{i=1}^{n} \epsilon_i^2 \right). \tag{3.59}$$

This equation means that the probability that $(\epsilon_1, \epsilon_2, \ldots, \epsilon_n)$ take values in the ranges $(\epsilon_1' \leq \epsilon_1 \leq \epsilon_1' + \Delta\epsilon_1', \ldots, \epsilon_n' \leq \epsilon_n \leq \epsilon_n' + \Delta\epsilon_n')$ is presented as

$$\int_{\epsilon_1'}^{\epsilon_1' + \Delta\epsilon_1'} \cdots \int_{\epsilon_n'}^{\epsilon_n' + \Delta\epsilon_n'} p(\epsilon_1, \epsilon_2, \ldots, \epsilon_n) d\epsilon_1 d\epsilon_2 \cdots d\epsilon_n. \tag{3.60}$$

Hence, the degree of the likelihood of the assumption "a data ($\{(X_i, Y_i)\}$) is generated by $y = \tilde{m}(x) + \tilde{\epsilon}$ (where $\tilde{\epsilon}$ is a realization of $N(0.0, \tilde{\sigma}^2)$ (normal distribution; average is 0.0, and variance is $\tilde{\sigma}^2$))" is expressed as

$$Like(\tilde{m}(\cdot), \tilde{\sigma}^2 | \{(X_i, Y_i)\}) = \left(\frac{1}{\sqrt{2\pi\tilde{\sigma}^2}}\right)^n \exp\left(-\frac{1}{2\tilde{\sigma}^2}\sum_{i=1}^{n}(\tilde{m}(X_i) - Y_i)^2\right).$$
(3.61)

This is a typical likelihood. Since eq(3.59) is a function to provide a value that is proportional to the frequency of the emergence of ($\epsilon_1, \epsilon_2, \ldots, \epsilon_n$), a large value of eq(3.61) means that the value of the target variable ($\{Y_i\}$) responding to the predictor value ($\{X_i\}$) is quite common under the condition that the functional relationship $y = \tilde{m}(x) + \tilde{\epsilon}$ exists. If so, the assumption of the functional relationship ($y = \tilde{m}(x) + \tilde{\epsilon}$) is fairly reasonable. In other words, the likelihood of this assumption is supported strongly by the data ($\{(X_i, Y_i)\}$). Therefore, the degree of cogency of the assumption of $y = \tilde{m}(x) + \tilde{\epsilon}$ ($\tilde{\epsilon}$ is a realization of $N(0.0, \tilde{\sigma}^2)$) is estimated in light of $\{(X_i, Y_i)\}$; the result is the right-hand side of eq(3.61). In other words, when $\{(X_i, Y_i)\}$ is fixed, and $\tilde{m}(\cdot)$ and $\tilde{\sigma}^2$ are given, the cogency of $\tilde{m}(\cdot)$ and $\tilde{\sigma}^2$ are represented by the right-hand side of eq(3.61). This is reflected by the expression $Like(\tilde{m}(\cdot), \tilde{\sigma}^2 | \{(X_i, Y_i)\})$.

Since eq(3.61) takes a positive value, the logarithm of eq(3.61) is usually used to discuss its characteristics. This is because, when the condition to maximize a function is derived, a transformation of the function by a monotonically increasing function does not affect the conditions of maximization of the function (here, the conditions are represented by $\tilde{m}(\cdot)$ and the value of $\tilde{\sigma}^2$), and a sum of terms is easier to treat than their multiplication. Hence, $L(\{Y_i\})$ is defined as log-likelihood:

$$L(\{Y_i\}) = -\frac{n}{2}\log(2\pi) - \frac{n}{2}\log(\tilde{\sigma}^2) - \frac{1}{2\tilde{\sigma}^2}\sum_{i=1}^{n}(\tilde{m}(X_i) - Y_i)^2.$$
(3.62)

Since this $L(\{Y_i\})$ is the cogency of $\tilde{m}(\cdot)$ and $\tilde{\sigma}^2$ when $\{Y_i\}$ is fixed, it could be more correctly expressed as $L(\tilde{m}(\cdot), \tilde{\sigma}^2 | \{(X_i, Y_i)\})$. However, to simplify the notation, $L(\{Y_i\})$ is used here. This expression is adopted hereafter. By augmenting the value of $L(\{Y_i\})$, the data ($\{(X_i, Y_i)\}$) becomes more common on the assumption of $\tilde{m}(x)$ and $\tilde{\sigma}^2$. Since it is rational for the accumulated data to be considered not rare ones but common ones, highly probable regression equations are obtained by maximizing this value ($L(\{Y_i\})$) to derive the values of $\tilde{m}(x)$ and $\tilde{\sigma}^2$. Thus, the derivation of assumptions on data-generating functions (e.g., regression equations or probability density functions which data is expected to obey) by maximizing likelihood (or log-likelihood) is called the maximum likelihood procedure (maximum likelihood method). Furthermore, the maximal value of log-likelihood is termed the maximum log-likelihood, and the estimate obtained by maximizing log-likelihood is called the maximum likelihood estimate.

Then, to maximize $L(\{Y_i\})$, $L(\{Y_i\})$ is differentiated with respect to $\tilde{\sigma}^2$ to be 0; we have

$$\frac{\partial L(\{Y_i\})}{\partial \tilde{\sigma}^2} = -\frac{n}{2\tilde{\sigma}^2} + \frac{1}{2(\tilde{\sigma}^2)^2}\sum_{i=1}^{n}(\tilde{m}(X_i) - Y_i)^2 = 0.$$
(3.63)

Hence, $\tilde{\sigma}^2$ is represented as

$$\tilde{\sigma}^2 = \frac{1}{n} \sum_{i=1}^{n} (\tilde{m}(X_i) - Y_i)^2. \tag{3.64}$$

This value, the maximum likelihood estimate of $\tilde{\sigma}^2$, is termed $\hat{\sigma}^2$, and its substitution into $\tilde{\sigma}^2$ in eq(3.62) leads to

$$L(\{Y_i\}) = -\frac{n}{2}\log(2\pi) - \frac{n}{2}\log(\hat{\sigma}^2) - \frac{n}{2}. \tag{3.65}$$

To maximize this value $\tilde{m}(x)$, which minimizes $\hat{\sigma}^2$, namely, the value of $\sum_{i=1}^{n}(\tilde{m}(X_i) - Y_i)^2$, is obtained and the resulting function is termed $\hat{m}(x)$. This is the least squares method. That is, the least squares method is a method of estimating a regression equation by the maximum likelihood procedure using the log-likelihood defined by eq(3.62).

Substitution of $\hat{\sigma}^2$ of the regression equation given by applying the least squares into eq(3.65) provides a maximum log-likelihood, which is the result of optimizing an assumption by using data at hand and estimating the cogency of the assumption in light of the same data. That is, the value of maximum log-likelihood obtained from the regression equation by the least squares using a data set collected in an actual experiment or observation is the cogency of the regression equation derived from the data; note that the cogency is estimated in light of the same data. Hence, the value of log-likelihood, which represents the cogency of data collected in another experiment or observation in the same setting, is expected to be smaller than that previously obtained. Therefore, analysts should estimate the degree of cogency of the assumption (i.e., $\hat{\sigma}^2$ and $\hat{m}(x)$) with respect to future data; otherwise, they cannot assess the real usefulness of $\hat{\sigma}^2$ and $\hat{m}(x)$.

Assume that future data are $\{(X_i, Y_i^*)\}$ $(1 \leq i \leq n)$ generated by the function $y = \hat{m}(x) + \epsilon^*$ (ϵ^* is also the realization of $N(0.0, \sigma^2)$); this is a procedure similar to that by which the data at hand were generated. Then, the cogency of $\hat{\sigma}^2$ and $\hat{m}(x)$ in light of $\{(X_i, Y_i^*)\}$ is represented as the log-likelihood:

$$L(\{Y_i\}, \{Y_i^*\}) = -\frac{n}{2}\log(2\pi) - \frac{n}{2}\log(\hat{\sigma}^2) - \frac{1}{2\hat{\sigma}^2} \sum_{i=1}^{n} (\hat{m}(X_i) - Y_i^*)^2. \tag{3.66}$$

$L(\{Y_i\}, \{Y_i^*\})$ is the log-likelihood that shows the cogency of $\hat{\sigma}^2$ and $\hat{m}(x)$ (i.e., an assumption on a function that generates $\{(X_i, Y_i)\}$) in light of $\{(X_i, Y_i^*)\}$.

The simplest way to estimate $L(\{Y_i\}, \{Y_i^*\})$ using the data at hand is cross-validation. σ^2 and $m(x)$, which are derived using the data obtained by omitting (X_j, Y_j) from $\{(X_i, Y_i)\}$, are named $\hat{\sigma}_{(j)}^2$ and $\hat{m}_{(j)}(x)$, respectively; the cogency of this model in light of (X_j, Y_j) is represented as the log-likelihood:

$$-\frac{1}{2}\log(2\pi) - \frac{1}{2}\log(\hat{\sigma}_{(j)}^2) - \frac{1}{2\hat{\sigma}_{(j)}^2}(\hat{m}_{(j)}(X_j) - Y_j)^2. \tag{3.67}$$

This procedure is repeated with each j $(1 \leq j \leq n)$, and the sum of the values can be employed as an approximation of $L(\{Y_i\}, \{Y_i^*\})$. Such a method is called likelihood cross-validation. When all of the values of $\hat{\sigma}^2_{(j)}$ in eq(3.67) are equal, the essential part of the multiplication of the result of likelihood cross-validation by $\hat{\sigma}^2_{(j)}$ is equivalent to the result of ordinary cross-validation (eq(3.33)). Even so, since the value of $\hat{\sigma}^2_{(j)}$ relies on the selection of the form of $m(x)$, the result of the choice of a regression equation by likelihood cross-validation is difficult to relate to that by ordinary cross-validation.

On the other hand, $L(\{Y_i\}, \{Y_i^*\})$ can be approximated without cross-validation. Namely, even if likelihood cross-validation is not expressed as eq(3.62), we can make use of the relationship between $E[L(\{Y_i\}, \{Y_i^*\})]$ and $E[L(\{Y_i\})]$ as below; this relationship holds in more general cases.

$$E[L(\{Y_i\}, \{Y_i^*\})] \approx E[L(\{Y_i\})] - K. \qquad (3.68)$$

K is the number of parameters in model and regression equations; when a linear estimator is used for calculating estimates, it is the sum of diagonal elements (trace) of a hat matrix. $E[L(\{Y_i\})]$ is the mean of $L(\{Y_i\})$ produced by data obtained as a result of the plurality of experiments or observations under the same conditions. $E[L(\{Y_i\}, \{Y_i^*\})]$ is the average (i.e., the average of the mean of log-likelihoods) calculated by repeating the calculation of the log-likelihood of the regression equation obtained in a single experiment or observation, in light of the data given by numerous experiments or observations under the same conditions. Then, eq(3.68) indicates that the log-likelihoods in light of new data are smaller by around K than the log-likelihood in light of the data used to produce the regression equation. This $E[L(\{Y_i\}, \{Y_i^*\})]$ is called the expected mean log-likelihood. Although the term "expected mean log-likelihood" may convey the impression that it stands for $E[L(\{Y_i\})]$, it is actually $E[L(\{Y_i\}, \{Y_i^*\})]$.

Eq(3.68) is based on the two relationships

$$E[L(\{Y_i\}, \{Y_i^*\})] \approx E[L^+(\{Y_i^*\})] - \frac{K}{2}, \qquad (3.69)$$

$$E[L^+(\{Y_i^*\})] \approx E[L(\{Y_i\})] - \frac{K}{2}, \qquad (3.70)$$

where $L^+(\{Y_i^*\})$ is the log-likelihood expressing the cogency of parameters (here, the values of regression coefficients and $\hat{\sigma}^2$) of a regression equation that is obtained using an infinite number of data (the form of the regression equation remains the same as that for n data) in light of $\{(X_i, Y_i^*)\}$. To derive eq(3.69) and eq(3.70), refer to [2].

If errors obey a normal distribution, we have

$$E[L^+(\{Y_i^*\})] = -\frac{n}{2}\log(2\pi) - \frac{n}{2}\log(\sigma^2) - \frac{n}{2}. \qquad (3.71)$$

Since we assume that data at hand and new data are generated by the same equation, we obtain the relationship

$$E[L^+(\{Y_i^*\})] = E[L^+(\{Y_i\})]. \qquad (3.72)$$

The main reason for the difficulty in choosing the optimal regression equation is that the value of σ^2 (the squared sum of the difference between the values of $m(x)$

and the data) is difficult to determine. However, the difference between $E[L(\{Y_i\})]$ and $E[L(\{Y_i\}, \{Y_i^*\})]$ does not depend on σ^2 in eq(3.68). Therefore, if the value of $L(\{Y_i\})$ given by data at hand is used as an approximation of $E[L(\{Y_i\})]$, the value of $E[L(\{Y_i\}, \{Y_i^*\})]$ can be estimated using eq(3.70), and the optimal regression equation can be selected by maximizing $E[L(\{Y_i\}, \{Y_i^*\})]$ without obtaining the value of σ^2.

Eq(3.68), eq(3.69), eq(3.70), and eq(3.72) hold when a probability density function and the like are estimated, as well as regression. Hence, these equations are useful in coping with diverse problems. It should be noted, however, that in common with cross-validation, these relationships are useful under the condition that data at hand and future data share almost the same statistical characteristics.

Multiplication of eq(3.68) by (-2) leads to Akaike's Information Criterion (AIC). That is,

$$AIC = -2E[L(\{Y_i\})] + 2K. \tag{3.73}$$

Maximization of $E[L(\{Y_i\}, \{Y_i^*\})]$ estimated using eq(3.68) is equivalent to the minimization of AIC.

If errors obey a normal distribution, substitution of eq(3.64) and eq (3.65) into eq(3.68) provides

$$E[L(\{Y_i\}, \{Y_i^*\})] \approx -\frac{n}{2}\log(2\pi) - \frac{n}{2}\log\left(\frac{1}{n}\sum_{i=1}^{n}(\hat{m}(X_i) - Y_i)^2\right) - \frac{n}{2} - K. \tag{3.74}$$

Normally, extraction of essential parts from this equation and multiplication by (-2) result in AIC for the selection of an optimal regression equation.

$$AIC = n \cdot \log\left(\frac{1}{n}\sum_{i=1}^{n}(\hat{m}(X_i) - Y_i)^2\right) + 2K. \tag{3.75}$$

Eq(3.55) is the AIC for the linear estimator. To estimate a regression equation, regression is carried out using the least squares to obtain several regression equations, and the values of AIC are calculated using eq(3.75). The regression equation that yields the smallest value of AIC is expected to gain the greatest cogency in light of future data.

When estimates are derived using a linear estimator, the expectation of the third term on the right-hand side of eq(3.66) (page 123) is calculated in an approximate manner. A hat matrix is written as $\mathbf{H}$ (the size is $n \times n$). $\mathbf{m} = (m(X_1), m(X_2), \dots, m(X_n))^t$, $\boldsymbol{\epsilon} = (\epsilon_1, \epsilon_2, \dots, \epsilon_n)^t$, and $\boldsymbol{\epsilon}^* = (\epsilon_1^*, \epsilon_2^*, \dots, \epsilon_n^*)^t$ are set out. Then, the mean of the third term on the right-hand side of eq(3.66) is expressed as

$$E\left[-\left(\frac{1}{2\hat{\sigma}^2}\sum_{i=1}^{n}(\hat{m}(X_i) - Y_i^*)^2\right)\right]$$
$$= -E\left[\frac{1}{2\hat{\sigma}^2}(\mathbf{Hm} + \mathbf{H\epsilon} - \mathbf{m} - \boldsymbol{\epsilon}^*)^t(\mathbf{Hm} + \mathbf{H\epsilon} - \mathbf{m} - \boldsymbol{\epsilon}^*)\right]. \tag{3.76}$$

Polynomial regression and multiple regression (multiple linear regression) satisfy $\mathbf{H} \cdot \mathbf{H} = \mathbf{H}$, $\mathbf{H}^t = \mathbf{H}$. Furthermore, if a linear estimator is unbiased, we gain

$\mathbf{Hm} = \mathbf{m}$. The use of these relationships yields

$$
\begin{aligned}
E&\left[-\left(\frac{1}{2\hat{\sigma}^2}\sum_{i=1}^{n}(\hat{m}(X_i)-Y_i^*)^2\right)\right] \\
&= -E\left[\frac{1}{2\hat{\sigma}^2}(\mathbf{Hm}+\mathbf{H}\epsilon-\mathbf{m}-\epsilon^*)^t(\mathbf{Hm}+\mathbf{H}\epsilon-\mathbf{m}-\epsilon^*)\right] \\
&= -E\left[\frac{1}{2\hat{\sigma}^2}(\mathbf{H}\epsilon-\epsilon^*)^t(\mathbf{H}\epsilon-\epsilon^*)\right] \\
&= -E\left[\frac{\epsilon^t\mathbf{H}\epsilon}{2\hat{\sigma}^2}\right]+2E\left[\frac{\epsilon^{*t}\mathbf{H}\epsilon}{2\hat{\sigma}^2}\right]-E\left[\frac{\epsilon^{*t}\epsilon^*}{2\hat{\sigma}^2}\right] \\
&\approx -\frac{(\sigma^2\text{trace}(\mathbf{H})+\sigma^2 n)}{2E[\hat{\sigma}^2]} \\
&= -\frac{(\sigma^2 K+\sigma^2 n)}{2E[\hat{\sigma}^2]} \\
&= -\frac{n(\sigma^2 K+\sigma^2 n)}{2(\sigma^2(n-K))} \\
&= -\frac{n(K+n)}{2(n-K)} \\
&= -\frac{n}{2}\frac{\left(1+\frac{K}{n}\right)}{1-\frac{K}{n}} \\
&\approx -\frac{n}{2}-K.
\end{aligned}
\tag{3.77}
$$

The equality between the sixth line and the seventh line is based on eq(3.48) (page 119). The approximation between the ninth line and the tenth line is appropriate when n is much larger than K. Furthermore, the relationship between the fourth line and the fifth line is due to the approximations

$$
E\left[\frac{\epsilon^t\mathbf{H}\epsilon}{2\hat{\sigma}^2}\right]\approx\frac{\sigma^2\text{trace}(\mathbf{H})}{2E[\hat{\sigma}^2]},
\tag{3.78}
$$

$$
E\left[\frac{\epsilon^{*t}\mathbf{H}\epsilon}{2\hat{\sigma}^2}\right]\approx 0,
\tag{3.79}
$$

$$
E\left[\frac{\epsilon^{*t}\epsilon^*}{2\hat{\sigma}^2}\right]\approx\frac{\sigma^2}{2E[\hat{\sigma}^2]}.
\tag{3.80}
$$

Namely, under the assumption that a linear estimator is unbiased, the approximations above result in eq(3.77); therefore, $-\frac{1}{2}AIC$, which eq(3.73) defines, is identical to $E[L(\{Y_i\},\{Y_i^*\})]$ (expected mean log-likelihood). However, it is not easy to examine the appropriateness of the assumptions and approximations here analytically.

Then, we show how $-\frac{1}{2}AIC$ works as an approximation of $E[L(\{Y_i\},\{Y_i^*\})]$ by means of a numerical calculation using an example of polynomials. The simulation data used here consist of 400 data given by

$$
X_i = 0.0025i \qquad (1\leq i\leq 400),
\tag{3.81}
$$

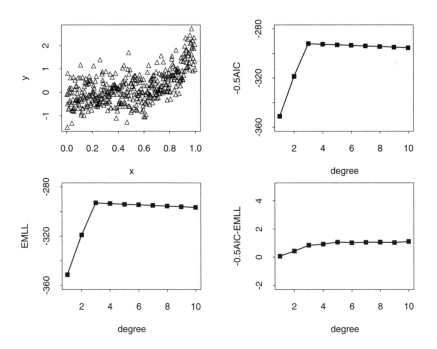

Figure 3.5   Some of the simulation data used here (400 data; standard deviation of the error is 0.5) (top left). Mean of $-\frac{1}{2}AIC$ (top right). Estimate of $E[L(\{Y_i\}, \{Y_i^*\})]$ (EMLL) (bottom left). Subtraction of the estimate of $E[L(\{Y_i\}, \{Y_i^*\})]$ (EMLL) from the mean of $-\frac{1}{2}AIC$ (bottom right).

$$Y_i = 10(X_i - 0.2)(X_i - 0.4)(X_i - 0.6) + \epsilon_i, \tag{3.82}$$

where $\{\epsilon_i\}$ are realizations of $N(0.0, 0.5^2)$ (normal distribution; mean is 0.0, and standard deviation is 0.5). Some of the simulation data obtained by eq(3.82) are illustrated in figure 3.5(top left). A total of 500 simulation data sets are generated by alternating the seed of a pseudorandom number. Then 1st-degree (linear equation) to 10th-degree polynomials are fitted to each data set to obtain $-\frac{1}{2}AIC$ for each. Then, the values of $-\frac{1}{2}AIC$ corresponding to the 500 simulation data sets are averaged to estimate the mean of $-\frac{1}{2}AIC$ for each degree of polynomials.

Next, an approximation of $E[L(\{Y_i\}, \{Y_i^*\})]$ based on eq(3.66) is calculated:

$$E[L(\{Y_i\}, \{Y_i^*\})] \approx \frac{1}{500} \sum_{j=1}^{500} \left(-\frac{n}{2}\log(2\pi) - \frac{n}{2}\log(\hat{\sigma}_j^2) \right.$$

$$\left. -\frac{1}{2\hat{\sigma}_j^2} \frac{1}{100} \sum_{k=1}^{100} \sum_{i=1}^{400} (\hat{m}_j(X_i) - Y_i^{(*jk)})^2 \right). \tag{3.83}$$

$\{Y_i^{(*jk)}\}$ $((1 \leq i \leq 400, \ 1 \leq j \leq 500, \ 1 \leq k \leq 100))$ is a simulation data set obtained using eq(3.82) with different seeds of random number. $\hat{m}_j(\cdot)$ is a regression equation derived when the jth simulation data set is used. This $E[L(\{Y_i\}, \{Y_i^*\})]$ is regarded as a mean of log-likelihood to show cogency of a regression equation created by the least squares using each simulation data in light of other simulation data. The average $-\frac{1}{2}AIC$ obtained above is diplayed in figure 3.5(top right), and the values of the estimates of $E[L(\{Y_i\}, \{Y_i^*\})]$ are illustrated in figure 3.5(bottom left). To compare the two estimates, subtractions of $E[L(\{Y_i\}, \{Y_i^*\})]$ from $-\frac{1}{2}AIC$ are shown in figure 3.5(bottom right). Both $E[L(\{Y_i\}, \{Y_i^*\})]$ and $-\frac{1}{2}AIC$ indicate that a cubic equation is the optimal regression equation.

Next, using eq(3.81) and eq(3.82), the results obtained using simulation data when the standard deviation of errors was set at 0.1 are displayed in the four graphs in figure 3.6. Although it is not very clear from these graphs, both $E[L(\{Y_i\}, \{Y_i^*\})]$ and $-\frac{1}{2}AIC$ give maximal values when a cubic equation is adopted.

Figure 3.5 and figure 3.6 show that $-\frac{1}{2}AIC$ acts as an appropriate approximation of $E[L(\{Y_i\}, \{Y_i^*\})]$. However, the augmentation of the value of K leads to the tendency that the values of $-\frac{1}{2}AIC$ surpass those of $E[L(\{Y_i\}, \{Y_i^*\})]$. Therefore, estimation of the usefulness of regression equations by the use of $-\frac{1}{2}AIC$ tends to favor regression equations with a large value of K, compared with the estimation using $E[L(\{Y_i\}, \{Y_i^*\})]$. That is, selection of a regression equation by using $-\frac{1}{2}AIC$ may yield a value of K that is slightly larger than the optimal one. Regarding smoothing, estimates are sometimes rougher than necessary. To reduce this tendency, $AICc$ (Corrected AIC) has been developed. When errors obey a normal distribution, $AICc$ is defined as [13].

$$AICc = \log\left(\frac{1}{n} \sum_{i=1}^{n} (\hat{m}(X_i) - Y_i)^2 \right) + \frac{n+K}{n-K-2}. \tag{3.84}$$

$AICc$ is useful when the number of data is small (Chapter 5 of [2]).

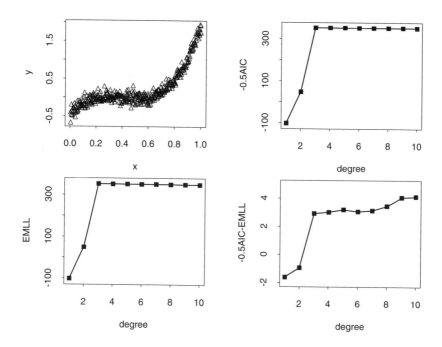

Figure 3.6 Simulation data used here (the number of simulation data is 400; standard deviation of error is 0.1) (top left). Averages of $-\frac{1}{2}AIC$ (top right). Estimates of $E[L(\{Y_i\}, \{Y_i^*\})]$ (EMLL) (bottom left). Subtraction of the estimates of $E[L(\{Y_i\}, \{Y_i^*\})]$ (EMLL) from the average of $-\frac{1}{2}AIC$ (bottom right).

3.4   NADARAYA-WATSON ESTIMATOR

All of the techniques described in Chapter 2 are considered to be some type of weighted average. Even if the values of the predictor of data are not equispaced, or the estimates of $m(x)$ at locations where data do not exist are required, weighted average plays an important role. Below, methods that directly utilize weighted averages of data to derive $m(x)$ are introduced first.

Both a predictor and a target variable are regarded as random variables (variate is a synonym for random variable). A random variable is a variable for which the probability of its taking a specific value obeys a certain probability distribution. For example, since a value gained by observing a particular value is expected to obey a certain probability distribution, the values obtained are looked upon as realizations of this probability distribution. Then, the values constituting the predictor of data and those of the target variable are considered to be random variables corresponding to the predictor and the target variable. That is, the random variable X yields the predictor variable part of data ($\{X_i\}$), and the random variable Y gives the target variable part of data ($\{Y_i\}$).

Then, a function $f_X(x)$ is defined. When the probability that X takes a value between x and $(x + \Delta x)$ (Δx is a small positive value) is $f_X(x)\Delta x$, $f_X(x)$ is called the probability density function that X obeys. $f_X(x)$ satisfies

$$\int_{-\infty}^{\infty} f_X(x)dx = 1. \tag{3.85}$$

On the other hand, if the probability that X takes a value between x and $(x + \Delta x)$ and Y takes a value between y and $(y + \Delta y)$ (Δy is also a small positive value) is written as $f(x, y)\Delta x \Delta y$, $f(x, y)$ is called the joint probability density function (or it can be called simply a joint density) of X and Y. It satisfies

$$\int_{-\infty}^{\infty} \int_{-\infty}^{\infty} f(x, y)dxdy = 1. \tag{3.86}$$

Furthermore, if the value of X is known to be x and the probability of Y taking a value between y and $(y+\Delta y)$ is written as $f(y|x)\Delta y$, $f(y|x)$ is called the conditional probability density function under the condition of $X = x$ (or simply, a conditional density); it satisfies the relationship

$$\int_{-\infty}^{\infty} f(y|x)dy = 1. \tag{3.87}$$

Then, there exists the following relationship among $f(x, y)$, $f(y|x)$, and $f_X(x)$:

$$f(x, y) = f(y|x)f_X(x). \tag{3.88}$$

This is named the multiplication rule; it can also be called the multiplication theorem. This equation is proved in the following manner.

The probability of X taking a value between x and $(x + \Delta x)$ is $f_X(x)\Delta x$. Furthermore, if X is known to take a value between x and $(x + \Delta x)$, the probability of Y taking a value between y and $(y + \Delta y)$ is $f(y|x)\Delta y$. Then, the probability that X

takes a value between x and $(x + \Delta x)$ and Y takes a value between y and $(y + \Delta y)$ is the product of the two probabilities. Namely, $f(y|x)f_X(x)\Delta x\Delta y$ is obtained; this is equal to $f(x, y)\Delta x\Delta y$ and hence eq(3.88) is derived.

Since the value of function $m(x)$ corresponding to a predictor value (x) is equivalent to the expectation of a target variable under the condition that the value of the predictor is fixed at x, the discussion above leads to

$$
\begin{aligned}
m(x) &= E(Y|X = x) \\
&= \int_{-\infty}^{\infty} yf(y|x)dy \\
&= \frac{1}{f_X(x)} \int_{-\infty}^{\infty} yf(x, y)dy.
\end{aligned}
\tag{3.89}
$$

The equality between the first line and the second line is the definition of $E(Y|X = x)$ (expectation of Y when X is fixed at x). Equality between the second line and the third line is derived from eq(3.88). This equation shows that the estimation of $m(x)$ is equivalent to that of $f(y|x)$ (i.e., the estimation of $f(x, y)$ and $f_X(x)$).

Then, let us adopt the kernel density estimator as one of the simplest methods of estimating $f(x, y)$ and $f_X(x)$. $f(x, y)$ and $f_X(x)$ estimated by this technique are termed $\hat{f}(x, y)$ and $\hat{f}_X(x)$, respectively,

$$
\hat{f}(x, y) = \frac{1}{nh_xh_y} \sum_{i=1}^{n} K_x \left(\frac{x - X_i}{h_x} \right) K_y \left(\frac{y - Y_i}{h_y} \right),
\tag{3.90}
$$

$$
\hat{f}_X(x) = \frac{1}{nh_x} \sum_{i=1}^{n} K_x \left(\frac{x - X_i}{h_x} \right),
\tag{3.91}
$$

where $K_x(\cdot)$ and $K_y(\cdot)$ are kernel functions. h_x and h_y are positive constants called bandwidths. Kernel functions usually satisfy

$$
\int_{-\infty}^{\infty} K_x(u)du = \int_{-\infty}^{\infty} K_y(u)du = 1,
\tag{3.92}
$$

$$
\int_{-\infty}^{\infty} uK_x(u)du = \int_{-\infty}^{\infty} uK_y(u)du = 0,
\tag{3.93}
$$

$$
\int_{-\infty}^{\infty} u^2 K_x(u)du < \infty,
\tag{3.94}
$$

$$
\int_{-\infty}^{\infty} u^2 K_y(u)du < \infty.
\tag{3.95}
$$

Eq(3.92) and eq(3.93) are equivalent, respectively, to the two equations below.

$$
\frac{1}{h_x} \int_{-\infty}^{\infty} K_x \left(\frac{x - X_i}{h_x} \right) dx = \frac{1}{h_y} \int_{-\infty}^{\infty} K_y \left(\frac{y - Y_i}{h_y} \right) dy = 1,
\tag{3.96}
$$

$$
\frac{1}{(h_x)^2} \int_{-\infty}^{\infty} xK_x \left(\frac{x}{h_x} \right) dx = \frac{1}{(h_y)^2} \int_{-\infty}^{\infty} yK_y \left(\frac{y}{h_y} \right) dy = 0.
\tag{3.97}
$$

Then, the Gaussian function below is, for example, used as $K_x(u)$ and $K_y(u)$.

$$\frac{1}{\sqrt{2\pi}}\exp\left(-\frac{u^2}{2}\right). \tag{3.98}$$

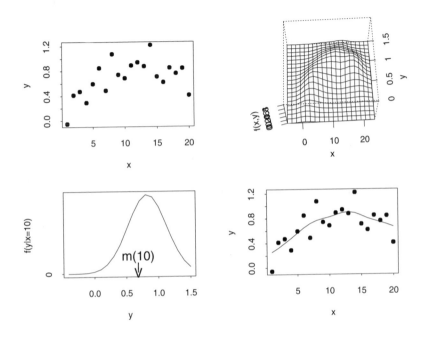

Figure 3.7   Procedure for estimating $m(x)$ using Nadaraya-Watson estimator. Data (top left). Estimation of $f(x, y)$ (top right). Estimation of $f(y|x = 10)$ and derivation of $\hat{m}(10)$ using $f(y|x = 10)$ (bottom left). Estimation of a regression equation $\hat{m}(x)$ by giving a variety of values for x (bottom right).

Figure 3.7 displays the processes of deriving $m(x)$ using data $(\{X_i, Y_i\})$ on the basis of the procedure described above, summarized as follows. The procedure is:
(1) Data is given (top left graph).
(2) $f(x, y)$ is estimated using eq(3.90) (top right graph).
(3) To calculate the estimate when $x = 10$, $f_X(x = 10)$ is estimated using eq(3.91). Furthermore, $f(y|x = 10)$ is estimated on the basis of eq(3.88); $f(x, y)$ obtained in process (2) is utilized (bottom left graph).
(4) The value $m(10)$ is estimated using $f(y|x = 10)$ derived in process (3); the second line in eq(3.89) is taken into account.
(5) Processes (3) and (4) are executed using various values of x to derive $\hat{m}(x)$ (bottom right graph).

In truth, $\hat{m}(x)$ is obtained through a simpler calculation. The substitution of eq(3.90) and eq(3.91) into eq(3.89) results in

$$\hat{m}(x) = \int_{-\infty}^{\infty} y \frac{\hat{f}(x, y)}{\hat{f}_X(x)} dy$$

$$
= \int_{-\infty}^{\infty} y \frac{\frac{1}{nh_x h_y} \sum_{i=1}^{n} K_x \left(\frac{x - X_i}{h_x} \right) K_y \left(\frac{y - Y_i}{h_y} \right)}{\frac{1}{nh_x} \sum_{i=1}^{n} K_x \left(\frac{x - X_i}{h_x} \right)} dy
$$

$$
= \frac{\sum_{i=1}^{n} K_x \left(\frac{x - X_i}{h_x} \right) \int_{-\infty}^{\infty} y K_y \left(\frac{y - Y_i}{h_y} \right) dy}{h_y \sum_{i=1}^{n} K_x \left(\frac{x - X_i}{h_x} \right)}. \tag{3.99}
$$

Then, if we set $\eta = \frac{y - Y_i}{h_y}$, we have

$$
\frac{1}{h_y} \int_{-\infty}^{\infty} y K_y \left(\frac{y - Y_i}{h_y} \right) dy = \int_{-\infty}^{\infty} (h_y \eta + Y_i) K_y(\eta) d\eta
$$
$$
= Y_i. \tag{3.100}
$$

The equality between the first line and the second line makes use of eq(3.92) and eq(3.93). Substitution of eq(3.100) into eq(3.99), and simplification of $K_x(\cdot)$ to $K(\cdot)$ and that of h_x to h yield

$$
\hat{m}(x) = \frac{\sum_{i=1}^{n} K \left(\frac{x - X_i}{h} \right) Y_i}{\sum_{k=1}^{n} K \left(\frac{x - X_k}{h} \right)} = \sum_{i=1}^{n} W_i(x) Y_i. \tag{3.101}
$$

This method is called the Nadaraya-Watson estimator. Then, we set

$$
W_i(x) = \frac{K \left(\dfrac{x - X_i}{h} \right)}{\displaystyle\sum_{k=1}^{n} K \left(\dfrac{x - X_k}{h} \right)}. \tag{3.102}
$$

$\{W_i(x)\}$ is called an equivalent kernel. Since the estimate is written as eq(3.101), a Nadaraya-Watson estimator is a linear estimator. $\{W_i(x)\}$ in a Nadaraya-Watson estimator has the characteristic

$$
\sum_{i=1}^{n} W_i(x) = 1. \tag{3.103}
$$

That is, an estimate provided by a Nadaraya-Watson estimator is a weighted average of $\{Y_i\}$. The fact that $\{W_i(x)\}$ satisfies eq(3.103) shows that one of the methods for smoothing by weighted averages when the values of the predictor are not equispaced is the Nadaraya-Watson estimator. The value of weight is determined from the shape of the kernel function ($K(\cdot)$) and its bandwidth (h). The form of eq(3.102) indicates that even when $K(\cdot)$ does not satisfy eq(3.92), it does not affect $\hat{m}(x)$. Therefore, when eq(3.98) is replaced, for example, with the equation below, the results remain the same.

$$
\exp \left(-\frac{u^2}{2} \right). \tag{3.104}
$$

Furthermore, when the estimates at x (in eq(3.101)), which coincide with X_i, are called $\hat{Y}_i$, we obtain

$$\hat{Y}_i = \sum_{j=1}^{n} W_j(X_i)Y_j. \tag{3.105}$$

Rewriting $W_i(X_j)$ as H_{ij}, we have

$$\hat{Y}_i = \sum_{j=1}^{n} H_{ij}Y_j, \tag{3.106}$$

$$H_{ij} = \frac{K\left(\dfrac{X_i - X_j}{h}\right)}{\displaystyle\sum_{k=1}^{n} K\left(\dfrac{X_i - X_k}{h}\right)}. \tag{3.107}$$

The construction of a matrix (the size is $n \times n$) in which H_{ij} is an element yields a hat matrix (**H**). Figure 3.8 illustrates an example of a hat matrix given by a Gaussian kernel function (eq(3.98) or eq(3.104)). Namely, the kernel function here is

$$K\left(\frac{x - X_i}{h}\right) = \exp\left(-\frac{1}{2}\left(\frac{x - X_i}{h}\right)^2\right), \tag{3.108}$$

where $n = 20$ and $\{X_i\} = \{1, 2, 3, \ldots, 20\}$; the bandwidths are $h = 1$ and $h = 2$. Eq(3.107) also shows that the shape of the hat matrix takes the form of a Gaussian function. Figure 3.9 illustrates an example where the values of a predictor are not equispaced but $\{X_i\} = \{1^{1.8}, 2^{1.8}, 3^{1.8}, \ldots, 20^{1.8}\}$. The bandwidths are $h = 10$ and $h = 20$. These results show that an estimation where data is dense uses a large number of data.

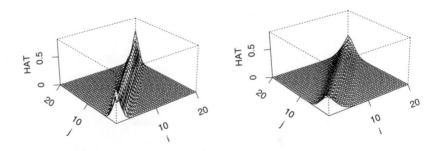

Figure 3.8   Hat matrix of a Nadaraya-Watson estimator using a Gaussian kernel function. $n = 20$, $\{X_i\} = \{1, 2, 3, \ldots, 20\}$. $h = 1$ (left). $h = 2$ (right).

There is another procedure for deriving a Nadaraya-Watson estimator: minimization of the value below to derive $\hat{m}(x)$.

$$E_{const}(x) = \sum_{i=1}^{n} (Y_i - a_0)^2 w\left(\frac{x - X_i}{h}\right). \tag{3.109}$$

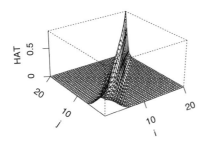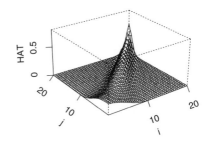

Figure 3.9   Hat matrix of a Nadaraya-Watson estimator using a Gaussian kernel function. $n = 20$, $\{X_i\} = \{1^{1.8}, 2^{1.8}, 3^{1.8}, \ldots, 20^{1.8}\}$. $h = 10$ (left). $h = 20$ (right).

Here, $w\left(\frac{x-X_i}{h}\right)$ takes a positive value or 0 similar to $K\left(\frac{x-X_i}{h}\right)$ described above; it takes the maximal value at $x = X_i$ and then decreases according to the absolute value of $(x - X_i)$. Then, eq(3.109) is used to derive a value of a_0 using data located in the neighborhood of $x = X_i$ under the condition that $m(x)$ takes a constant of a_0 around $x = X_i$. $w\left(\frac{x-X_i}{h}\right)$ determines the degree of importance to be placed on the ith data according to the distance from x when the estimate at x is obtained. Therefore, derivation of a_0 using various values of x leads to $\hat{m}(x)$. To calculate a_0 that minimizes the value which eq(3.109) provides, eq(3.109) is differentiated to be 0 with respect to a_0; it ends up with

$$a_0 = \frac{\sum\limits_{i=1}^{n} Y_i w\left(\dfrac{x - X_i}{h}\right)}{\sum\limits_{i=1}^{n} w\left(\dfrac{x - X_i}{h}\right)}. \tag{3.110}$$

We set $w\left(\frac{x-X_i}{h}\right)$ as $K\left(\frac{x-X_i}{h}\right)$; this yields the same equation as eq(3.101) (page 133). That is, the Nadaraya-Watson estimator is considered to be a method for estimating $m(x)$ by the use of $w\left(\frac{x-X_i}{h}\right)$ as a kernel function under the assumption that $m(x)$ consists of local constants. On the basis of this view, $K\left(\frac{x-X_i}{h}\right)$ in eq(3.101) determines the relative weights for $(Y_i - a_0)^2$. This is another justification for the needlessness of the limitation that the integration of $\frac{1}{h}K\left(\frac{x-X_i}{h}\right)$ in the whole region must be equal to 1.

Figure 3.10 displays estimates smoothed by the Nadaraya-Watson estimator using 30 data. The values of data are superimposed. A Gaussian function is utilized as a kernel function. The bandwidth is $h = 0.18$. The opimization procedure of the bandwidth is illustrated in figure 3.11. Since a hat matrix defined by eq(3.107) yields both CV and GCV, both of them are calculated. As a result, both statistics show that $h = 0.18$ is optimal, and hence we adopted this value in the calculation of estimates shown in figure 3.10.

Next, figure 3.12 and figure 3.13 display the results of the calculation of bias, variance, and mean squared error with smoothing by a Nadaraya-Watson estimator; the data are the same as those used in figure 3.1 (page 107) to figure 3.4. In the

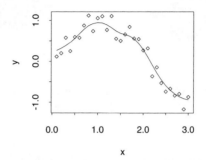

Figure 3.10 Estimates derived using the Nadaraya-Watson estimator ($h = 0.18$) (solid line) and using data ($\Diamond$).

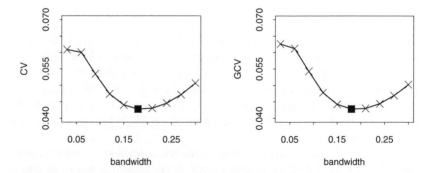

Figure 3.11 Results of calculation of CV and GCV with diverse values of bandwidth to optimize the bandwidth for deriving estimates illustrated in figure 3.10.

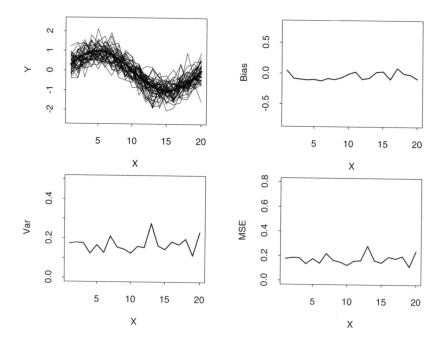

Figure 3.12 $m(x)$ and estimates smoothed by a Nadaraya-Watson estimator ($h = 0.5$) (top left). Bias ($Bias[\hat{Y}_i]$) (top right). Variance ($Var[\hat{Y}_i]$) (bottom left). Mean squared error ($MSE[\hat{Y}_i]$) (bottom right).

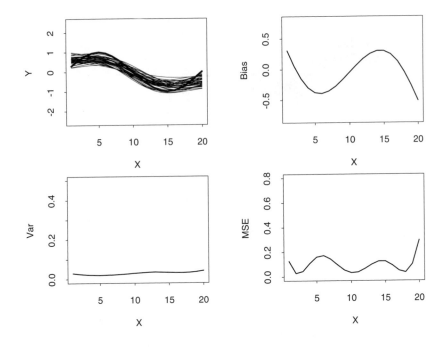

Figure 3.13 $m(x)$ and estimates smoothed by a Nadaraya-Watson estimator ($h = 3$) (top left). Bias ($Bias[\hat{Y}_i]$) (top right). Variance ($Var[\hat{Y}_i]$) (bottom left). Mean squared error ($MSE[\hat{Y}_i]$) (bottom right).

case of figure 3.12, $h = 0.5$, and in the case of figure 3.13, $h = 3$. Comparison of figure 3.1 to figure 3.4 indicates that the results of using a binomial filter with a small value of m and those obtained using a Nadaraya-Watson estimator with a small value of h have common features; the contrary is also true. This is inferred from the fact that a hat matrix of a binomial filter is analogous to that of a Nadaraya-Watson estimator. Therefore, while a binomial filter requires the limitation that the predictor be distributed at even intervals, elimination of this limitation yields a Nadaraya-Watson estimator with a kernel function, as represented by eq(3.104) (page 133).

3.5 LOCAL POLYNOMIAL REGRESSION

As eq(3.109) shows, the Nadaraya-Watson estimator is considered to be a method of fitting constants locally. Hence, this technique is sometimes called local constant estimation. On the other hand, since the local linear regression described in Chapter 1 is a regression to fit linear equations locally when the predictor of data is assumed to be equispaced, it is regarded as an extension of the local fitting of constants. Let us combine these two methods to develop the local linear regression into a method of treating data with a non-equispaced predictor. At the same time, equations for local fitting are not limited to a linear equation; pth-degree (p is a positive integer) polynomials are used.

A local polynomial equation ($m(x, x^*)$), when the value of predictor (x) is close to x^*, is defined as

$$m(x, x^*) = a_0(x^*) + \sum_{j=1}^{p} a_j(x^*)(x - x^*)^j. \tag{3.111}$$

This equation corresponds to an approximation using the first p terms of an equation provided by Taylor expansion of $m(x)$ around x^*. $\{a_0(x^*), a_1(x^*), a_2(x^*), \ldots, a_p(x^*)\}$ is derived by minimizing

$$\begin{aligned}
E_{local}(x^*) &= \sum_{i=1}^{n} \left(w\left(\frac{X_i - x^*}{h}\right)(m(X_i, x^*) - Y_i)^2 \right) \\
&= \sum_{i=1}^{n} \left(w\left(\frac{X_i - x^*}{h}\right)\left(a_0(x^*) + \sum_{j=1}^{p} a_j(x^*)(X_i - x^*)^j - Y_i\right)^2 \right).
\end{aligned}$$

$$\tag{3.112}$$

$w(\frac{X_i - x^*}{h})$ is a function for determining the degree of significance of data ((X_i, Y_i)) when a regression equation at x^* is obtained; this is a kind of kernel function. In eq(3.110) (page 135), $w\left(\frac{x - X_i}{h}\right)$ is regarded as a kernel density estimator, though this is not the case in local polynomial equations in general. However, $w(\frac{X_i - x^*}{h})$ is called a kernel function in the sense that it is a function that provides weight.

As is the case of the Nadaraya-Watson estimator, a kernel function is usually a function that takes 0 or a positive value and a maximum value when $X_i - x^* = 0$; it declines according to the absolute value of ($X_i - x^*$). Since $m(x, x^*)$ ought to give a good approximation when x is close to x^*, the data in the neighborhood of x^* should be emphasized to carry out the least squares. The estimate at $x = x^*$ through the use of $\hat{m}(x, x^*)$ obtained in this manner is $\hat{m}(x^*, x^*) = \hat{a}_0(x^*)$. That is, the derivation of $\{\hat{a}_0(x^*), \hat{a}_1(x^*), \hat{a}_2(x^*), \ldots, \hat{a}_p(x^*)\}$ by the least squares to minimize eq(3.112) using eq(3.111) yields a simple form of the estimate at $x = x^*$: $\hat{a}_0(x^*)$. $w(\frac{X_i - x^*}{h})$ is employed as a kernel function here. The form of $w(\frac{X_i - x^*}{h})$ implies that the shape of $w(\cdot)$ is unrelated to the value of either X_i or x^*. However, a kernel function can rely on these values.

As a kernel function $(w(\frac{X_i - x^*}{h}))$, we can use a Gaussian function again:

$$w\left(\frac{X_i - x^*}{h}\right) = \exp\left(-\frac{1}{2}\left(\frac{X_i - x^*}{h}\right)^2\right). \qquad (3.113)$$

h is a positive constant for determining the breadth of $w(\frac{X_i - x^*}{h})$; it is called a bandwidth here again. Other functions commonly used as kernel functions are the bisquare weight function (biweight, bisquare), the triweight, and the tricube weight function, as respectively shown below.

$$w\left(\frac{X_i - x^*}{h}\right) = \begin{cases} \left(1 - \left(\frac{X_i - x^*}{h}\right)^2\right)^2 & \text{if } \left(\frac{X_i - x^*}{h}\right)^2 \leq 1 \\ 0 & \text{if } \left(\frac{X_i - x}{h}\right)^2 > 1, \end{cases} \qquad (3.114)$$

$$w\left(\frac{X_i - x^*}{h}\right) = \begin{cases} \left(1 - \left(\frac{X_i - x^*}{h}\right)^2\right)^3 & \text{if } \left(\frac{X_i - x^*}{h}\right)^2 \leq 1 \\ 0 & \text{if } (\frac{X_i - x}{h})^2 > 1, \end{cases} \qquad (3.115)$$

$$w\left(\frac{X_i - x^*}{h}\right) = \begin{cases} \left(1 - \left(\frac{|X_i - x^*|}{h}\right)^3\right)^3 & \text{if } \left(\frac{|X_i - x^*|}{h}\right)^3 \leq 1 \\ 0 & \text{if } \left(\frac{|X_i - x|}{h}\right)^3 > 1. \end{cases} \qquad (3.116)$$

To carry out the minimization of eq(3.112), $\mathbf{X}$ (a matrix with size $n \times (p+1)$), which is called a design matrix, and $\mathbf{y}$ (a vector with n elements) are defined in a similar manner to those of polynomial regression (cf. Chapter 2 (page 33)). It should be noted, however, that $\mathbf{X}$ here is a function of x^*. While polynomial regression seeks a regression equation that is useful in the whole region where data exist, local polynomial regression creates a regression equation that fits the purpose in the neighborhood of x^*.

$$\mathbf{X} = \begin{pmatrix} 1 & (X_1 - x^*) & (X_1 - x^*)^2 & \cdots & (X_1 - x^*)^p \\ 1 & (X_2 - x^*) & (X_2 - x^*)^2 & \cdots & (X_2 - x^*)^p \\ \vdots & \vdots & \vdots & \ddots & \vdots \\ 1 & (X_n - x^*) & (X_n - x^*)^2 & \cdots & (X_n - x^*)^p \end{pmatrix}, \qquad (3.117)$$

$$\mathbf{y} = \begin{pmatrix} Y_1 \\ \vdots \\ Y_n \end{pmatrix}. \qquad (3.118)$$

Next, a diagonal matrix $\mathbf{W}$ (with size $n \times n$) is defined as

$$
\mathbf{W} =
\begin{pmatrix}
w\left(\dfrac{X_1 - x^*}{h}\right) & 0 & 0 & \cdots & 0 \\
0 & w\left(\dfrac{X_2 - x^*}{h}\right) & 0 & \cdots & 0 \\
0 & 0 & w\left(\dfrac{X_3 - x^*}{h}\right) & \cdots & 0 \\
\vdots & \vdots & \vdots & \ddots & \vdots \\
0 & 0 & 0 & \cdots & w\left(\dfrac{X_n - x^*}{h}\right)
\end{pmatrix}.
\tag{3.119}
$$

Furthermore, $\mathbf{a}$ is defined as

$$
\mathbf{a} =
\begin{pmatrix}
a_0(x^*) \\
\vdots \\
a_p(x^*)
\end{pmatrix}.
\tag{3.120}
$$

Then, eq(3.112) is rewritten as

$$
E_{local}(x^*) = (\mathbf{Xa} - \mathbf{y})^t \, \mathbf{W} \, (\mathbf{Xa} - \mathbf{y}).
\tag{3.121}
$$

Differentiations of $E_{local}(x^*)$ to 0 with respect to $\{a_0(x^*), a_1(x^*), a_2(x^*), \ldots, a_p(x^*)\}$ lead to $\hat{\mathbf{a}}$:

$$
\hat{\mathbf{a}} = (\mathbf{X}^t \mathbf{W} \mathbf{X})^{-1} \mathbf{X}^t \mathbf{W} \mathbf{y}.
\tag{3.122}
$$

This result is based on the calculation

$$
\begin{aligned}
E_{local}(x^*) &= \sum_{i=1}^{n}\left(\sum_{j=0}^{p}[\mathbf{X}]_{ij}[\mathbf{a}]_j - Y_i\right)[\mathbf{W}]_{ii}\left(\sum_{k=0}^{p}[\mathbf{X}]_{ik}[\mathbf{a}]_k - Y_i\right) \\
&= \sum_{i=1}^{n}\sum_{j=0}^{p}\sum_{k=0}^{p}[\mathbf{X}]_{ij}[\mathbf{a}]_j[\mathbf{W}]_{ii}[\mathbf{X}]_{ik}[\mathbf{a}]_k - \sum_{i=1}^{n}\sum_{k=0}^{p}Y_i[\mathbf{W}]_{ii}[\mathbf{X}]_{ik}[\mathbf{a}]_k \\
&\quad - \sum_{i=1}^{n}\sum_{j=0}^{p}[\mathbf{X}]_{ij}[\mathbf{a}]_j[\mathbf{W}]_{ii}Y_i + \sum_{i=1}^{n}Y_i[\mathbf{W}]_{ii}Y_i.
\end{aligned}
\tag{3.123}
$$

Hence, the $(p+1)$ equations as below must hold.

$$
\begin{aligned}
\frac{\partial E_{local}(x^*)}{\partial a_l} &= \sum_{i=1}^{n}\sum_{j=0}^{p}\sum_{k=0}^{p}[\mathbf{X}]_{ij}\frac{\partial [\mathbf{a}]_j}{\partial [\mathbf{a}]_l}[\mathbf{W}]_{ii}[\mathbf{X}]_{ik}[\mathbf{a}]_k \\
&\quad + \sum_{i=1}^{n}\sum_{j=0}^{p}\sum_{k=0}^{p}[\mathbf{X}]_{ij}[\mathbf{a}]_j[\mathbf{W}]_{ii}[\mathbf{X}]_{ik}\frac{\partial [\mathbf{a}]_k}{\partial [\mathbf{a}]_l} \\
&\quad - \sum_{i=1}^{n}\sum_{k=0}^{p}Y_i[\mathbf{W}]_{ii}[\mathbf{X}]_{ik}\frac{\partial [\mathbf{a}]_k}{\partial [\mathbf{a}]_l} - \sum_{i=1}^{n}\sum_{j=0}^{p}[\mathbf{X}]_{ij}\frac{\partial [\mathbf{a}]_j}{\partial [\mathbf{a}]_l}[\mathbf{W}]_{ii}Y_i \\
&= \sum_{i=1}^{n}\sum_{k=0}^{p}[\mathbf{X}]_{il}[\mathbf{W}]_{ii}[\mathbf{X}]_{ik}[\mathbf{a}]_k + \sum_{i=1}^{n}\sum_{j=0}^{p}[\mathbf{X}]_{ij}[\mathbf{a}]_j[\mathbf{W}]_{ii}[\mathbf{X}]_{il}
\end{aligned}
$$

$$-\sum_{i=1}^{n} Y_i [\mathbf{W}]_{ii}[\mathbf{X}]_{il} - \sum_{i=1}^{n}[\mathbf{X}]_{il}[\mathbf{W}]_{ii}Y_i$$

$$= 2\sum_{i=1}^{n}\sum_{j=0}^{p}[\mathbf{X}^t]_{li}[\mathbf{W}]_{ii}[\mathbf{X}]_{ij}[\mathbf{a}]_j - 2\sum_{i=1}^{n}[\mathbf{X}^t]_{li}[\mathbf{W}]_{ii}Y_i$$

$$= 0 \qquad (0 \le l \le p), \tag{3.124}$$

where the product rule of differentiation and the equation below are utilized:

$$\frac{\partial [\mathbf{a}]_i}{\partial [\mathbf{a}]_j} = \begin{cases} 1 & \text{if } i = j \\ 0 & \text{if } i \ne j. \end{cases} \tag{3.125}$$

Eq(3.124) leads to eq(3.122). Furthermore, taking account of $\hat{m}(x^*, x^*) = \hat{a}_0(x^*)$, we have

$$\hat{m}(x^*, x^*) = \mathbf{e}_1^t (\mathbf{X}^t \mathbf{W} \mathbf{X})^{-1} \mathbf{X}^t \mathbf{W} \mathbf{y}, \tag{3.126}$$

where $\mathbf{e}_1$ is a column vector $(1, 0, 0, \ldots, 0)^t$ (the number of elements is $(p + 1)$). This unit vector functions to extract only the first element of a vector.

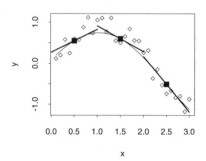

Figure 3.14   Concept of local linear regression. A local linear equation corresponding to each x resembles a tangent of the regression equation. Then, the point of tangency between the linear equation and the regression equation is the estimate at x. The three linear lines represent $\hat{m}(x, 0.5)$, $\hat{m}(x, 1.5)$, and $\hat{m}(x, 2.5)$, respectively, from left to right. The three ■ symbols signify estimates at $x = 0.5$, $x = 1.5$, and $x = 2.5$, respectively, from left to right; they are $\hat{a}_0(0.5)$ ($= \hat{m}(0.5, 0.5)$), $\hat{a}_0(1.5)$ ($= \hat{m}(1.5, 1.5)$), and $\hat{a}_0(2.5)$ ($= \hat{m}(2.5, 2.5)$).

Figure 3.14 illustrates the concept of local linear regression. Three linear equations are obtained from local fitting of linear equations at $x = 0.5, 1.5$, and 2.5 using eq(3.112) (page 139) ($p = 1$). That is, the linear equations are $\hat{m}(x, 0.5)$, $\hat{m}(x, 1.5)$, and $\hat{m}(x, 2.5)$, respectively. Hence, estimates at $x = 0.5, 1.5$, and 2.5 are $\hat{a}_0(0.5)$ ($= \hat{m}(0.5, 0.5)$), $\hat{a}_0(1.5)$ ($= \hat{m}(1.5, 1.5)$), and $\hat{a}_0(2.5)$ ($= \hat{m}(2.5, 2.5)$), respectively.

Next, a column vector $\mathbf{q}(x^*)$ ($= (q_1(x^*), q_2(x^*), \ldots, q_n(x^*))^t$) is defined as

$$\mathbf{q}(x^*)^t = \mathbf{e}_1^t (\mathbf{X}^t \mathbf{W} \mathbf{X})^{-1} \mathbf{X}^t \mathbf{W}. \tag{3.127}$$

$q(x^*)$ is called a weight diagram or a weight diagram vector. Taking this $q(x^*)$ into account, eq(3.126) is rewritten as

$$\hat{m}(x^*, x^*) = q(x^*)^t y = \sum_{j=1}^{n} q_j(x^*) Y_j. \tag{3.128}$$

Therefore, $\{q_j(x^*)\}$ is regarded as an equivalent kernel for local polynomial regression. Multiplication of $\mathbf{X}$ from the right on both sides of eq(3.127) leads to

$$\begin{aligned} q(x^*)^t \mathbf{X} &= e_1^t (\mathbf{X}^t \mathbf{W} \mathbf{X})^{-1} \mathbf{X}^t \mathbf{W} \mathbf{X} \\ &= e_1^t. \end{aligned} \tag{3.129}$$

Since all $[\mathbf{X}]_{i1}$ $(1 \le i \le n)$ are 1 (eq(3.117) (page 140)), we obtain

$$\sum_{i=1}^{n} q_i(x^*) = 1. \tag{3.130}$$

This is assured by the fact that if all $\{Y_j\}$ are 1, the estimate is 1 regardless of the value of x^*. Eq(3.128) and eq(3.130) indicate that an estimate given by local polynomial regression is a weighted average of $\{Y_j\}$.

When $x^* = X_k$, eq(3.128) becomes

$$\hat{m}(X_k, X_k) = \hat{Y}_k = \sum_{j=1}^{n} q_j(X_k) Y_j, \tag{3.131}$$

where $q_j(X_k)$ plays the role of the kj element of a hat matrix. That is, $q(x^*)^t$ is a generalized concept of a row vector constituting a hat matrix. While a hat matrix $(\mathbf{H}_{kj})$ represents the amount of influence of a datum (Y_j) on the estimate $\hat{Y}_k$, $q_j(x^*)$ shows the degree of the effect of a datum (Y_j) on the estimate $\hat{m}(x^*, x^*)$ $(= \hat{a}_0(x^*))$.

Using eq(3.127), an element of a hat matrix is

$$\begin{aligned} [\mathbf{H}]_{kj} &= q_j(X_k) \\ &= [e_1^t (\mathbf{X}^t \mathbf{W} \mathbf{X})^{-1} \mathbf{X}^t \mathbf{W}]_j \\ &= [(\mathbf{X}^t \mathbf{W} \mathbf{X})^{-1} \mathbf{X}^t \mathbf{W}]_{1j} \\ &= \sum_{i=1}^{n} [(\mathbf{X}^t \mathbf{W} \mathbf{X})^{-1} \mathbf{X}^t]_{1i} [\mathbf{W}]_{ij} \\ &= [(\mathbf{X}^t \mathbf{W} \mathbf{X})^{-1} \mathbf{X}^t]_{1j} [\mathbf{W}]_{jj} \\ &= [(\mathbf{X}^t \mathbf{W} \mathbf{X})^{-1} \mathbf{X}^t]_{1j} \cdot w \left(\frac{X_j - X_k}{h} \right) \\ &= e_1^t (\mathbf{X}^t \mathbf{W} \mathbf{X})^{-1} \mathbf{X}^t e_j' \cdot w \left(\frac{X_j - X_k}{h} \right), \end{aligned} \tag{3.132}$$

where e_j' is a column vector (a unit vector with n elements). Only the jth element is 1; the others are 0. It should be noted here that $\mathbf{W}$ and $\mathbf{X}$ in eq(3.132), eq(3.133), and eq(3.148) are those when $x^* = X_k$. Since the kth column of $\mathbf{X}^t$ is $\mathbf{e}'_1$, the diagonal elements of a hat matrix are

$$
\begin{aligned}
[\mathbf{H}]_{kk} &= q_k(X_k) \\
&= \mathbf{e}_1^t \left(\mathbf{X}^t \mathbf{W} \mathbf{X} \right)^{-1} \mathbf{X}^t \mathbf{e}'_k \cdot w(0) \\
&= [(\mathbf{X}^t \mathbf{W} \mathbf{X})^{-1} \mathbf{X}^t]_{1k} \cdot w(0) \\
&= \sum_{i=0}^{p} [(\mathbf{X}^t \mathbf{W} \mathbf{X})^{-1}]_{1i} [\mathbf{X}^t]_{ik} \cdot w(0) \\
&= \sum_{i=0}^{p} [(\mathbf{X}^t \mathbf{W} \mathbf{X})^{-1}]_{1i} [\mathbf{e}'_1]_i \cdot w(0) \\
&= [(\mathbf{X}^t \mathbf{W} \mathbf{X})^{-1}]_{11} \cdot w(0).
\end{aligned}
\tag{3.133}
$$

In ordinary local polynomial regression, $[\mathbf{H}]_{kk}$ takes a smaller value as $w(\frac{X_i - X_k}{h})$ broadens out. For instance, a large value of h in eq(3.113) (page 140), eq(3.114), eq(3.115), and eq(3,116) yields a small value of $[\mathbf{H}]_{kk}$. This implies that h has the characteristic of a smoothing parameter. To prove that the value of $[\mathbf{H}]_{kk}$ does not become larger when $w(\frac{X_i - X_k}{h})$ broadens out, $w(\frac{X_i - X_k}{h})$ with a certain breadth is set to be $[\mathbf{H}^{narrow}]_{kk}$ and $w(\frac{X_i - X_k}{h})$ with a wider breadth is defined as $[\mathbf{H}^{wide}]_{kk}$. Then, we prove

$$
[\mathbf{H}^{wide}]_{kk} \le [\mathbf{H}^{narrow}]_{kk}.
\tag{3.134}
$$

Without loss of generality, we assume that $w(0)$ which provides $[\mathbf{H}^{wide}]_{kk}$ is identical to $w(0)$ which gives $[\mathbf{H}^{narrow}]_{kk}$. This assumption allows us to rewrite eq(3.133) as

$$
\begin{aligned}
[\mathbf{H}^{wide}]_{kk} &= [(\mathbf{X}^t \mathbf{W} \mathbf{X} + \mathbf{X}^t \mathbf{\Delta} \mathbf{W} \mathbf{X})^{-1}]_{11} \cdot w(0), \\
[\mathbf{H}^{narrow}]_{kk} &= [(\mathbf{X}^t \mathbf{W} \mathbf{X})^{-1}]_{11} \cdot w(0),
\end{aligned}
\tag{3.135}
$$

where $\mathbf{W}$ is a diagonal matrix in which $w(\frac{X_i - X_k}{h})$ is a diagonal element; this $\mathbf{W}$ creates $[\mathbf{H}^{narrow}]_{kk}$. Furthermore, $\mathbf{\Delta}\mathbf{W}$ is $w(\frac{X_i - X_k}{h})$ which gives $[\mathbf{H}^{wide}]_{kk}$ minus $w(\frac{X_i - X_k}{h})$ which gives $[\mathbf{H}^{narrow}]_{kk}$. Hence, $\mathbf{\Delta}\mathbf{W}$ is a diagonal matrix in which a diagonal element is either a positive value or 0 and the kth diagonal element is 0.

To offer proof of eq(3.134), we prove the inequality

$$
\mathbf{v}^t (\mathbf{B} + \alpha_1 \mathbf{C})^{-1} \mathbf{v} \ge \mathbf{v}^t (\mathbf{B} + \alpha_2 \mathbf{C})^{-1} \mathbf{v},
\tag{3.136}
$$

where α_1 and α_2 are constants that satisfy $0 < \alpha_1 < \alpha_2$, and $\mathbf{v}$ is a column vector with n elements. Both $\mathbf{B}$ and $\mathbf{C}$ are symmetric matrices (with size $n \times n$), and each of them is a quasi-definite matrix (positive semidefinite matrix, nonnegative definite matrix) (i.e., a symmetric matrix in which all eigenvalues (characteristic values) are either positive or 0). Furthermore, it is assumed that there exist inverse matrices of $(\mathbf{B} + \alpha_1 \mathbf{C})$ and $(\mathbf{B} + \alpha_2 \mathbf{C})$.

Note that $\mathbf{B}$ is a quasi-definite matrix if and only if there exists $\mathbf{F}$ that satisfies

$$
\mathbf{B} = \mathbf{F}^t \mathbf{F}.
\tag{3.137}
$$

Since $\mathbf{C}$ is a symmetric matrix, $\mathbf{U}$ (orthogonal matrix), which satisfies $\mathbf{C} = \mathbf{U}\mathbf{A}\mathbf{U}^t$ ($\mathbf{A}$ is a diagonal matrix), is obtainable. The diagonal elements of $\mathbf{A}$ are positive or 0 because $\mathbf{C}$ is a quasi-definite matrix. Hence, the left-hand side of eq(3.136) is written as

$$\mathbf{v}^t(\mathbf{B} + \alpha_1\mathbf{U}\mathbf{A}\mathbf{U}^t)^{-1}\mathbf{v}. \tag{3.138}$$

Next, if we set $\mathbf{B}_2 = \mathbf{U}^t\mathbf{B}\mathbf{U}$, eq(3.138) becomes

$$\mathbf{v}^t\left(\mathbf{U}(\mathbf{B}_2 + \alpha_1\mathbf{A})\mathbf{U}^t\right)^{-1}\mathbf{v} = \mathbf{v}^t\mathbf{U}(\mathbf{B}_2 + \alpha_1\mathbf{A})^{-1}\mathbf{U}^t\mathbf{v}. \tag{3.139}$$

Furthermore, a diagonal matrix $\mathbf{A}_2$, whose diagonal elements are identical to the positive square roots of diagonal elements of $\mathbf{A}$, is set, and $\mathbf{B}_3$ is defined as $\mathbf{B}_3 = \mathbf{A}_2^{-1}\mathbf{B}_2\mathbf{A}_2^{-1}$. Then, eq(3.139) becomes

$$\mathbf{v}^t\mathbf{U}(\mathbf{A}_2\mathbf{B}_3\mathbf{A}_2 + \alpha_1\mathbf{A}_2^2)^{-1}\mathbf{U}^t\mathbf{v} = \mathbf{v}^t\mathbf{U}\mathbf{A}_2^{-1}(\mathbf{B}_3 + \alpha_1\mathbf{I})^{-1}\mathbf{A}_2^{-1}\mathbf{U}^t\mathbf{v}. \tag{3.140}$$

When the eigenvalues of $\mathbf{B}_3$ are defined as $\{b_i\}$, the eigenvalues of $(\mathbf{B}_3 + \alpha_1\mathbf{I})^{-1}$ are found to be $\{\frac{1}{b_i+\alpha_1}\}$. Hence, eq(3.140) is rewritten as

$$\mathbf{v}^t\mathbf{U}\mathbf{A}_2^{-1}\mathbf{U}' \begin{pmatrix} \dfrac{1}{b_1+\alpha_1} & 0 & \cdots & 0 \\ 0 & \dfrac{1}{b_2+\alpha_1} & \cdots & 0 \\ \vdots & \vdots & \ddots & \vdots \\ 0 & 0 & \cdots & \dfrac{1}{b_n+\alpha_1} \end{pmatrix} \mathbf{U}'^t\mathbf{A}_2^{-1}\mathbf{U}^t\mathbf{v}$$

$$= \mathbf{v}'^t \begin{pmatrix} \dfrac{1}{b_1+\alpha_1} & 0 & \cdots & 0 \\ 0 & \dfrac{1}{b_2+\alpha_1} & \cdots & 0 \\ \vdots & \vdots & \ddots & \vdots \\ 0 & 0 & \cdots & \dfrac{1}{b_n+\alpha_1} \end{pmatrix} \mathbf{v}', \tag{3.141}$$

where $\mathbf{U}'$ is an orthogonal matrix used to orthogonalize $(\mathbf{B}_3 + \alpha_1\mathbf{I})^{-1}$, and we set $\mathbf{v}' = \mathbf{U}'^t\mathbf{A}_2^{-1}\mathbf{U}^t\mathbf{v}$. If we define $\mathbf{v}'$ as $\mathbf{v}' = (v_1', v_2', \ldots, v_n')^t$, eq(3.141) becomes

$$\sum_{i=1}^{n} \frac{(v_i')^2}{b_i + \alpha_1}. \tag{3.142}$$

Since $\mathbf{B}$ is a quasi-definite matrix, $\mathbf{B}_2$ and $\mathbf{B}_3$ are represented in a similar manner to eq(3.137). Namely, $\mathbf{B}_3$ is a quasi-definite matrix. Hence, because all of $\{b_i\}$ are positive or 0, we have

$$\frac{(v_i')^2}{b_i + \alpha_1} \geq \frac{(v_i')^2}{b_i + \alpha_2}. \tag{3.143}$$

Therefore, we obtain the relationship

$$\sum_{i=1}^{n} \frac{(v_i')^2}{b_i + \alpha_1} \geq \sum_{i=1}^{n} \frac{(v_i')^2}{b_i + \alpha_2}. \tag{3.144}$$

Thus, eq(3.136) is derived (Q.E.D.).

Next, both $(\mathbf{X}^t\mathbf{WX})$ and $(\mathbf{X}^t\boldsymbol{\Delta}\mathbf{WX})$ are quasi-definite matrices; this is proved by

$$
\begin{aligned}
\mathbf{X}^t\mathbf{WX} &= \mathbf{X}^t\mathbf{W}'\mathbf{W}'\mathbf{X} \\
&= (\mathbf{W}'\mathbf{X})^t(\mathbf{W}'\mathbf{X}),
\end{aligned}
\tag{3.145}
$$

where $\mathbf{W}'$ is a diagonal matrix whose diagonal elements are square roots of those of $\mathbf{W}$. Through the use of eq(3.137), $\mathbf{X}^t\mathbf{WX}$ is found to be a quasi-definite matrix. Then, since the inverse matrix of a quasi-definite matrix is a quasi-definite matrix, $(\mathbf{X}^t\mathbf{WX})^{-1}$ is also a quasi-definite matrix.

The same is true of $(\mathbf{X}^t\boldsymbol{\Delta}\mathbf{WX})$, as is shown below:

$$
\begin{aligned}
\mathbf{X}^t\boldsymbol{\Delta}\mathbf{WX} &= \mathbf{X}^t\boldsymbol{\Delta}\mathbf{W}'\boldsymbol{\Delta}\mathbf{W}'\mathbf{X} \\
&= (\boldsymbol{\Delta}\mathbf{W}'\mathbf{X})^t(\boldsymbol{\Delta}\mathbf{W}'\mathbf{X}),
\end{aligned}
\tag{3.146}
$$

where $\boldsymbol{\Delta}\mathbf{W}'$ is a diagonal matrix whose diagonal elements are equal to the positive square roots of the diagonal elements of $\boldsymbol{\Delta}\mathbf{W}$. Hence, $(\mathbf{X}^t\boldsymbol{\Delta}\mathbf{WX})^{-1}$ is also a quasi-definite matrix.

Therefore, in eq(3.136) (page 144), we set $\alpha_1 = 0$, $\alpha_2 = 1$, $\mathbf{B} = \mathbf{X}^t\mathbf{WX}$, $\mathbf{C} = \mathbf{X}^t\boldsymbol{\Delta}\mathbf{WX}$, and $\mathbf{v} = \mathbf{e}'_1$. Thus, eq(3.134) (page 144) is obtained. Consequently, it is proved that broadening $w(X_i - X_k)$ does not increase the value of $\mathbf{H}_{kk}$.

Figure 3.15 displays a hat matrix of local linear regression; $n = 20$ and $\{X_i\} = \{1, 2, 3, \ldots, 20\}$. A Gaussian kernel is used and the bandwidths are $h = 3$ and $h = 6$. The influence of the data near the two ends on estimates is substantially different from that of the Nadaraya-Watson estimator. Figure 3.16 is based on the setup that a predictor is not equispaced but $\{X_i\} = \{1^{1.8}, 2^{1.8}, 3^{1.8}, \ldots, 20^{1.8}\}$. The bandwidths are $h = 30$ and $h = 60$, respectively. For figure 3.17, we use some of the predictor values as those in figure 3.16; some elements of a hat matrix with $h = 25$ are illustrated: $\{[\mathbf{H}]_{1j}\}$ $(1 \le j \le 20)$, $\{[\mathbf{H}]_{6j}\}$, $\{[\mathbf{H}]_{11j}\}$, $\{[\mathbf{H}]_{16j}\}$, and $\{[\mathbf{H}]_{20j}\}$. To represent relative weights, normalization is carried out to obtain $\mathbf{H}'$ (figure 3.17(right)) in which the maximal value of "a" becomes 1, that of "b" also becomes 1, and the same is true of "c," "d," and "e." As is expected from the definition of local polynomial regression, the broadness which $\mathbf{H}'$ shows is not affected by the density of the distribution of the predictor of data.

Thus, since a hat matrix corresponding to local polynomial regression can be defined, use of CV or GCV optimizes the bandwidth (h). Figure 3.18 displays the results of the calculations of CV and GCV using the data in figure 3.10 (page 136) and a Gaussian kernel function. The bandwidths optimized by CV and GCV are $h = 0.25$ and $h = 0.2$, respectively. However, it hardly matters which result is adopted. Comparison of figure 3.19(left) and figure 3.19(right) shows that the estimates are not greatly affected by the selection of the bandwidth. Furthermore, comparison of figure 3.19(left), figure 3.19(right), and figure 3.10 (page 136) indicates that, although most estimates are not very different, some estimates close to the two ends of the data region in figure 3.10 are slightly larger than those imagined intuitively from the data. This is called boundary bias. When the Nadaraya-Watson estimator is used, estimation near the left end of the data region, for example, provides estimates that take values close to those of data on the right-hand side of the estimation point,

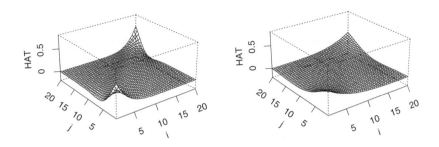

Figure 3.15   Hat matrix corresponding to local linear regression using a Gaussian kernel; $n = 20$, $\{X_i\} = \{1, 2, 3, \ldots, 20\}$. $h = 3$ (left). $h = 6$ (right).

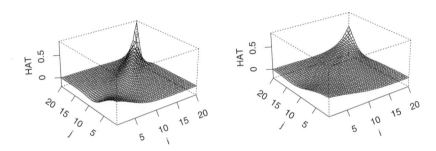

Figure 3.16   Hat matrix corresponding to local linear regression using a Gaussian kernel; $n = 20$, $\{X_i\} = \{1^{1.8}, 2^{1.8}, 3^{1.8}, \ldots, 20^{1.8}\}$. $h = 30$ (left). $h = 60$ (right).

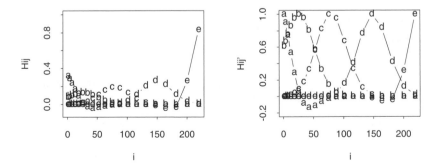

Figure 3.17   Some elements of a hat matrix when $h = 25$. "a" indicates $\{[\mathbf{H}]_{1j}\}$, "b" indicates $\{[\mathbf{H}]_{6j}\}$, "c" indicates $\{[\mathbf{H}]_{11j}\}$, "d" indicates $\{[\mathbf{H}]_{16j}\}$, and "e" indicates $\{[\mathbf{H}]_{20j}\}$ (left). The values of "a," "b," "c," "d," and "e" are normalized to make each maximal value 1 ($\mathbf{H}'$) (right).

because no data are available on the left-hand side of that point. On the other hand, this problem is alleviated by the use of local polynomial regression because it assumes that the trend of data to the right of the estimation point can be extended to data on the left-hand side. Small boundary bias is one of the beneficial characteristics of local polynomial regression.

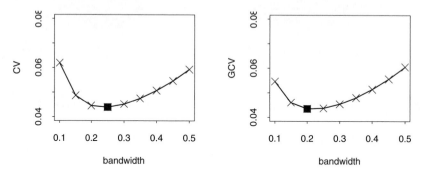

Figure 3.18   Results of calculations of CV and GCV using the same data as those in figure 3.10.

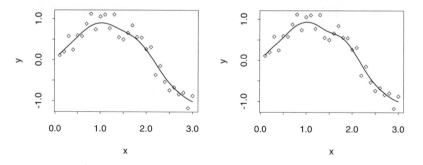

Figure 3.19   Estimates (solid line) and data ($\Diamond$) when local linear regression with a Gaussian kernel function is used. $h = 0.25$ (left). $h = 0.2$ (right).

Next, the variance of estimates obtained by local polynomial regression is discussed. Variance of $\hat{m}(x^*, x^*)$ is written as

$$
\begin{aligned}
Var[\hat{m}(x^*, x^*)] &= E[(\hat{m}(x^*, x^*) - E[\hat{m}(x^*, x^*)])^2] \\
&= E\left[\left(\sum_{i=1}^{n} q_i(x^*)(m(X_i) + \epsilon_i) - \sum_{i=1}^{n} q_i(x^*)m(X_i)\right) \right. \\
&\quad \left. \cdot \left(\sum_{j=1}^{n} q_j(x^*)(m(X_i) + \epsilon_j) - \sum_{j=1}^{n} q_j(x^*)m(X_i)\right)\right] \\
&= E\left[\sum_{i=1}^{n} \sum_{j=1}^{n} q_i(x^*)q_j(x^*)\epsilon_i\epsilon_j\right]
\end{aligned}
$$

$$= \sum_{i=1}^{n} \sum_{j=1}^{n} q_i(x^*) q_j(x^*) \delta_{ij} \sigma^2$$

$$= \sigma^2 \cdot \sum_{i=1}^{n} q_i(x^*)^2. \qquad (3.147)$$

This is the generalized version of eq(3.53) (page 120).

Now, let us prove $Var[\hat{m}(X_k, X_k)] \leq \sigma^2$. Since variance is σ^2 if Y_k is used as $\hat{Y}_k$, this inequality means that the variance of $\hat{Y}_k$ obtained by local polynomial regression is smaller than that of the estimates (intact $\{Y_k\}$); these two variances are identical in a special situation.

Eq(3.53) and eq(3.132) (page 143) yield the calculation

$$E[(\hat{m}(X_k) - E[\hat{m}(X_k)])^2]$$

$$= Var[\hat{m}(X_k)]$$

$$= \sigma^2 \cdot \sum_{j=1}^{n} ([\mathbf{H}]_{kj})^2$$

$$= \sigma^2 \sum_{j=1}^{n} \mathbf{e}_1^t \left((\mathbf{X}^t \mathbf{W} \mathbf{X})^{-1} \mathbf{X}^t \right) \mathbf{e}_j' \cdot w \left(\frac{X_j - X_k}{h} \right) \cdot$$

$$\mathbf{e}_1^t \left((\mathbf{X}^t \mathbf{W} \mathbf{X})^{-1} \mathbf{X}^t \right) \mathbf{e}_j' \cdot w \left(\frac{X_j - X_k}{h} \right)$$

$$= \sigma^2 \sum_{j=1}^{n} \mathbf{e}_1^t \left((\mathbf{X}^t \mathbf{W} \mathbf{X})^{-1} \mathbf{X}^t \right) \mathbf{e}_j' (\mathbf{e}_j')^t \left(\mathbf{X} (\mathbf{X}^t \mathbf{W} \mathbf{X})^{-1} \right) \mathbf{e}_1 \cdot w \left(\frac{X_j - X_k}{h} \right)^2$$

$$\leq \sigma^2 \sum_{j=1}^{n} \mathbf{e}_1^t \left((\mathbf{X}^t \mathbf{W} \mathbf{X})^{-1} \mathbf{X}^t \right) \mathbf{e}_j' (\mathbf{e}_j')^t \left(\mathbf{X} (\mathbf{X}^t \mathbf{W} \mathbf{X})^{-1} \right) \mathbf{e}_1 \cdot w \left(\frac{X_j - X_k}{h} \right)$$

$$\cdot w(0)$$

$$= \sigma^2 \mathbf{e}_1^t \left((\mathbf{X}^t \mathbf{W} \mathbf{X})^{-1} \mathbf{X}^t \mathbf{W} \mathbf{X} (\mathbf{X}^t \mathbf{W} \mathbf{X})^{-1} \right) \mathbf{e}_1 \cdot w(0)$$

$$= \sigma^2 \mathbf{e}_1^t \left((\mathbf{X}^t \mathbf{W} \mathbf{X})^{-1} \right) \mathbf{e}_1 \cdot w(0)$$

$$= \sigma^2 [(\mathbf{X}^t \mathbf{W} \mathbf{X})^{-1}]_{11} \cdot w(0)$$

$$= \sigma^2 [\mathbf{H}]_{kk}, \qquad (3.148)$$

where the equality between the third to fourth line and the fifth line is derived as a result of $\mathbf{e}_1^t \left((\mathbf{X}^t \mathbf{W} \mathbf{X})^{-1} \mathbf{X}^t \right) \mathbf{e}_j'$ being a scalar and hence unaffected by the transpose (^{t}) operation. The inequality between the fifth line and the sixth to seventh line holds because $w \left(\frac{X_k - X_k}{h} \right)$ ($= w(0)$) is largest among $\{w \left(\frac{X_j - X_k}{h} \right)\}$ ($1 \leq j \leq n$). The equality between the sixth to seventh line and the eighth line uses $\mathbf{W} = \sum_{j=1}^{n} \mathbf{e}_j' (\mathbf{e}_j')^t \cdot w \left(\frac{X_j - X_k}{h} \right)$. Eq(3.148) leads to the relationship

$$\sum_{j=1}^{n} ([\mathbf{H}]_{kj})^2 \leq [\mathbf{H}]_{kk}. \qquad (3.149)$$

Taking account of $([\mathbf{H}]_{kk})^2 \le \sum_{j=1}^{n}([\mathbf{H}]_{kj})^2$, we obtain

$$0 \le ([\mathbf{H}]_{kk})^2 \le [\mathbf{H}]_{kk}. \qquad (3.150)$$

Hence, we have the relationship

$$0 \le [\mathbf{H}]_{kk} \le 1. \qquad (3.151)$$

Substitition of this into eq(3.148) yields

$$Var[\hat{m}(X_k)] \le \sigma^2. \qquad (3.152)$$

Thus, it is proved that the variance of estimates smoothed by local polynomial regression is smaller than that of the error of data; the two variances are identical under a special circumstance.

Furthermore, bias of $\hat{m}(x^*, x^*)$ is calculated as follows.

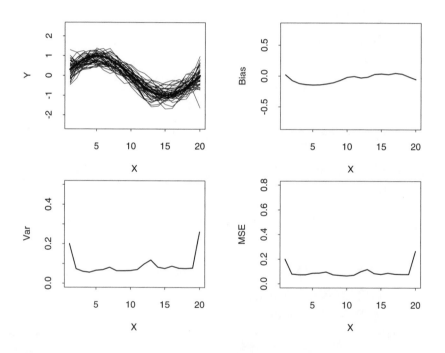

Figure 3.20   Trade-off of bias versus variance in local linear regression with a Gaussian kernel. Comparison of $m(x)$ and estimates (top left). Bias $(Bias[\hat{Y}_i])$ (top right). Variance $(Var[\hat{Y}_i])$ (bottom left). Mean squared error $(MSE[\hat{Y}_i])$ (bottom right).

Using eq(3.2) (page 105), bias of $\hat{m}(x^*, x^*)$ is defined as

$$Bias[\hat{m}(x^*, x^*)] = E[\hat{m}(x^*, x^*)] - m(x^*, x^*). \qquad (3.153)$$

Eq(3.126) (page 142) and $E[\epsilon_i] = 0 (1 \le i \le n)$ enable $E[\hat{m}(x^*, x^*)]$ to be rewritten as

$$E[m(x^*, x^*)] = \mathbf{e}_1^t (\mathbf{X}^t \mathbf{W} \mathbf{X})^{-1} \mathbf{X}^t \mathbf{W} \mathbf{m}, \qquad (3.154)$$

where $\mathbf{m} = (m(X_1), m(X_2), \ldots, m(X_n))^t$ is a column vector (with n elements) indicating the values of $m(x)$ at the locations of $\{X_i\}$. Next, Taylor expansion of $m(x)$ in the neighborhood of x^* results in

$$
\begin{aligned}
m(x) &= m(x^*) + m^{(1)}(x^*)(x - x^*) + \frac{m^{(2)}(x^*)}{2}(x - x^*)^2 + \cdots \\
&\quad + \frac{m^{(p)}(x^*)}{p!}(x - x^*)^p + r(x),
\end{aligned}
\tag{3.155}
$$

where $m^{(p)}(x)$ is the pth derivative of $m(x)$ with respect to x. $p!$ denotes the factorial of p. $r(x)$ is a residual which is $m(x)$ minus an approximation of terms up to the pth power produced by Taylor expansion. Therefore, eq(3.154) is transformed into

$$
\begin{aligned}
E[\hat{m}(x^*, x^*)] &= \mathbf{e}_1^t(\mathbf{X}^t\mathbf{W}\mathbf{X})^{-1}\mathbf{X}^t\mathbf{W}\mathbf{X}\beta + \mathbf{e}_1^t(\mathbf{X}^t\mathbf{W}\mathbf{X})^{-1}\mathbf{X}^t\mathbf{W}\mathbf{r} \\
&= m(x^*) + \mathbf{e}_1^t(\mathbf{X}^t\mathbf{W}\mathbf{X})^{-1}\mathbf{X}^t\mathbf{W}\mathbf{r},
\end{aligned}
\tag{3.156}
$$

where $\beta = \left(m(x^*), m^{(1)}(x^*), \frac{m^{(2)}(x^*)}{2}, \ldots, \frac{m^{(p)}(x^*)}{p!}\right)^t$, and $\mathbf{r} = (r(X_1), r(X_2), \ldots, r(X_n))^t$.

Then, we have

$$
Bias[\hat{m}(x^*, x^*)] = \mathbf{e}_1^t(\mathbf{X}^t\mathbf{W}\mathbf{X})^{-1}\mathbf{X}^t\mathbf{W}\mathbf{r}.
\tag{3.157}
$$

That is, $Bias[\hat{m}(x^*, x^*)]$ is $\hat{r}(x^*)$, which is obtained by local polynomial regression of the residuals: $m(x)$ minus approximations by terms up to the pth power produced by Taylor expansion around x^*. This local polynomial regression is carried out in the same way as the calculation of $\hat{m}(x^*, x^*)$.

Figure 3.20 displays variance, bias, and mean squared error obtained using local linear regression with a Gaussian kernel. The data and the numerical procedure are the same as those in the case of figure 3.13 (page 138). This graph is characterized by the marked tendency that the value of variance increases in the vicinity of the two ends. As was mentioned previously, it is the advantage of local linear regression that bias near the two ends is small. However, this is at the cost of increased variance close to the two ends. This viewpoint also exemplifies the trade-off of bias and variance.

3.6 NATURAL SPLINE AND SMOOTHING SPLINE

Let us define the function below in an infinite region $(-\infty < x < \infty)$:

$$
s_\infty(x) = \tilde{a}_0 + \tilde{a}_1 x + \tilde{a}_2 x^2 + \tilde{a}_3 x^3 + \sum_{j=1}^{n} \tilde{b}_j \cdot ((x - X_j)_+)^3,
\tag{3.158}
$$

where $\tilde{a}_0, \tilde{a}_1, \tilde{a}_2, \tilde{a}_3, \tilde{b}_1, \tilde{b}_2, \ldots,$ and $\tilde{b}_n$ are constants. $\{X_i\}$ $(X_1 < X_2 < \cdots < X_n)$ are knots (breakpoints). $((\cdot)_+)^3$ is a sort of truncated power function. $(x)_+$ is defined as

$$
(x)_+ = \begin{cases} x & \text{if} \quad x \geq 0 \\ 0 & \text{if} \quad x < 0. \end{cases}
\tag{3.159}
$$

$s_\infty(x)$ is expressed as a cubic function throughout the whole region (i.e., at every x in $-\infty < x < \infty$). At the points of knots, the value of $s_\infty(x)$, the first derivative function of $s_\infty(x)$, and the second derivative function of $s_\infty(x)$ are continuous (figure 3.21). Since this is a function represented by piecewise cubic functions, $s_\infty(x)$ is called the cubic spline function. A spline function with another degree can be defined. However, since the cubic spline function is most commonly used, it is sometimes simply called the spline function; we call it spline hereafter. $s_\infty(x)$ is rewritten as

$$s_\infty(x) = \bar{a}_0 + \bar{a}_1 x + \bar{a}_2 x^2 + \bar{a}_3 x^3 + \sum_{j=1}^{n} \bar{b}_j \cdot \mid x - X_j \mid^3 . \tag{3.160}$$

By the use of $((x)_+)^3 = \frac{1}{2}(x^3 + \mid x \mid^3)$, the relationship between eq(3.158) and eq(3.160) is found:

$$\tilde{a}_0 + \tilde{a}_1 x + \tilde{a}_2 x^2 + \tilde{a}_3 x^3 = \bar{a}_0 + \bar{a}_1 x + \bar{a}_2 x^2 + \bar{a}_3 x^3 - \sum_{j=1}^{n} \bar{b}_j \cdot (x - X_j)^3$$

$$\tilde{b}_j = 2\bar{b}_j. \tag{3.161}$$

That is, eq(3.158) is equivalent to eq(3.160).

Figure 3.21 exemplifies eq(3.160); $n = 6$ is set, namely, the number of knots is 6 (the number of piecewise cubic functions is 7). The first derivative of this function, the second derivative of this function, and the third derivative of this function are also drawn in figure 3.21. These graphs show the following features.

(1) The first derivative of this function consists of piecewise quadratic functions. The value of the derivative function and its first derivative are continuous at the knots.

(2) The second derivative of the function consists of piecewise linear functions. The value of the derivative function is continuous at the knots.

(3) The third derivative of the function consists of piecewise constants. The value of the derivative function is discontinuous at knots.

These features are clear from eq(3.160). The spline in figure 3.21 makes use of 10 coefficients $((\bar{a}_0, \bar{a}_1, \bar{a}_2, \bar{a}_3, \bar{b}_1, \ldots, \bar{b}_6))$ to interpolate 6 data. Hence, there are an infinite number of splines that have knots at the same places as this spline and whose values at knots are identical to those of this spline. Figure 3.22 shows an example of this. Although this spline has knots at the same places and with the same values as those in figure 3.21, the shape of the function is quite different. The first derivative of the function, the second derivative of the function, and the third derivative of the function are also profoundly different.

Furthermore, the behavior of the third derivative function in figure 3.21 (bottom right) implies that the spline in figure 3.21 uses 10 coefficients. This third derivative of the function is derived with 7 constants. Since the spline is a result of three integrations, 3 integral constants are added. Therefore, the spline in figure 3.21(top left) is depicted by 10 coefficients. Generalization of this discussion leads to the background to the descriptions of eq(3.158) and eq(3.160), both of which use $(n+4)$ coefficients.

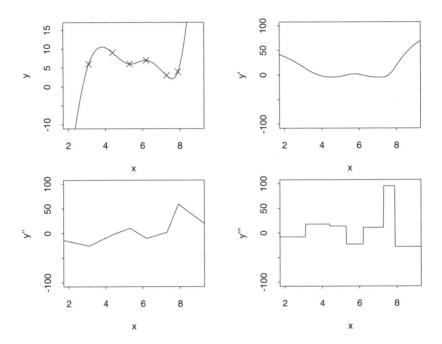

Figure 3.21   An example of a spline in an infinite region (number of knots is 6). In addition to the curve of the spline, × symbols are plotted to show the positions of knots and the values of the function at the knots (top left). First derivative of the function (top right), second derivative of the function (bottom left), and third derivative of the function (bottom right).

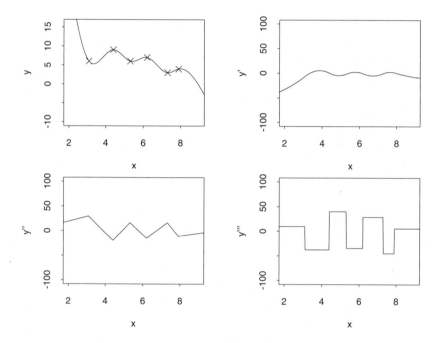

Figure 3.22 An example of a spline in an infinite region (number of knots is 6). In addition to the curve of the spline, × symbols are plotted to show the positions of knots and the values of the function at the knots (top left). First derivative of the function (top right), second derivative of the function (bottom left), and third derivative of the function (bottom right).

In accordance with the description above, a cubic spline in a finite region ($X_1 \leq x \leq X_n$) is defined as

$$s_{finite}(x) = \tilde{a}_0 + \tilde{a}_1 x + \tilde{a}_2 x^2 + \tilde{a}_3 x^3 + \sum_{j=2}^{n-1} \tilde{b}_j \cdot ((x - X_j)_+)^3, \qquad (3.162)$$

$$s_{finite}(x) = \bar{a}_0 + \bar{a}_1 x + \bar{a}_2 x^2 + \bar{a}_3 x^3 + \sum_{j=2}^{n-1} \bar{b}_j \cdot \mid x - X_j \mid^3, \qquad (3.163)$$

where the number of coefficients is reduced by 2 because the values of the third derivative of the function (i.e., the gradient of the second derivative of the function) outside the knots at the two ends (X_1 and X_n) no longer need to be given. Hence, eq(3.163) is rewritten as

$$s_{finite}(x) = a_0 + a_1 x + \frac{1}{12} \sum_{j=1}^{n} b_j \cdot \mid x - X_j \mid^3 . \qquad (3.164)$$

The sudden appearance of $\frac{1}{12}$ will be justified later. Eq(3.163) is transformed into eq(3.164) in the following manner:

$$\bar{a}_0 + \bar{a}_1 x + \bar{a}_2 x^2 + \bar{a}_3 x^3 = a_0 + a_1 x + \frac{1}{12} b_1 \cdot (x - X_1)^3 - \frac{1}{12} b_n \cdot (x - X_n)^3,$$

$$\bar{b}_j = \frac{1}{12} b_j \qquad (2 \leq j \leq n - 1). \qquad (3.165)$$

Eq(3.162), eq(3.163), and eq(3.164) are equivalent to each other. That is, a function that is represented by one of these three equations can be represented by either of the remaining two. Hence, discussion is essentially unaffected by the choice of the function. In the description hereafter, we adopt eq(3.164) to represent a cubic spline in the region of $X_1 \leq x \leq X_n$ because it facilitates the expansion of the natural spline in a finite region to that in an infinite region, it relates splines to physical phenomena easily, and it associates splines with those accompanying the plurality of predictors.

The first derivative, the second derivative, the third derivative, and the fourth derivative of the function in eq(3.164) are

$$\frac{ds_{finite}(x)}{dx} = a_1 + \frac{1}{4} \sum_{j=1}^{n} b_j \cdot sign(x - X_j) \cdot (x - X_j)^2, \qquad (3.166)$$

$$\frac{d^2 s_{finite}(x)}{dx^2} = \frac{1}{2} \sum_{j=1}^{n} b_j \cdot \mid x - X_j \mid, \qquad (3.167)$$

$$\frac{d^3 s_{finite}(x)}{dx^3} = \frac{1}{2} \sum_{j=1}^{n} sign(x - X_j) \cdot b_j, \qquad (3.168)$$

$$\frac{d^4 s_{finite}(x)}{dx^4} = \sum_{j=1}^{n} b_j \cdot \delta(x - X_j), \qquad (3.169)$$

where $\delta(\cdot)$ is a delta function, and $sign(\cdot)$ is defined as

$$sign(x) = \begin{cases} 1 & x > 0 \\ 0 & x = 0 \\ -1 & x < 0. \end{cases} \tag{3.170}$$

Figure 3.23 displays an example of a spline in a finite region. Expression of this curve using eq(3.164) yields $n = 6$. Thus, this spline consists of 5 piecewise cubic equations, and hence, 8 coefficients $((a_0, a_1, b_1, \ldots, b_6))$ are required to represent these cubic equations. Comparing figure 3.21 with figure 3.23, it is seen that while the number of knots is identical, the number of coefficients needed is reduced from 10 to 8 because the value of the function is not defined outside the two ends. However, in the region between the two ends, the curve in figure 3.21(top left) is identical to that in figure 3.23(top left), and the curve in figure 3.22(top left) is the same as that in figure 3.24(top left). Much the same is true for the first derivative, the second derivative, and the third derivative.

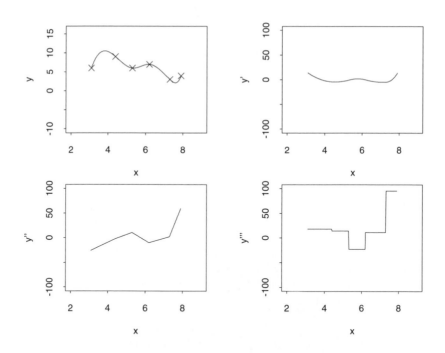

Figure 3.23   An example of a spline in a finite region (number of knots is 6). In addition to the curve of the spline, × symbols are plotted to show the positions of knots and the values of the function at knots (top left). First derivative of the function (top right). Second derivative of the function (bottom left). Third derivative of the function (bottom right).

In this context, the main target of dealing with the spline is the derivation of a smoothing spline. Hence, the predictor values of data are identical to the positions of data. However, the knots that are used to carry out interpolation using the spline can be

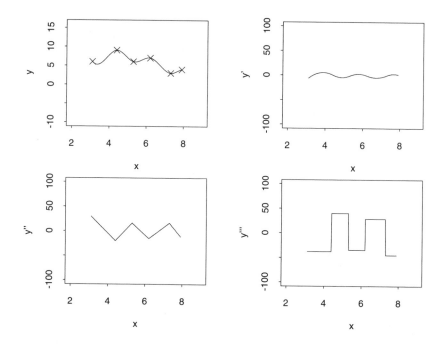

Figure 3.24 An example of a spline in a finite region (number of knots is 6). Although the values at knots coincide with those in figure 3.23, the entire shape of the function is different. In addition to the curve of the spline, × symbols are plotted to show the positions of knots and the values of the function at knots (top left). First derivative of the function (top right). Second derivative of the function (bottom left). Third derivative of the function (bottom right).

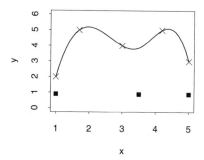

Figure 3.25 An example of interpolation using a spline in a finite region. The positions of knots do not coincide with the predictor values of data. The data (× in the graph) are $\{X_i\} = \{1, 1.7, 3, 4.2, 5\}$, $\{Y_i\} = \{2, 5, 4, 5, 3\}$. The positions of knots (■ in the graph) are $\{1, 3.5, 5\}$.

positioned in a different fashion. For example, figure 3.25 is a spline for interpolating the data ($\{X_i\} = \{1, 1.7, 3, 4.2, 5\}$, $\{Y_i\} = \{2, 5, 4, 5, 3\}$); the knot positions are $\{1, 3.5, 5\}$. In this instance, because $n = 3$ in eq(3.164), the number of coefficients needed to describe a spline is 5. That is to say, the number of data is the same as that of the coefficients of the spline. Therefore, a spline is determined uniquely by data. It should be noted, however, that the coincidence between the number of data and that of coefficients of the spline is not sufficient. Schoenberg-Whitney conditions must be met [14]; the Schoenberg-Whitney theorem indicates that a spline is derived under these conditions (page 236 of [14]).

Next, when the predictor variable of data is $\{X_i\}$ ($X_1 < X_2 < \cdots < X_n$, $\{X_i\}$ are not necessarily equispaced), and the target variable of data is $\{Y_i\}$ ($1 \le i \le n$), our task is the interpolation of this data to obtain a smooth function $m(x)$. For this purpose, a possible strategy is to adopt $m(x)$ with which to interpolate $\{X_i, Y_i\}$ and minimize

$$\int_{X_1}^{X_n} \left(\frac{d^2 m(x)}{dx^2} \right)^2 dx. \tag{3.171}$$

This value is an index representing the degree of global roughness of $m(x)$ in the region where data exist; this is based on the idea that the amount of roughness of $m(x)$ at each x is expressed as the square of curvature ($\left[\frac{d^2 m(x')}{dx'^2} \right]_{x'=x}$) of $m(x)$. Minimization of the above value under the condition that $m(x)$ is a C^2 class function (second derivative of the function is continuous in the region to be dealt with) results in a function that is represented as $s_{finite}(x)$ in eq(3.164) (page 155). Furthermore, $s_{finite}(x)$, which is used to interpolate $\{X_i, Y_i\}$ and which satisfies the conditions below, minimizes eq(3.171).

$$\sum_{j=1}^{n} b_j = 0, \tag{3.172}$$

$$\sum_{j=1}^{n} b_j X_j = 0. \tag{3.173}$$

$s_{finite}(x)$, which satisfies eq(3.172) and eq(3.173), is termed the natural cubic spline, or simply the natural spline. It is specified as $ns(x)$ hereafter.

$m(x)$, which minimizes eq(3.171), is interpreted from a physical standpoint; the shape of the elastic beam, which minimizes the elastic energy (in this instance, only bending of the beam is considered and hence it can be called bending elastic energy) accumulated in the beam under the condition that the beam is used to interpolate $\{X_i, Y_i\}$, is identical to the function that approximately minimizes eq(3.171). This is because when a straight beam (an elastic body) is bent into the shape that interpolates $\{X_i, Y_i\}$, the elastic energy accumulated in a beam is proportional to

$$\int_{X_1}^{X_n} \frac{\left(\frac{d^2 m(x)}{dx^2} \right)^2}{\left(1 + \left(\frac{dm(x)}{dx} \right)^2 \right)^{5/2}} dx. \tag{3.174}$$

If $\left(\frac{dm(x)}{dx}\right)^2$ in the integrand in this equation is negligible, that is, $m(x)$ does not vary dramatically, eq(3.174) can be approximated by eq(3.171).

According to eq(3.172) and eq(3.173), the second derivative of $ns(x)$ is 0 when $x = X_1$ and $x = X_n$. This is assured by calculating these values as follows. When $x = X_1$, $ns(x)$ is written as

$$ns(x) = a_0 + a_1 x - \frac{1}{12} \sum_{j=1}^{n} b_j \cdot (x - X_j)^3. \qquad (3.175)$$

Hence, the second derivative at $x = X_1$ is $\sum_{j=1}^{n}(-\frac{1}{2}b_j \cdot (X_1 - X_j))$. Eq(3.172) and eq(3.173) indicate that this value is 0. Furthermore, when $x = X_n$, $ns(x)$ is written as

$$ns(x) = a_0 + a_1 x + \frac{1}{12} \sum_{j=1}^{n} b_j \cdot (x - X_j)^3. \qquad (3.176)$$

Hence, the second derivative at $x = X_n$ is proved to be $\sum_{j=1}^{n}(\frac{1}{2}b_j(X_n - X_j))$. Eq(3.172) and eq(3.173) show that this value is also 0. On the contrary, if the second derivatives at $x = X_1$ and $x = X_n$ are 0, eq(3.172) and eq(3.173) are easily proved.

That is, the conditions of eq(3.172) and eq(3.173) are equivalent to

$$\left[\frac{d^2 ns(x)}{dx^2}\right]_{x=X_1} = 0, \qquad (3.177)$$

$$\left[\frac{d^2 ns(x)}{dx^2}\right]_{x=X_N} = 0. \qquad (3.178)$$

These conditions are called natural boundary conditions, or Neumann conditions. There is some literature in which natural boundary conditions are defined as the second derivatives, and the third derivative in the region of $x < X_1$ and $x > X_n$ is a constant of 0 (e.g., page 24 of [1], page 140 of [10]). In our context, however, $ns(x)$ is first defined in the region of $X_1 \leq x \leq X_n$, then the region of defining the function is extended. Hence, eq(3.177) and eq(3.178) define the natural boundary conditions. The two definitions are equivalent in essentials.

Eq(3.164) (page 155) contains $(n+2)$ unknown variables $(a_0, a_1, b_1, b_2, \ldots, b_n)$. Therefore, these $(n+2)$ variables are calculated using $(n+2)$ equations given by the following conditions: n conditions that $s_{finite}(x)$ interpolates $\{(X_i, Y_i)\}$, and the two conditions eq(3.172) and eq(3.173). Furthermore, $s_{finite}(x)$ belongs to the C^2 class. That is, the problem of deriving a function (C^2 class) which minimizes eq(3.171) under the condition that the function interpolates $\{(X_i, Y_i)\}$ is equivalent to the selection of $s_{finite}(x)$ (eq(3.164)) which satisfies eq(3.172) and eq(3.173) and interpolates $\{(X_i, Y_i)\}$. Therefore, the function we pursue is obtained by solving a simultaneous equation consisting of $(n+2)$ equations. $ns(x)$ obtained by this process minimizes eq(3.171). This is proved below.

Assume that a function (C^2 class) which interpolates $\{(X_i, Y_i)\}$ and minimizes eq(3.171) is not $ns(x)$; the function is termed $\tilde{s}(x)$. Then, $\Delta s(x)$ is defined as

$$\Delta s(x) = \tilde{s}(x) - ns(x). \qquad (3.179)$$

Our calculation proceeds as

$$
\int_{X_1}^{X_n} \left(\frac{d^2 ns(x)}{dx^2}\right)\left(\frac{d^2 \Delta s(x)}{dx^2}\right) dx
$$

$$
= \left[\left(\frac{d^2 ns(x)}{dx^2}\right)\left(\frac{d\Delta s(x)}{dx}\right)\right]_{X_1}^{X_n} - \int_{X_1}^{X_n}\left(\frac{d^3 ns(x)}{dx^3}\right)\left(\frac{d\Delta s(x)}{dx}\right) dx
$$

$$
= -\int_{X_1}^{X_n}\left(\frac{d^3 ns(x)}{dx^3}\right)\left(\frac{d\Delta s(x)}{dx}\right) dx
$$

$$
= -\sum_{j=1}^{n-1}\xi_j\int_{X_j}^{X_{j+1}}\left(\frac{d\Delta s(x)}{dx}\right) dx
$$

$$
= -\sum_{j=1}^{n-1}\xi_j(\Delta s(X_{j+1}) - \Delta s(X_j))
$$

$$
= 0, \tag{3.180}
$$

where ξ_j $(1 \leq j \leq n-1)$ is

$$
\xi_j = \frac{1}{2}\left(\sum_{i=1}^{j} b_j - \sum_{i=j+1}^{n} b_j\right), \tag{3.181}
$$

where the equality between the first line and the second line is based on integration by parts, and that between the second line and the third line is due to eq(3.177) and eq(3.178). "$= 0$" in the sixth line is obtained by $\Delta s(X_j) = \tilde{s}(X_j) - ns(X_j) = Y_j - Y_j = 0$.

Next, substitution of $\tilde{s}(x)$ into $m(x)$ in eq(3.171) gives

$$
\int_{X_1}^{X_n}\left(\frac{d^2 \tilde{s}(x)}{dx^2}\right)^2 dx
$$

$$
= \int_{X_1}^{X_n}\left(\frac{d^2(ns(x) + \Delta s(x))}{dx^2}\right)^2 dx
$$

$$
= \int_{X_1}^{X_n}\left(\frac{d^2 ns(x)}{dx^2}\right)^2 dx + 2\int_{X_1}^{X_n}\left(\frac{d^2 ns(x)}{dx^2}\right)\left(\frac{d^2 \Delta s(x)}{dx^2}\right) dx
$$

$$
+ \int_{X_1}^{X_n}\left(\frac{d^2 \Delta s(x)}{dx^2}\right)^2 dx
$$

$$
= \int_{X_1}^{X_n}\left(\frac{d^2 ns(x)}{dx^2}\right)^2 dx + \int_{X_1}^{X_n}\left(\frac{d^2 \Delta s(x)}{dx^2}\right)^2 dx
$$

$$
\geq \int_{X_1}^{X_n}\left(\frac{d^2 ns(x)}{dx^2}\right)^2 dx, \tag{3.182}
$$

where the equality between the third to fourth lines and the fifth line is derived using eq(3.180). This result contradicts the assumption of $\int_{X_1}^{X_n}\left(\frac{d^2 \tilde{s}(x)}{dx^2}\right)^2 dx <$

$\int_{X_1}^{X_n} \left(\dfrac{d^2 ns(x)}{dx^2} \right)^2 dx$. Hence, by *reductio ad absurdum*, it is proved that the function which minimizes eq(3.171) (page 158) is $ns(x)$. Furthermore, the equality of the sixth line of eq(3.182) holds only if $\frac{d^2 \Delta s(x)}{dx^2} \equiv 0$ ($X_1 \leq x \leq X_n$). This indicates that $\Delta s(x)$ is a linear function in this region. In addition, $\Delta s(X_i) = 0$ ($1 \leq i \leq n$) is required. Hence, since $\Delta s(x) \equiv 0$ if $n \geq 2$, the equality holds only if $\tilde{s}(x) \equiv ns(x)$. Therefore, it is proved that if $s_{finite}(x)$ defined by eq(3.164) (page 155) satisfies eq(3.172) and eq(3.173) (i.e., a natural spline), it is the only function that minimizes eq(3.171).

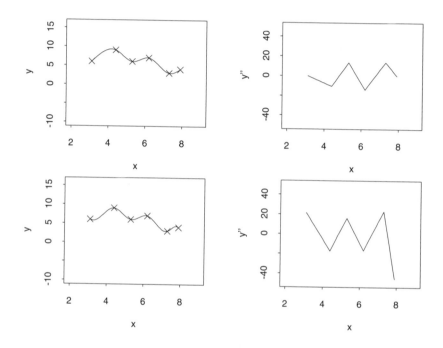

Figure 3.26   Natural spline and another spline. Natural spline obtained with values at six knots given (top left). Another spline with the same values at the knots (bottom left). The second derivative of each spline is plotted to the right of each graph.

Figure 3.26 shows the natural spline and another spline. The graph placed at the top left is the spline that satisfies the two conditions, and the values between knots are also shown. This is a natural spline that is obtained under the condition that the function interpolates the values given at the six knots. Therefore, the six coordinates determine the function uniquely. On the other hand, the graph at the bottom left shows a spline that does not satisfy the two conditions. There exists an infinite number of splines which take these values at the six knots and which do not satisfy the two conditions; the splines in this graph are an examples of such. The second derivative of each function is displayed on the right-hand side of each graph. As is clear from eq(3.177) and eq(3.178), the second derivative of the spline shown in the top left graph is 0 at the two ends. Furthermore, while the value defined by eq(3.171) is 263 for the function in the top left graph, that for the function in the bottom left graph is 695.

Next, we seek $m(x)$ (C^2 class), which minimizes the elastic energy in a wider region under the condition that it interpolates $\{(X_i, Y_i)\}$.

$$\int_{\gamma_1}^{\gamma_2} \left(\frac{d^2 m(x)}{dx^2} \right)^2 dx, \tag{3.183}$$

where, $\gamma_1 < X_1, \gamma_2 > X_n$. Hence, by dividing the integral region into three regions, eq(3.183) is written as

$$\int_{\gamma_1}^{\gamma_2} \left(\frac{d^2 m(x)}{dx^2} \right)^2 dx = \int_{\gamma_1}^{X_1} \left(\frac{d^2 m(x)}{dx^2} \right)^2 dx + \int_{X_1}^{X_n} \left(\frac{d^2 m(x)}{dx^2} \right)^2 dx$$
$$+ \int_{X_n}^{\gamma_2} \left(\frac{d^2 m(x)}{dx^2} \right)^2 dx. \tag{3.184}$$

Assume that $ns(x)$ (defined by eq(3.164) (page 155), eq(3.172) (page 158), and eq(3.173)) is employed as $m(x)$ in the region of $X_1 \leq x \leq X_n$. As was shown previously, the second term on the right-hand side of eq(3.184) is the minimized value among those that are given by a function that interpolates $\{(X_i, Y_i)\}$. Furthermore, the second derivatives of the function at $x = X_1$ and $x = X_n$ are 0. Then, $m(x)$ in the regions of $\gamma_1 \leq x < X_1$ and $X_n < x \leq \gamma_2$ is defined as follows: the values and the first derivatives of $m(x)$ at $x = X_1$ and $x = X_n$ (i.e., $ns(X_1)$ and $ns(X_n)$) are used to create a linear equation in the regions of $\gamma_1 \leq x < X_1$ and $X_n < x \leq \gamma_2$, respectively; the function derived by connecting the linear functions to $m(x)$ ($= ns(x)$) is termed $s^*(x)$. Then, $s^*(x)$ and its first and second derivatives are continuous at $x = X_1$ and $x = X_n$ (second derivatives at these positions are 0) and in the regions of $\gamma_1 \leq x < X_1$ and $X_n < x \leq \gamma_2$ (second derivatives $\equiv 0$). Hence, this function belongs to the C^2 class. Additionally, the first and third terms on the right-hand side of eq(3.184) are 0, and the second term is the result of the minimization of elastic energy in the region of $X_1 \leq x \leq X_n$ under the condition that the function interpolates $\{(X_i, Y_i)\}$. Therefore, $s^*(x)$ is the function that we need.

In fact, $s^*(x)$ obtained by the above procedure is the function derived by extending the domain of function ($ns(x)$), which is defined by eq(3.164) (page 155), eq(3.172) (page 158), and eq(3.173), to $\gamma_1 \leq x < X_1$ and $X_n < x \leq \gamma_2$. This is ensured by the characteristics of the function obtained by extending the domain.

$$\frac{d^2 ns(x)}{dx^2} = \sum_{j=1}^{n} \left(-\frac{1}{2} b_j (x - X_j) \right) = 0 \quad (\gamma_1 \leq x < X_1), \tag{3.185}$$

$$\frac{d^2 ns(x)}{dx^2} = \sum_{j=1}^{n} \left(\frac{1}{2} b_j (x - X_j) \right) = 0 \quad (X_n < x \leq \gamma_2). \tag{3.186}$$

That is, extension of the domain of $ns(x)$ to the regions of $\gamma_1 \leq x < X_1$ and $X_n < x \leq \gamma_2$ leads to linear functions in those regions. Furthermore, $ns(x)$ obtained by extension of the domain also belongs to the C^2 class. Therefore, derivation of $ns(x)$ by the use of eq(3.164), eq(3.172), and eq(3.173) and the extension of the domain simply yield a function which minimizes eq(3.183) under the condition that the function interpolates $\{(X_i, Y_i)\}$.

When we set $\gamma_1 \to -\infty$ and $\gamma_2 \to \infty$, the resulting function is a kind of spline in the infinite region ($s_\infty(x)$). Hence, this $ns(x)$ is regarded as a natural spline in the infinite region. Simple extension of the domain of a spline in a finite region produces a natural spline in a wider region because the natural spline is defined by eq(3.164), eq(3.172), and eq(3.173). When a natural spline is created using the B-spline (B-splines) (which will be explained later), the value of $ns(x)$ and its first derivatives at $x = X_1$ and $x = X_n$ must be calculated to derive linear functions outside the region of $X_1 \leq x \leq X_n$.

Next, let us consider $s_\infty(x)$ other than $ns(x)$. That is, assume that $\frac{d^2 s_\infty(x)}{dx^2} \equiv 0$ does not hold in the region of $x < X_1$, $x > X_n$ (i.e., $s_\infty(x)$ in this region is not a linear function). In this case, the elastic energy of a beam having the shape $s_\infty(x)$ is infinite. It is assured that if $\frac{d^2 s_\infty(x)}{dx^2} = \tau_0 + \tau_1 x$ in the region of $x > X_n$, the elastic energy in this region is proportional to

$$
\begin{aligned}
\int_{X_n}^\infty \left(\frac{d^2 s_\infty(x)}{dx^2} \right)^2 dx &= \int_{X_n}^\infty (\tau_0 + \tau_1 x)^2 dx \\
&= \left[\tau_0^2 x + \tau_0 \tau_1 x^2 + \frac{1}{3} \tau_1^2 x^3 \right]_{X_n}^\infty .
\end{aligned} \tag{3.187}
$$

To prevent this value from becoming infinite, it is required that $\tau_0 = 0$ and $\tau_1 = 0$. That is, $s_\infty(x)$ must be a linear equation in the region of $x > X_n$. The same is true of the region of $x < X_1$. Therefore, the elastic energy of a beam of shape $s_\infty(x)$ is one of only two kinds. When $s_\infty(x)$ is $ns(x)$, the elastic energy is finite. All others are situations where $s_\infty(x)$ is not identical to $ns(x)$; the elastic energy is infinite. Hence, a natural spline in an infinite region can be defined as the only function that has the property that the elastic energy of an infinitely long beam in the shape of this function is not infinite, and is a spline ($s_\infty(x)$) interpolating $\{(X_i, Y_i)\}$.

Figure 3.27 illustrates a natural spline and another spline in the region of $\gamma_1 \leq x \leq \gamma_2$. The top left graph shows a natural spline obtained by extending the domain of the natural spline in figure 3.26. The second derivative of the function is shown to the right of the graph. Values of the second derivatives are 0 at the two end knots and beyond the ends. On the other hand, the bottom left graph shows a spline whose second derivatives at the two ends are not 0. As is clear from the bottom right graph, the value of eq(3.183) becomes infinite if $\gamma_1 \to -\infty$ and $\gamma_2 \to \infty$. That is, since the second derivative beyond the two ends is described as a linear equation beyond the two sides, the second derivative of a spline beyond the two ends must be 0 (i.e., the spline must be a natural spline) in order to make the value of eq(3.183) finite.

$(n + 2)$ equations showing n conditions that $ns(x)$ interpolates $\{(X_i, Y_i)\}$, and two conditions represented by eq(3.172) (page 158) and eq(3.173) are derived as follows. First, $\mathbf{Q}$, $\mathbf{R}$, $\mathbf{a}$, and $\mathbf{b}$ are set as

$$
\mathbf{Q} = \begin{pmatrix} 1 & 1 & 1 & \cdots & 1 \\ X_1 & X_2 & X_3 & \cdots & X_n \end{pmatrix}, \tag{3.188}
$$

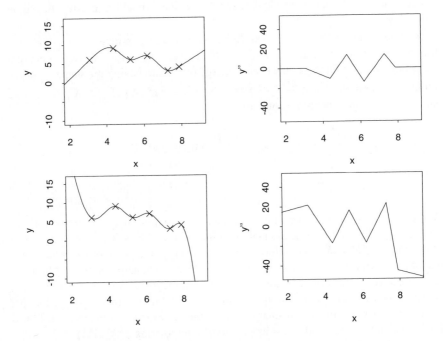

Figure 3.27   A natural spline and another spline in the region of $\gamma_1 \leq x \leq \gamma_2$. Natural spline when values at six knots are provided (top left). Another spline with the same values at the six knots (bottom left). The second derivative of each function is shown to the right of each graph.

$$\mathbf{R} =$$

$$
\begin{pmatrix}
0 & \dfrac{|X_1 - X_2|^3}{12} & \dfrac{|X_1 - X_3|^3}{12} & \cdots & \dfrac{|X_1 - X_n|^3}{12} \\[2mm]
\dfrac{|X_2 - X_1|^3}{12} & 0 & \dfrac{|X_2 - X_3|^3}{12} & \cdots & \dfrac{|X_2 - X_n|^3}{12} \\[2mm]
\vdots & \vdots & \vdots & \ddots & \vdots \\[2mm]
\dfrac{|X_n - X_1|^3}{12} & \dfrac{|X_n - X_2|^3}{12} & \dfrac{|X_n - X_3|^3}{12} & \cdots & 0
\end{pmatrix},
$$

$$\tag{3.189}$$

$$
\mathbf{a} = \begin{pmatrix} a_0 \\ a_1 \end{pmatrix}, \tag{3.190}
$$

$$
\mathbf{b} = \begin{pmatrix} b_1 \\ \vdots \\ b_n \end{pmatrix}. \tag{3.191}
$$

Then, the conditions that eq(3.164) (page 155) interpolates n data and eq(3.172) and eq(3.173) are satisfied are expressed by one equation:

$$
\begin{pmatrix} \mathbf{R} & \mathbf{Q}^t \\ \mathbf{Q} & 0 \end{pmatrix}
\begin{pmatrix} \mathbf{b} \\ \mathbf{a} \end{pmatrix}
= \begin{pmatrix} \mathbf{y} \\ 0 \end{pmatrix}. \tag{3.192}
$$

The solution of this simultaneous equation provides a natural spline $(ns(x))$ which interpolates n data.

The natural spline is interpreted in the context of the mechanics of materials as follows. Details of this topic are described in textbooks on the mechanics of materials.

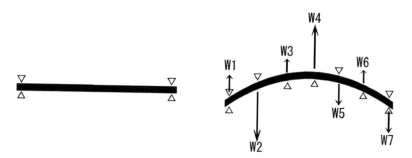

Figure 3.28   Simply supported beam without load (left). Simply supported beam with load (right).

Let us consider a beam supported by columns at two ends, such as that shown in figure 3.28(left). The beam is pinned at the left end to allow rotation; horizontal translational displacement is restrained at this point. Furthermore, the beam is supported by a roller at the right end to allow horizontal translation. The beam is loaded vertically at $\{X_j\}$ ($2 \le j \le n-1$); the forces at these positions are $\{W_i\}$ ($2 \le i \le n-1$). The coordinates of right and left columns are defined as (X_1, Y_1),

(X_n, Y_n) $(X_1 < X_2 < \cdots < X_{n-1} < X_n)$, and the forces at these two points are specified as W_1 and W_n, respectively. Since the beam is transformed by forces, each X_i is displaced horizontally. The influence of this transformation, however, is negligible when it is not large. This beam (figure 3.28(right)) is called a simply supported beam. In the context of the mechanics of materials, $\{W_j\}$ $(2 \le j \le n-1)$ are assumed to be downward forces (concentrated loads), and W_1 and W_n are assumed to be upward forces (reaction forces). Furthermore, $Y_1 = 0$ and $Y_n = 0$ are assumed when calculation is carried out. However, these limitations are not imposed here. Moreover, concentrated loads and reaction forces are not distinguished; this does not greatly affect our discussion. Let us now discuss the probelm of deriving $y(x)$ (a deflection curve) when the shape of the beam under these conditions is defined to be $y = y(x)$ $(Y_1 = y(X_1), Y_2 = y(X_2), \ldots, Y_{n-1} = y(X_{n-1}), Y_n = y(X_n))$. Furthermore, though the downward force is expressed as positive values and the upward force as negative values in the context of the mechanics of materials, here, the upward direction is described by positive values of $y(x)$ in the conventional system and, accordingly, upward forces are represented by positive values.

First, the two conditions below must be satisfied for the beam to be at rest. One is that the vertical forces on the beam sum up to 0. Therefore, the equation below must hold.

$$\sum_{j=1}^{n} W_j = 0. \tag{3.193}$$

The other condition is that when a point of the X coordinate on a beam is defined as x $(X_1 \le x \le X_n)$, the narrow interval in the beam must not rotate. The clockwise moment at this point is specified to be a positive, and the moment at x, due to forces applied on the left-hand side of x $(\{W_j\}$ $(1 \le j \le k))$, is called $M_L(x)$. Then $M_L(x)$ is represented as

$$M_L(x) = \sum_{j=1}^{k} W_j \cdot (x - X_j). \tag{3.194}$$

The moment due to the forces on the right-hand side of x is termed $M_R(x)$.

$$M_R(x) = \sum_{j=k+1}^{n} W_j \cdot (x - X_j). \tag{3.195}$$

To prevent rotation at x, we require that

$$M_L(x) + M_R(x) = \sum_{j=1}^{n} W_j \cdot (x - X_j) = 0. \tag{3.196}$$

Taking account of eq(3.193), eq(3.196) becomes

$$\sum_{j=1}^{n} W_j \cdot X_j = 0. \tag{3.197}$$

This describes the second condition. Eq(3.197) does not depend on x. It shows that if a moment at one point is 0, a moment at another point is also 0. Eq(3.193) also describes the condition that each point does not move vertically. Hence, in conjunction with eq(3.197), we gain the condition that all points on a beam neither move vertically nor rotate.

Since the moment required to bend the beam at point x is $M_L(x)$ ($M_R(x)$ is a reaction force), Hooke's law provides

$$\frac{d^2y(x)}{dx^2} = \frac{M_L(x)}{EI},$$

(3.198)

where E is Young's modulus, and I is the geometrical moment of inertia; these are constants representing the physical characteristics of the beam. EI is called the flexural rigidity. This equation can be understood by corresponding it to Hooke's law for the stretching of a spring:

$$\Delta y = \frac{P}{\kappa},$$

(3.199)

where the length of the spring is 0 when no forces are applied to it, and the length is Δy when a force is applied. κ is the spring constant, and P is the force applied to the spring. Comparison with eq(3.198) reveals that stretching of the spring corresponds to the second derivative of $y(x)$, and the spring constant corresponds to flexural rigidity.

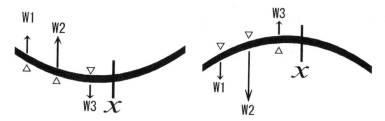

Figure 3.29   If the moment at x ($M_L(x)$) is positive, the curvature is positive (the forces applied on the left side of x are shown) (left). If the moment at x ($M_L(x)$) is negative, the curvature is negative (right).

Figure 3.29 illustrates the relationship between the moment at x ($M_L(x)$) and the curvature ($\left[\frac{d^2y(x')}{dx'^2}\right]_{x'=x}$). As expected from eq(3.198), the curvature is positive if $M_L(x)$ is positive. Because of the condition of eq(3.196), the curvature at x is determined by considering the forces on the left-hand side of x. In the context of the mechanics of materials, the graph of the values of $M_L(x)$ is called the bending moment diagram. Since figure 3.26(top right) (page 161) is regarded as a graph of the result of multiplying of $M_L(x)$ with a constant, a bending moment diagram can be obtained by changing the scale of the vertical axis. Furthermore, a graph that illustrates a function derived by differentiating $M_L(x)$ (i.e., the sum of forces on the left-hand side of x) is termed a shearing force diagram.

The beam is bent in the interval between x and $(x+\Delta x)$ (i.e., the second derivative of $y(x)$ is not 0). The elastic energy accumulated by this bending is

$$E_{bending}(x) = \frac{M_L(x)^2}{2EI}\Delta x = \frac{EI\left(\frac{d^2y(x)}{dx^2}\right)^2}{2}\Delta x.$$

(3.200)

This equation corresponds to the equation representing the elastic energy of a spring:

$$E_{sping}(x) = \frac{P^2}{2\kappa} = \frac{\kappa(\Delta y)^2}{2}.$$ (3.201)

Integration of eq(3.200) over the entire beam yields the total elastic energy accumulated in the beam:

$$\int_{X_1}^{X_n} E_{bending}(x)dx = \frac{1}{2EI} \int_{X_1}^{X_n} M_L(x)^2 dx = \frac{EI}{2} \int_{X_1}^{X_n} \left(\frac{d^2 y(x)}{dx^2}\right)^2 dx.$$ (3.202)

The principle of minimum potential energy leads to the solution of this simply supported beam problem. The resulting $y(x)$ minimizes the value of $\int_{X_1}^{X_n} E_{bending}(x)dx$. Since the derivation of $y(x)$ that minimizes $\frac{EI}{2} \int_{X_1}^{X_n} \left(\frac{d^2 y(x)}{dx^2}\right)^2 dx$ is equivalent to the derivation of $y(x)$ that minimizes $\int_{X_1}^{X_n} \left(\frac{d^2 y(x)}{dx^2}\right)^2 dx$, $y(x)$ that minimizes eq(3.202) under the condition that $y(x)$ interpolates $\{(X_i, Y_i)\}$ is identical to the function (i.e., $ns(x)$) that minimizes eq(3.171) (page 158) under the condition that $s_{finite}(x)$ interpolates $\{(X_i, Y_i)\}$.

Substitution of eq(3.194) (page 166) into eq(3.198) (page 167) results in

$$\frac{d^2 y(x)}{dx^2} = \frac{1}{EI} \sum_{j=1}^{k} W_j \cdot (x - X_j) \quad (X_k \le x < X_{k+1}).$$ (3.203)

Referring to eq(3.196) (page 166), we obtain

$$\frac{d^2 y(x)}{dx^2} = \frac{1}{2EI} \sum_{j=1}^{k} W_j \cdot (x - X_j) - \frac{1}{2EI} \sum_{j=k+1}^{n} W_j \cdot (x - X_j).$$ (3.204)

Therefore, $y(x)$ is a piecewise cubic equation defined in the region of $X_1 \le x \le X_n$, and the second derivative of the function is a continuous function (a piecewise linear equation) between the knots ($\{X_i\}$). The general form of the equation is written below (eq(3.164) (page 155)).

$$y(x) = a_0 + a_1 x + \frac{1}{12} \sum_{j=1}^{n} b_j \cdot |x - X_j|^3 .$$ (3.205)

The second derivative of this equation in the region of $X_k \le x < X_{k+1}$ ($1 \le k \le n - 1$) is

$$\frac{d^2 y(x)}{dx^2} = \frac{1}{2} \sum_{j=1}^{k} b_j \cdot (x - X_j) - \frac{1}{2} \sum_{j=k+1}^{n} b_j \cdot (x - X_j).$$ (3.206)

Comparison of eq(3.204) and eq(3.206) yields

$$b_j = \frac{1}{EI} W_j \quad (1 \le j \le n).$$ (3.207)

Hence, the $y(x)$ that we need is written as

$$y(x) = a_0 + a_1 x + \frac{1}{12EI} \sum_{j=1}^{n} W_j \cdot \mid x - X_j \mid^3 . \qquad (3.208)$$

The values of $\{\frac{1}{EI}W_i\}$ are derived under the conditions that this $y(x)$ interpolates $\{(X_i, Y_i)\}$ and satisfies eq(3.193) (page 166) and eq(3.197) (page 166). Then, when we set $b_j = \frac{1}{EI}W_j$, we obtain the same function as eq(3.192) (page 165). That is, eq(3.192) is the equation describing the shape of the beam ($y(x)$) under the condition that forces $\{\frac{1}{EI}W_i\}$ (values of $\{W_i\}$ are unknown) are applied at $\{X_i\}$ and $y(x)$ interpolates $\{(X_i, Y_i)\}$. Now, it is clear why eq(3.205) is used as a general form of a spline in a finite region. With the use of this equation, $\{b_j\}$ coincides with $\{W_j\}$ when $EI = 1$.

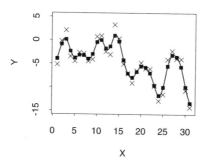

Figure 3.30   Interpolation of the estimates obtained using a binomial filter. × represents data. ■ is the result of smoothing using a binomial filter. The solid line shows the result of interpolation using a natural spline.

Next, let us consider a method of smoothing by applying the natural spline. The simplest smoothing method possible as an application of interpolation is one in which estimates ($\{\hat{Y}_i\}$) that are the results of smoothing of $\{Y_i\}$ are interpolated using the natural spline. For example, if $\{X_i\}$ is equispaced, $\{Y_i\}$ is smoothed, using a binomial filter, to obtain $\{\hat{Y}_i\}$, and a natural spline is derived to interpolate $\{X_i, \hat{Y}_i\}$. This procedure provides $\hat{m}(x)$, which is a smooth function that interpolates the estimates smoothed using a binomial filter. Figure 3.30 shows the estimates by smoothing using a binomial filter ($m = 2$, figure 2.4(left) (page 29)) and the result of the interpolation of the estimates using the natural spline. The values of a predictor of data ($\{X_i\}$) coincide with knots. Then, since the number of coefficients of a spline in a finite region is larger than that of data by 2 (eq(3.162) (page 155)), the conditions of eq(3.172) and (3.173), in addition to the data, lead to unique values of coefficients. Compared with figure 2.4(left) (interpolation of estimates using linear curves), the curve (interpolated values) in figure 3.30 is smoother.

However, if smoothing of Y_i to obtain $\hat{Y}_i$ and connection of these estimates to derive a continuous function are carried out simultaneously, the smoothing method is expected to be mathematically clearer and to have wider applications. From this point of view, we develop a smoothing spline described in Chapter 2. When the number of predictors is one, $m(x)$ that minimizes the value below is adopted as a function representing estimates.

$$E_{ss} = \sum_{i=1}^{n}(Y_i - m(X_i))^2 + \lambda \int_{-\infty}^{\infty} \left(\frac{d^2m(x)}{dx^2}\right)^2 dx. \qquad (3.209)$$

$\hat{m}(x)$, which minimizes the above value, is termed a cubic smoothing spline (cubic smoothing splines) or simply a smoothing spline. The second term on the right-hand side is called a roughness penalty; this value becomes larger as the roughness of $\hat{m}(x)$ increases. As is the smoothing spline discussed in Chapter 2, this λ is also a smoothing parameter (a positive constant). While the equations in Chapter 2 are in a form in which differences of a regression equation are used to represent the roughness penalty, derivatives instead of differences are utilized here because $m(x)$ is a continuous function. In addition, instead of the sum of squared differences, the integral of squared derivatives in the infinite region is adopted. Similar to the smoothing spline in Chapter 2, $\lambda \to 0$ leads to $\hat{m}(X_i)$ $(1 \leq i \leq n)$, which is very close to Y_i, and $\lambda \to \infty$ results in the linear equation $\hat{m}(x)$ in the entire region. Note that we assume that all the values of a predictor of data are distinct here. When there are tied data in values of the predictor of data, a peculiar skill is required ([1]; page 43 of [10]).

Figure 3.31 shows examples of $\hat{m}(x)$ obtained using 7 data and by minimizing eq(3.209), with four different values of smoothing parameters: $\lambda = 10^{-3}, 10^{-1}, 10^{1}, 10^{3}$. The regression equation becomes smoother and the curve gradually approaches a straight line as the value of the smoothing parameter is incremented. Figure 3.32 displays the results of interpolation using the natural spline and fitting a linear equation using the least squares. The estimates of the smoothing spline with $\lambda = 10^{-3}$ are located close to the curve obtained by interpolation using the natural spline, and those with $\lambda = 10^{3}$ are analogous to the least squares linear equation.

$\hat{m}(x)$ given by minimizing E_{ss} in eq(3.209) is always a natural spline ($s^*(x)$ (page 162), namely, $ns(x)$); this is not peculiar to the case of the value of λ being very small. This is proved as follows.

Assume that $\hat{m}(x)$ obtained by minimizing E_{ss} in eq(3.209) is not a natural spline. Then, a smoothing spline that interpolates $\{X_i, \hat{m}(X_i)\}$ $(1 \leq i \leq n)$, termed $\bar{m}(x)$, does not coincide with $\hat{m}(x)$. However, $\bar{m}(x)$ also interpolates $\{X_i, \hat{m}(X_i)\}$. Hence, the value obtained by substituting $\hat{m}(x)$ into $m(x)$ of the first term on the right-hand side of eq(3.209) is identical to the value calculated by substituting $\bar{m}(x)$ into $m(x)$ at the same place. On the other hand, since $\bar{m}(x)$ is a natural spline, $\bar{m}(x)$ is a unique function that interpolates $\{X_i, \hat{m}(X_i)\}$ and minimizes the second term on the right-hand side. Therefore, the value of E_{ss}, when $\hat{m}(x)$ is used as $m(x)$, is larger than that when $\bar{m}(x)$ is adopted as $m(x)$. This contradicts the assumption that $\hat{m}(x)$ is derived by minimizing E_{ss}. Therefore, by *reductio ad absurdum*, it is proved that $\hat{m}(x)$ is a natural spline.

Hence, $\hat{m}(x)$ is linear outside the region where data exist; that is, the second derivative is 0. As a consequence, as long as the integral region of the second term of eq(3.209) includes the area where data exist, the resulting $\hat{m}(x)$ remains the same.

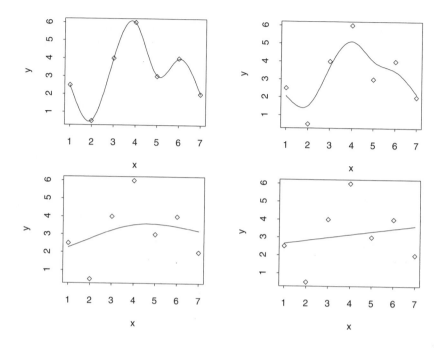

Figure 3.31   Estimates obtained by minimizing eq(3.209) using 7 data. The values of the smoothing parameter are $\lambda = 10^{-3}$ (top left), $\lambda = 10^{-1}$ (top right), $\lambda = 10^{1}$ (bottom left), and $\lambda = 10^{3}$ (bottom right).

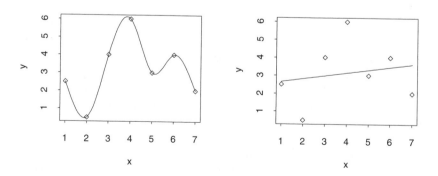

Figure 3.32   Result of smoothing by natural spline using the same data as in figure 3.31 (left). Result of fitting of a linear equation by the least squares (right).

Furthermore, since a smoothing spline derived by minimizing eq(3.209) is a natural spline that interpolates $\{X_i, \hat{m}(X_i)\}$ $(1 \le i \le n)$, it is written in the form of eq(3.164) (page 155). On the other hand, calculation based on the calculus of variations to minimize eq(3.209) produces the equation below. By this equation, though the form is the same as that of eq(3.164), a peculiar meaning is attached to each coefficient.

$$m(x) = c_0 + c_1 x + \sum_{j=1}^{n} \theta_j \cdot G(x - X_j). \qquad (3.210)$$

Here, $(c_0, c_1, \theta_1, \theta_2, \ldots, \theta_n)$ are constants. $G(x - X_j)$ is

$$G(x - X_j) = \frac{\Gamma(-1.5)}{16 \cdot \sqrt{\pi}} \mid x - X_j \mid^3, \qquad (3.211)$$

where $\Gamma(\cdot)$ is a gamma function. $G(x - X_j)$ can be written simply as $\mid x - X_j \mid^3$ so that the constant for multiplying $\mid x - X_j \mid^3$ will include θ_i. However, in the context of the problem of obtaining $m(x)$ by minimizing eq(3.209) via the calculus of variations, it is expressed as eq(3.210) because the function defined by eq(3.211) acts as a Green function (Green's function) (G of $G(x - X_j)$ is named from the initial letter of Green). $\Gamma(-1.5)$ takes the value of

$$\Gamma(-1.5) = \frac{4.0 \cdot \sqrt{\pi}}{3.0}. \qquad (3.212)$$

Hence, eq(3.210) becomes

$$m(x) = c_0 + c_1 x + \frac{1}{12} \sum_{j=1}^{n} \theta_j \cdot \mid x - X_j \mid^3 . \qquad (3.213)$$

The third term on the right-hand side includes $\frac{1}{12}$ in the same fashion as eq(3.164) (page 155). As mentioned above, in the simply supported beam problem, the shape of the beam when forces $\{\frac{W_i}{EI}\}$ are applied at $\{X_i\}$ is expressed as $m(x)$ in eq(3.213). When $EI = 1$, we have $\theta_j = W_j$. This means that $\theta_i = W_i$ is required to realize a simply supported beam with $EI = 1$; this beam has the shape of the smoothing spline (eq(3.213)) derived by minimizing eq(3.209). That is, the influence of the force (W_i) at X_i of a simply supported beam $(m(x))$ with $EI = 1$ is represented as $G(x - X_i)$. This is the origin of calling $G(x - X_i)$ the Green function. This $G(x - X_i)$ is an increasing function; the value of the function increases as that of $\mid x - X_i \mid$ is enhanced. On the other hand, the effect of each data (Y_i) upon $m(x)$ is depicted in figure 3.36 (page 178).

Figure 3.33 displays $\{\theta_j\}$ obtained using the same data as those in figure 3.31 and $\lambda = 10^1$ (the value is optimized by GCV). Data and estimates obtained using these $\{\theta_j\}$ and c_0 and c_1 are superimposed. Comparison of figure 3.33(left) and figure 3.34 indicates that the values of $\{\theta_j\}$ are proportional to the force applied to a simply supported beam at $\{X_j\}$. The sources of the forces applied to a simply supported beam correspond to the orientation at which the value of the first term of E_{ss} (eq(3.209) (page 170)), namely, the discrepancy between data and estimates, is reduced. In fact, comparison of figure 3.33(right) and figure 3.34 reveals that when

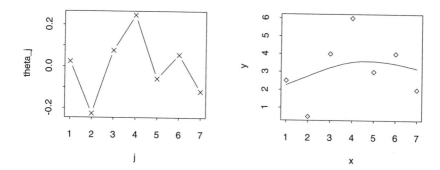

Figure 3.33   The values of $\{\theta_j\}$ when the same data as those in figure 3.31 were used and $\lambda = 10^1$ (left). Data and estimates (right).

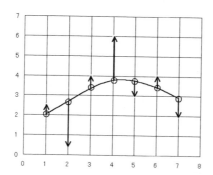

Figure 3.34   The forces (by data) applied to estimates in figure 3.33 are illustrated in the same manner as for forces applied to a simply supported beam.

an estimate is located beneath a datum, θ_j corresponding to that datum is positive. When the physical relationship between a datum and an estimate is the opposite, θ_j is negative. Additionally, the absolute value of each force is proportional to the distance between a datum and an estimate. That is to say, the distance between data and estimates is reduced by dragging estimates by data according to Hooke's law.

Eq(3.213) contains $(n+2)$ unknown coefficients: $(c_0, c_1, \theta_1, \theta_2, \ldots, \theta_n)$. A total of n data are available for solving this equation. Furthermore, these unknown coefficients are calculable, since the two equations below hold because of the fact that $m(x)$ is a natural spline.

$$\sum_{j=1}^{n} \theta_j = 0, \tag{3.214}$$

$$\sum_{j=1}^{n} \theta_j X_j = 0. \tag{3.215}$$

Then, in eq(3.209), let us introduce a method of deriving an equation in the form of eq(3.213), given the value of a smoothing paramter (λ) and data ($\{(X_i, Y_i)\}$). First, we set $\mathbf{c} = (c_0, c_1)^t$ and $\boldsymbol{\theta} = (\theta_1, \theta_2, \ldots, \theta_n)^t$; the definition of $\mathbf{Q}$ is eq(3.188) (page 163), and the definition of $\mathbf{R}$ is eq(3.189) (page 165). Then, taking account of $\mathbf{R} = \mathbf{R}^t$, eq(3.209) is transformed into

$$
\begin{aligned}
E_{ss} &= (\mathbf{y} - \mathbf{R}\boldsymbol{\theta} - \mathbf{Q}^t\mathbf{c})^t(\mathbf{y} - \mathbf{R}\boldsymbol{\theta} - \mathbf{Q}^t\mathbf{c}) + \lambda\boldsymbol{\theta}^t\mathbf{R}\boldsymbol{\theta} \\
&= (\boldsymbol{\theta}^t \quad \mathbf{c}^t) \begin{pmatrix} \mathbf{R}^2 + \lambda\mathbf{R} & \mathbf{R}\mathbf{Q}^t \\ \mathbf{Q}\mathbf{R} & \mathbf{Q}\mathbf{Q}^t \end{pmatrix} \begin{pmatrix} \boldsymbol{\theta} \\ \mathbf{c} \end{pmatrix} \\
&\quad -2 (\boldsymbol{\theta}^t \quad \mathbf{c}^t) \begin{pmatrix} \mathbf{R} \\ \mathbf{Q} \end{pmatrix} \mathbf{y} + \mathbf{y}^t\mathbf{y},
\end{aligned} \tag{3.216}
$$

where the roughness penalty in eq(3.209) (page 170) is represented as $\lambda\boldsymbol{\theta}^t\mathbf{R}\boldsymbol{\theta}$. This is proved as

$$
\begin{aligned}
&\int_{-\infty}^{\infty} \left(\frac{d^2m(x)}{dx^2} \right)^2 dx \\
&= \left[\left(\frac{d^2m(x)}{dx^2} \right) \left(\frac{dm(x)}{dx} \right) \right]_{-\infty}^{\infty} - \int_{-\infty}^{\infty} \left(\frac{d^3m(x)}{dx^3} \right) \left(\frac{dm(x)}{dx} \right) dx \\
&= - \left[\left(\frac{d^3m(x)}{dx^3} \right) m(x) \right]_{-\infty}^{\infty} + \int_{-\infty}^{\infty} \left(\frac{d^4m(x)}{dx^4} \right) m(x) dx \\
&= \int_{-\infty}^{\infty} \left(\frac{d^4m(x)}{dx^4} \right) m(x) dx \\
&= \left(\sum_{i=1}^{n} \theta_i \sum_{j=1}^{n} \frac{1}{12}\theta_j \mid X_i - X_j \mid^3 \right) + c_0 \sum_{j=1}^{n} \theta_j + c_1 \sum_{j=1}^{n} \theta_j X_j \\
&= \boldsymbol{\theta}^t\mathbf{R}\boldsymbol{\theta}, \tag{3.217}
\end{aligned}
$$

where both the first term of the second line and the first term of the third line are 0 because $m(x)$ is a natural spline. In addition, when the equality between the fourth line and the fifth line is derived, the transformation of $\frac{d^4m(x)}{dx^4}$, as below, is utilized. This is because the last equation of eq(3.166) (page 155) holds since $m(x)$ is a natural spline.

$$\frac{d^4m(x)}{dx^4} = \sum_{j=1}^{n}(\theta_j \cdot \delta(x - X_j)). \qquad (3.218)$$

Here $\delta(\cdot)$ is a delta function. Furthermore, both the second term and the third term of the fifth line are 0 because of eq(3.214) and eq(3.215).

Next, differentiation of eq(3.216) to be 0 with respect to each of $(\theta_1, \theta_2, \ldots, \theta_n, c_0, c_1)$ yields the $(n+2)$ equations:

$$\begin{pmatrix} \mathbf{R}^2 + \lambda\mathbf{R} & \mathbf{R}\mathbf{Q}^t \\ \mathbf{Q}\mathbf{R} & \mathbf{Q}\mathbf{Q}^t \end{pmatrix} \begin{pmatrix} \boldsymbol{\theta} \\ \mathbf{c} \end{pmatrix} = \begin{pmatrix} \mathbf{R} \\ \mathbf{Q} \end{pmatrix} \mathbf{y}. \qquad (3.219)$$

This equation is justified by the fact that, for instance, the result of differentiation of $\boldsymbol{\theta}^t\mathbf{R}\boldsymbol{\theta}$ with respect to $\boldsymbol{\theta}$ is written as

$$\begin{aligned} \frac{\partial \boldsymbol{\theta}^t\mathbf{R}\boldsymbol{\theta}}{\partial \theta_k} &= \frac{\partial}{\partial \theta_k}\sum_{i=1}^{n}\sum_{j=1}^{n}\theta_i[\mathbf{R}]_{ij}\theta_j \\ &= \sum_{i=1}^{n}\sum_{j=1}^{n}\left(\frac{\partial\theta_i}{\partial\theta_k}[\mathbf{R}]_{ij}\theta_j + \theta_i[\mathbf{R}]_{ij}\frac{\partial\theta_j}{\partial\theta_k}\right) \\ &= \sum_{j=1}^{n}[\mathbf{R}]_{kj}\theta_j + \sum_{i=1}^{n}\theta_i[\mathbf{R}]_{ik} \\ &= 2\sum_{j=1}^{n}[\mathbf{R}]_{kj}\theta_j, \qquad (3.220) \end{aligned}$$

where the derivation of the second line from the first one involves the use of the product rule of differentiation. The equality between the second line and the third line is given by

$$\frac{\partial\theta_i}{\partial\theta_k} = \begin{cases} 1 & \text{if } i = k \\ 0 & \text{if } i \neq k. \end{cases} \qquad (3.221)$$

The equality between the third line and the fourth line is due to the fact that $\mathbf{R}$ is a symmetric matrix.

Eq(3.219) is transformed into

$$\begin{pmatrix} \mathbf{R} & 0 \\ \mathbf{Q} & -\lambda\mathbf{I}_2 \end{pmatrix}\begin{pmatrix} \mathbf{R}+\lambda\mathbf{I}_n & \mathbf{Q}^t \\ \mathbf{Q} & 0 \end{pmatrix}\begin{pmatrix} \boldsymbol{\theta} \\ \mathbf{c} \end{pmatrix} = \begin{pmatrix} \mathbf{R} & 0 \\ \mathbf{Q} & -\lambda\mathbf{I}_2 \end{pmatrix}\begin{pmatrix} \mathbf{y} \\ 0 \end{pmatrix}, \qquad (3.222)$$

where $\mathbf{I}_n$ is an identity matrix (the size is $n \times n$), and $\mathbf{I}_2$ is also an identity matrix (the size is 2×2). Since eq(3.219) is a simultaneous equation that yields a unique solution, $\begin{pmatrix} \mathbf{R} & 0 \\ \mathbf{Q} & -\lambda\mathbf{I}_2 \end{pmatrix}$ is a regular matrix. Hence, the solution of eq(3.219) (i.e., eq(3.222)) to obtain $\mathbf{c}$ and $\boldsymbol{\theta}$ is equivalent to calculations of $\mathbf{c}$ and $\boldsymbol{\theta}$ using

$$\begin{pmatrix} \mathbf{R}+\lambda\mathbf{I}_n & \mathbf{Q}^t \\ \mathbf{Q} & 0 \end{pmatrix}\begin{pmatrix} \boldsymbol{\theta} \\ \mathbf{c} \end{pmatrix} = \begin{pmatrix} \mathbf{y} \\ 0 \end{pmatrix}. \qquad (3.223)$$

c and θ, which are given by solving this simultaneous equation, yield the function that minimizes eq(3.209) (page 170). This equation contains $\mathbf{Q}\theta = \mathbf{0}$. Hence, c and θ obtained using eq(3.223) satisfy eq(3.214) and eq(3.215). In addition, eq(3.223) becomes close to eq(3.192) (page 165) when $\lambda \to 0$ and hence, the resulting $\hat{m}(x)$ (natural spline) under the condition of $\lambda \to 0$ is close to the function that interpolates data.

Eq(3.223) yields

$$
\begin{pmatrix} \theta \\ c \end{pmatrix} = \begin{pmatrix} \mathbf{R} + \lambda \mathbf{I}_n & \mathbf{Q}^t \\ \mathbf{Q} & 0 \end{pmatrix}^{-1} \begin{pmatrix} \mathbf{y} \\ \mathbf{0} \end{pmatrix}.
\tag{3.224}
$$

Therefore, when a vector consisting of estimates derived using the smoothing spline is represented as $\hat{\mathbf{y}} = (\hat{Y}_1, \hat{Y}_2, \ldots, \hat{Y}_n)^t \; (= (\hat{m}(X_1), \hat{m}(X_2), \ldots, \hat{m}(X_n))^t)$, we have

$$
\hat{\mathbf{y}} = (\mathbf{R} \quad \mathbf{Q}^t) \begin{pmatrix} \theta \\ c \end{pmatrix} = (\mathbf{R} \quad \mathbf{Q}^t) \begin{pmatrix} \mathbf{R} + \lambda \mathbf{I}_n & \mathbf{Q}^t \\ \mathbf{Q} & 0 \end{pmatrix}^{-1} \begin{pmatrix} \mathbf{y} \\ \mathbf{0} \end{pmatrix}.
\tag{3.225}
$$

Therefore, the hat matrix (**H**) is

$$
\mathbf{H} = \left[(\mathbf{R} \quad \mathbf{Q}^t) \begin{pmatrix} \mathbf{R} + \lambda \mathbf{I}_n & \mathbf{Q}^t \\ \mathbf{Q} & 0 \end{pmatrix}^{-1} \right]_{(n,n)},
\tag{3.226}
$$

where $[\;]_{(n,n)}$ is a submatrix (i.e., a matrix (size $n \times n$) that makes up the left-hand part of the main matrix).

In addition, eq(3.224) yields

$$
\begin{aligned}
\theta &= \mathbf{T}\mathbf{y} \\
&= \mathbf{T}\mathbf{H}^{-1}\hat{\mathbf{y}},
\end{aligned}
\tag{3.227}
$$

where **T** is

$$
\mathbf{T} = \left[\begin{pmatrix} \mathbf{R} + \lambda \mathbf{I}_n & \mathbf{Q}^t \\ \mathbf{Q} & 0 \end{pmatrix}^{-1} \right]_{(n,n)},
\tag{3.228}
$$

and $[\;]_{(n,n)}$ is the left-top submatrix (size $n \times n$) of the main matrix. Substitution of eq(3.227) into eq(3.216) (page 174) results in

$$
E_{ss} = (\mathbf{y} - \hat{\mathbf{y}})^t(\mathbf{y} - \hat{\mathbf{y}}) + \lambda \hat{\mathbf{y}}^t(\mathbf{T}\mathbf{H}^{-1})^t \mathbf{R}\mathbf{T}\mathbf{H}^{-1}\hat{\mathbf{y}}.
\tag{3.229}
$$

When we set $\mathbf{L} = (\mathbf{T}\mathbf{H}^{-1})^t \mathbf{R}\mathbf{T}\mathbf{H}^{-1}$, we have

$$
E_{ss} = (\mathbf{y} - \hat{\mathbf{y}})^t(\mathbf{y} - \hat{\mathbf{y}}) + \lambda \hat{\mathbf{y}}^t \mathbf{L}\hat{\mathbf{y}}.
\tag{3.230}
$$

When this equation is differentiated to be 0 with respect to each element of $\hat{\mathbf{y}}$, the following result is obtained using, for example, eq(3.220) (page 175).

$$
\hat{\mathbf{y}} = (\mathbf{I}_n + \lambda \mathbf{L})^{-1}\mathbf{y}.
\tag{3.231}
$$

This equation indicates that the use of eq(3.209) in place of eq(2.59) (page 51) replaces **S** in eq(2.63) (page 52, chapter 2) by **L**. A matrix such as **S** or **L** is called a penalty

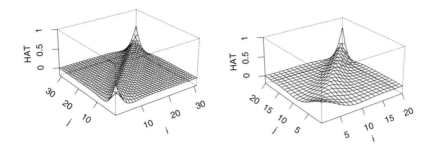

Figure 3.35   Values of elements of the hat matrix provided by the smoothing spline ($\lambda =$ 10.0) when the predictor values are $\{1, 2, 3, \ldots, 31\}$ (left). Values of elements of the hat matrix provided by the smoothing spline ($\lambda = 10^4$) when the predictor values are $\{1^{1.8}, 2^{1.8}, 3^{1.8}, \ldots, 20^{1.8}\}$ (right).

matrix. There are other procedures to derive this **L** (e.g., Chapter 2 of [10]; page 29 of [1]; page 61 of [11]; page 181 of [8]).

Figure 3.35(left) shows an example of a hat matrix calculated using eq(3.226). The predictor values of data are $\{1, 2, 3, \ldots, 31\}$; the number of data is 31. The smoothing parameter (λ) is 10.0. This hat matrix is analogous to that obtained using the roughness penalty defined using differences (figure 2.29(right) (page 53)). Figure 3.35(right) displays the hat matrix when the predictor is not equispaced: $\{1^{1.8}, 2^{1.8}, 3^{1.8}, \cdots, 20^{1.8}\}$. The value of λ here is 10^4. This hat matrix is characterized by a small wideness in the region where data are sparse. It is due to the smaller number of neighboring data in an area with lower density of data. This feature is similar to that of the Nadaraya-Watson estimator and local polynomial regression. However, there is some difference. Figure 3.36(left) illustrates some elements of a hat matrix ($\lambda = 10^4$, the predictor values are the same as the previous ones): $\{[\mathbf{H}]_{1j}\}$ ($1 \leq j \leq 20$), $\{[\mathbf{H}]_{6j}\}$, $\{[\mathbf{H}]_{11j}\}$, $\{[\mathbf{H}]_{16j}\}$, and $\{[\mathbf{H}]_{20j}\}$. To represent relative weights, normalization is carried out to obtain $\mathbf{H}'$ (figure 3.36(right)) in which the maximal value of "a" becomes 1, that of "b" also becomes 1, and the same is true of "c," "d," and "e." The shape that $\mathbf{H}'$ forms is broader in the region where the predictor values are distributed sparsely. That is, the number of data to be used in calculating an estimate is prevented from being very small in a sparse-data area. It is known that the broadness of the hat matrix is proportional to the $-\frac{1}{4}$ power of the density of data, except in areas close to the two ends (page 48 of [10]; [17]). LOESS, which is discussed later, also yields an estimate using data in a wide area in a region of sparse data. While the smoothing spline is similar to LOESS in terms of the feature that both methods refer to data located in a wider area for lower density of data, this trend is not marked in the smoothing spline. In this sense, the smoothing spline falls between local polynomial regression (figure 3.17 (page 147)) and LOESS (figure 3.43 (page 189)).

In addition, eq(3.223) (page 175) is derived by applying the simply supported beam discussed earlier. As seen in figure 3.34, springs are attached at $\{X_i, Y_i\}$ to bend the beam by pulling it up or down. The resulting shape of the beam is $y(x)$.

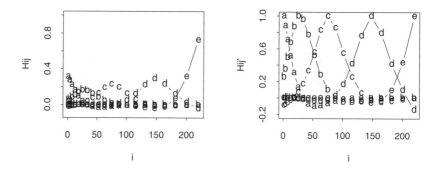

Figure 3.36   Some elements of a hat matrix when $\lambda = 10^4$. "a" indicates $\{[\mathbf{H}]_{1j}\}$. "b" indicates $\{[\mathbf{H}]_{6j}\}$. "c" indicates $\{[\mathbf{H}]_{11j}\}$. "d" indicates $\{[\mathbf{H}]_{16j}\}$. "e" indicates $\{[\mathbf{H}]_{20j}\}$ (left). "a," "b," "c," "d," and "e" are normalized to make each maximal value 1 ($\mathbf{H}'$) (right).

Since the springs obey Hooke's law (eq(3.199) (page 167)), we obtain

$$
\begin{aligned}
y(X_i) &= Y_i + \Delta Y_i \\
&= Y_i - \frac{W_i}{\kappa} \\
&= Y_i - \frac{EI \cdot \theta_i}{\kappa},
\end{aligned} \tag{3.232}
$$

where $\{W_i\}$ indicate the forces of springs exerted on the beam at $\{X_i\}$; the upward direction is positive. On the other hand, the force applied by the ith spring is $-W_i$ ($= -EI \cdot \theta_i$) because of the law of action and reaction. Hence, we arrive at eq(3.232). By taking account of eq(3.214) (page174) and eq(3.215), we have

$$
\begin{pmatrix} \mathbf{R} + \dfrac{EI}{\kappa}\mathbf{I}_n & \mathbf{Q}^t \\ \mathbf{Q} & \mathbf{0} \end{pmatrix} \begin{pmatrix} \boldsymbol{\theta} \\ \mathbf{c} \end{pmatrix} = \begin{pmatrix} \mathbf{y} \\ \mathbf{0} \end{pmatrix}. \tag{3.233}
$$

Hence, λ in eq(3.224) (page 176) corresponds to the flexural rigidity of a beam divided by the spring constant. This is ensured by the fact that the total elastic energy accumulated in the beam and the spring is written as below using eq(3.201) (page 168) and eq(3.202).

$$
E_{bending+spring} = \frac{\kappa}{2} \sum_{i=1}^{n} (m(X_i) - Y_i)^2 + \frac{EI}{2} \int_{X_1}^{X_n} \left(\frac{d^2 m(x)}{dx^2} \right)^2 dx. \tag{3.234}
$$

The principle of minimum potential energy states that $\hat{m}(x)$, which represents the shape of a beam, should minimize $E_{bending+spring}$. Derivation of $\hat{m}(x)$ that minimizes this energy is equivalent to the minimization of eq(3.209) (page 170) after $\lambda = \frac{EI}{\kappa}$ is set. Hence, $\hat{m}(x)$ given by the smoothing spline indicates the shape of a beam in which the total elastic energy of the beam and springs is minimized when these objects are derived as in figure 3.37; that is, this system is in stable equilibrium. In addition, while λ determines the trade-off of bias and variance in the smoothing spline, in the problem of the system of a beam and springs, $\frac{EI}{\kappa}$ determines the trade-off of the elasticity of springs and that of the beam.

Furthermore, if we set $\frac{EI}{\kappa} \to \infty$ in eq(3.233) (i.e., $\lambda \to \infty$ in eq(3.224)), all $\{\theta_i\}$ become close to 0. This corresponds to the fact that when the elasticity of the beam is very high or that of springs is low, the shape of the beam comes close to a straight line. On the other hand, if the elasticity of the beam is very low or that of the spring is very high, the shape of the beam becomes similar to that of a natural spline that interpolates data. Additionally, if we set each $\{Y_i\}$ as 1, all estimates are 1 regardless of the elasticity of springs; this is obvious from physical intuition (the value of $E_{bending+spring}$ in eq(3.234) is 0 in this case). Therefore, an estimate obtained using the smoothing spline is a weighted average of $\{Y_i\}$.

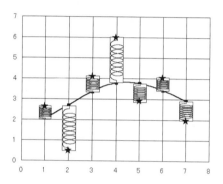

Figure 3.37   Smoothing spline is realized by springs and a beam.

Direct use of eq(3.192) (page 165) and eq(3.225) (page 176) to compute estimates sometimes results in an ill-conditioned numerical calculation. Two methods of avoiding this difficulty are well known. One is the use of the B-spline. The other is a method that utilizes QR decomposition, Cholesky decomposition, and singular value decomposition. For the latter, refer to [1] or page 165 of [10]. Here we outline the method making use of the B-spline as bases.

First, $m(x)$ is set as

$$m(x) = \sum_{j=1}^{P} \beta_j \cdot g_j(x). \tag{3.235}$$

This equation means that $m(x)$ is approximated by a linear combination of P bases: $\{g_j(x)\}$. Various bases are possible. A typical choice is the B-spline or a Gaussian function. To derive a smoothing spline based on a linear combination of bases using eq(3.235), we set

$$\mathbf{G} = \begin{pmatrix} g_1(X_1) & g_2(X_1) & \cdots & g_P(X_1) \\ g_1(X_2) & g_2(X_2) & \cdots & g_P(X_2) \\ g_1(X_3) & g_2(X_3) & \cdots & g_P(X_3) \\ \vdots & \vdots & \ddots & \vdots \\ g_1(X_n) & g_2(X_n) & \cdots & g_P(X_n) \end{pmatrix}, \tag{3.236}$$

$$\boldsymbol{\beta} = \begin{pmatrix} \beta_1 \\ \vdots \\ \beta_P \end{pmatrix}, \tag{3.237}$$

$$\mathbf{K} = \begin{pmatrix} k_{11} & k_{12} & \cdots & k_{1P} \\ k_{21} & k_{22} & \cdots & k_{2P} \\ \vdots & \vdots & \ddots & \vdots \\ k_{P1} & k_{P2} & \cdots & k_{PP} \end{pmatrix}. \tag{3.238}$$

k_{ij} in eq(3.238) is

$$k_{ij} = \int_{-\infty}^{\infty} \left(\frac{d^2 g_i(x)}{dx^2} \right) \left(\frac{d^2 g_j(x)}{dx^2} \right) dx. \tag{3.239}$$

Using eq(3.236), eq(3.237), and eq(3.238), eq(3.209) (page 170) is rewritten as

$$E_{ss} = (\mathbf{y} - \mathbf{G}\boldsymbol{\beta})^t (\mathbf{y} - \mathbf{G}\boldsymbol{\beta}) + \lambda \boldsymbol{\beta}^t \mathbf{K} \boldsymbol{\beta}. \tag{3.240}$$

If this equation is differentiated to be 0 with respect to each $\{\beta_i\}$ and the resultant $\boldsymbol{\beta}$ that minimizes this value is set to be $\hat{\boldsymbol{\beta}}$, then $\hat{\boldsymbol{\beta}}$ is expressed as

$$\hat{\boldsymbol{\beta}} = (\mathbf{G}^t \mathbf{G} + \lambda \mathbf{K})^{-1} \mathbf{G}^t \mathbf{y}. \tag{3.241}$$

Hence, a hat matrix is written as

$$\mathbf{H} = \mathbf{G}(\mathbf{G}^t \mathbf{G} + \lambda \mathbf{K})^{-1} \mathbf{G}^t. \tag{3.242}$$

This $\mathbf{H}$ is identical to that defined in eq(3.226) (page 176).

Then, we set $P = n + 2$, and $\{g_j(x)\}$ $(1 \le j \le n + 2)$ is assumed to be bases ($\{B_j(x)\}$) of the cubic B-spline whose knots are positioned at $\{X_i\}$ $(1 \le i \le n)$. Since we need $(n + 6)$ knots to create $(n + 2)$ bases of the cubic B-spline, 6 knots must be added to the knots at $\{X_i\}$ $(1 \le i \le n)$. X_1 and X_n are usually transformed into quadruple knots. However, as long as 3 knots out of 6 are placed on the left-hand side of X_1 or at X_1 and the remaining 3 are located on the right-hand side of X_n or at X_n, estimates positioned in the region of $X_1 \le x \le X_n$ are not affected.

Then, if $X_1 \le x \le X_n$, the cubic spline defined in eq(3.205) (page 168) is represented as

$$a_0 + a_1 + \frac{1}{12} \sum_{j=1}^{n} b_j \mid x - X_j \mid^3 = \sum_{j=1}^{n+2} \beta_j B_j(x). \tag{3.243}$$

$(n + 2)$ coefficients $((a_0, a_1, b_1, b_2, \ldots, b_n))$ on the left-hand side are transformed into $(n + 2)$ coefficients $((\beta_1, \beta_2, \ldots, \beta_{n+2}))$ on the right-hand side. For this reason, $\hat{m}(x)$ derived using this $\{B_j(x)\}$ $(1 \le j \le n+2)$ and by the calculation of eq(3.241) after setting $P = n + 2$ in eq(3.235) is equivalent to $\hat{m}(x)$ obtained through the direct use of eq(3.209) (page 170). That is, if a regression equation of the form of eq(3.235) in which cubic B-splines are employed as bases with knots located at $\{X_i\}$ $(1 \le i \le n)$ is derived, estimates that minimize eq(3.209) are obtained in the region of $X_1 \leq x \leq X_n$. An estimate outside $X_1 \leq x \leq X_n$ is calculated by extrapolation using linear equations created using values of $\hat{m}(x)$ and the first derivatives of $\hat{m}(x)$ at X_1 and X_n. Thus, the estimates are the same as those obtained by minimizing eq(3.216). Furthermore, $\mathbf{H}$ in eq(3.242) is identical to $\mathbf{H}$ in eq(3.226) (page 176). Additionally, the B-spline possesses an advantage over eq(3.210) (page 172); direct solution of the simultaneous equation (eq(3.241)) leads to estimates with high accuracy.

However, data with certain characteristics lead to unstable numerical calculation even when the B-spline is employed. In this case, the number of bases of the B-spline (i.e., the number of knots) can be reduced, which stabilizes the calculation and also reduces the amount of numerical calculation. Some instead of all $\{X_i\}$ are used as knots, which means that eq(3.213) (page 172) is approximated using an equation with a smaller number of knots. However, estimates given using commonplace data are negligibly affected by even a considerable reduction of the number of knots.

In the previous section, the value of $[\mathbf{H}]_{kk}$ was shown to decline as $w(X_i - X_k)$ used to carry out local polynomial regression broadened (eq(3.134) (page 144)). A similar relationship holds for the smoothing spline. That is to say, as the value of λ in eq(3.209) increases, the value of $[\mathbf{H}]_{kk}$ decreases. Since eq(3.209), eq(3.216) (page 174), and eq(3.241) are equivalent, the proof of this characteristic using $[\mathbf{H}]_{kk}$ given by eq(3.242) indicates that the smoothing spline in general has this relationship.

By the use of eq(3.242), $[\mathbf{H}]_{kk}$ is expressed as

$$[\mathbf{H}]_{kk} = \sum_{i=1}^{n}\sum_{j=1}^{n}[\mathbf{G}]_{ki}\left[(\mathbf{G}^t\mathbf{G} + \lambda\mathbf{K})^{-1}\right]_{ij}[\mathbf{G}^t]_{jk}. \qquad (3.244)$$

Comparison of this equation with eq(3.136) (page 144) shows that if both $\mathbf{G}^t\mathbf{G}$ and $\mathbf{K}$ are quasi-definite matrices, the increase of the value of λ causes the reduction of $\mathbf{H}_{kk}$. First, for $\mathbf{K}$, since $\boldsymbol{\beta}^t\mathbf{K}\boldsymbol{\beta}$ is a roughness penalty, $\boldsymbol{\beta}^t\mathbf{K}\boldsymbol{\beta}$ is positive or 0, regardless of $\boldsymbol{\beta}$. $\mathbf{K}$ is apparently a symmetric matrix by virtue of its definition. Hence, $\mathbf{K}$ is a quasi-definite matrix. Next, $\mathbf{G}^t\mathbf{G}$ is also a quasi-definite matrix because of eq(3.137) (page 144). Thus, it is proved that the value of $\mathbf{H}_{kk}$ decreases as the value of λ increases.

In addition, since $\mathbf{K}$ and $\mathbf{G}^t\mathbf{G}$ are quasi-definite matrices, a hat matrix corresponding to the smoothing spline is expressed as

$$\begin{aligned}
\mathbf{H} &= \mathbf{G}\bar{\mathbf{U}}\mathbf{A}\bar{\mathbf{U}}^t\mathbf{G}^t \\
&= \mathbf{U}^+\mathbf{A}\mathbf{U}^{+t}, \qquad (3.245)
\end{aligned}$$

where $\bar{\mathbf{U}}$ is an orthogonal matrix for diagonalizing $(\mathbf{G}^t\mathbf{G}+\lambda\mathbf{K})^{-1}$ and $\mathbf{A}$ is a diagonal matrix in which all diagonal elements are either positive or 0. In addition, $\mathbf{U}^+ = \mathbf{G}\bar{\mathbf{U}}$. By the use of a similar form to eq(2.71) (page 56), eq(3.245) is substantialized to be

$$\begin{aligned}
\mathbf{U}^+\mathbf{A}\mathbf{U}^{+t} &= \sum_{i=1}^{n}\alpha_i\mathbf{u}_i^+ \cdot \mathbf{u}_i^{+t} \\
&= \sum_{i=1}^{n}\frac{1}{1+l_i\lambda}\mathbf{u}_i^+ \cdot \mathbf{u}_i^{+t}, \qquad (3.246)
\end{aligned}$$

where $\{l_i\}$ are eigenvalues of $\mathbf{L}$ (eq(3.230) (page 176)). $\{\mathbf{u}_i^+\}$ are column vectors constituting $\mathbf{U}^+$; they are also eigenvectors of $\mathbf{L}$, and eigenvectors of $(\mathbf{I}_n + \lambda\mathbf{L})^{-1}$. Since $\mathbf{L}$ is a penalty matrix, all $\{l_i\}$ are positive or 0. Hence, all diagonal elements of the diagonal matrix obtained by diagonalizing $\mathbf{H}$ are positive. Furthermore, two of $\{\frac{1}{1+l_i\lambda}\}$ are 1 and two of $\{l_i\}$ are 0 because eq(2.93) (page 68) holds even when the predictor of data is not equispaced.

It is a favorable feature that all eigenvalues of a hat matrix are positive or 0 (cf. Chapter 2). The smoothing spline satisfies this condition. Furthermore, the results of numerical calculation in Chapter 2 imply that the hat matrix $((\mathbf{I}_n + \lambda\mathbf{S})^{-1}$ (eq(2.63) (page 52))) given by the smoothing spline with a roughness penalty defined using differences contains this feature. This mathematical justification is given as follows.

$\mathbf{S}$ is a quasi-definite matrix because eq(2.59) (page 51) is written as

$$
\begin{aligned}
E_{ss-equi} &= \| \mathbf{y} - \hat{\mathbf{y}} \|^2 + \lambda \| \mathbf{D}\mathbf{y} \|^2 \\
&= \| \mathbf{y} - \hat{\mathbf{y}} \|^2 + \lambda \mathbf{y}^t \mathbf{D}^t \mathbf{D}\mathbf{y},
\end{aligned}
\tag{3.247}
$$

where $\| \ \|$ stands for the length of a vector. $\mathbf{D}$ is

$$
\begin{pmatrix}
1 & -2 & 1 & 0 & \cdots & \cdots & \cdots & \cdots & \cdots & \cdots & \cdots & 0 \\
0 & 1 & -2 & 1 & 0 & \cdots & \cdots & \cdots & \cdots & \cdots & \cdots & 0 \\
0 & 0 & 1 & -2 & 1 & 0 & \cdots & \cdots & \cdots & \cdots & \cdots & 0 \\
\vdots & \vdots & \vdots & \vdots & \vdots & \vdots & \vdots & \vdots & \vdots & \vdots & \vdots & \vdots \\
0 & \cdots & \cdots & \cdots & \cdots & \cdots & 0 & 1 & -2 & 1 & 0 \\
0 & \cdots & \cdots & \cdots & \cdots & \cdots & \cdots & 0 & 1 & -2 & 1
\end{pmatrix}.
\tag{3.248}
$$

Calculation similar to eq(3.220) (page 175) yields

$$
\hat{\mathbf{y}} = (\mathbf{I}_n + \lambda\mathbf{D}^t\mathbf{D})^{-1}\mathbf{y}.
\tag{3.249}
$$

That is, $\mathbf{S}$ in eq(2.63) is equal to $\mathbf{D}^t\mathbf{D}$. Hence, eq(3.137) (page 144) shows that $\mathbf{S}$ is also a quasi-definite matrix. The fact that $\mathbf{D}^t\mathbf{D}$ in eq(3.247) is a penalty matrix is further evidence of $\mathbf{D}^t\mathbf{D}$ being a quasi-definite matrix. Consequently, $(\mathbf{I}_n + \lambda\mathbf{S})^{-1}$ becomes

$$
\begin{aligned}
\mathbf{H} &= (\mathbf{I}_n + \lambda\mathbf{S})^{-1} \\
&= \tilde{\mathbf{U}}\tilde{\mathbf{A}}\tilde{\mathbf{U}}^t \\
&= \sum_{i=1}^{n} \tilde{\alpha}_i \tilde{\mathbf{u}}_i \cdot \tilde{\mathbf{u}}_i^t \\
&= \sum_{i=1}^{n} \frac{1}{1 + s_i\lambda} \tilde{\mathbf{u}}_i \cdot \tilde{\mathbf{u}}_i^t,
\end{aligned}
\tag{3.250}
$$

where the result of the diagonalization of $(\mathbf{I}_n + \lambda\mathbf{S})^{-1}$ using $\tilde{\mathbf{U}}$ is diagonal matrix $\tilde{\mathbf{A}}$; $\{\tilde{\alpha}_i\}$ are diagonal elements of $\tilde{\mathbf{A}}$. The eigenvalues of $\mathbf{S}$ are defined as $\{s_i\}$ $(1 \leq i \leq n)$. $\{\tilde{\mathbf{u}}_i\}$ is a column vector constituting $\tilde{\mathbf{U}}$; it is also eigenvectors of $\mathbf{S}$, and eigenvectors of $(\mathbf{I}_n + \lambda\mathbf{S})^{-1}$. Since $\mathbf{S}$ is a quasi-definite matrix, all s_i are positive or 0. Hence, all of the diagonal elements of the matrix derived by diagonalizing $\mathbf{H}$

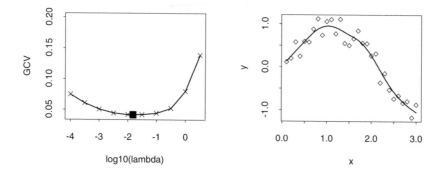

Figure 3.38   Optimization of a smoothing parameter using GCV (after GCV for 10 smoothing parameters are calculated, the smoothing parameter was optimized by the golden ratio search to obtain ■) (left). Estimates obtained using the resulting value of the smoothing parameter (right).

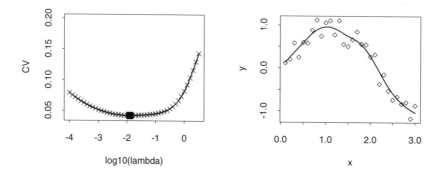

Figure 3.39   Optimization of a smoothing parameter using CV (CV for 46 smoothing parameters were calculated to obtain the optimal value (■)) (left). Estimates obtained using the resulting smoothing parameter (right).

given by eq(3.250) are also positive. In particular, allowing for eq(2.93) (page 68), two of $\{\frac{1}{1+s_i\lambda}\}$ are 1 and hence two of $\{s_i\}$ are 0.

Since estimates using the smoothing spline are obtained with a linear estimator, the hat matrix defined in eq(3.226) (page 176) or eq(3.242) (page 180) leads to, for example, CV, GCV, and AIC. However, these equations are not utilized as is in most cases; some techniques are applied to reduce the amount of computation and/or to stabilize the numerical calculation. Figure 3.38(left) shows the process by which, after 10 values $(10^{-4}, 10^{-3.5}, 10^{-3}, \ldots, 10^{0.5})$ are adopted as λ to obtain GCV and to elucidate the rough relationship between λ and GCV, the optimal $\lambda \, (= 10^{-1.814706})$ is found by conducting the golden ratio search around λ values corresponding to small values of GCV. Resulting estimates are displayed in figure 3.38(right). Figure 3.39 illustrates the values of CV derived using 46 λ values $(10^{-4}, 10^{-3.9}, 10^{-3.8}, \ldots, 10^{0.5})$ with the same data as those in figure 3.38, and estimates obtained by smoothing with the resulting value of $\lambda \, (= 10^{-1.9})$.

The cubic natural spline and cubic smoothing spline have been treated hitherto. This is because they are commonly used in practice and are strongly associated with physical problems such as the simply supported beam. However, alternatives with other degrees are possible. They are useful for several purposes, and hence an outline of them is presented next.

The degree of a natural spline must be an odd number. A $(2k-1)$-degree (k is a positive integer) natural spline ($m(x)$) that interpolates $\{X_i, Y_i\}$ minimizes the value below, under the condition of $m(X_i) = Y_i \, (1 \leq i \leq n)$.

$$\int_{\gamma_1}^{\gamma_2} \left(\frac{d^k m(x)}{dx^k} \right)^2 dx. \tag{3.251}$$

Here, $\gamma_1 < X_1$ and $\gamma_2 > X_n$. The resulting $s_{(2k-1)}(x)$ is

$$s_{(2k-1)}(x) = \tilde{a}_0 + \tilde{a}_1 x + \cdots + \tilde{a}_{k-1} x^{k-1} + \sum_{j=1}^{n} \tilde{b}_j \cdot ((x - X_j)_+)^{2k-1}. \tag{3.252}$$

$\tilde{a}_0, \tilde{a}_1, \ldots, \tilde{a}_{k-1}, \tilde{b}_1, \tilde{b}_2, \ldots, \tilde{b}_n$ are constants. Since $m(x)$ must be a $(k-1)$th-degree polynomial in the region of $x > X_n$, $\{\tilde{b}_j\}$ needs k conditions:

$$\sum_{j=1}^{n} \tilde{b}_j X_j^l = 0 \quad (l = 0, 1, \ldots, k - 1). \tag{3.253}$$

When $k = 2$, a function represented by eq(3.252) under the condition of eq(3.253) is equivalent to that represented by eq(3.164) (page 155) under the conditions of eq(3.172) (page 158) and eq(3.173). That is, the former equations can be represented by the latter equations; the other way around is also possible.

A $(2k-1)$-degree smoothing spline given by the application of such a $(2k-1)$-degree natural spline minimizes

$$E_{ss(2k-1)} = \sum_{i=1}^{n} (Y_i - m(X_i))^2 + \lambda \int_{-\infty}^{\infty} \left(\frac{d^k m(x)}{dx^k} \right)^2 dx. \tag{3.254}$$

As in the case of the cubic smoothing spline, the region of integration for obtaining a roughness penalty does not affect the results, as long as it contains the region where data exist. If $s_{(2k-1)}(x)$ is written in a similar manner to eq(3.210) (page 172), we have

$$s_{(2k-1)}(x) = c_0 + c_1 x + \cdots + c_{k-1} x^{k-1} + \sum_{i=1}^{n} \theta_i \cdot G(x - X_i), \qquad (3.255)$$

where $c_0, c_1, \ldots, c_{k-1}, \theta_1, \theta_2, \ldots, \theta_n$ are constants. $G(x - X_i)$ is

$$G(x - X_i) = \frac{\Gamma(0.5 - k)}{2^{2k} \cdot \sqrt{\pi} \cdot (k-1)!} \, |x - X_i|^{2k-1}, \qquad (3.256)$$

where $(k-1)!$ stands for the factorial of $(k-1)$. Furthermore, $\{\theta_j\}$ must satisfy the k conditions:

$$\sum_{j=1}^{n} \theta_j X_j^l = 0 \quad (l = 0, 1, \ldots, k-1). \qquad (3.257)$$

GCVPACK ([1]) enables the calculation of a smoothing spline with a degree other than 3. S-Plus objects to carry out this sort of calculation are available (page 60 of [15]).

3.7 LOESS

Predictor values of data ($\{X_i\}$) are sometimes not equispaced. When local polynomial regression is applied to such data, the bandwidth (h) should vary depending on the estimation point (x) in order to obtain beneficial estimates. This is illustrated using a simple simulation data:

$$Y_i = \sin(0.1\pi X_i) + 0.7\sin(1.23\pi X_i) + 0.5\cos(2.7\pi X_i) + 0.2\sin(0.6\pi X_i) + \epsilon_i, \qquad (3.258)$$

where the number of data is 120 and $\{X_i\} = \{0.2, 0.4, 0.6, \ldots, 19.6, 19.8, 20, 21, 22, 23, \ldots, 40\}$. $\{\epsilon_i\}$ are realizations of $N(0.0, 0.2^2)$ (normal distribution; average is 0.0 and standard deviation is 0.2). Figure 3.40 displays the results of local linear regression with the bandwidths (h in eq(3.104) (page 133)) of $h = 0.15$ and $h = 1.5$ using these simulation data. To test the validity of the resulting estimates, E^l and E^r defined below are calculated.

$$E^l = \sum_{i=1}^{397} (\hat{Y}_i^+ - \tilde{Y}_i)^2, \qquad (3.259)$$

$$E^r = \sum_{i=398}^{797} (\hat{Y}_i^+ - \tilde{Y}_i)^2. \qquad (3.260)$$

$\hat{Y}_i^+$ is an estimate at X_i^+, and X_i^+ is

$$X_i^+ = 0.15 + 0.05i \quad (1 \le i \le 797). \qquad (3.261)$$

$\tilde{Y}_i$ is obtained by omitting ϵ_i from Y_i (eq(3.258)). When $h = 0.15$, $E^l = 22.30757$ and $E^r = 180.9858$ are obtained. $E^l = 156.0787$ and $E^r = 157.5099$ are obtained when $h = 1.5$. That is, a narrow bandwidth is better adapted to the region where data are densely distributed, while a broad bandwidth is better suited for the region where data are sparsely distributed. This indicates that in the region where data are sparse, quick oscillation of data should be regarded as aliasing errors (Chapter 1 (page 15)) that must be eliminated by smoothing data with a wide bandwidth. On the other hand, in the region where data are dense, fast movements of data are required to be looked upon as an intrinsic property of data, and hence a narrow bandwidth should be adopted for smoothing. Estimations with a constant bandwidth in the entire region do not handle this situation correctly. The bandwidth should be varied according to the density of data. For local linear regression, a method of varying the bandwidth for each data is to use the equation below in place of eq(3.112) (page 139):

$$E_{LOESS}(x^*) = \sum_{i=1}^{n} \left(w\left(\frac{X_i - x^*}{h_k(x^*)}\right) \left(a_0(x^*) + \sum_{j=1}^{p} a_i(x^*)(X_i - x^*)^j - Y_i\right)^2 \right),$$
(3.262)

where h is a function of x^* and k (i.e., $h_k(x^*)$). This $h_k(x^*)$ is the distance between the estimation points (x^*) and the kth-nearest data from x^*. That is, a high density of data at x^* is represented by the distance between x^* and kth nearest-data being short. $h_k(x^*)$ is called the k-nearest-neighbor distance (for short, k-NN). Apart from eq(3.262), an estimate derived using the k-nearest-neighbor distance is generally called a k-nearest-neighbor estimate.

A tricube weight function (eq(3.116) (page 128)) is usually used as $w\left(\frac{X_i - x^*}{h_k(x^*)}\right)$. It provides weights that are 0 for the data located at the k-nearest-neighbor distance or further. Since $\frac{k}{n}$ clearly indicates the broadness of $w\left(\frac{X_i - x^*}{h_k(x^*)}\right)$, $\frac{k}{n}$ is called the span; this is used as an index to show the broadness of a kernel function of LOESS instead of k. A span given by a tricube weight function is an approximate ratio of data used to calculate an estimate. When the span (s) is specified as an index of the broadness of a kernel function, $k = [ns]$ (n is the number of data, $[\cdot]$ is the smallest integer that is larger than or equal to the value within $[\ ]$). However, $[\cdot]$ is defined as the closest integer to the number within $[\ ]$ in some literature (e.g., page 24 of [7]). In addition, if the span is larger than 1, $k = n$ is set and $w\left(\frac{X_i - x^*}{h_k(x^*)}\right)$ in eq(3.262) is replaced with $w\left(\frac{X_i - x^*}{s \cdot h_k(x^*)}\right)$ (page 101 of [5]).

This technique is termed LOESS. LOESS is originally a word that means yellow ocher; it is pronounced like "low-" of lowland and "Is" of Islamism (however, this word is pronounced "low S" on page 77 of [6]). It was adopted after the fact that LOESS is slightly different from LOWESS, described later, and there is a geological formation of curved lines of yellow ocher in a soil profile ([4]). This word originates from the German word "löss" (corresponding to "loose" in English) (page 314 of [3]); the soil alongside the Rhine River has been called löss. In addition, loess as a nonparametric regression technique implies an abbreviation of "local regression" (page 314 of [3]; page 94 of [5]; page 331 of [16]; page 19 of [9]). To avoid confusion in the use of the term LOESS, a better definition should be used: nearest-neighbor local-linear regression.

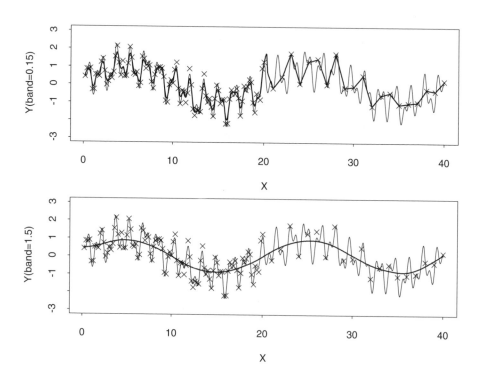

Figure 3.40   Data ($\times$), $m(x)$ (thin line), and estimates derived by local linear regression (thick line) (top). A different bandwidth is adopted ($h = 1.5$) (bottom). The data are the same as those in eq(3.258).

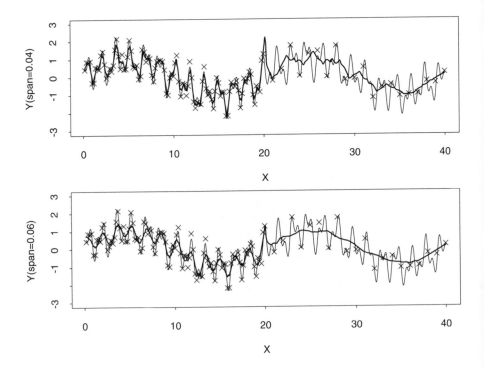

Figure 3.41   Data ($\times$), $m(x)$ (thin solid line), and estimates given by LOESS (thick solid line) ($s = 0.04$), (top). A different span is used ($s = 0.06$), (bottom). The data were created using eq(3.258).

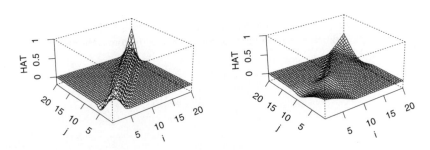

Figure 3.42   A hat matrix for $\{X_i\} = \{1^{1.8}, 2^{1.8}, 3^{1.8}, \dots, 20^{1.8}\}$. $s = 0.3$ (left). $s = 0.6$ (right).

Figure 3.41 shows the result of examining the features of LOESS using the same data as those in figure 3.40. The span is 0.04 or 0.06. A tricube weight function (eq(3.116) (page 128)) is used as a kernel function. Linear equations are used for local fitting in the implementation of LOESS. This figure indicates that the use of LOESS produces the favorable effect that a narrow bandwidth leaves quick movements of data as they are in the region where data are dense, and a broad bandwidth removes fast oscillation of data in the region where data are sparse. While the span of 0.04 yields $E^l = 19.89175$ and $E^r = 200.8386$, that of 0.06 results in $E^l = 67.18008$ and $E^r = 157.1754$.

When the number of data is 20 and the predictor values of data are $\{X_i\} = \{1^{1.8}, 2^{1.8}, 3^{1.8}, \ldots, 20^{1.8}\}$, the hat matrix is as illustrated in figure 3.42. The span is 0.3 or 0.6. As expected from the definition of LOESS, the broadness of the hat matrix is not narrow in the region where data are sparse (i.e., the value of i is large). This is in contrast to figure 3.16 (page 147). Figure 3.43(left) displays part of the hat matrix, $\{[\mathbf{H}]_{1j}\}$ $(1 \le j \le 20)$, $\{[\mathbf{H}]_{6j}\}$, $\{[\mathbf{H}]_{11j}\}$, $\{[\mathbf{H}]_{16j}\}$, and $\{[\mathbf{H}]_{20j}\}$, where the predictor values are the same as the previous ones and the span is 0.4. Normalization is carried out to obtain $\mathbf{H}'$ (figure 3.43(right)) in which the maximal value of "a" becomes 1, that of "b" also becomes 1, and the same is true of "c," "d," and "e." The values of elements of $\mathbf{H}'$ are distributed widely in the region where predictor values of data are sparse. Additionally, the variation of the broadness is more substantial than that of the smoothing spline (figure 3.36 (page 178)).

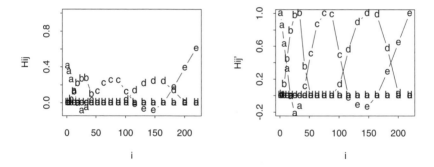

Figure 3.43   A part of the hat matrix when the span is 0.4. "a" stands for $\{[\mathbf{H}]_{1j}\}$, "b" stands for $\{[\mathbf{H}]_{6j}\}$, "c" stands for $\{[\mathbf{H}]_{11j}\}$, "d" stands for $\{[\mathbf{H}]_{16j}\}$, "e" stands for $\{[\mathbf{H}]_{20j}\}$ (left). The values of "a," "b," "c," "d," and "e" are normalized to make each maximal value 1 ($\mathbf{H}'$) (right).

Thus, a hat matrix is obtainable for LOESS because LOESS is an extension of local polynomial regression achieved by adjusting the broadness of a kernel function according to the density of data. However, when cross-validation is carried out using the definition (eq(3.33) (page 115)) as it is, $\{h_1(x^*), \ldots, h_{k-1}(x^*), h_{k+1}(x^*), \ldots, h_n(x^*)\}$ which is obtained by omitting the kth data, is slightly different from those gained by omitting $h_k(x^*)$ from $\{h_1(x^*), \ldots, h_n(x^*)\}$ (obtained using a whole set of data). Consequently, the use of eq(3.41) (page 117) to calculate CV of LOESS and eq(3.46) (page 118) to compute GCV yield only approximations.

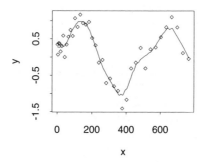

Figure 3.44 Example of smoothing by LOESS. The span is 0.18.

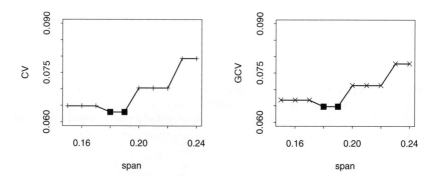

Figure 3.45 Calculations of CV and GCV for obtaining the span to be used in figure 3.44.

Simulation data are produced to show the procedure of smoothing by LOESS. The predictor value is $\{X_i\} = \{1^{1.8}, 2^{1.8}, 3^{1.8}, \ldots, 40^{1.8}\}$, and the values of the target variable are generated by

$$Y_i = \sin(0.004\pi X_i) + \epsilon_i, \tag{3.263}$$

where ϵ_i is a realization of $N(0.0, 0.3^2)$ (normal distribution; average is 0.0 and standard deviation is 0.3). Figure 3.44 shows the result when the span is 0.18. Figure 3.45 depicts the values of CV (eq(3.41) (page 117)) and GCV (eq(3.46) (page 118)) when the values of the span are $\{0.15, 0.16, \ldots, 0.24\}$. This figure shows that the optimal span is 0.18 (the result remains the same when the span is 0.19).

3.8 SUPERSMOOTHER

LOESS is a method aimed at coping with the variation of the density of data depending on the region through local adjustment of the bandwidth. However, there are two more reasons for modifying the local bandwidth. The one is the local variation of the variance of error ($\{\epsilon_i\}$). Although we have assumed hitherto that the variance of $\{\epsilon_i\}$ does not depend on area, some data do not satisfy this assumption. In some cases, a large value of a predictor may unavoidably result in a large amount of error in an experiment. The other reason is that the behavior of the variation of data may alter locally. In the region where $m(x)$ varies considerably and the variation has essential meanings, analysts are tempted to reduce the bandwidth (or the smoothing parameter) in order to reproduce the variation clearly (i.e., to reduce the bias). On the other hand, in the region where the variation of data is small, the focus is shifted to the reduction of variance to broaden the bandwidth because bias is not greatly enlarged by a broader bandwidth.

To treat the variation of bandwidth motivated by these three reasons, cross-validation should be carried out using the data located in a limited region instead of all data, in order to estimate prediction errors. By adjusting the bandwidth to reduce the value of prediction errors obtained in this manner, a bandwidth suited to the characteristics of each region is calculated to realize smoothing according to the local features of data. For this purpose, supersmoother was developed. The basic concept of supersmoother is LOESS in which $p = 1$ in eq(3.262) (page 186). However, it does not simply derive a linear equation using data with the k-nearest-neighbor distance; it also adjusts the number of data on the right-hand side of an estimation point to be the same as that on the left-hand side, except at estimation points located near the two ends. In addition, while LOESS determines, by a certain method, the span used to smooth the whole data set using the value, supersmoother uses three values of k to derive estimates corresponding to each value of data and calculates the final estimates based on these three sets of estimates.

This procedure is concretely shown as follows.
(1) Three values of k are set as $k_1 = 0.05n$, $k_2 = 0.2n$, and $k_3 = 0.5n$ (n is the number of data, fractions below the decimal point are truncated). LOESS based on local linear regression is carried out for estimation. Estimates using k_1, k_2, and k_3 are termed tweeter, midrange, and woofer, respectively.

(2) To determine the value of k at $\{X_i\}$, local cross-validation is performed. By the computation above, local cross-validated residuals ($\{r_{(i)}(s)\}$ ($1 \leq i \leq n$)) corresponding to k_1, k_2, and k_3 are calculated. A local cross-validated residual at i is defined as

$$r_{(i)}(k) = \frac{Y_i - \hat{m}_k(X_i)}{1 - h_{ii}}, \tag{3.264}$$

where n is the number of data. $\hat{m}_k(X_i)$ ($k = k_1, k_2, k_3$) are estimates obtained by using k. h_{ii} is the ith diagonal elements of a hat matrix given by local linear regression.

(3) The values of $|r_{(i)}(k)|$ at $\{X_i\}$ are compared, and the value, among k_1, k_2, and k_3, which gives the smallest $|r_{(i)}(k)|$ is named $k^*(X_i)$ ($\{|r_{(i)}(k)|\}$ may be smoothed before (3)).

(4) The data of $\{X_i, k^*(X_i)\}$ is smoothed using k_2; the result is termed $\{\tilde{k}^*(X_i)\}$. $\hat{Y}_i$ is calculated using $\tilde{k}^*(X_i)$ as k; this process is repeated for all i to obtain $\{\hat{Y}_i\}$.

A simple example shows that this method realizes smoothing adaptable to the local behavior of data.

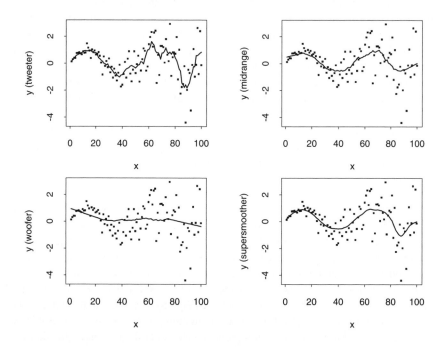

Figure 3.46   The amount of error depends heavily upon the region. Tweeter (top left). Midrange (top right). Woofer (bottom left). Estimates obtained using supersmoother (bottom right).

First, figure 3.46 exemplifies data in which error depends heavily upon the region. $n = 100$ and $X_i = i$ are set, and $\{Y_i\}$ is

$$Y_i = \sin(0.04\pi X_i) + X_i \cdot \epsilon_i. \tag{3.265}$$

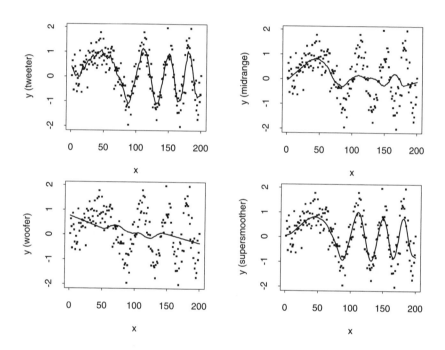

Figure 3.47 Width of oscillation depends upon the region. Tweeter (top left). Midrange (top right). Woofer (bottom left). Estimates obtained using supersmoother (bottom right).

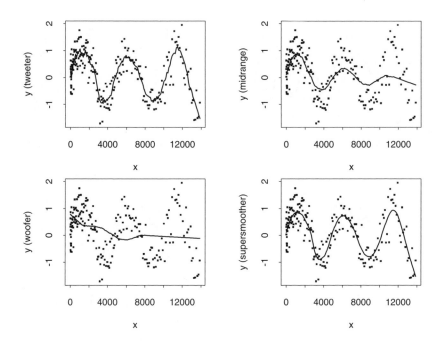

Figure 3.48   Density of data depends on the region. Tweeter (top left). Midrange (top right). Woofer (bottom left). Estimates obtained with supersmoother (bottom right).

ϵ_i is a realization of $N(0.0, 0.02^2)$ (normal distribution; mean is 0.0, and standard deviation is 0.02); that is, the standard deviation of error is proportional to X_i. While one of tweeter, midrange, and woofer gives good estimates in a specific region (bottom left), estimates obtained through the realization of local adaptation based on these three sets of estimates (given by tweeter, midrange, and woofer) are valid throughout the whole region. Figure 3.47 displays an example where the amount of variation of $m(x)$ greatly depends upon the region. $n = 200$ and $X_i = i$ are used; $\{Y_i\}$ is

$$Y_i = \sin(0.0002\pi X_i^2) + \epsilon_i. \tag{3.266}$$

ϵ_i is a realization of $N(0.0, 0.5^2)$ (normal distribution; mean is 0.0, and standard deviation is 0.5). Although the behavior of data dramatically alters depending on the region, supersmoother yields beneficial estimates throughout the whole region. Figure 3.48 shows a result of smoothing using a supersmoother; the predictor values of data are distributed inhomogeneously. $n = 200$ and $\{X_i\} = \{1^{1.8}, 2^{1.8}, \ldots, 200^{1.8}\}$ are used. $\{Y_i\}$ is generated from

$$Y_i = \sin(0.0004\pi X_i) + \epsilon_i. \tag{3.267}$$

ϵ_i is a realization of $N(0.0, 0.5^2)$ (normal distribution; mean is 0.0, and standard deviation is 0.5). The estimates given with supersmoother are valid throughout the whole region.

As is inferred from the algorithm, supersmoother yields estimates merely at the predictor values of data ($\{(X_i, \hat{Y}_i)\}$). This feature prompts the need for the interpolation of $\{(X_i, \hat{Y}_i)\}$ in one way or another.

3.9 LOWESS

Outliers sometimes cause critical problems because a small number of outliers threaten to seriously impair the intrinsic content of data by dramatically changing the trend of estimates. Therefore, some smoothing methods to curb the influence of outliers (i.e., methods robust to outliers) have been developed. LOWESS is well known among them. LOWESS is called a "robust version of LOESS" on occasion. In fact, S-Plus realizes this technique by altering an argument of the object loess(). It may sound slightly strange that LOWESS is renamed a robust version of LOESS since LOESS is coined to indicate a slight difference from LOWESS. However, a robust version of LOESS can be a more useful terminology than LOWESS because modification of LOESS to be robust to outliers produces LOWESS, whereas the term "LOcal WEighted Scatterplot Smoother (LOWESS)" does not contain the meaning that the method is robust to outliers.

The following is the algorithm of LOWESS.

(1) The result of smoothing of data by LOESS is termed $\hat{Y}_i$. Residuals ($\{r_i\}$ ($= \{Y_i - \hat{Y}_i\}$)) are computed.

(2) Robustness weights ($\{\sigma_i\}$ ($i = 1, \ldots, n$)) are derived as below:

$$\sigma_i = B(r_i/(6\hat{s})), \tag{3.268}$$

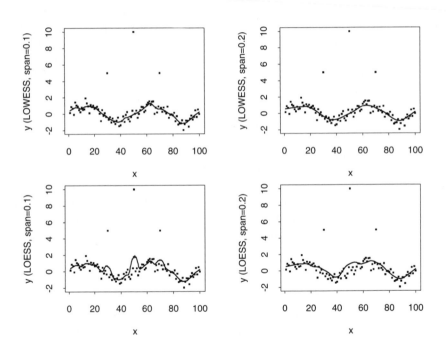

Figure 3.49   Smoothing data with outliers. LOWESS (span is 0.1) (top left). LOWESS (span is 0.2) (top right). LOESS (span is 0.1) (bottom left). LOESS (span is 0.2) (bottom right).

where $B(u)$ is a bisquare weight function (eq(3.114) (page 140)):

$$B(u) = \begin{cases} \frac{15}{16}(1-u^2)^2 & 0 \le |u| \le 1 \\ 0 & 1 < |u|. \end{cases} \quad (3.269)$$

$\hat{s}$ (scale estimate) is

$$\hat{s} = \text{med}(\{|r_i|\}). \quad (3.270)$$

med() indicates median.

(3) Smoothing by LOESS is carried out with weight of σ_i on ith data. That is, $w\left(\frac{X_i - x^*}{h_k(x)}\right)$ in eq(3.262) (page 186) is replaced with $\left(\sigma_i \cdot w\left(\frac{X_i - x^*}{h_k(x)}\right)\right)$ to obtain estimates. Then, residuals ($\{r_i\}$ ($= \{Y_i - \hat{Y}_i\}$)) are calculated.

(4) (2) and (3) above are repeated three more times.

This procedure results in smoothing that is robust to outliers. The number of repetitions in (4) can be another one.

Figure 3.49 illustrates the result of smoothing by LOWESS or LOESS (both are based on a linear equation); $n = 100$ and $X_i = i$ were set. $\{Y_i\}$ is generated in the following manner.

$$Y_i = \sin(0.04\pi X_i) + \epsilon_i. \quad (3.271)$$

ϵ_i is a realization of $N(0.0, 0.5^2)$ (normal distribution; mean is 0.0, and standard deviation is 0.5). Note that $Y_{30} = 5$, $Y_{50} = 10$, and $Y_{70} = 5$ are given as outliers. It is clearly apparent that estimates supplied by LOWESS are less influenced by outliers. As expected, while estimates by LOESS with a narrow span are substantially influenced by outliers, LOWESS successfully reduces the effects of outliers.

3.10 EXAMPLES OF S-PLUS OBJECT

(A) Smoothing by using the Nadaraya-Watson estimator (construction of figure 3.10 (page 136). Use of `ksmooth()`)

```
function()
{
#   (1)
    d1 <- read.table("c:\\datasets\\test31.csv", sep = ",")
    xx <- d1[, 1.]
    yy <- d1[, 2.]
#   (2)
    nd <- length(xx)
#   (3)
    bw <- 0.18
#   (4)
    ex <- seq(from = 0.1, to = 3., by = 0.1)
#   (5)
    bwsplus <- bw/0.3708159
    fit.ks <- ksmooth(xx, yy, kernel = "normal", bandwidth =
    bwsplus, x.points = ex)
```

```
#   (6)
    ey <- fit.ks$y
#   (7)
    par(mfrow = c(1., 1.), mai = c(2., 3., 2., 3.),
      oma = c(2., 2., 2., 2.))
    par(mfrow = c(1., 1.), mai = c(2., 3., 2., 3.),
      oma = c(2., 2., 2., 2.))
    plot(xx, yy, type = "n", xlab = "x", ylab = "y")
    points(xx, yy, pch = 5., cex = 0.8)
    lines(ex, ey)
}
```

(1) Data are retrieved; the predictor values are stored as xx, and the values of a target variable are saved as yy. The file test31.csv has the following form.

0.1, 0.1151
0.2, 0.19762
0.3, 0.578587
0.4, 0.249493

$$\vdots \qquad \vdots$$

(2) The number of data (nd) is defined as length(xx).
(3) The bandwidth (bw) is configured as 0.18.
(4) The x-coordinate value of the estimation point (ex) is set to be $\{0.1, 0.2, \ldots, 3.0\}$.
(5) The estimates are calculated using the Nadaraya-Watson estimator, and the results are saved in fit.ks. The bandwidth is given as bandwidth, and the x-coordinate value of the estimation point is given as x.points. Since kernel = "normal" is assigned, a Gaussian kernel function is adopted. It should be noted that if the value for bandwidth = in ksmooth() is assumed to be h_{splus}, the kernel function below is employed.

$$K\left(\frac{x - X_i}{c_{splus}h_{splus}}\right) = \exp\left(-\frac{1}{2}\left(\frac{x - X_i}{c_{splus}h_{splus}}\right)^2\right) \tag{3.272}$$

c_{splus} is approximately 0.3708159. Using this value, the result of integration of eq(3.272) in the region from $(X_i - 0.25h_{splus})$ to $(X_i + 0.25h_{splus})$ is half of that from $-\infty$ to ∞. Hence, if the value of bw is provided as h of a kernel function in eq(3.108) (page 134), the transformation of bwsplus <- bw/0.3708159 must be used as a value for bandwidth = in ksmooth().

If kernel = "normal" in ksmooth() is omitted or kernel = "box" is designated, the kernel function below is adopted.

$$K\left(\frac{x - X_i}{h_{splus}}\right) = \begin{cases} 1 & \text{if } \frac{|x - X_i|}{h} \leq 0.5 \\ 0 & \text{if } \frac{|x - X_i|}{h} > 0.5. \end{cases} \tag{3.273}$$

The integration of this kernel function in the region from $(X_i - 0.25h_{splus})$ to $(X_i + 0.25h_{splus})$ is also half of that from $-\infty$ to ∞. This characteristic remains when "triangle" or "parzen" is assigned as bandwidth =.
(6) The estimates stored in fit.ks$y are reproduced as ey.

(7) The relationship between data and estimates is plotted in a graph.

(B) Calculation of CV and GCV of smoothing by Nadaraya-Watson estimator
Object: kscv1()

```
function(x1, y1, nd, bandw, ntrial)
{
#   (1)
    kscvgcv <- function(bw, x1, y1)
    {
#   (2)
        nd <- length(x1)
        bwsplus <- bw/0.3708159
        fit.ks <- ksmooth(x1, y1, "normal",
         bandwidth = bwsplus)
        res <- y1 - fit.ks$y
#   (3)
        dhat1 <- function(x2, bw)
        {
            nd2 <- length(x2)
            diag1 <- diag(nd2)
            bwsplus <- bw/0.3708159
            dhat <- rep(0, length = nd2)
#   (4)
            for(jj in 1.:nd2) {
                y2 <- diag1[, jj]
                fit.ks <- ksmooth(x2, y2, "normal",
                 bandwidth = bwsplus)
                dhat[jj] <- fit.ks$y[jj]
            }
            return(dhat)
        }
#   (5)
        dhat <- dhat1(x1, bw)
        trhat <- sum(dhat)
        sse <- sum(res^2.)
#   (6)
        cv <- sum((res/(1. - dhat))^2.)/nd
        gcv <- sse/(nd * (1. - (trhat/nd))^2.)
#   (7)
        return(cv, gcv)
    }
#   (8)
    cvgcv <- lapply(as.list(bandw), kscvgcv,
     x1 = x1, y1 = y1)
    cvgcv <- unlist(cvgcv)
    cv <- cvgcv[attr(cvgcv, "names") == "cv"]
    gcv <- cvgcv[attr(cvgcv, "names") == "gcv"]
```

```
#  (9)
    return(cv, gcv)
}
```

(1) The function `kscvgcv()` is defined. This function is used to compute the values of CV and GCV of smoothing using a Nadaraya-Watson estimator with a certain value of bandwidth.

(2) The number of data is defined as nd. Smoothing using a Nadaraya-Watson estimator is conducted with a data (the values of a predictor are x1, and the values of a target variable are y1). The bandwidth is bw. Since the form of a kernel function is assigned as "normal," a Gaussian kernel is employed. In addition, box (a rectangular box, this is the default form if `kernel` = is not assigned), `triangle` (a triangular box), and `parzen` (parzen kernel) are available. Then, residuals are obtained by subtraction of estimates (`fit.ks$y`) from values of a target variable of data (y1); they are stored as res.

(3) The function `dhat1()` is defined to derive diagonal elements of a hat matrix corresponding to a Nadaraya-Watson estimator. The number of data is named nd2, and a diagonal matrix (the size is nd2 × nd2) is defined as `diag1`. Then, the bandwidth (`bwsplus`) to be used in `ksmooth()` is computed. The object dhat is prepared for accommodating diagonal elements of a hat matrix.

(4) Estimates are calculated using a Nadaraya-Watson estimator; the predictor value is x2, and each column vector constituting `diag1` corresponds to values of a target variable of each data. The object dhat, which is given by connecting estimates, constitutes diagonal elements of a hat matrix. This object is an output argument.

(5) Using `dhat1()`, diagonal elements (dhat) of a hat matrix when the bandwidth is bw are derived. The sum of diagonal elements of a hat matrix is saved as `trhat`. In addition, the sum of previously obtained squared residuals (res) is computed. The resultant value (residual sum of squares) is saved as sse.

(6) The values of CV and GCV when the bandwidth is bw are saved as cv and gcv, respectively.

(7) The objects cv and gcv are designated as output arguments.

(8) Through the use of `kscvgcv()`, the values of CV and GCV corresponding to ntrial bandwidths stored in bandw are computed. The expression of `as.list`
(bandw) in the first line can be simplified to bandw. By carrying out `unlist(cvgcv)`, a vector is obtained; the labels cv and gcv are attached to the elements of the vector. These labels enable the extraction of cv (a vector) and gcv (a vector).

(9) The objects cv and gcv are output arguments.

An example of an object to use `kscv1()` (construction of figure 3.11 (page 136))

```
function()
{
#  (1)
    mm <- read.table("c:\\datasets\\test31.csv", sep = ",")
    xx <- mm[[1]]
    yy <- mm[[2]]
#  (2)
```

```
        nd <- length(xx)
#    (3)
        ntrial <- 10.
#    (4)
        bandw <- seq(from = 0.03, by = 0.03, length = ntrial)
#    (5)
        output.ks <- kscv1(xx, yy, nd, bandw, ntrial)
        cv <- output.ks$cv
        gcv <- output.ks$gcv
#    (6)
        par(mfrow = c(1., 2.), mai = c(2., 1., 2., 1.),
          oma = c(2., 2., 1., 1.))
        plot(bandw, cv, type = "n", ylim = c(0.04, 0.07),
          xlab = "bandwidth", ylab = "CV")
        points(bandw, cv, cex = 1.2, pch = 4.)
        lines(bandw, cv, lwd = 2.)
#    (7)
        cvmin <- min(cv)
#    (8)
        icvmin <- (1.:ntrial)[cv == cvmin]
        bandcv <- bandw[icvmin]
        points(bandcv, cvmin, cex = 1.2, pch = 15.)
#    (9)
        plot(bandw, gcv, type = "n", ylim = c(0.04, 0.07),
          xlab = "bandwidth", ylab = "GCV")
        points(bandw, gcv, cex = 1.2, pch = 4.)
        lines(bandw, gcv, lwd = 2.)
#    (10)
        gcvmin <- min(gcv)
#    (11)
        igcvmin <- (1.:ntrial)[gcv == gcvmin]
        bandgcv <- bandw[igcvmin]
        points(bandgcv, gcvmin, cex = 1.2, pch = 15.)
}
```

(1) Data are retrieved. The predictor values are named xx and those of a target variable are named yy.

(2) The number of data (nd) is set at length(xx).

(3) The number of bandwidths (ntrial) is fixed at 10.

(4) The ntrial bandwidths are given as $\{0.03, 0.06, \ldots, 0.3\}$, and they are stored as bandw.

(5) The values of CV and GCV are calculated using the object "kscv1()" and are saved as cv and gcv, respectively.

(6) The graph to show the relationship between the values of the bandwidth (bandw) and those of CV (cv) is plotted.

(7) The minimum value of cv is saved as cvmin.

(8) The value of icvmin is selected such that the bandwidth (bandw(icvmin)) leads to cvmin. The value of bandw(icvmin) is named bandcv. Furthermore, the point of the optimal bandwidth (bandcv) in terms of CV versus the value of the resulting CV (cvmin) is plotted in the graph.

(9) The relationship between bandwidth (bandw) and GCV (gcv) is shown in the graph.

(10) The minimum value of gcv is saved as gcvmin.

(11) The value of igcvmin is chosen to determine the optimal bandwidth (bandw(igcvmin)), giving gcvmin. The value of bandw(igcvmin) is stored as bandgcv. Furthermore, the point of the optimal bandwidth (bandgcv) in terms of GCV versus the resulting value of GCV (gcvmin) is plotted in the graph.

(C) Calculation of CV and GCV in the case of smoothing by local linear regression

Object: llcv1()

```
function(x1, y1, nd, bandw, ntrial)
{
#   (1)
    llinest <- function(ex1, xdata, ydata, band)
    {
#   (2)
        x2 <- xdata - ex1
        wts <- exp((-0.5 * x2^2.)/band^2.)
#   (3)
        data1 <- data.frame(x = x2, y = ydata, www = wts)
#   (4)
        fit.lm <- lm(y ~ x, data = data1, weights = www)
#   (5)
        est <- fit.lm$coef[1.]
        names(est) <- NULL
#   (6)
        return(est)
    }
#   (7)
    dhat2 <- function(xdata, band)
    {
#   (8)
        dhat3 <- function(ydata, xdata, band)
        {
#   (9)
            x2 <- xdata - xdata[ydata == 1.]
            wts <- exp((-0.5 * x2^2.)/band^2.)
            data1 <- data.frame(x = x2, y = ydata, www = wts)
#   (10)
            fit.lm <- lm(y ~ x, data = data1, weights = www,
              x = T)
            inf1 <- lm.influence(fit.lm)
```

```
                        dhatj <- inf1$hat[ydata == 1.]
# (11)
                    return(dhatj)
                }
# (12)
                diag1 <- diag(length(xdata))
# (13)
                dhat <- apply(diag1, 2., dhat3, xdata = xdata,
                  band = band)
                dhat <- as.vector(dhat)
# (14)
                return(dhat)
            }
# (15)
        cv <- numeric(ntrial)
        gcv <- numeric(ntrial)
# (16)
        for(kk in 1.:ntrial) {
# (17)
                ey <- lapply(as.list(x1), llinest, xdata = x1,
                  ydata = y1, band = bandw[kk])
                ey <- unlist(ey)
# (18)
                res <- y1 - ey
                sse <- sum(res^2.)
# (19)
                dhat <- dhat2(x1, bandw[kk])
                trhat <- sum(dhat)
# (20)
                cv[kk] <- sum((res/(1. - dhat))^2.)/nd
                gcv[kk] <- sse/(nd * (1. - (trhat/nd))^2.)
        }
# (21)
        return(cv, gcv)
}
```

(1) The function `llinest()` to carry out local linear regression is defined. The object ex1 (a scalar) is the coordinate of the position where estimation is executed. The object xdata shows predictor values of data. The object ydata represents target values of data. The object band denotes h in eq(3.113) (page 140).

(2) The values corresponding to $\{X_i - x^*\}$ $(1 \le i \le n)$ $(x^* = \text{ex1}, \{X_i\} = \text{xdata})$ in eq(3.113) are stored as one vector (x2). Furthermore, the values ($\{w(\frac{X_i - x^*}{h})\}$ $(1 \le i \le n)$) of a kernel function (eq(3.113)) are computed to be saved as wts.

(3) The data for using `lm()` are put together in a data frame (data1). In the data frame, x2 is called x, ydata is named y, and wts is defined as www.

(4) $x^* = $ ex1 (eq(3.112) (page 139)) is set to derive a regression equation. When y ~ poly (x, 1) is assigned in lm() instead of y ~ x, (5) should be replaced by the following procedure.

```
#  (5)'
    assign("data1", data1, frame = 1)
    data2 <- data.frame(x = 0)
    est <- predict.gam(fit.lm, newdata = data2)
```

When local quadratic regression is carried out, fit.lm <- lm(y ~ x, data = data1, weights = www) should be replaced by

```
fit.lm <- lm(y ~ x + x^2, data = data1, weights = www)
```

For local cubic regression, it should be

```
fit.lm <- lm(y ~ x + x^2 + x^3, data = data1, weights = www)
```

(5) The constant term ($a_0(x^*)$ in eq(3.112) (page 139)) is extracted from fit.lm to be an estimate (est). names(est) <- NULL aims to omit names ((Intercept)) from est.

(6) The object est is outputted from llinest().

(7) The function dhat2() for obtaining a hat matrix is defined. The object xdata is the vector consisting of predictor values of data; band is a bandwidth.

(8) The function dhat3() for deriving a diagonal element of a hat matrix is defined. The object ydata is a vector to store values of a target variable (one value is 1 and others are 0). The object xdata is a vector to save values of a predictor; band is bandwidth. Eq(2.6) (page 24) is used for the calculation of diagonal elements of a hat matrix. Eq(3.133) (page 144) is also applicable.

(9) Since the value of a predictor when the value of a target variable is 1 is xdata[ydata == 1], a data frame (data1) is constructed to carry out local linear regression at that point.

(10) A local linear equation is derived using lm() and the result is saved in fit.lm. The object lm.influence() is executed using fit.lm as an argument and the result is stored in inf1; the values of diagonal elements of a hat matrix are to be saved as inf1$hat. The value in inf1$hat that corresponds to the value 1 in ydata gives a diagonal element of a hat matrix of the local linear regression. The assignment x = T is set in lm() because fit.lm, which is used for lm.influence(), is assured to include the values of a predictor. These three lines can be replaced with the following two lines.

```
    fit.lm <- lm(y ~ x, data = data1, weights = www)
    dhatj <- fit.lm$coef[1]
```

The object lm.influence() is not used above and hence x = T does not need to be assigned in fit.lm.

(11) The object dhatj is outputted from dhat3().

(12) An identity matrix (size is nd×nd) is defined as diag1.

(13) The object dhat3() is carried out; each column of diag1 is used as values of a target variable, xdata is used as values of a predictor, and band is a bandwidth. The

resultant values are connected to constitute dhat; dhat comprises diagonal elements of a hat matrix by local linear regression.

(14) The object dhat is outputted from dhat2().

(15) The object cv is prepared for storing CV, and the object gcv is for saving GCV.

(16) A total of ntrial bandwidths stored in bandw are utilized to calculate CV and GCV.

(17) Estimates corresponding to each element of x1 is carried out using llinest() to save the result as ex (a vector). The expression of as.list(x1) can be simply x1.

(18) Residuals obtained by subtracting ey from y1 are saved as res. The residual sum of squares is defined as sse.

(19) Diagonal elements of a hat matrix are obtained using dhat2() and are defined as dhat. Their sum is saved as trhat.

(20) The value of CV is computed using eq(3.41) (page 117) and is saved as cv. The value of GCV is calculated using eq(3.46) (page 118) and is stored as gcv.

(21) The objects cv and gcv are outputted.

An example of an object to use llcv1() (construction of figure 3.18 (page 148))

To calculate the values of CV and GCV corresponding to smoothing by local linear regression through llcv1() and to obtain figure 3.18, procedures of (4) and (5) in "an example of an object to use kscv1()" are replaced with the following procedures.

```
#   (4)'
    bandw <- seq(from = 0.1, by = 0.05, length = ntrial)
#   (5)'
    output.ll <- llcv1(xx, yy, nd, bandw, ntrial)
    cv <- output.ll$cv
    gcv <- output.ll$gcv
```

(D) Smoothing by local linear regression
Object: lll()

```
function(ex1, xdata, ydata, band)
{
#   (1)
    x2 <- xdata - ex1
    wts <- exp((-0.5 * x2^2.)/band^2.)
#   (2)
    data1 <- data.frame(x = x2, y = ydata, www = wts)
#   (3)
    fit.lm <- lm(y ~ x, data = data1, weights = www)
#   (4)
    est <- fit.lm$coef[1.]
    names(est) <- NULL
#   (5)
    return(est)
```

```
}
```

This object is identical to `llinest()` in `llcv1()` (presented as (C)).

In `lline()` (page 89), `llcv1()`, and `lll()`, all data are used to derive a linear equation for calculating each estimate. However, weights of data located far from the estimation points are positive values close to 0 or exactly 0; these data have little impact on regression coefficients of polynomials. This consideration leads to the regression in which data near estimation points are used. For this purpose, `lll()` is, for example, replaced with the object below. In this object, data with weights more than or equal to 0.00001 are selected. This implementation reduces the amount of calculation required to derive a linear equation. However, when the calculation time for unsubstantial parts is long, this effect is difficult to realize.

```
function(ex1, xdata, ydata, band)
{
#   (1)
    x2 <- xdata - ex1
    wts <- exp((-0.5 * x2^2.)/band^2.)
    set1 <- rep(1., length = length(xdata))
    set1[wts < 1e-005] <- 0.
#   (2)
    data1 <- data.frame(x = x2, y = ydata, www = wts,
      set2 = set1)
#   (3)
    fit.lm <- lm(y ~ x, data = data1, weights = www,
      subset = set2 == 1.)
#   (4)
    est <- fit.lm$coef[1.]
    names(est) <- NULL
#   (5)
    return(est)
}
```

An example of an object to use `lll()` (construction of figure 3.19(left) (page 148))

```
function()
{
#   (1)
    mm <- read.table("c:\\datasets\\test31.csv", sep = ",")
    xx <- mm[[1]]
    yy <- mm[[2]]
#   (2)
    band1 <- 0.25
#   (3)
    ne <- 59.
#   (4)
    ex <- seq(from = 0.1, to = 3., length = ne)
```

```
    ey <- sapply(as.list(ex), lll, xdata = xx, ydata = yy,
      band = band1)
#   (5)
    par(mfrow = c(1., 1.), mai = c(2., 3., 2., 3.),
      oma = c(2., 2., 2., 2.))
    plot(xx, yy, type = "n", xlab = "x", ylab = "y")
    points(xx, yy, cex = 0.8, pch = 5.)
    lines(ex, ey, lwd = 2.)
}
```

(1) Data are retrieved.

(2) The bandwidth is set at 0.25.

(3) The number of estimation points is given as ne. The values of a predictor at estimation points are set as ex.

(4) Estimates corresponding to ex are derived using lll() and the results are stored as ey. Since sapply() is used, ey is a vector. The expression of as.list(ex) can be ex.

(5) Data and estimates are plotted in the graph.

(E) Interpolation using B-spline
Object: intsp1()

```
function(x1, y1, intkn, ex)
{
#   (1)
    data1 <- data.frame(x = x1, y = y1)
    assign("data1", data1, frame = 1.)
    formula.name <- substitute(y ~ bs(x, knots = kn,
      degree = 3.), list(kn = intkn))
    fit.lm <- lm(formula.name, data = data1)
#   (2)
    data2 <- data.frame(x = ex)
    ey <- predict(fit.lm, newdata = data2)
#   (3)
    return(ey)
}
```

(1) A cubic B-spline is fitted by the least squares. When the number of data is equal to that of regression coefficients (eq(3.164) (page 155)), interpolation is realized. If the number of data is more than that, smoothing is accomplished. The knots at the two boundaries are omitted in the knots designated as "knots =" (intkn in this example). Knots at the two boundaries are set at the maximum and minimum values of a predictor of data. Knots at the two boundaries are called boundary knots (boundary breakpoint), and remaining knots are called internal knots (internal breakpoint). Regarding substitute(), refer to Chapter 6.

(2) Estimates are calculated at ex. The maximum and minimum values of ex are identical to those of data.

(3) The object ey is outputted.

An example of an object to use `intsp1()` (construction of figure 3.25 (page 157))

```
function()
{
  (1)
    xx <- c(1., 1.7, 3., 4.2, 5.)
    yy <- c(2., 5., 4., 5., 3.)
    intkn <- c(3.5)
  (2)
    ne <- 41.
    ex <- seq(from = 1., by = 0.1, length = ne)
  (3)
    ey <- intsp1(xx, yy, intkn, ex)
  (4)
    par(mfrow = c(1., 1.), mai = c(2., 3., 2., 3.),
      oma = c(2., 2., 2., 2.))
    plot(ex, ey, type = "n", xlab = "x", ylab = "y",
      ylim = c(0., 6.))
    points(xx, yy, pch = 4., cex = 1.3)
    lines(ex, ey, lwd = 2.)
  (5)
    points(c(min(xx), intkn, max(xx)), rep(0.9,
      length = length(intkn) + 2.), pch = 15., cex = 1.)
}
```

(1) Data are given as xx (predictor) and yy (target variable). The coordinates of internal knots are represented as a vector (intkn).
(2) The number of estimates to be obtained is set at 41. The predictor values at estimation points are given as ex.
(3) Estimates (ey) are obtained using intsp1().
(4) Data and estimates are illustrated.
(5) The locations of knots are displayed.

(F) Interpolation using natural spline
Object: intns1() (use of spline())

```
function(x1, y1, ne)
{
#  (1)
    exy <- spline(x1, y1, n = ne, boundary = 0.)
#  (2)
    ex <- exy$x
    ey <- exy$y
#  (3)
    return(ex, ey)
}
```

(1) Interpolation using the natural spline is executed using `spline()` (an object of interpolation with assigned boundary conditions). The second derivatives at both boundaries are set at 0 by assigning boundary = 0. The expression of n = ne designates the approximate number of points at which estimates are derived.
(2) The coordinates of estimation points are extracted from the output of `spline()` (exy) to be defined as ex, and estimates are extracted to be saved as ey.
(3) The objects ex and ey are outputted.

Object: `intns2()` (use of `smooth.spline()`)

```
function(x1, y1, ex)
{
#    (1)
    fit.sp <- smooth.spline(x1, y1, spar = 1e-010, all.knot = T)
#    (2)
    exy <- predict.smooth.spline(fit.sp, ex)
#    (3)
    ey <- exy$y
#    (4)
    return(ey)
}
```

(1) A smoothing spline is derived by the use of `smooth.spline()`. The value of spar (a smoothing parameter, λ in eq(3.209) (page 170)) is set to be a very small positive value (10^{-10}). Hence, interpolation using the natural spline is substantially realized.
(2) Estimates are calculated at ex.
(3) The values of estimates are extracted from exy to define them as ey.
(4) The object ey is outputted.

Object: `intns3()` (use of `lm()` and `ns()`)

```
function(x1, y1, ex)
{
#    (1)
    data1 <- data.frame(x = x1, y = y1)
#    (2)
    kn1 <- x1[2.:(length(x1) - 1.)]
#    (3)
    formula.name <- substitute(y ~ ns(x, knots = kn),
      list(kn = kn1))
    fit.lm <- lm(formula.name, data = data1)
#    (4)
    data2 <- data.frame(x = ex)
    ey <- predict(fit.lm, newdata = data2)
#    (5)
    return(ey)
}
```

(1) Data are put together in a data frame (data1) to use `lm()`.

(2) The positions of internal knots are given as kn1. Use of `attach()` and `detach()` is an alternative to make kn function in the same manner. However, this is not advisable because of potential problems.

(3) Fitting of the natural spline by the least squares is carried out and the result is stored in `fit.lm`. Regarding `substitute()`, refer to Chapter 6.

(4) Estimates at ex are derived and the result is saved as ey. It should be noted that the maximum and minimum values of ex are assumed to be identical to those of data.

(5) The object ey is outputted.

An example of object to use `intns1()`, `intns2()`, and `intns3()` (construction of figure 3.30 (page 169))

```
function()
{
#   (1)
    yd <- scan("c:\\datasets\\wak2.csv")
    yy <- scan("c:\\datasets\\wak2s.txt")
    nd <- length(yy)
    xx <- seq(from = 1., by = 1., length = nd)
#   (2)
    ne <- 301.
#   (3)
    exy <- intns1(xx, yy, ne)
    ex <- exy$ex
    ey <- exy$ey
#   (4)
    par(mfrow = c(1., 1.), mai = c(2.5, 2.5, 2.5, 2.5),
      oma = c(1., 1., 1., 1.))
    plot(xx, yy, type = "n", ylim = c(-15., 5.), xlab = "X",
      ylab = "Y")
    lines(ex, ey, lwd = 2.)
    points(xx, yy, pch = 15., cex = 0.8)
    points(xx, yd, pch = 4.)
}
```

(1) Data (a target variable) is retrieved and it is named yd. Estimates given by binomial filter smoothing are retrieved and they are defined as yy. The number of elements of yy (i.e., the number of data) is named nd and predictor values are stored as xx.

(2) The number of estimation points is defined as ne and the coordinates of estimation points are stored as ex.

(3) Estimates are calculated by the use of `intns1()` and are called ey. To use `intns2()`, procedure (3) above is replaced by the following.

```
#   (3)'
    ex <- seq(from = 1, length = ne, by = 0.1)
    ey <- intns2(xx, yy, ex)
```

Furthermore, to make use of `intns3()`, this part is again replaced with

```
#   (3)"
```

```
ex <- seq(from = 1, length = ne, by = 0.1)
ey <- intns3(xx, yy, ex)
```

Since `intns3()` utilizes the result of interpolation by `lm()` and `ns()` so as to obtain estimates using `predict()`, the values of ex at both ends are required to be equal to those of data, as in the case of `intsp1()` ((E)).

(4) Data, estimates using a binomial filter, and estimates obtained by interpolation are illustrated in a graph.

(G) An object to realize the same calculation as that by dtpss (a subroutine of GCVPACK) in S-Plus
Object: `dtpss()`

```
function(des, lddes, nobs, dim, m, s, lds, ncov, y, ntbl,
 adiag, lamlim, dout, iout, coef, svals, tbl, ldtbl,
 auxtbl, work, lwa, iwork, liwa, job, info)
{
#  (1)
   dll.load("c:\\datasets\\gcvpack.dll",
    symbols = "dtpss_")
#  (2)
   storage.mode(auxtbl) <- "double"
#  (3)
   fit <- .C("dtpss_",
    des = as.double(des),
    lddes = as.integer(lddes),
    nobs = as.integer(nobs),
    dim = as.integer(dim),
    m = as.integer(m),
    s = as.double(s),
    lds = as.integer(lds),
    ncov = as.integer(ncov),
    y = as.double(y),
    ntbl = as.integer(ntbl),
    adiag = as.double(adiag),
    lamlim = as.double(lamlim),
    dout = as.double(dout),
    iout = as.integer(iout),
    coef = as.double(coef),
    svals = as.double(svals),
    tbl = as.double(tbl),
    ldtbl = as.integer(ldtbl),
    auxtbl = auxtbl,
    work = as.double(work),
    lwa = as.integer(lwa),
    iwork = as.integer(iwork),
    liwa = as.integer(liwa),
    job = as.integer(job),
```

```
      info = as.integer(info))
#  (4)
      return(fit)
}
```

(1) A DLL file ("d:\\x-2nd\\datasets\\gcvpack.dll") is made accessible in S-Plus. The program (gcvpack.f) is processed by MinGW (a free software available on the Internet. Version 2.0.0-3 was used in this example) to create gcvpack.dll. The above-mentioned gcvpack.f is obtained by the addition of LINPACK to GCVPACK (second release). Both GCVPACK and LINPACK are a collection of subroutines written in FORTRAN77; they are available free of charge. GCVPACK is provided, for example, on the web page of Professor Bates in the Department of Statistics, at Wisconsin University. LINPACK is a collection of subroutines to carry out fundamental computation on matrix and simultaneous equations; it is available on the site "netlib" (where programs and data for numerical calculation are distributed), for instance.

To construct gcvpack.dll, gcvpack.f is placed at "\MinGW\bin" ("\MinGW" is the folder where MinGW is installed). Next, DOS prompt is activated to execute "g77 -c -o gcvpack.o gcvpack.f" at "\MinGW\bin". Furthermore, "dllwrap –export-all-symbols -o gcvpack.dll gcvpack.o" is executed. Should the message "no export definition file provided. Creating one, but that may not be what you want" be displayed, it can be disregarded. The file "gcvpack.dll" is created herewith at "\MinGW\bin". (2) If auxtbl = as.double(auxtbl) is assigned in .C(), auxtbl, which is output from .C(), and takes the form of a vector instead of a matrix. This is caused by as.double(). To prevent this, storage.mode(auxtbl) <- "double" is executed before .C() and auxtbl = auxtbl is assigned in .C(). (3) Calculation by dtpss (part of gcvpack.f) is carried out. For this purpose, all arguments required by dtpss (input argument, output argument, input output argument) are transformed appropriately to the form which dtpss needs; if it has a double precision form, as.double() is used, and if it has an integer form, as.integer() is used. No transformations or declarations are required to make distinctions among a scalar, a vector, and a matrix or to specify the number of elements of a vector or the size of a matrix, because the forms used in S-Plus are inherited in dtpss. All results are stored in fit, which has the form of a list. More detailed explanations are available in comment lines in GCVPACK or [1]. (4) The object fit is outputted. The object fit contains all results given by calculation using dtpss. In this example, fit is the only argument of return(). Hence, the output from this object is an object (in the form of a list), which has the same components as those of fit. When the arguments in return() are plural objects in the form of a list or an object in the form of a list accompanied by objects in other forms, the output is an object in the form of a list, which contains those items as components.

An example of object to use dtpss() (construction of figure 3.38 (page 183))

```
function()
{
#  (1)
```

```
    mm <- read.table("c:\\datasets\\test31.csv", sep = ",")
    xx <- mm[[1]]
    yy <- mm[[2]]
#   (2)
    nobs <- 30.
    if(length(xx) != nobs | length(yy) != nobs)
    stop("The number of data is not equal to nobs.")
    ne <- 146.
    ex <- seq(from = 0.1, by = 0.02, length = ne)
    ey <- NULL
#   (3)
    dim <- 1.
    m <- 2.
    ntbl <- 10.
    lamlim <- c(-4., 0.5)
    job <- 1110.
    des <- xx
    y <- yy
    maxobs <- nobs
    maxtbl <- ntbl
    mxncts <- m + dim - 1.
    maxuni <- nobs
    maxpar <- maxuni + mxncts
    lwa <- maxuni * (2. + mxncts + maxuni) + maxobs
    liwa <- 2. * maxobs + maxuni
    info <- 0.
    iwork <- rep(0., liwa)
    iout <- rep(0., 4.)
    adiag <- rep(0., maxobs)
    tbl2 <- rep(0., (maxtbl * 3.))
    tbl <- matrix(tbl2, ncol = 3.)
    coef <- rep(0., maxpar)
    auxtbl <- matrix(rep(0., length = 9.), ncol = 3.)
    svals <- rep(0., maxpar)
    dout <- rep(0., 5.)
    work <- rep(0., lwa)
    lddes <- maxobs
    lds <- maxobs
    ldtbl <- maxtbl
    ncov <- 0.
    s <- rep(0., nobs)
#   (4)
    fit <- dtpss(des, lddes, nobs, dim, m, s, lds,
     ncov, y, ntbl, adiag, lamlim, dout, iout, coef,
     svals, tbl, ldtbl, auxtbl, work, lwa, iwork, liwa,
     job, info)
#   (5)
```

```
    auxtbl <- fit$auxtbl
#   (6)
    print("info=")
    print(fit$info)
#   (7)
    gammav <- gamma(0.5 - m)
    const <- 2.^(-2. * m)/sqrt(pi)/gamma(m)
    m1 <- m + 1.
    nobsm <- nobs + m
    for(i in 1.:ne) {
        absv <- abs(ex[i] - xx)^(2. * m - 1.)
        ey[i] <- gammav * const * sum(fit$coef[m1:nobsm]
         * absv) + fit$coef[1.]
        for(j in 1.:(m - 1.)) {
            ey[i] <- ey[i] + fit$coef[j + 1.] * ex[i]^(j)
        }
    }
#   (8)
    par(mfrow = c(1., 2.), mai = c(2.5, 1.5, 2., 0.5),
     oma = c(1., 1., 1., 1.))
    tblm <- matrix(fit$tbl, ncol = 3.)
    sparam <- tblm[, 1.]
    gcv <- tblm[, 2.]
    plot(sparam, gcv, type = "n", xlab =
     "log10(lambda)", ylab = "GCV", ylim = c(0.04, 0.2))
    lines(sparam, gcv, lwd = 2.)
    points(sparam, gcv, pch = 4.)
    sparagcv <- auxtbl[1., 1.]
    gcvmin <- auxtbl[1., 2.]
    points(sparagcv, gcvmin, type = "p", pch = 15., cex = 1.2)
#   (9)
    plot(xx, yy, type = "n", xlab = "x", ylab = "y")
    points(xx, yy, pch = 5.)
    lines(ex, ey, lwd = 2.)
}
```

(1) Data are retrieved. The values of a predictor are saved as xx and those of a target variable are stored as yy. If the elements of xx are not aligned in ascending order, the three lines below should be added.

```
    order1 <- order(xx)
    xx <- xx[order1]
    yy <- yy[order1]
```

(2) The number of data (nobs) is set at 30 and the number of estimation points (ne) is set at 146. The values of a predictor at estimation points are stored as ex (a vector). The object ey is prepared for storing estimates.

(3) The values of arguments to use dtpss() are given. In this object, the number of elements dim is fixed at 1, and the degree of differentiation to obtain a roughness

penalty (m, the value of k in eq(3.251) (page 184)) is fixed at 2 (i.e., eq(3.209) (page 170) is used). By assigning `ntbl <- 10` and `lamlim <- c(-4, 0.5)`, the value of the smoothing parameter (λ in eq(3.209)) is altered from 10^{-4} to $10^{0.5}$ in 10 steps (i.e., $\{10^{-4}, 10^{-3.5}, \ldots, 10^{0.5}\}$) to calculate the values of GCV. It should be noted that the equation below is adopted in `dtpss()`, instead of eq(3.209).

$$E'_{ss} = \frac{1}{n} \sum_{i=1}^{n} (Y_i - m(X_i))^2 + \lambda' \int_{-\infty}^{\infty} \left(\frac{d^2 m(x)}{dx^2} \right)^2 dx \qquad (3.274)$$

In addition, the value of $(n \cdot \lambda')$ is assumed to be defined as `lamlim`. However, since $n \cdot E'_{ss} = E_{ss}$, it is equivalent to defining the value of λ in eq(3.209) because we can set $\lambda = n \cdot \lambda'$.

(4) Calculation by `dtpss()` is carried out and the result is saved as `fit`. Diagonal elements of a hat matrix are output as `fit$adiag`; they lead to the values of CV and AIC.

(5) The component `fit$auxtbl` is extracted from `fit` and is defined as `auxtbl`.

(6) The value of the argument `info`, which indicates whether procedure (4) stopped in an orderly manner, is displayed. The value of `info` has the following.

`info=` 0: The calculation ended normally.

`info=` -2: The value of the second element of `lamlim` (the maximum value of $\log_{10}$(smoothing parameter), i.e., $\log_{10}(\lambda)$, $= \log_{10}(\lambda' \cdot n)$, to be used for optimizing the value of a smoothing parameter) is too small and hence the result obtained using the second value of `lamlim` was regarded as the best one. To obtain the optimal value of a smoothing parameter, the value of the second element of `lamlim` is increased for recalculation. If the optimal regression equation leads to an almost straight line, the optimal value of $\log_{10}$ (smoothing parameter) ($\log_{10}(\lambda)$) is very large and hence the result with the value of `info` ($= -2$) is adopted as the optimal one.

`info=` -1: The value of the first element of `lamlim` (the minimum value of $\log_{10}(\lambda)$ to be used for obtaining the optimal value of a smoothing parameter) is too large. Therefore, the result given by the first element of `lamlim` was considered to be the best one. To optimize the value of the smoothing parameter, the value of the first element of `lamlim` is reduced for recalculation.

`info=` 1: The arguments for `dtpss()` contain errors; the number of elements of vectors or the size of matrices is wrong. Since GCVPACK and LINPACK are written in FORTRAN77, these programs do not function if the number of elements of vectors or the size of matrices is not designated properly.

`info=` 2: If `ncov` is set at a positive value and predictor values other than those in `xx` are saved as `s`, a partial spline (Chapter 4 (page 270)) is realized. If the size of `s` is different from `des`, this error occurs. When the assignment of `ncov=0` is employed for simple smoothing, this error does not arise.

`info=` 3: `lwa` (the number of elements of `work` (a vector for work space)) is too small. In this example, the value of `lwa` is provided by values of other arguments. Hence, errors of this kind do not arise as long as the values of those arguments are appropriate.

`info=` 4: The value of `liwa` (the number of elements of the vector `iwork` (a vector to be used as work space)) is too small. In this object, the value of `liwa` is given by values of other arguments and hence this error does not arise unless the values of those arguments are inappropriate.

$10 <$ info < 20: An error arises in dsetup (a subroutine used by dtpss).

$100 <$ info < 200: An error arises in dsgdc1 (a subroutine used by dtpss).

$200 <$ info < 300: An error arises in dgcv1 (a subroutine used by dtpss).

(7) The estimates are calculated using eq(3.210) (page 172); eq(3.255) (page 185) is used if m $\neq$ 2. The value of the smoothing parameter which gives the minimum value of GCV (the smoothing parameter is optimized by a golden ratio search) is utilized.

(8) A graph is drawn to represent the relationship between the smoothing parameter $(\log_{10}(\lambda) = \log_{10}(\lambda' \cdot n))$ and GCV. The object dtpss() calculates the values of GCV using the value of a smoothing parameter designated by ntbl and lamlim, and then optimizes the value of the smoothing parameter by a golden ratio search in the area where the value of the smoothing parameter gives a small value of GCV. The resultant value of the smoothing parameter (a logarithmic transformation to base 10 is performed) is saved in auxtbl[1, 1], and the value of GCV corresponding to this value of the smoothing parameter is stored in auxtbl[1, 2]. These two values are named sparagcv and gcvmin, respectively. Furthermore, the point given by sparagcv and gcvmin is plotted in the graph.

(9) The estimates computed using the value of the smoothing parameter which gives the smallest GCV value are plotted in a graph.

When some predictor values of data are tied, job <- 1110 in (3) is replaced with job <- 1111 and (7) is substituted with the following procedure.

```
#   (7)'
    gammav <- gamma(0.5 - m)
    const <- 2^(-2 * m)/sqrt(pi)/gamma(m)
    m1 <- m + 1
    nobsm <- nobs + m
    xxu <- unique(xx)
    coef1 <- fit$coef[m1:nobsm]
    coefu <- coef1[1:length(xxu)]
    for(i in 1:ne) {
        absv <- abs(ex[i] - xxu)^(2 * m - 1)
        ey[i] <- gammav * const * sum(coefu * absv) + fit$coef[1]
        for(j in 1:(m - 1)) {
            ey[i] <- ey[i] + fit$coef[j + 1] * ex[i]^(j)
        }
    }
```

(H) Smoothing using smoothing spline (construction of figure 3.39 (page 183), use of smooth.spline())

```
function()
{
#   (1)
    ll <- list(r1 = 0, r2 = 0)
```

```
    mm <- scan("c:\\datasets\\test31.csv", ll, sep = ",")
    xx <- mm$r1
    yy <- mm$r2
    nd <- length(xx)
#   (2)
    ntrial <- 46
    tran1 <- (max(xx) - min(xx))^(-3)
    sparam <- 10^seq(from = -4, by = 0.1, length = ntrial)
    cv <- rep(0, length = ntrial)
#   (3)
    for(k in c(1:ntrial)) {
#   (4)
        sparsplus <- tran1 * sparam[k]
        fit.sp <- smooth.spline(xx, yy, cv = T,
         spar = sparsplus, all.knots = T)
#   (5)
        cv[k] <- fit.sp$cv
    }
#   (6)
    par(mfrow = c(1, 2), mai = c(2.5, 1.5, 2, 0.5),
     oma = c(1, 1, 1, 1))
    plot(log10(sparam), cv, ylim = c(0.04, 0.2),
     xlab = "log10(lambda)", ylab = "CV", type = "n")
    points(log10(sparam), cv, pch = 3)
    lines(log10(sparam), cv, lwd = 2)
    cvmin <- min(cv)
    icvmin <- (1:ntrial)[cv == cvmin]
    sparacv <- sparam[icvmin]
    points(log10(sparam[icvmin]), cvmin, pch = 15)
#   (7)
    sparsplus <- tran1 * sparacv
    fit.sp <- smooth.spline(xx, yy, spar = sparsplus,
     all.knot = T)
    ex <- seq(from = 0.1, to = 3, by = 0.02)
    exy <- predict.smooth.spline(fit.sp, ex)
    ey <- exy$y
    plot(xx, yy, xlab = "x", ylab = "y", type = "n")
    points(xx, yy, pch = 5)
    lines(ex, ey, lwd = 2)
}
```

(1) Data are retrieved. The data are xx (values of predictor) and yy (values of a target variable). The number of data is named nd.

(2) The value of ntrial is fixed at 46 to apply 46 values of the smoothing parameter. When the weight of all data is 1, the value of the smoothing parameter for smooth.spline() (λ_{splus}) is given after a transformation:

$$\lambda_{splus} = (X_n - X_1)^{-3} \cdot \lambda. \qquad (3.275)$$

When a weight of ith data is w_i $(1 \leq i \leq n)$, the transformation equation above is replaced with

$$\lambda_{splus} = \frac{n}{\sum_{i=1}^{n} w_i} (X_n - X_1)^{-3} \cdot \lambda. \tag{3.276}$$

Therefore, tran1 is defined to give sparam the value of λ as is. Furthermore, an object (cv) is prepared for storing CV.

(3) The values of CV are obtained using ntrial values of the smoothing parameter.

(4) Using the smoothing parameter values (λ) saved in sparam, calculation of the smoothing spline is carried out with smooth.spline(), and the values of CV corresponding to these values of the smoothing parameter are derived. The assignment cv = T indicates the calculation of CV. If cv = T is omitted or cv = F is designated, the value of GCV is obtained. The object sparsplus ($= \lambda_{splus}$) is obtained, by multiplying sparam[k] and tran1, and is used as the value for spar. When all.knots = T is assigned, calculation of the smoothing spline is executed following eq(3.243) (page 180). If this assignment is omitted or all.knots = F is assigned, the number of B-spline bases becomes smaller than that of an ordinary smoothing spline when the number of data is large. This adjustment reduces the amount of calculation and stabilizes the numerical computation, and hence it deserves to be attempted when the number of data is large. Furthermore, if the degree of smoothing is given by the value of the degrees of freedom ($= \text{trace}(\mathbf{H})$), spar = is replaced by df =.

(5) The value of CV (fit.sp$cv) is extracted from the result of smooth.spline() (fit.sp) to be saved as cv. Even if smooth.spline() yields the value of GCV, the value is outputted as fit.sp$cv. Furthermore, the diagonal elements of a hat matrix (namely, leverage (leverage value)) is obtained as fit.sp$lev (for other componets of fit.sp, refer to the S-Plus manual, for example).

(6) A graph is created to represent the relationship between the smoothing parameter (λ) and CV, in the same manner as that described in "an example of an object to use kscv1()."

(7) Calcuation of the smoothing spline is carried out using the value of the smoothing parameter (λ_{splus} = tran1 * sparacv) which gives the minimal value of CV. Estimates at ex are derived using the result (fit.sp) and are stored as ey. A graph is constructed to show the relationship between data and estimates.

The value of the smoothing parameter is optimized using the capability of automatic optimization of a smoothing parameter implemented in smooth.spline(). When the result is used for calculation of the smoothing spline, fit.sp <- smooth.spline(xx, yy, cv = T, spar = sparsplus, all.knots = T) is replaced by

```
fit.sp <- smooth.spline(xx, yy, cv = T, all.knots = T)
```

The above line executes the optimization of the smoothing parameter and the calculation of the smoothing parameter using the result of optimization. If GCV is used instead of CV, cv = T is deleted or cv = F is designated. Note that when some predictor values (xx in this example) are tied, the function of automatic optimization of a smoothing parameter implemented in smooth.spline() may not work as intended. This seems to be caused by a flaw in smooth.spline() (as was pointed out in an article in s-news mailing list written by Professor Ripley (Oxford University)).

(I) Calculation of CV and GCV for LOESS

Object: locv1()

```
function(x1, y1, nd, span1, ntrial)
{
#   (1)
    locvgcv <- function(sp, x1, y1)
    {
        nd <- length(x1)
#   (2)
        assign("data1", data.frame(xx1 = x1, yy1 = y1),
         frame = 1.)
        fit.lo <- loess(yy1 ~ xx1, data = data1, span = sp,
         family = "gaussian", degree = 1., surface = "direct")
        res <- residuals(fit.lo)
#   (3)
        dhat2 <- function(x1, sp)
        {
            nd2 <- length(x1)
            diag1 <- diag(nd2)
            dhat <- rep(0, length = nd2)
#   (4)
            for(jj in 1.:nd2) {
                y2 <- diag1[, jj]
                assign("data1", data.frame(xx1 = x1, yy1 = y2),
                 frame = 1.)
                fit.lo <- loess(yy1 ~ xx1, data = data1,
                 span = sp, family = "gaussian", degree = 1.,
                 surface = "direct")
                ey <- fitted.values(fit.lo)
                dhat[jj] <- ey[jj]
            }
            return(dhat)
        }
#   (5)
        dhat <- dhat2(x1, sp)
        trhat <- sum(dhat)
        sse <- sum(res^2.)
#   (6)
        cv <- sum((res/(1. - dhat))^2.)/nd
        gcv <- sse/(nd * (1. - (trhat/nd))^2.)
#   (7)
        return(cv, gcv)
    }
#   (8)
```

```
    cvgcv <- lapply(as.list(span1), locvgcv, x1 = x1, y1 = y1)
    cvgcv <- unlist(cvgcv)
    cv <- cvgcv[attr(cvgcv, "names") == "cv"]
    gcv <- cvgcv[attr(cvgcv, "names") == "gcv"]
#   (9)
    return(cv, gcv)
}
```

(1) A function `locvgcv()` is defined to calculate CV and GCV corresponding to smoothing by LOESS with a specific value of bandwidth.
(2) The number of data is defined as nd. Smoothing by LOESS is carried out using given data (the values of a predictor are x1 and the values of a target variable are y1). The assignment surface = "direct" is adopted, and hence, estimates are computed using the definition of LOESS. If this assignment is omitted or surface = "interpolate" is designated, estimation is carried out at a small number of points and the estimates are interpolated to derive other estimates. This strategy reduces the amount of computation. The span is defined as sp. The values of residuals derived by subtracting the estimates (ey) from the values of the target variable of data (y1) are saved as res.
(3) The function dhat2() is defined to calculate the values of diagonal elements of a hat matrix of LOESS; the values of a predictor and the value of bandwidth are used in the function. The number of data is saved as nd2, and the values of the elements of a diagonal matrix (the size is nd2 × nd2) are defined as diag1. Furthermore, an object (dhat) is prepared for saving values of diagonal elements of a hat matrix.
(4) Calculation of LOESS is carried out; x1 is used as values of a predictor and each column vector constituting diag1 is used as values of a target variable. The values of diagonal elements (dhat) of a hat matrix are given by connecting the resultant estimates; dhat is output. If degree = is not assigned in loess(), a quadratic equation is utilized (i.e., degree = 2 is assumed).
(5) Using dhat2(), the values of diagonal elements of a hat matrix (dhat) are obtained; the span is fixed at sp. The sum of diagonal elements of a hat matrix is defined as trhat. Furthermore, the sum of squares of the residuals (res) is obtained and the resultant value (residual sum of squares) is defined as sse.
(6) The values of CV and GCV, which are given by sp (the value of span), are saved as cv (a scalar) and gcv (a scalar).
(7) The objects cv and gcv are outputted.
(8) Using locvgcv(), the values of CV and GCV are calculated corresponding to ntrial values of the smoothing parameter (stored as span1). Each resultant value is saved as cv (a vector) and gcv (a vector).
(9) The objects cv and gcv are outputted.

An example of an object to use locv1() (construction of figure 3.45 (page 190))

```
function()
{
#   (1)
    set.seed(193.)
```

```
    nd <- 40.
    xx <- seq(from = 1., by = 1., length = nd)^1.8
    yy <- sin(0.004 * pi * xx) + rnorm(nd, mean = 0., sd = 0.3)
#   (2)
    ntrial <- 10.
    span1 <- seq(from = 0.15, by = 0.01, length = ntrial)
#   (3)
    output.lo <- locv1(xx, yy, nd, span1, ntrial)
    cv <- output.lo$cv
    gcv <- output.lo$gcv
#   (4)
    par(mfrow = c(1., 2.), mai = c(2., 1., 2., 1.),
     oma = c(2., 2., 1., 1.))
    plot(span1, cv, type = "n", ylim = c(0.06, 0.09),
     xlab = "span", ylab = "CV")
    points(span1, cv, pch = 3.)
    lines(span1, cv, lwd = 2.)
    pcvmin <- seq(along = cv)[cv == min(cv)]
    spancv <- span1[pcvmin]
    cvmin <- cv[pcvmin]
    points(spancv, cvmin, cex = 1., pch = 15.)
#   (5)
    plot(span1, gcv, type = "n", ylim = c(0.06, 0.09),
     xlab = "span", ylab = "GCV")
    points(span1, gcv, pch = 4.)
    lines(span1, gcv, lwd = 2.)
    pgcvmin <- seq(along = gcv)[gcv == min(gcv)]
    spangcv <- span1[pgcvmin]
    gcvmin <- gcv[pgcvmin]
    points(spangcv, gcvmin, cex = 1., pch = 15.)
}
```

(1) A total of 40 simulation data are generated; values of a predictor are xx and those of a target value are yy.

(2) A total of ntrial values are used as the values of span. They are defined as span1.

(3) The values of CV and GCV are calculated using locv1() and the results are saved as cv and gcv.

(4) A graph is created to represent the relationship between the value of span and that of CV in the same manner as that in "an example of an object to use kscv1()." Optimization of the value of span in LOESS yields the same optimal values of CV or GCV for different values of span; the object above handles this problem appropriately.

(5) A graph is created to represent the relationship between the value of span and that of GCV.

(J) Smoothing by supersmoother (construction of figure 3.46 (page 192), use of supsmu())

```
function()
{
#   (1)
    nd <- 100.
    set.seed(123.)
    xx <- seq(from = 1., by = 1., length = nd)
    yy <- sin(0.04 * pi * xx) + rnorm(nd, mean = 0.,
      sd = 0.02) * xx
#   (2)
    par(mfrow = c(2., 2.), mai = c(1., 1., 0.5, 0.5),
      oma = c(2., 2., 2., 2.))
#   (3)
    fit.su <- supsmu(xx, yy, span = 0.05)
    plot(xx, yy, type = "n", xlab = "x", ylab = "y (tweeter)")
    points(xx, yy, pch = 15., cex = 0.3)
    lines(fit.su$x, fit.su$y, lwd = 2.)
#   (4)
    fit.su <- supsmu(xx, yy, span = 0.2)
    plot(xx, yy, type = "n", xlab = "x", ylab = "y (midrange)")
    points(xx, yy, pch = 15., cex = 0.3)
    lines(fit.su$x, fit.su$y, lwd = 2.)
#   (5)
    fit.su <- supsmu(xx, yy, span = 0.5)
    plot(xx, yy, type = "n", xlab = "x", ylab = "y (woofer)")
    points(xx, yy, pch = 15., cex = 0.3)
    lines(fit.su$x, fit.su$y, lwd = 2.)
#   (6)
    fit.su <- supsmu(xx, yy, span = "cv")
    plot(xx, yy, type = "n", xlab = "x", ylab =
      "y (supersmoother)")
    points(xx, yy, pch = 15., cex = 0.3)
    lines(fit.su$x, fit.su$y, lwd = 2.)
}
```

(1) A total of 100 simulation data are generated.

(2) The total of 4 ($= 2 \cdot 2$) graphs are to be drawn.

(3) Smoothing with the use of supsmu() is carried out; the span is fixed at 0.05. The result is illustrated along with data in a graph. These estimates correspond to tweeter in supersmoother.

(4) Smoothing with the use of supsmu() is carried out; the span is set at 0.2. The result is illustrated along with data. These estimates correspond to midrange in supersmoother.

(5) Smoothing with the use of supsmu() is executed; the span is set at 0.5. The result is illustrated along with data in a graph. These estimates correspond to woofer in supersmoother.

(6) The result of smoothing using supersmoother (supsmu()) is illustrated along with data. The assignment of span = "cv" indicates that the value of span to obtain each estimate is determined by local cross-validation.

(K) Smoothing by LOWESS (robust version of LOESS) and by LOESS (construction of figure 3.49 (page 196), use of loess())

```
function()
{
#   (1)
    nd <- 100.
    set.seed(123.)
    xx <- seq(from = 1., by = 1., length = nd)
    yy <- sin(0.04 * pi * xx) + rnorm(nd, mean = 0., sd = 0.5)
    yy[30.] <- 5.
    yy[50.] <- 10.
    yy[70.] <- 5.
#   (2)
    ex <- seq(from = 1., to = nd, by = 0.1)
#   (3)
    assign("data1", data.frame(x = xx, y = yy), frame = 1.)
    data2 <- data.frame(x = ex)
#   (4)
    par(mfrow = c(2, 2), mai = c(1, 1, 0.5, 0.5),
      oma = c(2, 2, 1, 1))
#   (5)
    fit.low <- loess(y ~ x, data = data1, span = 0.1,
      family = "symmetric", degree = 1., surface = "direct")
    ey <- predict.loess(fit.low, newdata = data2)
    plot(xx, yy, type = "n", xlab = "x",
      ylab = "y (LOWESS, span=0.1)")
    points(xx, yy, pch = 15., cex = 0.3)
    lines(ex, ey, lwd = 2.)
#   (6)
    fit.low <- loess(y ~ x, data = data1, span = 0.2,
      family = "symmetric", degree = 1., surface = "direct")
    ey <- predict.loess(fit.low, newdata = data2)
    plot(xx, yy, type = "n", xlab = "x",
      ylab = "y(LOWESS, span=0.2)")
    points(xx, yy, pch = 15., cex = 0.3)
    lines(ex, ey, lwd = 2.)
#   (7)
    fit.lo <- loess(y ~ x, data = data1, span = 0.1,
      family = "gaussian", degree = 1., surface = "direct")
    ey <- predict.loess(fit.lo, newdata = data2)
    plot(xx, yy, type = "n", xlab = "x",
      ylab = "y (LOESS, span=0.1)")
    points(xx, yy, pch = 15., cex = 0.3)
```

```
    lines(ex, ey, lwd = 2.)
#   (8)
    fit.lo <- loess(y ~ x, data = data1, span = 0.2,
     family = "gaussian", degree = 1., surface = "direct")
    ey <- predict.loess(fit.lo, newdata = data2)
    plot(xx, yy, type = "n", xlab = "x",
     ylab = "y, (LOESS, span=0.2)")
    points(xx, yy, pch = 15., cex = 0.3)
    lines(ex, ey, lwd = 2.)
}
```

(1) A total of 100 simulation data are generated.

(2) The coordinates of the points at estimation points are saved as ex.

(3) Since data for loess() are assumed to take the form of a data frame, the data for regression are stored in data1 and the values of a predictor to derive estimates are saved in data2.

(4) Four graphs are to be drawn on the screen.

(5) Smoothing by LOWESS with span = 0.1 is carried out, and the results are illustrated in a graph. Since degree = 1 is designated, local linear regression is employed. If degree = is not designated, degree = 2 is adopted by default (i.e., local quadratic equation is utilized). Note that in the case of scatter.smooth(), which is an object to calculate estimates using LOWESS or LOESS and draw the results, nonexistence of degree = means degree=1 and the local linear equation is employed. Furthermore, since family = "symmetric" is assigned in loess(), calculation of LOWESS is carried out. If family = is not designated or family = "gaussian" is assigned, LOESS is executed.

(6) Smoothing by LOWESS with span = 0.2 is carried out and the results are illustrated in a graph.

(7) Smoothing by LOESS with span = 0.1 is executed and the results are displayed in a graph.

(8) Smoothing by LOESS with span = 0.2 is executed and the results are shown in a graph.

REFERENCES

1. D.M. Bates, M.J. Lindstorm, G. Wahba, and B.S. Yandell (1987). GCVPACK — routines for generalized cross validation. *Communication in Statistics, Series B*, Vol. 16, No. 1, pp. 263-297.

2. K.P. Burnhan and D.R. Anderson (1998). *Model Selection and Inference — A Practical Information-Theoretic Approach*, Springer-Verlag.

3. J.M. Chambers and T.J. Hastie (1993). *Statistical Models in S*, Chapman & Hall/CRC.

4. W.S.Cleveland and S.J. Devlin (1988). Locally weighted regression: an approach to regression analysis by local fitting, *Journal of the American Statistical Association*, Vol. 83, pp. 596-610.

5. W.S. Cleveland (1993). *Visualizing Data*, Hobart Press.

6. B. Efron and R.J. Tibshirani (1993). *An Introduction to the Bootstrap*, Chapman & Hall/CRC.

7. J. Fan and I. Gijbels (1996). *Local Polynomial Modelling and Its Application*, Chapman & Hall/CRC.

8. L. Fahrmeir and G. Tutz (2001). *Multivariate Statistical Modelling Based on Generalized Linear Models* (Springer Series in Statistics), 2nd edition, Springer-Verlag.

9. J. Fox (2000). *Nonparametric Simple Regression* (Series: Quantitative Applications in the Social Sciences), Sage Publications.

10. P.J. Green and B.W. Silverman (1994). *Nonparametric Regression and Generalized Linear Models, a Roughness Penalty Approach*, Chapman & Hall/CRC.

11. C. Gu (2002). *Smoothing Spline ANOVA Models*, Springer-Verlag.

12. T.J. Hastie and R.J. Tibshirani (1990). *Generalized Additive Model*, Chapman & Hall/CRC.

13. C.M. Hurvich, J.S. Simonoff, and C.-L. Tsai (1998). Smoothing parameter selection in nonparametric regression using an improved Akaike information criterion. *Journal of the Royal Statistical Society*, Series B, Vol. 60, Part 2, pp. 271-293.

14. M.J. Powell (1981). *Approximation Theory and Methods*, Cambridge University Press.

15. Ramsay and Silverman (1997). *Functional Data Analysis*, Springer-Verlag.

16. T.P. Ryan (1997). *Modern Regression Methods*, John Wiley & Sons.

17. B.W. Silverman (1984). Spline smoothing: the equivalent variable kernel method. *Annals of Statistics*, Vol.12, pp. 898-916.

Problems

3.1   Consider the trade-off of bias versus variance of a binomial filter by the simulation data obtained using eq(3.13) (page 110).
(a) Create an S-Plus or R object for drawing graphs similar to that in figure 3.2 (page 108) and figure 3.3 (page 109) for a binomial filter.
(b) Run the object produced in **(a)** and describe the influence of m on bias ($Bias[\hat{Y}_i]$), variance ($Var[\hat{Y}_i]$), and mean squared error ($MSE[\hat{Y}_i]$). Consider the relationship between the appearance of data and these three values, and the behavior of these three values near the two ends of the data region.

3.2   Consider the trade-off of bias versus variance of local linear regression by the simulation data obtained using eq(3.13) (page 110).
(a) Create an S-Plus or R object for constructing graphs similar to those in figure 3.2 (page 108) and figure 3.3 (page 109) for local linear regression.
(b) Run the object produced in **(a)** and observe the influence of the value of the smoothing parameter on bias ($Bias[\hat{Y}_i]$), variance ($Var[\hat{Y}_i]$), and mean squared error ($MSE[\hat{Y}_i]$). Consider the relationship between the appearance of data and these three values, and the behavior of these three values near the two ends of the data region.

3.3   Consider the trade-off of bias versus variance of the smoothing spline by the simulation data obtained using eq(3.13) (page 110).
(a) Create an S-Plus or R object for drawing graphs similar to those in figure 3.2 (page 108) and figure 3.3 (page 109) for the smoothing spline.
(b) Run the object produced in **(a)** and observe the influence of the smoothing parameter on bias ($Bias[\hat{Y}_i]$), variance ($Var[\hat{Y}_i]$), and mean squared error ($MSE[\hat{Y}_i]$). Consider the relationship between the appearance of data and these three values, and the behavior of these three values near the two ends of the data region.

3.4   Prove eq(3.41) (page 117) in a case in which a local linear regression is used for deriving estimates.

3.5   Prove eq(3.41) (page 117) in a case in which a polynomial regression is employed for deriving estimates; use the Sherman-Morrison-Woodbury theorem.

3.6   Consider the trade-off of bias versus variance of the Nadaraya-Watson estimator by simulation data obtained using eq(3.13) (page 110).
(a) Create an S-Plus or R object for drawing graphs similar to those in figure 3.2 (page 108) and figure 3.3 (page 109) by utilizing the object presented in **(A)** of this chapter.
(b) Run the object produced in **(a)** and observe the influence of the bandwidth on bias ($Bias[\hat{Y}_i]$), variance ($Var[\hat{Y}_i]$), and mean squared error ($MSE[\hat{Y}_i]$). Consider the relationship between the appearance of data and these three values, and the behavior of these three values near the two ends of the data region.
(c) Estimate the optimal bandwidth in terms of CV or GCV using the object presented in **(B)** of this chapter. Construct graphs of bias ($Bias[\hat{Y}_i]$), variance ($Var[\hat{Y}_i]$), and mean squared error ($MSE[\hat{Y}_i]$) using the object created in **(a)**, and verify that the optimal bandwidth in terms of CV or GCV yields satisfactory bias, variance, and mean squared error values. If the optimal bandwidth does not provide a sufficient result, discuss the reason and attempt the use of other statistics for choosing the optimum bandwidth.

3.7   Consider the relationship between a local linear regression and the tangent lines of the function that generates data.

(a) Construct graphs similar to those in figure 3.14 (page 142) using the object presented in (D) of this chapter.

(b) Create an S-Plus or R object for drawing two graphs on a screen. One graph shows data and estimates obtained by local linear regression. The other graph displays the values of the gradient estimated by local linear regression and those derived using a real function that generates the data. Use simulation data such as those generated by

$$X_i = i \qquad (1 \le i \le 20), \qquad\qquad (3.277)$$

$$Y_i = \sin(0.2\pi X_i) + \epsilon_i, \qquad\qquad (3.278)$$

where $\{\epsilon_i\}$ are realizations of $N(0.0, 0.5^2)$ (normal distribution; the mean is 0.0, and the standard deviation is 0.5). Then, the real value of a target variable is calculated using $Y_i = \sin(0.2\pi X_i)$, and the first derivatives (i.e., gradients) of the function at each $\{X_i\}$ are derived analytically.

(c) Run the object created in (b). The bandwidth is optimized visually. Then, confirm that when the appropriate estimates are obtained, the gradients derived by local linear regression are close to those obtained using real values. Note, however, that this observation does not guarantee that good estimates always result in good estimates of the gradient; for more detailed information on this matter, refer to more advanced textbooks on smoothing.

3.8   Construct an S-Plus or R object for drawing weight diagrams (i.e., equivalent kernels) of local linear regression on the basis of eq(3.127) (page 142) and eq(3.128) (page 143). Assume that $\{X_i\}$ are $\{1, 2, 3, \ldots, 20\}$, and compare the values of the equivalent kernels with those of the Nadaraya-Watson estimator (in particular, the equivalent kernels for calculating estimates near two ends of the data region). Furthermore, verify eq(3.130) (page 143) numerically.

3.9   Examine the effect of the degree of the polynomial equation on the estimates derived by local polynomial regression.

(a) Modify the object 111() shown in (D) of this chapter to employ a cubic equation for local polynomial regression.

(b) Check that the estimates supplied by local cubic regression are more flexible than those supplied by local linear regression when the span is the same.

(c) Examine the characteristics of bias, variance, and mean squared error of these two smoothing methods (local linear regression and local cubic regression) using the technique adopted in Problem **3.1**.

3.10   Check numerically that the natural boundary conditions (eq(3.172) (page 158) and eq(3.173) (page 158)) minimize the value defined by eq(3.171) (page 158), which the cubic spline function gives.

(a) Create an S-Plus or R object for constructing a cubic spline function under diverse boundary conditions. When the natural boundary conditions are employed, the function is realized by solving eq(3.192) (page 165). When the natural boundary conditions are not employed, the right-hand side of eq(3.192) is modified.

(b) Produce an S-Plus or R object for calculating numerically the value defined by eq(3.171) using the cubic spline function.

(c) Using the object constructed in **(b)**, confirm that the natural boundary conditions minimize the value defined by eq(3.171) of the cubic spline function.

3.11   Calculate the values of the elements of $\mathbf{L}$ and $(\mathbf{I}_n + \lambda\mathbf{L})^{-1}$ defined in eq(3.231) (page 176) on the assumption that the predictor values are written as $\{X_i\} = \{1, 2, 3, \ldots, 20\}$. Eq(3.229) (page 176) can be used for this purpose. If possible, however, attempt the use of previously reported procedures in other literature to derive them and compare the results.

3.12   Consider the fact that, among various natural cubic splines, the natural cubic spline given by eq(3.223) (page 175) minimizes the value defined by eq(3.209) (page 170).
(a) Create an S-Plus or R object for computing the value of eq(3.209) on the assumption that the regression equation is derived using eq(3.192) (page 165) (hint: if the regression equation is a natural spline, eq(3.217) (page 174) holds).
(b) Run the object created in **(a)** with various y values on the right-hand side of eq(3.192) to check that the value of eq(3.209) is minimized by eq(3.223) (page 175). If possible, use an optimization technique such as the generic algorithm to minimize the value defined by eq(3.209).

3.13   Consider the knots of the smoothig spline.
(a) Construct an S-Plus or R object for calculating the estimates of the smoothig spline on the basis of eq(3.242) (page 180). B-spline bases are used and knots are placed on all data points.
(b) Modify the object constructed in **(a)** to compute the estimates of the smoothig spline when some knots are deleted.
    Calculate the estimates using the objects produced in **(a)** and **(b)** with various simulation data. Determine the case in which a marked difference between the values of estimates of the two objects is observed.

3.14   On the basis of eq(3.243) (page 180), construct an S-Plus or R object for transforming the values of $\{a_0, a_1, b_1, b_2, \ldots, b_n\}$ into $\{\beta_1, \beta_2, \ldots, \beta_{n+2}\}$. Description in Appendix A will be useful.

3.15   The cross-validation (for LOESS) implemented in the object presented in (l) of this chapter is based on the assumption that the estimate obtained by deleting one data is represented by eq(3.41) (page 117); however, this is an approximation. Compare the results obtained using the object presented in (l) with those obtained by conducting the cross-validation based on its real definition.

3.16   Consider the fact that the supersmoother sometimes does not function well when the number of data is large.
(a) Generate data sets (the number of data is n) based on the equation below.

$$Y_i = \sin(0.2\pi X_i) + \epsilon_i, \tag{3.279}$$

where $\{X_i\}$ is $\{0.1, 0.2, 0.3, \ldots, 0.1n\}$ and ϵ_i is a realization of $N(0.0, 0.1^2)$ (a normal distribution; the mean is 0.0 and the standard deviation is 0.1). The numbers of data (n) are 100, 500, and 2000.

(b) Confirm that the estimates are not satisfactory when the number of data is large. Discuss the mechanism behind this phenomenon and determine countermeasures.

3.17   Produce an S-Plus or R object for the robust version of LOESS without using the objects for the robust version of LOESS included in S-Plus or R by default, and illustrate the iterative procedure in a graph to show the process of reducing the effect of outliers.

3.18   Assume that the eigenvalue of $\mathbf{A}$ is $\{a_i\}$; $\mathbf{A}$ is a symmetric matrix. Prove that the eigenvalues of $(\mathbf{A} - \beta\mathbf{I})^{-1}$ ($\mathbf{I}$ is an identity matrix) are $\{\frac{1}{a_i+\beta}\}$.

3.19   Prove eq(3.32) (page 115).
(a) Assume that a random variable X obeys $p(x)$ (a probability density function) and a random variable Y obeys $q(y)$ (a probability density function). Moreover, assume the equation below.

$$E[Y] = \int yq(y)dy = 0. \tag{3.280}$$

In addition, assume that X and Y are independent of each other; that is, the equation below holds

$$r(x,y) = p(x)q(y), \tag{3.281}$$

where $r(x,y)$ is a joint probability density function of X and Y. Prove the equation below using eq(3.280) and eq(3.281). $f(\cdot)$ is an arbitrary integrable function.

$$E[f(X)Y] = 0. \tag{3.282}$$

(b) Prove eq(3.32) (page 115) by the use of eq(3.282) in **(a)**.

3.20   Prove that in the case of linear regression, eq(3.32) (page 115) holds under the assumption of $E[\epsilon_i\epsilon_j^*] = 0$ $(i \neq j)$; this assumption is weaker than that of Problem **3.19**.

3.21   When a data set contains data in which predictor values are the same or very close, the use of the least squares as is may not be the best strategy. A slightly different value should be minimized to obtain regression coefficients; this method can be preferable in terms of computational cost or computational stability. Then, it is assumed that a data set is $\{(X_i, Y_{ij})\}$ $(1 \le i \le n, 1 \le j \le t_i)$. The conventional least squares minimizes the value below to derive regression coefficients.

$$E_1 = \sum_{i=1}^{n}\sum_{j=1}^{t_i}(Y_{ij} - m(X_i))^2, \tag{3.283}$$

where $m(X_i)$ is a regression equation. Prove that the derivation of regression coefficients by minimizing the value defined in eq(3.283) is equivalent to that by minimizing

$$E_2 = \sum_{i=1}^{n}(t_i(\bar{Y}_i - m(X_i)^2), \tag{3.284}$$

where $\bar{Y}_i$ is defined as

$$\bar{Y}_i = \frac{1}{t_i}\sum_{j=1}^{t_i}Y_{ij}. \tag{3.285}$$

(Hint: Substitute eq(3.285) into eq(3.284), and compare the result with eq(3.283); the essential parts are expected to be the same.)

CHAPTER 4

MULTIDIMENSIONAL SMOOTHING

4.1 INTRODUCTION

This chapter concerns nonparametric regression when the number of predictors is plural. When the number of predictors is unity, our aim is to make a smooth line pass through the data in a graph; the line should resemble one that a person would draw. Although the line generated by nonparametric regression is different from a hand-drawn one in that the former has a mathematical background, nonparametric regression actually provides lines mimicking those drawn by hand. Hence, nonparametric regression does not satisfy the expectation that calculations by a computer, based on mathematical techniques, should bring about results that are impossible by freehand. Therefore, the significance of nonparametric regression for data with one predictor might be nothing more than the realization of labor-saving using a computer, or mathematical justification of the intuition of a human being. However, when the number of predictors is more than one, it is difficult to plot data on a graph and derive estimates (estimated values) manually on the basis of the graph. Consequently, by developing techniques built up for one predictor into those for plural predictors, we

Introduction to Nonparametric Regression, Kunio Takezawa.
Copyright © 2006 John Wiley & Sons, Inc.

can live up to the expectation that a computer should achieve something that a human being would find difficult to do.

In this chapter, therefore, several techniques of nonparametric regression are presented to cope with data with plural predictors. These techniques are separated into two categories: extensions of the method treated in Chapter 3 and methods peculiar to data with plural predictors. Note that the methods introduced hereafter are confined to fundamental ones. For information on advanced techniques, refer to other literature (e.g., [10]). There are a myriad of methods for deriving a functional relationship from data with plural predictors. Even now, various innovative techniques continue to be developed, and their characteristics are under investigation. This is attributable to the fact that this is fertile territory for motivating a variety of challenges and has highly practical demands driven by diverse applications.

4.2   LOCAL POLYNOMIAL REGRESSION FOR MULTIDIMENSIONAL PREDICTOR

Local polynomial regression for data with one predictor is developed into that for data with plural predictors as follows. For instance, when the number of predictors is two, we can make use of the local linear equation around the position where the values of predictors are (x_1^*, x_2^*).

$$m(x_1, x_2, x_1^*, x_2^*) = a_0(x_1^*, x_2^*) + a_1(x_1^*, x_2^*)(x_1 - x_1^*) + a_2(x_1^*, x_2^*)(x_2 - x_2^*). \quad (4.1)$$

This equation corresponds to that for data with one predictor: eq(3.111) (page 139) with $p = 1$. When datum $\{X_{i1}, X_{i2}, Y_i\}$ $(1 \le i \le n)$ is given, $a_0(x_1^*, x_2^*)$, $a_1(x_1^*, x_2^*)$, and $a_2(x_1^*, x_2^*)$ are obtained by minimizing

$$
\begin{aligned}
& E_{2D-linear}(x_1^*, x_2^*) \\
&= \sum_{i=1}^{n} w\left(\frac{X_{i1} - x_1^*}{h_1}, \frac{X_{i2} - x_2^*}{h_2}\right)\left(m(X_{i1}, X_{i2}, x_1^*, x_2^*) - Y_i\right)^2 \\
&= \sum_{i=1}^{n} w\left(\frac{X_{i1} - x_1^*}{h_1}, \frac{X_{i2} - x_2^*}{h_2}\right) \cdot \Big(a_0(x_1^*, x_2^*) \\
&\qquad + a_1(x_1^*, x_2^*)(x_1 - x_1^*) + a_2(x_1^*, x_2^*)(x_2 - x_2^*) - Y_i\Big)^2. \quad (4.2)
\end{aligned}
$$

One predictor version of this equation is eq(3.112) (page 139) $(p = 1)$. Then, the estimate at (x_1^*, x_2^*) is $\hat{m}(x_1^*, x_2^*, x_1^*, x_2^*)$ $(= \hat{a}_0(x_1^*, x_2^*))$.

To utilize a local quadratic equation, we use

$$
\begin{aligned}
m(x_1, x_2, x_1^*, x_2^*) &= a_0(x_1^*, x_2^*) + a_1(x_1^*, x_2^*)(x_1 - x_1^*) + a_2(x_1^*, x_2^*)(x_2 - x_2^*) \\
&\quad + a_3(x_1^*, x_2^*)(x_1 - x_1^*)^2 + a_4(x_1^*, x_2^*)(x_2 - x_2^*)^2 \\
&\quad + a_5(x_1^*, x_2^*)(x_1 - x_1^*)(x_2 - x_2^*). \quad (4.3)
\end{aligned}
$$

To obtain $a_0(x_1^*, x_2^*)$, $a_1(x_1^*, x_2^*)$, $a_2(x_1^*, x_2^*)$, $a_3(x_1^*, x_2^*)$, $a_4(x_1^*, x_2^*)$, and $a_5(x_1^*, x_2^*)$, the value below is minimized.

$$E_{2D-quad}(x_1^*, x_2^*)$$

$$= \sum_{i=1}^{n} w \left(\frac{X_{i1} - x_1^*}{h_1}, \frac{X_{i2} - x_2^*}{h_2} \right) \left(m(X_{i1}, X_{i2}, x_1^*, x_2^*) - Y_i \right)^2$$

$$= \sum_{i=1}^{n} w \left(\frac{X_{i1} - x_1^*}{h_1}, \frac{X_{i2} - x_2^*}{h_2} \right) \cdot \Big(a_0(x_1^*, x_2^*)$$

$$+ a_1(x_1^*, x_2^*)(x_1 - x_1^*) + a_2(x_1^*, x_2^*)(x_2 - x_2^*)$$

$$+ a_3(x_1^*, x_2^*)(x_1 - x_1^*)^2 + a_4(x_1^*, x_2^*)(x_2 - x_2^*)^2$$

$$+ a_5(x_1^*, x_2^*)(x_1 - x_1^*)(x_2 - x_2^*) - Y_i \Big)^2. \tag{4.4}$$

This equation corresponds to an approximation by Taylor expansion of $m(x_1, x_2)$ up to the degree of 2 at (x_1^*, x_2^*).

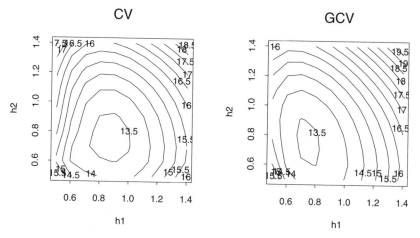

Figure 4.1 Smoothing data with two predictors. Optimization of bandwidth of local linear regression (h_1, h_2). Values of CV (left). Values of GCV (right).

As a kernel function $\left(w\left(\frac{X_{i1} - x_1^*}{h_1}, \frac{X_{i2} - x_2^*}{h_2} \right) \right)$, the Gaussian function below, for example, can be used.

$$w \left(\frac{X_{i1} - x_1^*}{h_1}, \frac{X_{i2} - x_2^*}{h_2} \right) = \exp\left(-\frac{1}{2} \left(\frac{X_{i1} - x_1^*}{h_1} \right)^2 - \frac{1}{2} \left(\frac{X_{i2} - x_2^*}{h_2} \right)^2 \right). \tag{4.5}$$

If the characteristics of x_1 are not greatly different from those of x_2, $h = h_1 = h_2$ is assumed. Consequently, we use the kernel function

$$w \left(\frac{X_{1i} - x_1^*}{h}, \frac{X_{2i} - x_2^*}{h} \right) = \exp\left(-\frac{1}{2} \left(\frac{X_{1i} - x_1^*}{h} \right)^2 - \frac{1}{2} \left(\frac{X_{2i} - x_2^*}{h} \right)^2 \right). \tag{4.6}$$

Differentiation of eq(4.2) to be 0 with respect to $a_0(x_1^*, x_2^*)$, $a_1(x_1^*, x_2^*)$, and $a_2(x_1^*, x_2^*)$ constitutes a simultaneous linear equation. The same is true of the differentiation of eq(4.4) to be 0 with respect to $a_0(x_1^*, x_2^*)$, $a_1(x_1^*, x_2^*)$, $a_2(x_1^*, x_2^*)$,

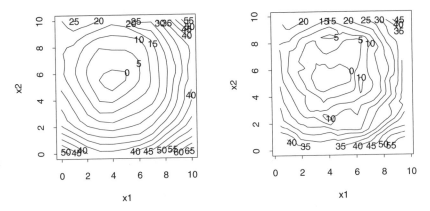

Figure 4.2   Estimates given by smoothing using local linear regression; the bandwidths (h_1, h_2) obtained from figure 4.1 (left). Estimates obtained by interpolation using Akima's method (right).

$a_3(x_1^*, x_2^*)$, $a_4(x_1^*, x_2^*)$, and $a_5(x_1^*, x_2^*)$. Therefore, a hat matrix is derived in a similar manner to that for data with one predictor to compute CV, GCV, and so forth.

Figure 4.1 and figure 4.2 illustrate the results of local linear regression using simulation data with two predictors. The number of simulation data is 100. $\{X_{i1}\}$ ($1 \leq i \leq n$) and $\{X_{i2}\}$ are realizations of a uniform pseudorandom number that takes a value between 0 and 10. $\{Y_i\}$ is generated as

$$Y_i = (X_{i1} - 4)^2 + (X_{i2} - 6)^2 + \epsilon_i, \tag{4.7}$$

where $\{\epsilon_i\}$ is a realization of $N(0.0, 3.0^2)$ (a normal distribution; the mean is 0.0 and standard deviation is 3.0). The relationships between h_1 ($= (0.5, 0.6, \ldots, 1.4)$) and h_2 ($= (0.5, 0.6, \ldots, 1.4)$) for CV and GCV are displayed in figure 4.1; eq(4.5) is employed as a kernel function. The bandwidth that gives the smallest GCV ($= 13.34$) is found to be $h_1 = 0.7$ and $h_2 = 0.8$. Estimates obtained using these values are shown in figure 4.2(left). Figure 4.2(right) illustrates interpolated estimates given by Akima's method.

4.3   THIN PLATE SMOOTHING SPLINES

The natural cubic spline and cubic smoothing spline (cubic smoothing splines) are also developed into methods for data with two predictors. In the case of more than one predictor, a concept similar to the natural cubic spline for data with one predictor is termed the thin plate splines (thin plate spline) or natural thin plate splines (natural thin plate spline). When the number of predictors is two, it is $m(x_1, x_2)$, which is obtained using the value below under the condition of $m(X_{1i}, X_{2i}) = Y_i$ ($1 \leq i \leq n$).

$$\int_{-\infty}^{\infty} \int_{-\infty}^{\infty} \left(\left(\frac{\partial^2 m(x_1, x_2)}{\partial x_1^2} \right)^2 + 2 \left(\frac{\partial^2 m(x_1, x_2)}{\partial x_1 \partial x_2} \right)^2 + \left(\frac{\partial^2 m(x_1, x_2)}{\partial x_2^2} \right)^2 \right) dx_1 dx_2. \tag{4.8}$$

The estimates given by data with two predictors are regarded as bent elastic thin plates in the same way that those given by the natural cubic spline using data with one predictor are considered as bent elastic beams. This is the origin of the term thin plate splines. Thin plate splines with more than two variables are looked upon as bent thin elastic plates in a higher dimensional space.

Furthermore, in analogy with a cubic smoothing spline, which is developed as a smoothing method for one predictor based on the natural cubic spline, $m(x_1, x_2)$ obtained by minimizing the value below can be used as a smoothed estimate for data with two predictors.

$$
\begin{aligned}
E_{2D-thin} &= \sum_{i=1}^{n}(Y_i - m(X_{i1}, X_{i2}))^2 \\
&+ \lambda \int_{-\infty}^{\infty} \int_{-\infty}^{\infty} \left(\left(\frac{\partial^2 m(x_1, x_2)}{\partial x_1^2} \right)^2 + 2\left(\frac{\partial^2 m(x_1, x_2)}{\partial x_1 \partial x_2} \right)^2 \right. \\
&+ \left. \left(\frac{\partial^2 m(x_1, x_2)}{\partial x_2^2} \right)^2 \right) dx_1 dx_2
\end{aligned}
\tag{4.9}
$$

Here, λ is a smoothing parameter. The resulting $\hat{m}(x_1, x_2)$ is called a thin plate smoothing spline. A thin plate smoothing spline with two predictors is a thin plate spline that interpolates $\{X_{i1}, X_{i2}, \hat{m}(X_{i1}, X_{i2})\}$ $(1 \le i \le n)$ in a similar fashion to a cubic smoothing spline with one predictor which is a natural spline that interpolates $\{X_i, \hat{m}(X_i)\}$ $(1 \le i \le n)$. This is proved in an analogous manner to the case with one predictor.

Furthermore, by analogy with physical objects, estimates with two predictors are compared to an infinitely wide elastic plate that is pulled by springs; the number of springs is the same as that of data. That is, the analogy with the case of one predictor is carried over; estimates with one predictor are associated with an elastic beam pulled by springs of the same number as that of data. Multiplication of eq(4.8) by a constant to yields the elastic energy of a plate with constant thickness when ν (the Poisson ratio) is 0; the elastic energy is written as

$$
\begin{aligned}
U &= \frac{D}{2} \int \int \left(\frac{\partial^2 m(x_1, x_2)}{\partial x_1^2} + \frac{\partial^2 m(x_1, x_2)}{\partial x_2^2} \right)^2 dx_1 dx_2 \\
&- D(1-\nu) \int \int \left(\frac{d^2 m(x_1, x_2)}{\partial x_1^2} \frac{\partial^2 m(x_1, x_2)}{\partial x_2^2} - \left(\frac{\partial^2 m(x_1, x_2)}{\partial x_1 \partial x_2} \right)^2 \right) dx_1 dx_2,
\end{aligned}
\tag{4.10}
$$

where D denotes the flexural rigidity of the plate. The domain of integration is the whole region of the plate. On the other hand, the domain in eq(4.8) and eq(4.9) is the entire two-dimensional space. When the number of predictors is one, the resulting regression equation (regression function) remains the same as long as the domain of integration contains the area where data exist. However, when the number of predictors is more than one, the result depends weakly on the domain of integration (page 150 of [5]). Hence, the commonly defined thin plate smoothing splines and thin plate splines utilize integration in the whole of two-dimensional space.

Minimization of eq(4.9) yields the solution ([6])

$$m(x_1, x_2) = c_0 + c_1 x_1 + c_2 x_2 + \sum_{i=1}^{n} \theta_i \cdot G_i(x_1 - X_{i1}, x_2 - X_{i2}), \qquad (4.11)$$

where the definition below is used:

$$G_i(x_1 - X_{i1}, x_2 - X_{i2}) = \begin{cases} \dfrac{t_i^2}{8\pi} \cdot \log(t_i) & \text{if} \quad t > 0 \\ 0 & \text{if} \quad t = 0, \end{cases} \qquad (4.12)$$

$$t_i = \sqrt{(x_1 - X_{i1})^2 + (x_2 - X_{i2})^2}. \qquad (4.13)$$

When the number of data is n, eq(4.11) contains $(n+3)$ unknown coefficients. The 3 lacking data correspond to the three conditions

$$\sum_{i=1}^{n} \theta_i = 0, \qquad \sum_{i=1}^{n} \theta_i X_{i1} = 0, \qquad \sum_{i=1}^{n} \theta_i X_{i2} = 0. \qquad (4.14)$$

The form of these three conditions is the straightforward extension of that in a smoothing spline (smoothing splines) with one predictor (eq(3.214) (page 174), eq(3.215)).

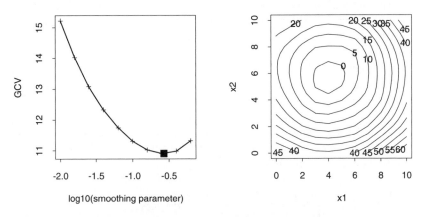

Figure 4.3   Smoothing using thin plate smoothing splines with the same data as in figure 4.1 and figure 4.2. After values of GCV (+) are calculated at $\lambda = \{10^{-2.0}, 10^{-1.8}, \ldots, 10^{-0.2}\}$, a golden ratio search is carried out in the region where a small value of GCV is observed; the result is plotted with ■ (left). Estimates given by the optimal smoothing parameter ($\lambda = 10^{-0.5724401}$) (right).

Figure 4.3 displays the result of two-dimensional smoothing using thin plate smoothing splines; the data are the same as those used for figure 4.1 and figure 4.2. The process of optimizing a smoothing parameter using GCV is illustrated in figure 4.3(left). Estimates using the optimal smoothing parameter are shown in figure 4.3(right). GCVPACK [1] is utilized for this calculation. It is assumed that the values of the two smoothing parameters are identical.

The object "`intns2()`"(page 209) introduced in Chapter 3 satisfactorily realizes interpolation using a natural spline if "`smooth.spline()`" (an object for carrying out smoothing spline) with a very small positive value of the smoothing parameter is utilized. Similarly, when the number of predictors is plural, interpolation using the thin plate splines is accomplished by using the thin plate smoothing splines with a very small positive number for the smoothing parameter (λ in eq(4.9) (page 235)). S-Plus contains `interp()` for interpolation to illustrate the shape of data with two predictors. This object makes use of Akima's method. However, substantial interpolation using the thin plate smoothing splines with very small positive values for the smoothing parameters, which is realized by GCVPACK, provides better estimates [2].

4.4   LOESS AND LOWESS WITH PLURAL PREDICTORS

For LOESS and LOWESS (LOcally WEighted Scatterplot Smoother), the equation for data with one predictor (eq(3.262) (page 186)) is simply extended to cope with data with plural predictors. When a local linear equation is used in LOESS, eq(4.2) (page 232) is transformed to use a kernel function with the k-nearest-neighbor distance. When the number of predictors is two and the characteristics of the two predictors are similar, we use

$$
\begin{aligned}
&E_{2D-LOESS}(x_1^*, x_2^*) \\
&= \sum_{i=1}^{n} w\left(\frac{X_{i1} - x_1^*}{h_k(x)}, \frac{X_{i2} - x_2^*}{h_k(x)}\right) \cdot \Bigg(a_0(x_1^*, x_2^*) \\
&\quad + a_1(x_1^*, x_2^*)(x_1 - x_1^*) + a_2(x_1^*, x_2^*)(x_2 - x_2^*) - Y_i \Bigg)^2 .
\end{aligned} \tag{4.15}
$$

If the predictor variable that is the k-nearest-neighbor distance from (x_1^*, x_2^*) is set to be $(X_1^{(k)}, X_2^{(k)})$, $h_k(x)$ is $\sqrt{(x_1^* - X_1^{(k)})^2 + (x_2^* - X_2^{(k)})^2}$. The values of $a_0(x_1^*, x_2^*)$, $a_1(x_1^*, x_2^*)$, and $a_2(x_1^*, x_2^*)$ are calculated by minimizing $E_{2D-LOESS}(x_1^*, x_2^*)$, and an estimate at (x_1^*, x_2^*) is defined as $\hat{a}_0(x_1^*, x_2^*)$. When a local quadratic equation is employed, a similar measure is applied to eq(4.4) (page 232).

A kernel function for plural predictors is, for example, a simple extension of a tricube weight function (eq(3.116) (page 128) for one predictor. When the value of the span (s) is set as $s \leq 1$, we have

$$
w\left(\frac{X_{i1} - x_1^*}{h_k(x)}, \frac{X_{i2} - x_2^*}{h_k(x)}\right) =
\begin{cases}
\left(1 - \left(\frac{((X_{i1} - x_1^*)^2 + (X_{i2} - x_2^*)^2)^{1/2}}{h_k(x)}\right)^3\right)^3 \\
\qquad \text{if } \left(\frac{((X_{i1} - x_1^*)^2 + (X_{i2} - x_2^*)^2)^{1/2}}{h_k(x)}\right)^3 \leq 1 \\
0 \\
\qquad \text{if } \left(\frac{((X_{i1} - x_1^*)^2 + (X_{i2} - x_2^*)^2)^{1/2}}{h_k(x)}\right)^3 > 1 .
\end{cases}
\tag{4.16}
$$

If $s > 1$, we set $w\left(\frac{X_{i1} - x_1^*}{s \cdot h_n(x)}, \frac{X_{i2} - x_2^*}{s \cdot h_n(x)}\right)$.

Using these settings, LOWESS (robust version of LOESS) for plural predictors is also constructed.

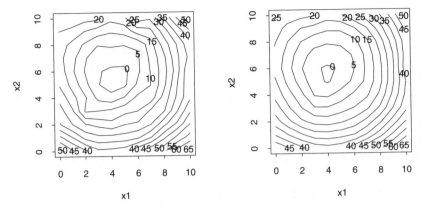

Figure 4.4   Example of smoothing by LOESS for two predictors. $s = 0.1$ (left). $s = 0.2$ (right).

Figure 4.4 displays estimates by LOESS with eq(4.16) as a kernel function; the data are the same as that in figure 4.1, figure 4.2, and figure 4.3, and the span is either 0.1 or 0.2.

4.5   KRIGING

Kriging interpolation, which has been developed mainly in the field of geostatistics, is a kind of interpolation method. The term kriging is named after D.G. Kringe [9]. This is not a smoothing method but an interpolation method, and hence, a line or a surface exhibiting the estimates created by this technique interpolates data points. As was described in Chapter 1, since smoothing is preferable to interpolation even if data do not contain errors, interpolation methods may not deserve attention. However, while we need somewhat complicated methods to smooth data that have errors with correlation, interpolation has the advantage that correlation of errors is easily incorporated by adopting kriging. The meaningfulness of sketching the outlines of kriging interpolation is also justified by the fact that it has long been practically used as an estimation method with weighted averages. Furthermore, interpolation has another merit: estimation at the data point by smoothing followed by interpolation by a smooth line or surface (figure 3.30 (page 169)) is a useful procedure.

Although kriging is treated as a nonparametric smoothing method for plural predictors on the basis of the historical background that kriging was chiefly developed in geostatistics, kriging is formalized and exemplified first for one predictor. This is because the essence of kriging interpolation does not contain this limitation on the number of predictors and the fundamental concepts are the same however many predictors (one or two) data may have, though it is usually applied to data with two variables for practical uses.

First, we discuss a method of regressing data ($\{X_i, Y_i\}$ ($1 \leq i \leq n$)) to the regression equation below for interpolation and derive estimates without bias.

$$y = \beta_0 + \epsilon. \qquad (4.17)$$

Then, the data are written as

$$Y_i = \beta_0 + \epsilon_i, \tag{4.18}$$

where β_0 is a regression coefficient, and ϵ is a random variable that satisfies $E[\epsilon] = 0$. This regression differs from those described hitherto with respect to the fact that if the two predictor values (x_a and x_b) are not identical ($x_a \neq x_b$), correlation between ϵ_a and ϵ_b is considered; ϵ_a and ϵ_b are values of ϵ at x_a and x_b, respectively. That is, even if $x_a \neq x_b$ holds, $Cov(\epsilon_a, \epsilon_b) = 0$ is not guaranteed. Then, $\mathbf{C}$ is defined as

$$[\mathbf{C}]_{ij} = Cov(Y_i, Y_j) = Cov(\epsilon_i, \epsilon_j) \qquad (1 \leq i \leq n, \ 1 \leq j \leq n). \tag{4.19}$$

In this setting, the estimation of a target variable (object variable) (y_0) at x_0 (a value of a predictor) is the estimation of

$$\hat{y}_0 = \hat{\beta}_0 + \hat{\epsilon}_0. \tag{4.20}$$

If $Cov(\epsilon_a, \epsilon_b) = 0$ ($x_a \neq x_b$) is assumed, $\hat{y}_0 = \hat{\beta}_0$ is obtained, but the assumption here is different. Hence, the elements of $\mathbf{d}_0$ ($= (d_1, d_2, \ldots, d_n)^t$) (superscript t stands for a transposed matrix) is defined as

$$d_i = Cov(Y_i, Y_0) = Cov(\beta_0 + \epsilon_i, \beta_0 + \epsilon_0) = Cov(\epsilon_i, \epsilon_0) \qquad (1 \leq i \leq n). \tag{4.21}$$

Next, we assume that $\hat{y}_0$ is represented as

$$\hat{y}_0 = \alpha_0 + \sum_{i=1}^{n} \alpha_i Y_i. \tag{4.22}$$

To derive $\{\alpha_0, \alpha_1, \alpha_2, \ldots, \alpha_n\}$ in eq(4.22) using the least squares (least square method), we maximize

$$
\begin{aligned}
E_{ordinary}(x_0) &= E\left[\left(y_0 - \alpha_0 - \sum_{i=1}^{n} \alpha_i Y_i\right)^2\right] \\
&= E\left[\left(\beta_0 + \epsilon_0 - \alpha_0 - \sum_{i=1}^{n} \alpha_i(\beta_0 + \epsilon_i)\right)^2\right] \\
&= E\left[\left((\beta_0 - \alpha_0 - \beta_0 \sum_{i=1}^{n} \alpha_i) + (\epsilon_0 - \sum_{i=1}^{n} \alpha_i \epsilon_i)\right)^2\right] \\
&= \left(\beta_0 - \alpha_0 - \beta_0 \sum_{i=1}^{n} \alpha_i\right)^2 + E[\epsilon_0^2] - 2\sum_{i=1}^{n} \alpha_i d_i \\
&\quad + \sum_{i=1}^{n}\sum_{j=1}^{n} \alpha_i [\mathbf{C}]_{ij} \alpha_j.
\end{aligned}
\tag{4.23}
$$

$(\beta_0 - \alpha_0 - \beta_0 \sum_{i=1}^{n} \alpha_i)^2$ becomes 0 if α_0 is

$$\alpha_0 = \beta_0 - \beta_0 \sum_{i=1}^{n} \alpha_i. \tag{4.24}$$

By this procedure, estimates without bias are obtained, because, since eq(4.22) and eq(4.18) give the equations below, respectively, the equality between $E[\hat{y}_0]$ and $E[y_0]$ is equivalent to eq(4.24).

$$\begin{aligned}
E[\hat{y}_0] &= E\left[\alpha_0 + \sum_{i=1}^{n} \alpha_i Y_i\right] \\
&= E\left[\alpha_0 + \sum_{i=1}^{n} \alpha_i (\beta_0 + \epsilon_i)\right] \\
&= \alpha_0 + \beta_0 \sum_{i=1}^{n} \alpha_i, \quad (4.25) \\
E[y_0] &= \beta_0. \quad (4.26)
\end{aligned}$$

Consequently, when estimates that satisfy eq(4.24) are employed, only minimization of variance leads to the estimates obtained using the least squares. In other words, by differentiating $(E[\epsilon_0^2] - 2\sum_{i=1}^{n} \alpha_i d_i + \sum_{i=1}^{n}\sum_{j=1}^{n} \alpha_i [\mathbf{C}]_{ij} \alpha_j)$ to be 0 with respect to $\{\alpha_1, \alpha_2, \ldots, \alpha_n\}$, the value of eq(4.23) is minimized. Then, we have

$$d_k = \sum_{i=1}^{n} [\mathbf{C}]_{ki} \alpha_i, \quad (4.27)$$

where it should be noted that $\mathbf{C}$ is a symmetric matrix. If $\mathbf{C}$ is a regular matrix (in other words, an invertible matrix), setting $\boldsymbol{\alpha} = (\alpha_1, \alpha_2, \ldots, \alpha_n)^t$ yields

$$\boldsymbol{\alpha} = \mathbf{C}^{-1} \mathbf{d}_0. \quad (4.28)$$

The value of $\hat{y}_0$ becomes as below because of eq(4.22) and eq(4.24); substituting eq(4.28) into this equation results in $\hat{y}_0$ expressed in terms of $\{Y_i\}$.

$$\hat{y}_0 = \beta_0 + \sum_{i=1}^{n} \alpha_i (Y_i - \beta_0). \quad (4.29)$$

That is, $\hat{\epsilon}_0$ in eq(4.20) is written as $\sum_{i=1}^{n} \alpha_i (Y_i - \beta_0)$. However, β_0 must be estimated. To estimate β_0 when $Cov(\epsilon_a, \epsilon_b) = 0$ ($x_a \neq x_b$) is not necessarily satisfied, generalized least squares is used. The value below is minimized to estimate β_0.

$$E_{general} = \sum_{i=1}^{n}\sum_{j=1}^{n} (\beta_0 - Y_i)[\mathbf{C}^{-1}]_{ij} (\beta_0 - Y_j). \quad (4.30)$$

Differentiation of this equation with respect to β_0 and consideration of the fact that $\mathbf{C}^{-1}$ is a symmetric matrix result in

$$\hat{\beta}_0 = \frac{\sum_{i=1}^{n}\sum_{j=1}^{n} [\mathbf{C}^{-1}]_{ij} Y_j}{\sum_{i=1}^{n}\sum_{j=1}^{n} [\mathbf{C}^{-1}]_{ij}}. \quad (4.31)$$

Substitution of this equation into eq(4.29) and the use of eq(4.28) lead to $\hat{y}_0$, which is a weighted average of $\{Y_i\}$:

$$\hat{y}_0 = \sum_{k=1}^{n}\sum_{l=1}^{n} [\mathbf{C}^{-1}]_{kl} \left(\frac{1 - \sum_{i=1}^{n}\sum_{j=1}^{n} [\mathbf{C}^{-1}]_{ij} d_j}{\sum_{i=1}^{n}\sum_{j=1}^{n} [\mathbf{C}^{-1}]_{ij}} + d_l \right) Y_k. \quad (4.32)$$

$$\sum_{l=1}^{n}[\mathbf{C}^{-1}]_{kl}\left(\frac{1-\sum_{i=1}^{n}\sum_{j=1}^{n}[\mathbf{C}^{-1}]_{ij}d_j}{\sum_{i=1}^{n}\sum_{j=1}^{n}[\mathbf{C}^{-1}]_{ij}}+d_l\right) \text{ is a weight of } Y_k \text{ to calculate } \hat{y}_0.$$

The sum of these weights with respect to k equals 1; this is proved by

$$\sum_{k=1}^{n}\sum_{l=1}^{n}[\mathbf{C}^{-1}]_{kl}\left(\frac{1-\sum_{i=1}^{n}\sum_{j=1}^{n}[\mathbf{C}^{-1}]_{ij}d_j}{\sum_{i=1}^{n}\sum_{j=1}^{n}[\mathbf{C}^{-1}]_{ij}}+d_l\right)$$

$$= \frac{1}{\sum_{i=1}^{n}\sum_{j=1}^{n}[\mathbf{C}^{-1}]_{ij}}\left(\sum_{k=1}^{n}\sum_{l=1}^{n}[\mathbf{C}^{-1}]_{kl}-\sum_{k=1}^{n}\sum_{l=1}^{n}\sum_{i=1}^{n}\sum_{j=1}^{n}[\mathbf{C}^{-1}]_{ij}[\mathbf{C}^{-1}]_{kl}d_j\right.$$

$$\left.+\sum_{k=1}^{n}\sum_{l=1}^{n}\sum_{i=1}^{n}\sum_{j=1}^{n}[\mathbf{C}^{-1}]_{ij}[\mathbf{C}^{-1}]_{kl}d_l\right)$$

$$= \frac{1}{\sum_{i=1}^{n}\sum_{j=1}^{n}[\mathbf{C}^{-1}]_{ij}}\left(\sum_{i=1}^{n}\sum_{j=1}^{n}[\mathbf{C}^{-1}]_{ij}-\sum_{k=1}^{n}\sum_{l=1}^{n}\sum_{i=1}^{n}\sum_{j=1}^{n}[\mathbf{C}^{-1}]_{ij}[\mathbf{C}^{-1}]_{kl}d_j\right.$$

$$\left.+\sum_{i=1}^{n}\sum_{j=1}^{n}\sum_{k=1}^{n}\sum_{l=1}^{n}[\mathbf{C}^{-1}]_{kl}[\mathbf{C}^{-1}]_{ij}d_j\right)$$

$$= 1, \tag{4.33}$$

where the equality between the second to third lines and the fourth to fifth lines are obtained by substitution between i and k and between j and l.

Next, $\mathbf{D}$ is defined as a matrix (size of $n \times n$):

$$\mathbf{D} = (\tilde{\mathbf{d}}_1, \tilde{\mathbf{d}}_2, \ldots, \tilde{\mathbf{d}}_n)^t = (\tilde{\mathbf{d}}_1, \tilde{\mathbf{d}}_2, \ldots, \tilde{\mathbf{d}}_n), \tag{4.34}$$

where $\tilde{\mathbf{d}}_i$ is $\mathbf{d}_0$ when $x_0 = x_i$. $\mathbf{D}$ coincides with $\mathbf{C}$. Hence, eq(4.32) results in

$$\hat{Y}_p = \sum_{k=1}^{n}\sum_{l=1}^{n}[\mathbf{C}^{-1}]_{kl}\left(\frac{1-\sum_{i=1}^{n}\sum_{j=1}^{n}[\mathbf{C}^{-1}]_{ij}[\tilde{\mathbf{d}}_p]_j}{\sum_{i=1}^{n}\sum_{j=1}^{n}[\mathbf{C}^{-1}]_{ij}}+[\tilde{\mathbf{d}}_p]_l\right)Y_k$$

$$(1 \leq p \leq n). \tag{4.35}$$

$\sum_{i=1}^{n}\sum_{j=1}^{n}[\mathbf{C}^{-1}]_{ij}[\tilde{\mathbf{d}}_p]_j$ becomes

$$\sum_{i=1}^{n}\sum_{j=1}^{n}[\mathbf{C}^{-1}]_{ij}[\tilde{\mathbf{d}}_p]_j = \sum_{i=1}^{n}\sum_{j=1}^{n}[\mathbf{C}^{-1}]_{ij}[\mathbf{C}]_{jp}$$

$$= \sum_{i=1}^{n}[\mathbf{I}_n]_{ip}$$

$$= 1 \qquad (1 \leq p \leq n), \tag{4.36}$$

where $\mathbf{I}_n$ is an identity matrix (the size is $n \times n$). Therefore, setting $\mathbf{y} = (Y_1, Y_2, \ldots, Y_n)^t$ and $\hat{\mathbf{y}} = (\hat{Y}_1, \hat{Y}_2, \ldots, \hat{Y}_n)^t$ and considering eq(4.35) and eq(4.36) lead to

$$\hat{\mathbf{y}} = \mathbf{C}^{-1}\mathbf{C}\mathbf{y}$$

$$= \mathbf{y}. \tag{4.37}$$

That is, a hat matrix is an identity matrix. This assures that this method implements an interpolation.

Another procedure also yields eq(4.32). As mentioned above, calculation of estimates using a linear estimator reduces to the derivation of the weights ($\{w_i\}$ $(1 \le i \le n)$) to obtain $\hat{y}_0$ using

$$\hat{y}_0 = \sum_{i=1}^{n} w_i Y_i. \tag{4.38}$$

The solution of the problem to obtain these weights is eq(4.32). The problem is as follows.

"Assume that the relationship between a predictor and a target variable is written as

$$y = m(x) + \epsilon, \tag{4.39}$$

where $m(x)$ is a function that represents a functional relationship that does not contain random factors, ϵ is a random variable, and $E[\epsilon] = 0$. Furthermore, when the two predictor values are x_a and x_b, and ϵ corresponding to these two values are ϵ_a and ϵ_b, $x_a \ne x_b$ does not necessarily mean $Cov(\epsilon_a, \epsilon_b) = 0$. Assume that $Cov(\epsilon_a, \epsilon_b)$ is known for any combination of x_a and x_b. Under these conditions, derive $\{w_i\}$ that minimizes the variance of the difference between $\hat{y}_0$ (given by eq(4.38)) and y_0. Note that $\sum_{i=1}^{n} w_i = 1$ is assumed."

This problem is solved as follows.

To minimize the variance of the difference between $\hat{y}_0$ and y_0, $\{w_i\}$ that minimizes the equation below is derived:

$$
\begin{aligned}
Var_{ordinary} &= E\left[\left(y_0 - \sum_{i=1}^{n} w_i Y_i - E[y_0 - \sum_{i=1}^{n} w_i Y_i]\right)^2\right] \\
&= E\left[\left(m_0 + \epsilon_0 - \sum_{i=1}^{n} w_i m_i - \sum_{i=1}^{n} w_i \epsilon_i - m_0 + \sum_{i=1}^{n} w_i m_i\right)^2\right] \\
&= E\left[\left(\epsilon_0 - \sum_{i=1}^{n} w_i \epsilon_i\right)^2\right] \\
&= \sum_{i=1}^{n}\sum_{j=1}^{n} w_i w_j [C]_{ij} - 2\sum_{i=1}^{n} w_i d_i + E[\epsilon_0^2], \tag{4.40}
\end{aligned}
$$

where we set $y_0 = m_0 + \epsilon_0$ ($m_0 = m(x_0)$: ϵ_0 is the value of ϵ when $x = x_0$) and $Y_i = m_i + \epsilon_i$ ($1 \le i \le n$). Using the method of Lagrange multipliers (Lagrange multiplier method), eq(4.40) is minimized under the condition of $\sum_{i=1}^{n} w_i = 1$. That is, the equation below is differentiated to be 0 with respect to each $\{w_i\}$ and the Lagrange multiplier.

$$E_{Lagrange} = \sum_{i=1}^{n}\sum_{j=1}^{n} w_i w_j [C]_{ij} - 2\sum_{i=1}^{n} w_i d_i + E[\epsilon_0^2] + \chi\left(\sum_{i=1}^{n} w_i - 1\right). \tag{4.41}$$

χ is the Lagrange multiplier. Hence, we have

$$\frac{\partial E_{Lagrange}}{\partial w_k} = 2 \sum_{i=1}^{n} w_i [\mathbf{C}]_{ik} - 2d_k + \chi = 0 \quad (1 \le k \le n), \qquad (4.42)$$

$$\sum_{i=1}^{n} w_i = 1, \qquad (4.43)$$

where we make use of the fact that $\mathbf{C}$ is a symmetric matrix. When we set $\mathbf{w} = (w_1, w_2, \ldots, w_n)^t$, eq(4.42) is rewritten as below under the condition that $\mathbf{C}$ is a regular matrix:

$$\mathbf{w} = \mathbf{C}^{-1}(\mathbf{d}_0 - \frac{1}{2}\chi), \qquad (4.44)$$

where χ is a column vector $(\chi, \chi, \ldots, \chi)^t$ (the number of elements is n). Comparison of this equation with eq(4.32) (page 240) indicates that the two equations are identical when the equation below holds.

$$-\frac{1}{2}\chi = \left(\frac{1 - \sum_{i=1}^{n} \sum_{j=1}^{n} [\mathbf{C}^{-1}]_{ij} d_j}{\sum_{i=1}^{n} \sum_{j=1}^{n} [\mathbf{C}^{-1}]_{ij}} \right). \qquad (4.45)$$

Furthermore, eq(4.33) shows that eq(4.43) holds. Hence, interpolation by $\mathbf{w}$ obtained here is identical to that by eq(4.32). It should be noted that $m(x)$ is found to be constant even though we do not assume that $m(x)$ is a constant; $m(x)$ does not depend on x_0. Interpolation using this $\mathbf{w}$ is called ordinary kriging.

Interpolation by ordinary kriging requires $\mathbf{C}$ defined by eq(4.19) (page 239) and $\mathbf{d}_0$ defined by eq(4.21). These values are derived using data unless prior knowledge concerning the phenomena that generate the data provides these values. For this purpose, we introduce the concept of a correlogram:

$$\begin{aligned} Correlo(\{X_i, X_j\}) &= \frac{E\left[(\epsilon_i - E[\epsilon_i])(\epsilon_j - E[\epsilon_j])\right]}{E\left[(\epsilon_i - E[\epsilon_i])^2\right]} \\ &= \frac{Cov(\epsilon_i, \epsilon_j)}{sd[\{\epsilon_i\}]sd[\{\epsilon_j\}]}, \end{aligned} \qquad (4.46)$$

where $sd[\{\epsilon_i\}]$ means the standard deviation of $\{\epsilon_i\}$; the definition of the standard deviation is

$$sd[\{\epsilon_i\}] = \sqrt{\frac{\sum_{i=1}^{n} (\epsilon_i - \bar{\epsilon})^2}{n}}. \qquad (4.47)$$

$\bar{\epsilon}$ stands for the mean of $\{\epsilon_i\}$. Additionally, $Correlo(\epsilon_i, \epsilon_i) = 1$ holds. The word correlogram is often used in the context of krging and time series analysis. The coefficient of correlation between ϵ_i and ϵ_j is a more common phrase in other fields. The value of correlation is between -1 and 1. In addition, a covariogram is defined as

$$\begin{aligned} Covario(\{X_i, X_j\}) &= E[(\epsilon_i - E[\epsilon_i])(\epsilon_j - E[\epsilon_j])] \\ &= Cov(\epsilon_i, \epsilon_j). \end{aligned} \qquad (4.48)$$

This is equivalent to covariance in contexts other than kriging.

A correlogram estimated using data is called an empirical correlogram (sample correlogram). To derive an empirical correlogram, it is usually assumed that $Correlo$ (X_i, X_j) is a function of only the distance between X_i and X_j. That is, when the distance between X_i and X_j ($\mid X_i - X_j \mid$ if the number of predictors is one, or $\sqrt{(X_{1i} - X_{1j})^2 + (X_{2i} - X_{2j})^2}$ if the number of predictors is two) is defined as τ, we assume that the empirical correlogram is written as $Correlo(\tau)$ calculated as

$$Correlo(\tau) = \frac{\sum_{\tau}^{\tau+\Delta\tau}\left((\epsilon_i - \bar{\epsilon}^\tau)(\epsilon_j - \bar{\epsilon}^\tau)\right)}{\sum_{\tau}^{\tau+\Delta\tau}(\epsilon_i - \bar{\epsilon}^\tau)^2}, \tag{4.49}$$

where $\sum_{\tau}^{\tau+\Delta\tau}$ means that the sum is taken if the distantce between ϵ_i and ϵ_j is equal to or more than τ and less than $(\tau + \Delta\tau)$. Furthermore, $\bar{\epsilon}^\tau$ is an average of ϵ_i, which is selected using $\sum_{\tau}^{\tau+\Delta\tau}$ (an average of ϵ_j selected using $\sum_{\tau}^{\tau+\Delta\tau}$ takes the same value as this average). In practice, τ is usually discretized to calculate $\{Correlo(\tau_k)\}$ corresponding to $\{\tau_k\}$ $(1 \leq k \leq K)$. In doing this, $\tau_k = (k-1)\Delta\tau$ is set out, and values of $\Delta\tau$ and K are determined to satisfy the condition that τ_K is larger than the maximum distance between X_i and X_j.

The concept of the correlogram has a meaning equivalent of that of the variogram. Variogram is more commonly used than correlogram in the context of kriging; variogram is defined as

$$Vario(X_i, X_j) = \frac{E[(\epsilon_i - \epsilon_j)^2]}{2}. \tag{4.50}$$

This value was originally termed semi-variogram, and the doubled value was called variogram. However, it is not unusual that eq(4.50) is used as the definition of the varigram. A variogram that is estimated using data is called an empirical variogram (sample variogram); it is defined in a similar manner to the empirical correlogram:

$$Vario(\tau) = \frac{\sum_{\tau}^{\tau+\Delta\tau}(\epsilon_i - \epsilon_j)^2}{2}. \tag{4.51}$$

There is a relationship between $Vario(\tau)$ in eq(4.51) and $Correlo(\tau)$ in eq(4.49):

$$
\begin{aligned}
Vario(\tau) &= \frac{\sum_{\tau}^{\tau+\Delta\tau}(\epsilon_i^2 + \epsilon_j^2 - 2\epsilon_i\epsilon_j)}{2} \\
&= \sum_{\tau}^{\tau+\Delta\tau}(\epsilon_i^2 - \epsilon_i\epsilon_j) \\
&= \left(\sum_{\tau}^{\tau+\Delta\tau}(\epsilon_i - \bar{\epsilon}^\tau)^2\right)\left(1 - Correlo(\tau)\right) \\
&= \left(\sum_{\tau}^{\tau+\Delta\tau}(\epsilon_i - \bar{\epsilon}^\tau)^2\right)\left(Correlo(0) - Correlo(\tau)\right). \quad (4.52)
\end{aligned}
$$

To calculate $\mathbf{C}$ and $\mathbf{d}_0$, simple forms to represent the correlogram and variogram are of great use and hence the empirical correlogram and empirical variogram are replaced with equations of specific forms. The functions obtained in this way are called the theoretical correlogram (correlogram model) and theoretical variogram (variogram model), respectively. Typical forms of a function to be used as a theoretical correlogram are

$$Correlo(\tau) = \exp\left(-\frac{\tau}{\delta}\right), \tag{4.53}$$

$$Correlo(\tau) = \exp\left(-\left(\frac{\tau}{\delta}\right)^2\right), \tag{4.54}$$

$$Correlo(\tau) = \left(1 - \frac{2}{\pi}\left(\frac{\tau}{\delta}\sqrt{1 - \frac{\tau^2}{\delta^2}} + \sin^{-1}\left(\frac{\tau}{\delta}\right)\right)\right). \tag{4.55}$$

Eq(4.53) is called the exponential correlogram, eq(4.54) is termed the Gaussian correlogram, and eq(4.55) is named the spherical correlogram. The correlogram and variogram created from real data sometimes cannot be depicted by functions of only τ. For instance, while the theoretical correlograms presented here are based on the assumption that data has isotropy (spatial autocorrelation of data does not rely on direction), data that has anisotropy (spatial autocorrelation of data does rely on direction) needs suitable measures such as transformation of data. Furthermore, to derive a theoretical correlogram from an empirical one, simple least squares is not appropriate. No values of the empirical correlogram can be regarded as a value of an intrinsic function plus a mutually independent error.

The procedure to implement interpolation using ordinary kriging is summarized as follows:

(1) An empirical correlogram (or empirical variogram) is obtained using data. In this case, ϵ_i, ϵ_j, and $\bar{\epsilon}^\tau$ in eq(4.49) can be replaced with Y_i, Y_j, and $\bar{Y}^\tau$ (an average of Y_i selected by $\sum_\tau^{\tau + \Delta\tau}$), respectively. This is because the addition of a constant to ϵ_i does not affect $Correlo(\tau)$.

(2) The form of a function to approximate the empirical correlogram is chosen so as to determine the values of the coefficients. That is to say, an analyst draws the shapes of empirical correlograms and those of a theoretical correlograms on a graph, and selects a theoretical correlogram that is nearly identical to an empirical correlogram.

(3) Using the theoretical correlogram identified in (2), d_0 at an estimation point (x_0) and C are calculated. By the use of these, α is calculated by eq(4.28) (page 240).

(4) $\hat{\beta}_0$ is computed using eq(4.31).

(5) $\hat{y}_0$ is computed using eq(4.29).

By interpolating data with one predictor using ordinary kriging, the effect of the broadness of the theoretical variogram on estimates can be examined; an example of the result is shown in figure 4.5. Three theoretical correlograms are attempted. It is assumed that $\delta = 0.3, 1$, and 3 in eq(4.54). When the area in which the correlation relationship expands is narrow, an estimate at a point different from predictor values of data ($\{X_i\}$) is close to a constant ($\hat{\beta}_0$). For comparison, this figure also shows the result of interpolation using a natural spline whose knots (break points) are located at predictor values of data.

If $\beta_0 = 0$ is always true in eq(4.17) (page 238), it is called a simple kriging; it utilizes the regression equation

$$y = \epsilon. \tag{4.56}$$

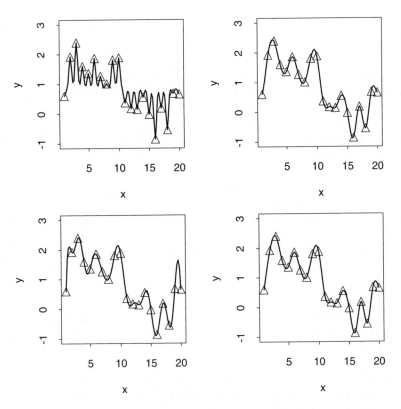

Figure 4.5   Examples of ordinary kriging (the number of predictors is one). Eq(4.54) (page 245) is used for theoretical correlograms with $\delta = 0.3, 1$, and 3 (top left, top right, and bottom left, respectively). The bottom right graph displays the result of interpolation using a natural spline with knots at data points.

$\hat{y}_0$ under this condition is

$$\hat{y}_0 = \hat{\epsilon}_0 = \sum_{i=1}^{n} \alpha_i Y_i. \tag{4.57}$$

This $\{\alpha_i\}$ is defined in eq(4.28) (page 240). These conditions transform eq(4.23) (page 239) into

$$E_{simple} = \alpha_0^2 + E[\epsilon_0^2] - 2\sum_{i=1}^{n} \alpha_i d_i + \sum_{i=1}^{n}\sum_{j=1}^{n} \alpha_i [\mathbf{C}]_{ij}\alpha_j. \tag{4.58}$$

We set $\alpha_0 = 0$ to obtain estimates without bias. Then, $\boldsymbol{\alpha} = (\alpha_1, \alpha_2, \dots, \alpha_n)^t$ given by eq(4.28) realizes interpolation by simple kriging.

Figure 4.6 exemplifies an interpolation of data with one predictor by simple kriging. Eq(4.54) is used under the condition of $\delta = 0.1, 0.3, 1$, and 3. When simple kriging is adopted, the area in which the correlation relationship expands is narrow, and an estimate not located at data points ($\{X_i\}$) is close to 0.

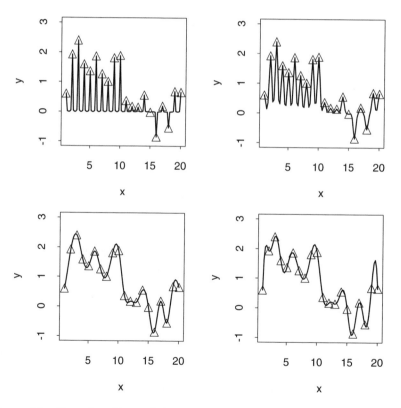

Figure 4.6   Examples of interpolation by simple kriging (the number of predictors is one). Eq(4.54) (page 245) is used for a correlogram with $\delta = 0.1, 0.3, 1$, and 3. Each value of δ corresponds to a graph located at the top left, top right, bottom left, and botttom right, respectively.

Smoothing is realized if ϵ in simple kringing (eq(4.56)) is assumed to be

$$\epsilon_0 = \epsilon_0^c + \epsilon_0^u, \tag{4.59}$$

$$\epsilon_i = \epsilon_i^c + \epsilon_i^u, \tag{4.60}$$

where $E[\epsilon_0^c] = 0$, $E[\epsilon_0^u] = 0$, $Cov(\epsilon_i^c, \epsilon_j^c) = [\mathbf{C}']_{ij}$, $Cov(\epsilon_0^c, \epsilon_i^c) = d_i'$, $Cov(\epsilon_0^c, \epsilon_0^u) = 0$, $Cov(\epsilon_i^c, \epsilon_j^u) = 0$, $Cov(\epsilon_i^u, \epsilon_j^u) = 0$ (if $i \neq j$), $Cov(\epsilon_i^u, \epsilon_i^u) = \sigma^2$, and $Cov(\epsilon_0^u, \epsilon_0^u) = \sigma^2$ ($1 \leq i \leq n$, $1 \leq j \leq n$). That is, it is assumed that ϵ can be divided into two parts: one (ϵ_0^c and ϵ_i^c) correlated with ϵ at another position (i.e., covariance is not 0), and one (ϵ_0^u and ϵ_i^u) that has no correlation with ϵ at another position. Then, to calculate $\hat{y}_0$ ($= \hat{\epsilon}_0^c$), $\boldsymbol{\alpha}$ ($= (\alpha_1, \alpha_2, \ldots, \alpha_n)^t$) is obtained by minimizing

$$
\begin{aligned}
Var_{simple,smoothing} &= E[(\epsilon_0^c - \sum_{i=1}^{n} \alpha_i(\epsilon_i^c + \epsilon_i^u))^2] \\
&= E[(\epsilon_0^c)^2] - 2\sum_{i=1}^{n} \alpha_i d_i' + \sum_{i=1}^{n}\sum_{j=1}^{n} \alpha_i [\mathbf{C}']_{ij} \alpha_j \\
&+ \sigma^2 \sum_{i=1}^{n} \alpha_i^2.
\end{aligned}
\tag{4.61}
$$

Differentiating this equation to be 0 with respect to α_k ($1 \leq k \leq n$) gives

$$\boldsymbol{\alpha} = (\mathbf{C}' + \sigma^2 \mathbf{I}_n)^{-1} \mathbf{d}_0', \tag{4.62}$$

where $\mathbf{d}_0' = (d_1', d_2', \ldots, d_n')^t$. An estimate ($\hat{\mathbf{y}} = (\hat{Y}_1, \hat{Y}_2, \ldots, \hat{Y}_n)^t$) at ($X_1, X_2, \ldots, X_n$) is derived, using eq(4.62), as

$$\hat{\mathbf{y}} = (\mathbf{C}' + \sigma^2 \mathbf{I}_n)^{-1} \mathbf{D}' \mathbf{y}, \tag{4.63}$$

where $\mathbf{D}'$ (a matrix; the size is $n \times n$), which is identical to $\mathbf{C}'$, is as follows.

$$\mathbf{D}' = (\tilde{\mathbf{d}}_1', \tilde{\mathbf{d}}_2', \ldots, \tilde{\mathbf{d}}_n')^t = (\tilde{\mathbf{d}}_1', \tilde{\mathbf{d}}_2', \ldots, \tilde{\mathbf{d}}_n'). \tag{4.64}$$

This $\tilde{\mathbf{d}}_i'$ is defined as $\mathbf{d}'_i$, where $x_0 = x_i$ is assumed. Consequently, we have

$$\hat{\mathbf{y}} = (\mathbf{I}_n + \sigma^2 \mathbf{D}'^{-1})^{-1} \mathbf{y}. \tag{4.65}$$

That is, a hat matrix for this regression is $(\mathbf{I}_n + \sigma^2 \mathbf{D}'^{-1})^{-1}$. This has the same form as that of a hat matrix of a smoothing spline (eq(3.231) (page 176)).

Figure 4.7 shows an example of smoothing by simple kriging; the number of predictors is one, and eq(4.57) and eq(4.62) are utilized. Eq(4.61) with $\sigma^2 = 0.25$, and eq(4.54) (page 245) with $\delta = 0.1, 0.3, 1, 3$ are employed. We assume that the value of σ^2 is known. However, it is usually unknown, whereby adjustment of the value of σ^2 and selection of a theoretical variogram allow a diversity of smoothing.

On the other hand, interpolation by universal kriging creates the equation

$$y = \sum_{j=1}^{P} \beta_j g_j(x) + \epsilon. \tag{4.66}$$

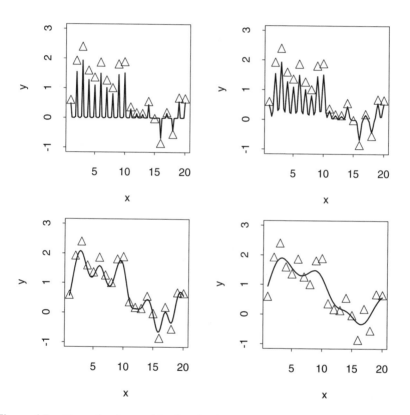

Figure 4.7 Example of smoothing by simple kriging (the number of predictors is one). $\sigma^2 = 0.25$ and eq(4.54) (page 245) are used for a theoretical correlogram. $\delta = 0.1, 0.3, 1,$ and 3 corresponds to graphs at the top left, top right, bottom left, and bottom right.

Using this equation, an estimate ($\hat{y}_0$) at x_0 (the value of predictor) is

$$\hat{y}_0 = \sum_{j=1}^{P} \beta_j g_j(x_0) + \hat{\epsilon}_0. \tag{4.67}$$

On this occasion, eq(4.23) (page 239) is transformed into

$$
\begin{aligned}
E_{universal} &= E[(y_0 - \alpha_0 - \sum_{i=1}^{n} \alpha_i Y_i)^2] \\
&= (m_0 - \alpha_0 - \sum_{i=1}^{n} \alpha_i m_i)^2 + E[\epsilon_0^2] - 2\sum_{i=1}^{n} \alpha_i d_i \\
&\quad + \sum_{i=1}^{n}\sum_{j=1}^{n} \alpha_i [\mathbf{C}]_{ij} \alpha_j, \tag{4.68}
\end{aligned}
$$

where we assume

$$m_0 = \sum_{j=1}^{P} \beta_j g_j(x_0), \tag{4.69}$$

$$m_i = \sum_{j=1}^{P} \beta_j g_j(X_i). \tag{4.70}$$

To realize universal kriging, eq(4.24) (page 239), which is used for ordinary kriging, is converted to

$$\alpha_0 = m_0 - \sum_{i=1}^{n} \alpha_i m_j. \tag{4.71}$$

Eq(4.27) (page 240) and eq(4.28) remain intact in universal kriging. Eq(4.30) (page 240) is replaced with

$$E_{general} = \sum_{i=1}^{n}\sum_{j=1}^{n}\left(\sum_{k=1}^{P} \beta_k g_k(X_i) - Y_i\right)[\mathbf{C}^{-1}]_{ij}\left(\sum_{k=1}^{P} \beta_k g_k(X_j) - Y_j\right), \tag{4.72}$$

where we set

$$\boldsymbol{\beta} = (\beta_1, \beta_2, \ldots, \beta_P)^t, \tag{4.73}$$

$$\mathbf{y} = (Y_1, Y_2, Y_3, \ldots, Y_n)^t, \tag{4.74}$$

$$\mathbf{G} = \begin{pmatrix} g_1(X_1) & g_2(X_1) & \cdots & g_P(X_1) \\ g_1(X_2) & g_2(X_2) & \cdots & g_P(X_2) \\ g_1(X_3) & g_2(X_3) & \cdots & g_P(X_3) \\ \vdots & \vdots & \ddots & \vdots \\ g_1(X_n) & g_2(X_n) & \cdots & g_P(X_n) \end{pmatrix}. \tag{4.75}$$

Hence, eq(4.72) is rewritten as

$$E_{general} = (\mathbf{G}\boldsymbol{\beta} - \mathbf{y})^t \mathbf{C}^{-1}(\mathbf{G}\boldsymbol{\beta} - \mathbf{y}). \tag{4.76}$$

Since $\mathbf{C}$ is a positive definite matrix, it is decomposed as

$$\mathbf{C} = \mathbf{L}\mathbf{L}^t, \tag{4.77}$$

where $\mathbf{L}$ is a lower triangular matrix (the size is $n \times n$). Decomposition of a positive definite matrix, such as that above, is termed Cholesky decomposition. Hence, $\mathbf{C}^{-1}$ is written as

$$\mathbf{C}^{-1} = (\mathbf{L}\mathbf{L}^t)^{-1} = (\mathbf{L}^t)^{-1}\mathbf{L}^{-1} = (\mathbf{L}^{-1})^t\mathbf{L}^{-1}. \tag{4.78}$$

Consequently, eq(4.76) becomes

$$\begin{aligned} E_{general} &= (\mathbf{L}^{-1}(\mathbf{G}\boldsymbol{\beta} - \mathbf{y}))^t(\mathbf{L}^{-1}(\mathbf{G}\boldsymbol{\beta} - \mathbf{y})) \\ &= \| \mathbf{L}^{-1}(\mathbf{G}\boldsymbol{\beta} - \mathbf{y}) \|^2, \end{aligned} \tag{4.79}$$

where $\| \ \|$ stands for the length of a vector.

Next, we carry out the transformation

$$\mathbf{L}^{-1}\mathbf{G} = \mathbf{Q}\mathbf{R}. \tag{4.80}$$

$\mathbf{Q}$ is an orthogonal matrix (the size is $n \times n$). $\mathbf{R}$ is an upper triangular matrix; the size of the matrix is $n \times P$ ($P \leq n$) and all elements between the $(P+1)$th row and nth row are 0. Transformation of this sort is called QR decomposition. Then, eq(4.79) is written as

$$E_{general} = \| \mathbf{Q}(\mathbf{R}\boldsymbol{\beta} - \mathbf{Q}^t\mathbf{L}^{-1}\mathbf{y}) \|^2, \tag{4.81}$$

where the elements between the $(P+1)$th one and nth one of $\mathbf{R}\boldsymbol{\beta}$ (a vector with n elements) are 0 and hence they are unaffected by $\boldsymbol{\beta}$. Then, to minimize the value of eq(4.81), $\boldsymbol{\beta}$ is adjusted to satisfy

$$[\mathbf{Q}^t\mathbf{L}^{-1}\mathbf{y}]_{(1:P)} = [\mathbf{R}\boldsymbol{\beta}]_{(1:P)}, \tag{4.82}$$

where $[\ ]_{(1:P)}$ is a vector resulting from extraction of elements between the 1st one and Pth one from a vector with n elements. Substitution of the resulting $\boldsymbol{\beta}$ into eq(4.71) (page 250) yields α_0. Hence, by incorporating $(\alpha_1, \alpha_2, \ldots, \alpha_n)$ given by eq(4.28) (page 240), eq(4.67) (page 250) leads to $\hat{y}_0$:

$$\hat{y}_0 = m_0 + \sum_{i=1}^{n} \alpha_i(Y_i - m_i). \tag{4.83}$$

This implies that $\hat{\epsilon}_0$ in eq(4.67) is identical to $\sum_{i=1}^{n} \alpha_i(Y_i - m_i)$. While m_0 is a constant (β_0) that does not depend on x_0 in ordinary kriging, m_0 derived by universal kriging depends on x_0 such as eq(4.71). It should be noted that universal kriging includes ordinary kriging as a special case because $P = 1$ and $g_1(x) = \beta_0$ (β_0 is a constant) in universal kriging result in ordinary kriging. However, use of the computational procedure of universal kriging to implement ordinary kriging is not efficient because ordinary kriging does not require an iteration method.

The procedure for interpolation by universal kriging is as follows.

(1) A function with a specific form is assumed to be the theoretical correlogram. The coefficients of the function are determined visually. $\mathbf{C}$ is derived using this function.

(2) Using the data and $\mathbf{C}$ derived in (1), a generalized least squares (eq(4.76)) is carried out to obtain β.

(3) $\{\hat{\epsilon}_i\}$ is calculated by eq(4.83).

(4) An empirical correlogram (or empirical variogram) for $\{\hat{\epsilon}_i\}$ derived in (3) is created. A theoretical correlogram is fitted to it.

(5) $\mathbf{C}$ is computed using the theoretical correlogram given by (4).

(6) Using the data and $\mathbf{C}$ obtained in (5), β is derived by implementing generalized least squares (eq(4.76)).

(7) Processes (3) to (6) are repeated until β is considered to have converged to a certain point.

(8) By means of β and the theoretical correlogram gained through (7), estimates using eq(4.83) are derived.

As described above, the procedure for interpolation by universal kriging contains an iterative calculation. The reason is that while ordinary kriging and simple kriging use $\{Y_i\}$ to derive an empirical correlogram for $\{\hat{\epsilon}_i\}$, universal kriging, apart from the special case that reduces to ordinary kriging, generates the following relationships: (1) β is needed to obtain $\{\hat{\epsilon}_i\}$. (2) A theoretical correlogram is requisite to computing β. (3) $\{\hat{\epsilon}_i\}$ is indispensable for deriving a theoretical correlogram. Additionally, when more than one form of a function is applied as a theoretical correlogram, coefficients for each function are estimated using this procedure to choose a function that fits an empirical correlogram well.

Figure 4.8 shows the result of the examination regarding the influence of the broadness of the theoretical variogram on the estimates given by universal kriging with one predictor. We set $P = 4$, $g_1(x) = 1$, $g_2(x) = x$, $g_3(x) = x^2$, and $g_4(x) = x^3$. That is, a cubic equation is adopted to represent a rough trend of estimates. Four kinds of correlogram (i.e., $\delta = 0.1, 0.3, 1, 3$ in eq(4.54) (page 245)) are utilized. When the area in which the correlation relationship expands is narrow, estimates that are not located at $\{X_i\}$ (predictor values of data) form a shape close to that for a cubic equation ($\sum_{j=1}^{4} \beta_j g_j(x)$).

The contents of simple kriging, ordinary kriging, and universal kriging, described so far, remain intact when the number of predictors is more than one. Hence, the reason for frequent practical use of kriging interpolation with two predictors is that, along with the historical background, there are not many alternative techniques for interpolation for two predictors compared to those for one predictor.

Simple examples of kriging with two predictors are illustrated. Figure 4.9 exemplifies an interpolation by ordinary kriging with 10 data; when the data are represented as $\{(X_{i1}, X_{i2}, Y_i)\}$, they are $\{(1, 7, 2), (2, 5, 7), (3, 4, 4), (4, 2, 3), (5, 9, 5), (6, 5, 2),$ $(7, 3, 9), (8, 4, 4), (9, 2, 1), (10, 9, 8)\}$. Eq(4.54) (page 245) is used as a theoretical correlogram; $\delta = 0.2, 0.5, 1, 2$.

Next, a result of universal kriging is displayed in figure 4.10; the data and the theoretical correlogram are the same as those in figure 4.9. Furthermore, $P = 6$, $g_1(x_1, x_2) = 1$, $g_2(x_1, x_2) = x_1$, $g_3(x_1, x_2) = x_2$, $g_4(x_1, x_2) = x_1^2$, $g_5(x_1, x_2) = x_1 x_2$, and $g_6(x_1, x_2) = x_2^2$ are set.

Figure 4.11 shows the results of using the thin plate splines and Akima's method for comparison with figure 4.9 and figure 4.10. Estimates are obtained using the thin plate smoothing splines with a very small positive value of a smoothing parameter. In

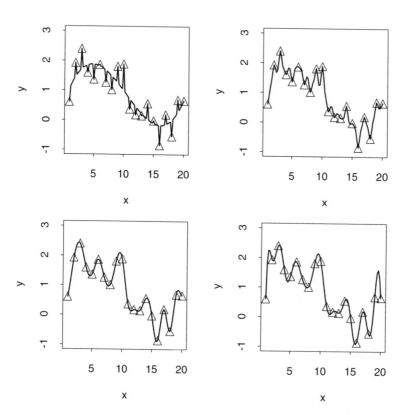

Figure 4.8 Example of universal kriging with one predictor (a cubic equation is used to represent a rough trend). Eq(4.54) (page 245) is employed as the theoretical correlogram; $\delta = 0.1, 0.3, 1,$ and 3 in graphs at the top left, top right, bottom left, and bottom right, respectively.

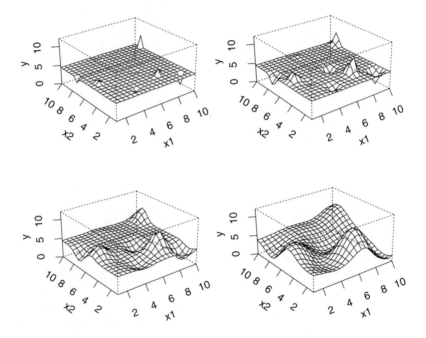

Figure 4.9   Example of ordinary kriging for two predictors. The theoretical correlogram is eq(4.54) (page 245) with $\delta = 0.2, 0.5, 1, 2$. Each value of δ corresponds to graphs located at the top left, top right, bottom left, and bottom right, respectively.

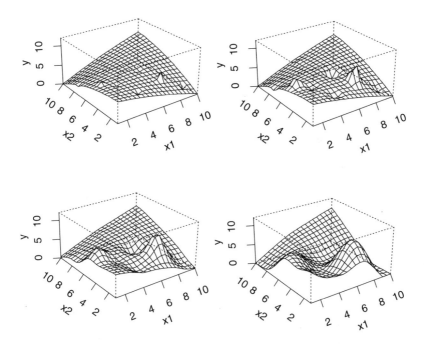

Figure 4.10   Example of ordinary kriging for two predictors. The data are the same as that used in the case of figure 4.9. The theoretical correlogram is eq(4.54) (page 245) with $\delta = 0.2, 0.5, 1, 2$. A rough trend is represented using a quadratic equation. Each value of δ corresponds to graphs at the top left, top right, bottom left, and botttom right, respectively.

addition, since Akima's method is for interpolation only, the values where estimates are not available are not shown.

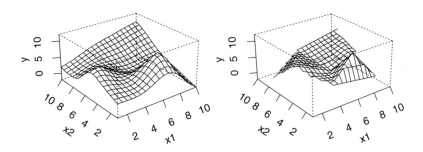

Figure 4.11   Result of interpolation with thin plate splines; the data are the same as those in figure 4.9 and figure 4.10 (left). Result of interpolation by Akima's method (right).

Following the procedure of universal kriging, the theoretical correlogram and interpolated values converge on the iterative calculation; figure 4.12 illustrates the results. The predictor values of the data (the number of data is 100) are defined as $\{X_{i1}\}$ $(1 \leq i \leq 100)$ and $\{X_{i2}\}$ $(1 \leq i \leq 100)$. Both predictors are generated as uniform pseudorandom numbers between 0 and 10. The values of the target variable are calculated as

$$Y_i = 8 + 5 \cdot \sin(0.1 \cdot X_{i1} \cdot X_{i1}) + 3 \cdot \cos(X_{i2}) + 3 \cdot \sin(2.5 \cdot X_{i2}) + 0.6 \cdot X_{i2}. \quad (4.84)$$

$P = 3$, $g_1(x_1, x_2) = 1$, $g_2(x_1, x_2) = x_1$, and $g_3(x_1, x_2) = x_2$ are set. Furthermore, eq(4.54) is employed as a theoretical correlogram. The goodness of fit of an empirical correlogram to a theoretical correlogram is measured visually; the value of δ is modified by an iterative calculation. As a result, a suitable value of δ for approximating the empirical correlogram is found to be 0.8 (figure 4.12(bottom left)). The empirical variogram obtained here is shown in figure 4.12(right top). The estimates obtained using the resulting theoretical correlogram are illustrated in figure 4.12(bottom left). For comparison, approximate estimates obtained with the thin plate splines are shown in figure 4.12(bottom right). These estimates are calculated using the thin plate smoothing splines with a very small positive smoothing parameter and should be almost identical to those obtained using thin plate splines.

Kriging was developed to infer the appearance throughout the entire domain from a limited number of data when the number of predictors is two. However, the use of a variety of theoretical correlograms in terms of the form of the function and the values of coefficients enables us to design a surface with a diversity of expressions under the condition that the surface is an interpolation of the given data. This technique is expected to be applied to tasks such as industrial design and creation of animation. Furthermore, a simpler and more flexible method for the same purpose has been developed [11].

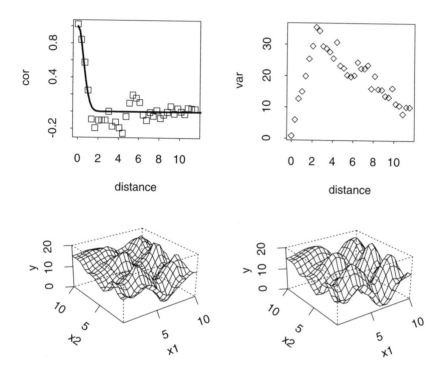

Figure 4.12   Interpolation of the simulation data generated by eq(4.84).   Empirical correlogram (□) and theoretical correlogram (solid line), (top left). Empirical variogram (top right).   Estimates obtained by interpolation using theoretical correlogram (top left), (bottom left). Estimates obtained with thin plate smoothing splines (bottom right).

4.6 ADDITIVE MODEL

Ordinary multiple regression involves the regression of data to

$$y = \sum_{i=1}^{r} a_i x_i + b. \tag{4.85}$$

This equation is based on the assumption that each predictor has a linear relationship with a target variable. This is a severe assumption. It may hamper satisfactory extraction of the contents of data. However, smoothing with plural predictors sometimes leads to unreliable results unless the number of data is huge, particularly in the case of a large number of predictors. Hence, the form of multiple regression is relaxed to some extent by making the most of the characteristics of nonparametric regression. One strategy is to use

$$y = \sum_{i=1}^{r} m_i(x_i) + \epsilon. \tag{4.86}$$

Each $\{m_i(x_i)\}$ is a smooth nonparametric function, and ϵ is error; the mean is 0 and the variance is constant. This is called an additive model.

Some algorithms have been suggested so far to derive regression equations of this form. The Gauss-Seidel method is a typical tool for this purpose. Derivation of an additive model mainly by this method is explained here. For more general and mathematical discussions, refer to [3] or [7].

To make notations simple, we assume $r = 2$. That is, we discuss the derivation of an additive model, as below, when data are $\{(X_{i1}, X_{i2}, Y_i)\}$ $(1 \leq i \leq n)$.

$$y = m_1(x_1) + m_2(x_2). \tag{4.87}$$

We assume that $m_1(x_1)$ is a Pth-degree polynomial and $m_2(x_2)$ is a Qth-degree one. Then, a hat matrix corresponding to the derivation of $\{\hat{m}_1(X_{i1})\}$ from $\{X_{i1}\}$ using a Pth-degree polynomial is termed $\mathbf{H}^{(1)}$; that corresponding to the derivation of $\{\hat{m}_2(X_{i2})\}$ from $\{X_{i2}\}$ using a Qth-degree polynomial is called $\mathbf{H}^{(2)}$. Eigenvalues (characteristic values) of $\mathbf{H}^{(1)}$ consist of 1 (the number is $(P+1)$) and 0 (the number is $(n - P - 1)$) (eq(2.77) (page 57)). The same is true of a Qth-degree polynomial. Then, eq(2.87) (page 58) results in

$$[\mathbf{H}^{(1)}]_{ij} = \sum_{l=1}^{P+1} u_{il} u_{jl}, \tag{4.88}$$

$$[\mathbf{H}^{(2)}]_{ij} = \sum_{l=1}^{Q+1} v_{il} v_{jl}, \tag{4.89}$$

where $\mathbf{u}_l = (u_{1l}, u_{2l}, \ldots, u_{nl})^t$ is an eigenvector of $\mathbf{H}^{(1)}$. $\mathbf{u}_l$ $(1 \leq l \leq P+1)$ are eigenvectors when the eigenvalue is 1. $\mathbf{u}_l$ $(P+2 \leq l \leq n)$ are eigenvectors when the eigenvalue is 0. Similarly, $\mathbf{v}_l = (v_{1l}, v_{2l}, \ldots, v_{nl})^t$ is an eigenvector of $\mathbf{H}^{(2)}$. $\mathbf{v}_l$ $(1 \leq l \leq Q+1)$ are eigenvectors when the eigenvalue is 1. $\mathbf{v}_l$ $(Q+2 \leq l \leq n)$ are eigenvectors when the eigenvalue is 0.

Hence, to derive $m_1(x_1) + m_2(x_2)$, we minimize the value

$$
\begin{aligned}
E_{additive} &= \sum_{i=1}^{n} (m_1(X_{i1}) + m_2(X_{i2}) - Y_i)^2 \\
&= \| \mathbf{m}_1 + \mathbf{m}_2 - \mathbf{y} \|^2 \\
&= \| \mathbf{H}^{(1)}\mathbf{m}_1 + \mathbf{H}^{(2)}\mathbf{m}_2 - \mathbf{y} \|^2 \\
&= \sum_{i=1}^{n} \left(\sum_{j=1}^{n} \sum_{l=1}^{P+1} u_{il}u_{jl}m_1(X_{j1}) + \sum_{j=1}^{n} \sum_{l=1}^{Q+1} v_{il}v_{jl}m_2(X_{j2}) - Y_i \right)^2 \\
&= \sum_{i=1}^{n} \left(\sum_{l=1}^{P+1} a_l u_{il} + \sum_{l=1}^{Q+1} b_l v_{il} - Y_i \right)^2,
\end{aligned}
\tag{4.90}
$$

where $\mathbf{m}_1 = (m_1(X_{11}), m_1(X_{21}), \ldots, m_1(X_{n1}))^t$, $\mathbf{m}_2 = (m_2(X_{12}), m_2(X_{22}), \ldots, m_2(X_{n2}))^t$, and $\mathbf{y} = (Y_1, Y_2, \ldots, Y_n)^t$. Additionally, a_l and b_l are

$$
a_l = \sum_{j=1}^{n} u_{jl}m_1(X_{j1}) = < \mathbf{u}_l \cdot \mathbf{m}_1 >,
\tag{4.91}
$$

$$
b_l = \sum_{j=1}^{n} v_{jl}m_2(X_{j2}) = < \mathbf{v}_l \cdot \mathbf{m}_2 >,
\tag{4.92}
$$

where $< \cdot >$ is an inner product of vectors.

Differentiation of $E_{additive}$ to be 0 with respect ot $\{a_k\}$ ($1 \le k \le P+1$) and $\{b_k\}$ ($1 \le k \le Q+1$) yields

$$
\sum_{i=1}^{n} u_{ik} \left(\sum_{l=1}^{P+1} a_l u_{il} + \sum_{l=1}^{Q+1} b_l v_{il} - Y_i \right) = 0 \quad (1 \le k \le P+1),
\tag{4.93}
$$

$$
\sum_{i=1}^{n} v_{ik} \left(\sum_{l=1}^{P+1} a_l u_{il} + \sum_{l=1}^{Q+1} b_l v_{il} - Y_i \right) = 0 \quad (1 \le k \le Q+1).
\tag{4.94}
$$

By solving the simultaneous equation consisting of these $(P+Q+2)$ linear equations, $\{a_k\}$ and $\{b_k\}$ can be derived. Eq(4.93) and eq(4.94) are simplified by setting

$$
\Delta Y_i = \sum_{l=1}^{P+1} a_l u_{il} + \sum_{l=1}^{Q+1} b_l v_{il} - Y_i \quad (1 \le i \le n).
\tag{4.95}
$$

Hence, eq(4.93) and eq(4.94) are transformed into

$$
\sum_{i=1}^{n} u_{ik} \Delta Y_i = < \mathbf{u}_k \cdot \Delta \mathbf{y} > = 0 \quad (1 \le k \le P+1),
\tag{4.96}
$$

$$
\sum_{i=1}^{n} v_{ik} \Delta Y_i = < \mathbf{v}_k \cdot \Delta \mathbf{y} > = 0 \quad (1 \le k \le Q+1),
\tag{4.97}
$$

where $\Delta\mathbf{y} = (\Delta Y_1, \Delta Y_2, \cdots, \Delta Y_n)^t$. In Chapter 2, we showed that a vector consisting of residuals given by fitting a linear equation by the least squares is orthogonal to the plane spanned by the eigenvectors corresponding to the eigenvalues with the value of 1. This concept extends to eq(4.96) and eq(4.97).

This method consists of three steps: (1) derivation of eigenvectors of a hat matrix, (2) solution of a simultaneous equation to obtain $\{a_l\}$ and $\{b_l\}$, and (3) calculation of $\mathbf{m}_1$ and $\mathbf{m}_2$ using $\{a_l\}$ and $\{b_l\}$. When the number of data (n) is large, the computation to obtain eigenvectors is particularly laborious. Therefore, let us pursue methods to derive an additive model without obtaining eigenvectors of a hat matrix.

$\{\mathbf{u}_k\}$ $(1 \le k \le P+1)$ constitutes an orthogonal system, as does $\{\mathbf{v}_k\}$ $(1 \le k \le Q+1)$. Hence, eq(4.96) and eq(4.97), respectively, are equivalent to

$$\sum_{i=1}^{n}\sum_{k=1}^{P+1} u_{jk}u_{ik}\Delta Y_i = \sum_{k=1}^{P+1} u_{jk} < \mathbf{u}_k \cdot \Delta\mathbf{y} >= 0 \quad (1 \le j \le n), \qquad (4.98)$$

$$\sum_{i=1}^{n}\sum_{k=1}^{Q+1} v_{jk}v_{ik}\Delta Y_i = \sum_{k=1}^{Q+1} v_{jk} < \mathbf{v}_k \cdot \Delta\mathbf{y} >= 0 \quad (1 \le j \le n). \qquad (4.99)$$

By the use of eq(4.88) and eq(4.89), the two equations above are rewritten as

$$\sum_{i=1}^{n}[\mathbf{H}^{(1)}]_{ji}\Delta Y_i = 0 \quad (1 \le j \le n), \qquad (4.100)$$

$$\sum_{i=1}^{n}[\mathbf{H}^{(2)}]_{ji}\Delta Y_i = 0 \quad (1 \le j \le n). \qquad (4.101)$$

These two equations are arranged as

$$\begin{pmatrix} \mathbf{I}_n & \mathbf{H}^{(1)} \\ \mathbf{H}^{(2)} & \mathbf{I}_n \end{pmatrix} \begin{pmatrix} \mathbf{m}_1 \\ \mathbf{m}_2 \end{pmatrix} = \begin{pmatrix} \mathbf{H}^{(1)}\mathbf{y} \\ \mathbf{H}^{(2)}\mathbf{y} \end{pmatrix}, \qquad (4.102)$$

where $\mathbf{I}_n$ is an identity matrix (size of $n \times n$). We utilized the relationships $\mathbf{H}^{(1)}\mathbf{m}_1 = \mathbf{m}_1$ and $\mathbf{H}^{(2)}\mathbf{m}_2 = \mathbf{m}_2$. They are based on the fact that $\mathbf{H}^{(1)}$ and $\mathbf{H}^{(2)}$ are idempotent. A smoother whose hat matrix is idempotent is sometimes called a projection-type smoother.

Using eq(4.102) and solving a simultaneous equation after the derivation of a hat matrix results in $\mathbf{m}_1$ and $\mathbf{m}_2$. Since $\mathbf{m}_1$ and $\mathbf{m}_2$ are good enough to easily derive underlying equations (a Pth polynomial and a Qth polynomial), an additive model in the form of eq(4.87) results. An equation in the form of eq(4.102) to derive an additive model, but not restricted to an additive model consisting of regression equations in the form of polynomials, is called a normal equation. Solving eq(4.102) is equivalent to deriving $\{a_l\}$ and $\{b_l\}$ in eq(4.90) using the normal equation (eq(2.28) (page 33)).

However, if ordinary polynomials are used as $m_1(x_1)$ and $m_2(x_2)$, the resulting $\mathbf{m}_1$ and $\mathbf{m}_2$ are not unique. This is because an additive model obtained in this way is

$$y = m_1(x_1) + m_2(x_2) = c_0 + \sum_{j=1}^{P} c_j x_1^j + d_0 + \sum_{j=1}^{Q} d_j x_2^j. \qquad (4.103)$$

Since, under this situation, there are an infinite number of combinations of values of c_0 and d_0 that give a specific value of $(c_0 + d_0)$, neither $m_1(x_1)$ nor $m_2(x_2)$ is uniquely determined even when $m_1(x_1) + m_2(x_2)$ is unique. To solve this problem, either $\mathbf{H}^{(1)}$ or $\mathbf{H}^{(2)}$ must be modified. For example, $\mathbf{H}^{(1)}$ is replaced with $\mathbf{H}'^{(1)}$:

$$\mathbf{H}'^{(1)} = \left(\mathbf{I}_n - \frac{1}{n}\mathbf{J}_n\right)\mathbf{H}^{(1)}. \tag{4.104}$$

$\mathbf{J}_n$ is a matrix with the size of $n \times n$; all of the elements are 1. A smoothing method that yields a hat matrix such as $\mathbf{H}'^{(1)}$ is called a centered smoother. This is based on the fact that if a hat matrix such as $\mathbf{H}'^{(1)}$ is employed as $\mathbf{H}^{(1)}$, the average of $\{m_1(X_{j1})\}$ $(1 \le j \le n)$ is sure to be 0. Addition of the same value to $\{m_1(X_{j1})\}$ does not yield another solution. Hence, the solution of a normal equation is uniquely determined. This means that eq(4.102) is rewritten as

$$\begin{pmatrix} \mathbf{m}_1 \\ \mathbf{m}_2 \end{pmatrix} = \begin{pmatrix} \mathbf{I}_n & \mathbf{H}^{(1)} \\ \mathbf{H}^{(2)} & \mathbf{I}_n \end{pmatrix}^{-1} \begin{pmatrix} \mathbf{H}^{(1)}\mathbf{y} \\ \mathbf{H}^{(2)}\mathbf{y} \end{pmatrix}. \tag{4.105}$$

Next, to derive a hat matrix, we set

$$\mathbf{\Phi} = \begin{pmatrix} \mathbf{I}_n & \mathbf{H}^{(1)} \\ \mathbf{H}^{(2)} & \mathbf{I}_n \end{pmatrix}^{-1} = \begin{pmatrix} \mathbf{\Phi}_{11} & \mathbf{\Phi}_{12} \\ \mathbf{\Phi}_{21} & \mathbf{\Phi}_{22} \end{pmatrix}, \tag{4.106}$$

$$\mathbf{\Phi}_{11} = \mathbf{\Phi}_{(1:n,1:n)}, \tag{4.107}$$

$$\mathbf{\Phi}_{12} = \mathbf{\Phi}_{(1:n,(n+1):(2n))}, \tag{4.108}$$

$$\mathbf{\Phi}_{21} = \mathbf{\Phi}_{((n+1):(2n),1:n)}, \tag{4.109}$$

$$\mathbf{\Phi}_{22} = \mathbf{\Phi}_{((n+1):(2n),(n+1):(2n))}, \tag{4.110}$$

where $\mathbf{\Phi}_{(1:n,1:n)}$ is the top-left quarter of $\mathbf{\Phi}$. Similarly, $\mathbf{\Phi}_{(1:n,(n+1):(2n))}$ is the top-right quarter of $\mathbf{\Phi}$, $\mathbf{\Phi}_{((n+1):(2n),1:n)}$ is the bottom-left quarter, and $\mathbf{\Phi}_{((n+1):(2n),(n+1):(2n))}$ is the bottom-right quarter. Hence, eq(4.105) is transformed into

$$\mathbf{m}_1 = (\mathbf{\Phi}_{11}\mathbf{H}^{(1)} + \mathbf{\Phi}_{12}\mathbf{H}^{(2)})\mathbf{y}, \tag{4.111}$$

$$\mathbf{m}_2 = (\mathbf{\Phi}_{21}\mathbf{H}^{(1)} + \mathbf{\Phi}_{22}\mathbf{H}^{(2)})\mathbf{y}. \tag{4.112}$$

Therefore, we have

$$\begin{aligned} \mathbf{m}_1 + \mathbf{m}_1 &= (\mathbf{\Phi}_{11}\mathbf{H}^{(1)} + \mathbf{\Phi}_{12}\mathbf{H}^{(2)} + \mathbf{\Phi}_{21}\mathbf{H}^{(1)} + \mathbf{\Phi}_{22}\mathbf{H}^{(2)})\mathbf{y} \\ &= \mathbf{H}\mathbf{y}. \end{aligned} \tag{4.113}$$

That is, a hat matrix corresponding to this formation is

$$\mathbf{H} = \mathbf{\Phi}_{11}\mathbf{H}^{(1)} + \mathbf{\Phi}_{12}\mathbf{H}^{(2)} + \mathbf{\Phi}_{21}\mathbf{H}^{(1)} + \mathbf{\Phi}_{22}\mathbf{H}^{(2)}. \tag{4.114}$$

The process of creating an additive model consisting of polynomials is shown using simulated data. The number of data is 40. Uniform pseudorandom numbers between 0 and 10 are used as $\{X_{i1}\}$ and $\{X_{i2}\}$. $\{Y_i\}$ are produced by

$$Y_i = \sin(0.2\pi X_{i1}) + 0.05X_{i2}^2 - 0.4X_{i2} + 3 + \epsilon_i, \tag{4.115}$$

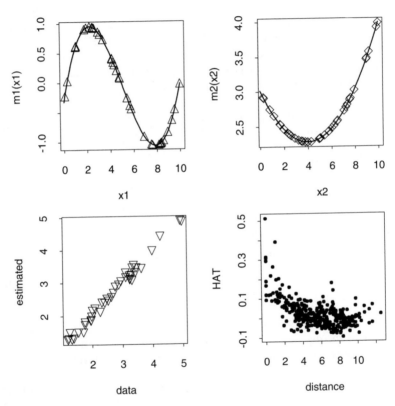

Figure 4.13   An additive model given by direct solution of a normal equation; the simulated data are generated by eq(4.115). Resultant quartic equation $(\hat{m}_1(x_1))$ is represented by a solid line. $\triangle$ stands for $\{(X_{i1}, \hat{m}_1(X_{i1}))\}$ (top left). Resultant quadratic equation $(\hat{m}_2(x_2))$ is represented by a solid line. $\diamond$ corresponds to $\{(X_{i2}, \hat{m}_2(X_{i2}))\}$ (top right). $\triangledown$ corresponds to $\{(Y_i, \hat{Y}_i)\}$ (bottom left). Relationship between the values of a hat matrix and the distance between data (400 points extracted from 1950 points are plotted) (bottom right).

where $\{\epsilon_i\}$ are realizations of $N(0.0, 0.1^2)$ (normal distribution with mean $= 0.0$, and standard deviation $= 0.1$). A quartic equation is used for $m_1(x_1)$ and a quadratic equation is used for $m_2(x_2)$; a normal equation (eq(4.102)) is solved to obtain a regression equation. To make the solution of a simultaneous equation unique, $\mathbf{H}'^{(1)}$ (eq(4.104) (page 261)) is used for $\mathbf{H}^{(1)}$. Thus, $m_1(x_1)$ and $m_2(x_2)$ are derived. They are depicted in figure 4.13(top left) and figure 4.13(top right). The relationship between the estimates ($\{\hat{Y}_i\}$) and the data ($\{Y_i\}$) is displayed in figure 4.13(bottom left). Furthermore, a hat matrix is obtained using eq(4.114) to find the relationship between the values of elements of the hat matrix and the distance between data (distance between data (X_{i1}, X_{i2}) and (X'_{i1}, X'_{i2}) is defined as $\sqrt{(X_{i1} - X'_{i1})^2 + (X_{i2} - X'_{i2})^2}$). The result is shown in figure 4.13(bottom right). A total of 400 points extracted from $1950 \, (= 40 \cdot 39 \cdot 0.5)$ points are plotted here. This graph shows approximately the fundamental characteristics that a hat matrix should acquire: the value of an element of a hat matrix should become smaller as the distance between data becomes longer.

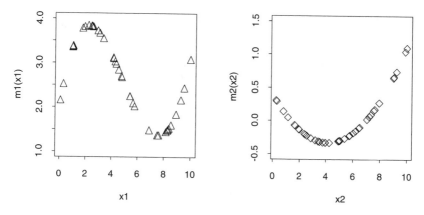

Figure 4.14 $\hat{m}_1$ (left) and $\hat{m}_2$ (right). A quartic equation and a quadratic equation are employed as $m_1(x_1)$ and $m_2(x_2)$, respectively. The Gauss-Seidel method is adopted. The number of iterations is 1.

On the other hand, when polynomial equations are employed as $m_1(x_1)$ and $m_2(x_2)$, applying the equation below, $\{c_0, c_1, \ldots, c_P, d_1, d_2, \ldots, d_Q\}$ can be derived directly by the least squares; a hat matrix need not be obtained.

$$y = m_1(x_1) + m_2(x_2) = c_0 + \sum_{j=1}^{P} c_j x_1^j + \sum_{j=1}^{Q} d_j x_2^j. \qquad (4.116)$$

While eq(4.102) (page 260) requires the solution of a $(2n)$-degree simultaneous equation, eq(4.116) only requires the solution of a $(P + Q + 1)$-degree multiple equation; the cost of calculation is usually reduced. However, if an iterative method for converging the solutions for a simultaneous equation is adopted to solve eq(4.102) instead of a universal method of multiple regression, this comparison is not so simple. For instance, the Gauss-Seidel method is a typical one for solving eq(4.102) by iteration.

Derivation of $\mathbf{m}_1$ and $\mathbf{m}_2$ by the Gauss-Seidel method is as follows: all initial values of $\mathbf{m}_2$ ($= \mathbf{m}_2^0$) are set to 0 (the average value of $\{Y_i\}$ is an alternative), and the

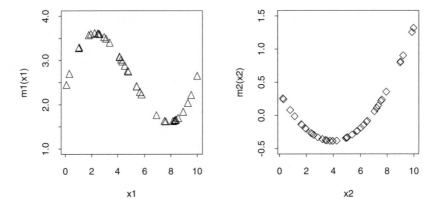

Figure 4.15 $\hat{m}_1$ (left) and $\hat{m}_2$ (right). A quartic equation and a quadratic equation are employed as $m_1(x_1)$ and $m_2(x_2)$, respectively. The Gauss-Seidel method is adopted. The number of iterations is 20.

calculation described below is repeated sequentially ($j = 1, 2, \ldots$) until $\mathbf{m}_1^j$ and $\mathbf{m}_2^j$ no longer change substantially. More accurately, however, this method is considered a block Gauss-Seidel method because $\mathbf{m}_1$ and $\mathbf{m}_2$ are vectors.

$$\mathbf{m}_1^j = \mathbf{H}^{(1)}(\mathbf{y} - \mathbf{m}_2^{j-1}), \tag{4.117}$$

$$\mathbf{m}_2^j = \mathbf{H}^{(2)}(\mathbf{y} - \mathbf{m}_1^j). \tag{4.118}$$

The merit of employing the Gauss-Seidel method to solve eq(4.102) lies not only in the possible saving of computational cost. When an additive model in the form of eq(4.103) is obtained using eq(4.102), the solution of the simultaneous equation becomes nonunique unless a hat matrix based on a centered smoother is adopted. On the other hand, while the Gauss-Seidel method causes $\mathbf{m}_1$ and $\mathbf{m}_2$ to converge to points that depend on the initial values, the values of $(\mathbf{m}_1 + \mathbf{m}_2)$ converge to the same points regardless of the initial values. That is to say, although the solution of the simultaneous equation (eq(4.102)) is not unique, the values of $(\mathbf{m}_1 + \mathbf{m}_2)$ are determined uniquely. Note that the method of executing eq(4.117) and eq(4.118) alternately is regarded as backfitting.

Using the same data as those in figure 4.13, a quartic equation and a quadratic equation are assumed for $m_1(x_1)$ and $m_2(x_2)$, respectively, and an iterative calculation based on the Gauss-Seidel method produces an additive model. The results are illustrated in figure 4.14, figure 4.15, and figure 4.16. The numbers of iterations are 1, 20, and 30, respectively. The similarity between figure 4.15 and figure 4.16 indicates numerical convergence.

When the B-spline (B-splines) is utilized for both $m_1(x_1)$ and $m_2(x_2)$, hat matrices for both regressions are also idempotent. Therefore, if $m_1(x_1) + m_2(x_2)$ is set out as follows, an additive model is created in a similar manner to that of polynomial

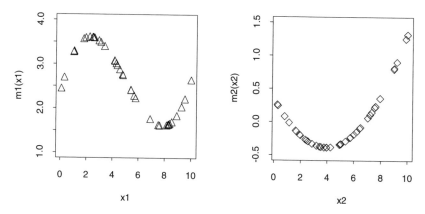

Figure 4.16   m̂$_1$(left) and m̂$_2$ (right). A quartic equation and a quadratic equation are assumed for $m_1(x_1)$ and $m_2(x_2)$, respectively, and the Gauss-Seidel method is employed to obtain them. The number of iterations is 30.

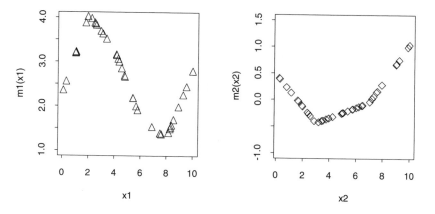

Figure 4.17   m̂$_1$ (left) and m̂$_2$ (right). The B-spline is employed to represent $m_1(x_1)$ and $m_2(x_2)$, and the Gauss-Seidel method is used to derive them. The number of iterations is 1.

equations.

$$y = m_1(x_1) + m_2(x_2) = \sum_{j=1}^{P} \beta_{j1} \cdot g_{j1}(x_1) + \sum_{j=1}^{Q} \beta_{j2} \cdot g_{j2}(x_2). \qquad (4.119)$$

The functions of $\{g_{1j}(x_1)\}$ and $\{g_{2j}(x_2)\}$ are the bases of the B-spline.

Figure 4.17, figure 4.18, and figure 4.19 illustrate additive models that are obtained using a one-degree B-spline (broken line) instead of polynomial equations and that are converged by the Gauss-Seidel method. The positions of knots of $m_1(x_1)$ are $(X_{min,1}, 2, 4, 6, 8, X_{max,1})$ ($X_{min,1}$ is the minimum value of $\{X_{i1}\}$; $X_{max,1}$ is the maximum value of $\{X_{i1}\}$). Those of $m_2(x_2)$ are $(X_{min,2}, 3, 7, X_{max,2})$ ($X_{min,2}$ is the minimum value of $\{X_{i2}\}$; $X_{max,2}$ is the maximum value of $\{X_{i2}\}$). The number of iterations are 1, 20, and 30, respectively.

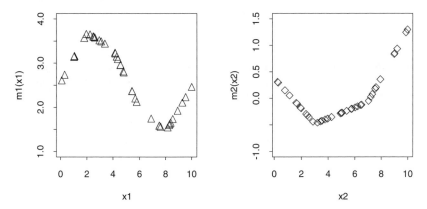

Figure 4.18   One-degree B-spline is used for $m_1(x_1)$ and $m_2(x_2)$, and $\hat{m}_1$ (left) and $\hat{m}_2$ (right) are obtained by the Gauss-Seidel method. The number of iterations is 20.

Next, we discuss the derivation of an additive model to which the smoothing spline is applied; this is a typical example in which $\mathbf{m}_1$ and $\mathbf{m}_2$ are not idempotent. The value below is minimized for this derivation.

$$\begin{aligned} E_{ss-add} = \ & \sum_{i=1}^{n} \Big(Y_i - m_1(X_{i1}) - m_2(X_{i2}) \Big)^2 \\ & + \lambda_1 \int_{-\infty}^{\infty} \left(\frac{d^2 m_1(x_1)}{dx_1^2} \right)^2 dx_1 \\ & + \lambda_2 \int_{-\infty}^{\infty} \left(\frac{d^2 m_2(x_2)}{dx_2^2} \right)^2 dx_2. \end{aligned} \qquad (4.120)$$

λ_1 and λ_2 are smoothing parameters for the two predictors, respectively. To minimize E_{ss-add}, a procedure similar to that described from eq(3.235) (page 179) to eq(3.239) is used. That is, the equations below are set.

$$m_1(x) = \sum_{j=1}^{P} \beta_{j1} \cdot g_{j1}(x), \qquad (4.121)$$

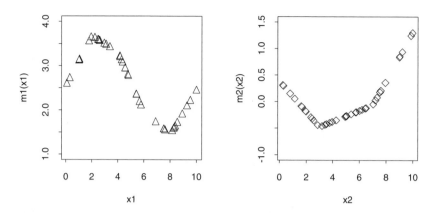

Figure 4.19 One-degree B-spline is used for $m_1(x_1)$ and $m_2(x_2)$, and $\hat{m}_1$ (left) and $\hat{m}_2$ (right) are obtained by the Gauss-Seidel method. The number of iterations is 30.

$$m_2(x) = \sum_{j=1}^{Q} \beta_{j2} \cdot g_{j2}(x), \qquad (4.122)$$

$$\mathbf{G}_1 = \begin{pmatrix} g_{11}(X_{11}) & g_{21}(X_{11}) & \cdots & g_{P1}(X_{11}) \\ g_{11}(X_{21}) & g_{21}(X_{21}) & \cdots & g_{P1}(X_{21}) \\ g_{11}(X_{31}) & g_{21}(X_{31}) & \cdots & g_{P1}(X_{31}) \\ \vdots & \vdots & \ddots & \vdots \\ g_{11}(X_{n1}) & g_{21}(X_{n1}) & \cdots & g_{P1}(X_{n1}) \end{pmatrix}, \qquad (4.123)$$

$$\mathbf{G}_2 = \begin{pmatrix} g_{12}(X_{12}) & g_{22}(X_{12}) & \cdots & g_{Q2}(X_{12}) \\ g_{12}(X_{22}) & g_{22}(X_{22}) & \cdots & g_{Q2}(X_{22}) \\ g_{12}(X_{32}) & g_{22}(X_{32}) & \cdots & g_{Q2}(X_{32}) \\ \vdots & \vdots & \ddots & \vdots \\ g_{12}(X_{n2}) & g_{22}(X_{n2}) & \cdots & g_{Q2}(X_{n2}) \end{pmatrix}, \qquad (4.124)$$

$$\boldsymbol{\beta}_1 = \begin{pmatrix} \beta_{11} \\ \vdots \\ \beta_{P1} \end{pmatrix}, \qquad (4.125)$$

$$\boldsymbol{\beta}_2 = \begin{pmatrix} \beta_{12} \\ \vdots \\ \beta_{Q2} \end{pmatrix}, \qquad (4.126)$$

$$\mathbf{K}_1 = \begin{pmatrix} k_{111} & k_{121} & \cdots & k_{1P1} \\ k_{121} & k_{221} & \cdots & k_{2P1} \\ \vdots & \vdots & \ddots & \vdots \\ k_{P11} & k_{P21} & \cdots & k_{PP1} \end{pmatrix}, \qquad (4.127)$$

$$k_{ij1} = \int_{-\infty}^{\infty} \left(\frac{d^2 g_{i1}(x_1)}{dx_1^2} \right) \left(\frac{d^2 g_{j1}(x)}{dx_1^2} \right) dx_1, \tag{4.128}$$

$$\mathbf{K}_2 = \begin{pmatrix} k_{112} & k_{212} & \cdots & k_{1Q2} \\ k_{212} & k_{222} & \cdots & k_{2Q2} \\ \vdots & \vdots & \ddots & \vdots \\ k_{Q12} & k_{Q22} & \cdots & k_{QQ2} \end{pmatrix}, \tag{4.129}$$

$$k_{ij2} = \int_{-\infty}^{\infty} \left(\frac{d^2 g_{i2}(x_2)}{dx_2^2} \right) \left(\frac{d^2 g_{j2}(x)}{dx_2^2} \right) dx_2. \tag{4.130}$$

Then, eq(4.120) is rewritten as

$$\begin{aligned} E_{ss-add} &= (\mathbf{y} - \mathbf{G}_1\boldsymbol{\beta}_1 - \mathbf{G}_2\boldsymbol{\beta}_2)^t (\mathbf{y} - \mathbf{G}_1\boldsymbol{\beta}_1 - \mathbf{G}_2\boldsymbol{\beta}_2) + \lambda_1 \boldsymbol{\beta}_1^t \mathbf{K}_1 \boldsymbol{\beta}_1 \\ &\quad + \lambda_2 \boldsymbol{\beta}_2^t \mathbf{K}_2 \boldsymbol{\beta}_2 \\ &= (\mathbf{y} - \mathbf{m}_1 - \mathbf{m}_2)^t (\mathbf{y} - \mathbf{m}_1 - \mathbf{m}_2) + \lambda_1 \boldsymbol{\beta}_1^t \mathbf{K}_1 \boldsymbol{\beta}_1 \\ &\quad + \lambda_2 \boldsymbol{\beta}_2^t \mathbf{K}_2 \boldsymbol{\beta}_2. \end{aligned} \tag{4.131}$$

In a similar manner to the procedure for obtaining eq(3.229) (page 176), $\mathbf{L}_1$ and $\mathbf{L}_2$ ($\mathbf{L}_1$ and $\mathbf{L}_2$ are symmetric matrices) are derived.

$$E_{ss-add} = (\mathbf{y} - \mathbf{m}_1 - \mathbf{m}_2)^t (\mathbf{y} - \mathbf{m}_1 - \mathbf{m}_2) + \lambda_1 \mathbf{m}_1^t \mathbf{L}_1 \mathbf{m}_1 + \lambda_2 \mathbf{m}_2^t \mathbf{L}_2 \mathbf{m}_2. \tag{4.132}$$

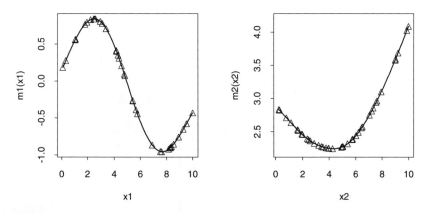

Figure 4.20   An additive model derived using a smoothing spline and the Gauss-Seidel method. $\triangle$ indicates $\hat{\mathbf{m}}_1$ (left) and $\hat{\mathbf{m}}_2$ (right). Solid lines represent $\hat{m}_1(x_1)$ (left) and $\hat{m}_2(x_2)$ (right).

Differentiation of this equation to be 0 with respect to $\mathbf{m}_1$ and $\mathbf{m}_2$ yields

$$\mathbf{y} - \mathbf{m}_2 = (\mathbf{I}_n + \lambda_1 \mathbf{L}_1)\mathbf{m}_1, \tag{4.133}$$

$$\mathbf{y} - \mathbf{m}_1 = (\mathbf{I}_n + \lambda_2 \mathbf{L}_1)\mathbf{m}_2. \tag{4.134}$$

These two equations are rewritten as

$$\mathbf{H}^{(1)}(\mathbf{y} - \mathbf{m}_2) = \mathbf{m}_1, \tag{4.135}$$

$$\mathbf{H}^{(2)}(\mathbf{y} - \mathbf{m}_1) = \mathbf{m}_2. \tag{4.136}$$

where $\mathbf{H}^{(1)}$ and $\mathbf{H}^{(2)}$ are

$$\mathbf{H}^{(1)} = (\mathbf{I}_n + \lambda_1 \mathbf{L}_1)^{-1}, \tag{4.137}$$

$$\mathbf{H}^{(2)} = (\mathbf{I}_n + \lambda_2 \mathbf{L}_2)^{-1}. \tag{4.138}$$

Eq(4.135) and eq(4.136) have the same form as eq(4.102) (page 260). That is, eq(4.102) can be used for smoothing when eq(4.120) is adopted. A more general proof, which shows that even if a hat matrix is not idempotent, an additive model can be derived using eq(4.102) (Chapter 4 of [7]), is possible; this is not limited to the derivation of an additive model using the smoothing spline. Furthermore, hat matrices based on the smoothing spline and on LOESS generally share the feature that when the Gauss-Seidel method is applied to solve this simultaneous equation, $(\mathbf{m}_1 + \mathbf{m}_2)$ converges to stable values even when the solution of the multiple equation is not unique; this is not confined to an idempotent hat matrix.

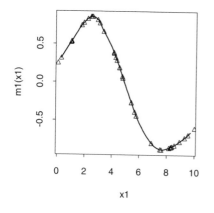 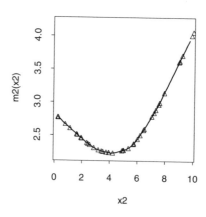

Figure 4.21 Additive model created by LOESS and the Gauss-Seidel method. $\triangle$ shows $\hat{\mathbf{m}}_1$ (left) and $\hat{\mathbf{m}}_2$ (right). The solid lines of the graphs represent $\hat{m}_1(x_1)$ (left) and $\hat{m}_2(x_2)$ (right).

Figure 4.20 illustrates the result of the derivation of an additive model based on the smoothing spline using the Gauss-Seidel method; the data are the same as that for figure 4.13 (page 262). The values of smoothing parameters (λ_1, λ_2) are adjusted to fix the values of the equivalent degrees of freedom at 4 for both predictors. Here, the term "equivalent degrees of freedom" denotes $\sum_{i=1}^{n}[\mathbf{H}^{(1)}]_{ii} - 1 \, (= \text{trace}(\mathbf{H}^{(1)}) - 1)$ for x_1, and $\sum_{i=1}^{n}[\mathbf{H}^{(2)}]_{ii} - 1 \, (= \text{trace}(\mathbf{H}^{(2)}) - 1)$ for x_2. The reason why the degrees of freedom of one of the nonparametric functions (page 118) constituting an additive model is so defined is that no nonparametric function contains a contant term, whereas an additive model as a whole has a constant term. Figure 4.21 shows the result of using LOESS in place of the smoothing spline. The span (page 186) for both predictors is set at 0.5.

Even when the number of predictors is r ($r \geq 3$), a similar method to that for two predictors can be used. In this instance, eq(4.102) is replaced with

$$
\begin{pmatrix}
\mathbf{I}_n & \mathbf{H}^{(1)} & \mathbf{H}^{(1)} & \cdots & \mathbf{H}^{(1)} & \mathbf{H}^{(1)} \\
\mathbf{H}^{(2)} & \mathbf{I}_n & \mathbf{H}^{(2)} & \cdots & \mathbf{H}^{(2)} & \mathbf{H}^{(2)} \\
\vdots & \vdots & \vdots & \vdots & \ddots & \vdots \\
\mathbf{H}^{(r)} & \mathbf{H}^{(r)} & \mathbf{H}^{(r)} & \cdots & \mathbf{H}^{(r)} & \mathbf{I}_n
\end{pmatrix}
\begin{pmatrix}
\mathbf{m}_1 \\
\mathbf{m}_2 \\
\vdots \\
\mathbf{m}_r
\end{pmatrix}
=
\begin{pmatrix}
\mathbf{H}^{(1)}\mathbf{y} \\
\mathbf{H}^{(2)}\mathbf{y} \\
\vdots \\
\mathbf{H}^{(r)}\mathbf{y}
\end{pmatrix}.
\tag{4.139}
$$

Furthermore, eq(4.117) and eq(4.118) are substituted with

$$
\mathbf{m}_1^j = \mathbf{H}^{(1)} \sum_{k=2}^{r} (\mathbf{y} - \mathbf{m}_k^{j-1}),
$$

$$
\mathbf{m}_2^j = \mathbf{H}^{(2)} \sum_{k=1}^{1} (\mathbf{y} - \mathbf{m}_k^j) + \mathbf{H}^{(2)} \sum_{k=3}^{r} (\mathbf{y} - \mathbf{m}_k^{j-1}),
$$

$$
\vdots \qquad \vdots
$$

$$
\mathbf{m}_r^j = \mathbf{H}^{(r)} \sum_{k=1}^{r-1} (\mathbf{y} - \mathbf{m}_k^j).
\tag{4.140}
$$

As a special case of an additive model, some $\{m_i(x_i)\}$ in eq(4.86) (page 258) can be parametric functions. Regression of this sort is called semiparametric regression on some occasions. Only when the regression equation employed in an additive model is derived using a linear estimator, is a regression equation obtained through the same procedure of replacing a hat matrix with that for parametric functions. Furthermore, when one among $\{m_i(x_i)\}$ is a nonparametric function, the smoothing spline is extended to realize semiparametric regression. This is called a partial spline. The subroutine of dtpss (or dptpss) included in GCVPACK enables this calculation. GCVPACK uses an algorithm in which iterative calculation is not required. Partial spline is used for calculating rice yields in average years in the Ministry of Agriculture, Forestry and Fisheries of Japan.

4.7 ACE

While an additive model transforms each predictor via a smooth function, regression that transforms a target variable via a smooth function along with predictors is possible. That is, the regression equation below is assumed.

$$
\eta(y) = \sum_{i=1}^{r} m_i(x_i) + \epsilon.
\tag{4.141}
$$

The functions of $\eta(y)$ and $\{m_i(x_i)\}$ are smooth regression equations, and ϵ is a random error with zero mean and constant variance. Note that the regression equations treated in this section have the form of eq(4.141); it is not the form

$$y = \eta^{-1}\left(\sum_{i=1}^{r} m_i(x_i)\right) + \epsilon, \tag{4.142}$$

where $\eta^{-1}(\cdot)$ is an inverse function of $\eta(\cdot)$. While for most regression equations we assume that the essential behavior of a target variable with added errors with zero mean and constant variance leads to values of a target variable of data, for eq(4.141), we assume that an error which is added to a transformed target variable has zero mean and constant variance. Since this assumption is part of the essence of the algorithm dealt with in this section, it considerably affects the characteristics of regression equations with this form. As for this point and more detailed description on the topics in this section, refer to Chapter 7 of [7] and to [4].

To obtain a regression equation with the form of eq(4.141), canonical correlation analysis is available. Let us cite an example where $\eta(y)$ and $\{m_i(x_i)\}$ are polynomial equations. Assume that the data are $\{(A_{i1}, A_{i2}, \ldots, A_{ip}, B_{i1}, B_{i2}, \ldots, B_{iq})\}$ $(1 \leq i \leq n)$ $((A_{i1}, A_{i2}, \ldots, A_{ip})$ are values of predictors of data and $(B_{i1}, B_{i2}, \ldots, B_{iq})$ are target variables of data). Canonical correlation analysis yields $\{a_1, a_2, \ldots, a_p\}$ and $\{b_1, b_2, \ldots, b_q\}$, which maximize

$$Cor(\{A_i^*\}, \{B_i^*\}) = \frac{\sum_{i=1}^{n} A_i^* B_i^*}{\sqrt{\sum_{i=1}^{n} (A_i^*)^2 \sum_{i=1}^{n} (B_i^*)^2}}, \tag{4.143}$$

where $Cor(\{A_i^*\}, \{B_i^*\})$ is a coefficient of correlation between $\{A_i^*\}$ and $\{B_i^*\}$. A_i^* and B_i^* are

$$A_i^* = \sum_{j=1}^{p} a_j(A_{ij} - \bar{A}_j), \qquad B_i^* = \sum_{j=1}^{q} b_j(B_{ij} - \bar{B}_j). \tag{4.144}$$

$\bar{A}_j$ and $\bar{B}_j$ are

$$\bar{A}_j = \frac{\sum_{i=1}^{n} A_{ij}}{n}, \qquad \bar{B}_j = \frac{\sum_{i=1}^{n} B_{ij}}{n}. \tag{4.145}$$

A_i^* and B_i^* are sometimes defined as those in which $\bar{A}_j$ and $\bar{B}_j$ are not subtracted, respectively. However, the definition described by eq(4.144) is adopted here because much emphasis is placed on the relationship between canonical correlation analysis and regression equations. Furthermore, A_i^* and B_i^* satisfy the conditions

$$\sum_{i=1}^{n} (A_i^*)^2 = n, \qquad \sum_{i=1}^{n} (B_i^*)^2 = n, \tag{4.146}$$

$$\sum_{i=1}^{n} A_i^* = 0, \qquad \sum_{i=1}^{n} B_i^* = 0. \tag{4.147}$$

If $\{A_i^*\}$ and $\{B_i^*\}$ are defined by eq(4.144), eq(4.147) holds. Then, using $\{A_i^*\}$ and $\{B_i^*\}$, two vectors are defined: $\alpha^* = (A_1^*, A_2^*, \ldots, A_n^*)^t$ and $\beta^* = (B_1^*, B_2^*, \ldots, B_n^*)^t$. Eq(4.146) ensures that both vectors have a length of $\sqrt{n}$. Therefore, an inner product between α^* and β^* is $(n \cdot Cor(\{A_i^*\}, \{B_i^*\}))$ $(-1 \leq Cor(\{A_i^*\}, \{B_i^*\}) \leq 1)$. Then, the maximum value of $Cor(\{A_i^*\}, \{B_i^*\})$ is defined as $\cos(\theta)$ $(0 \leq \theta \leq \pi)$. In the context of canonical correlation analysis, $\cos(\theta)$ is the maximum value among canonical correlation coefficients. Hence, $\cos(\theta)$ is called the first canonical correlation coefficient. Furthermore, $\{A_i^*\}$ and $\{B_i^*\}$, which give the first canonical correlation coefficient, are one set among canonical variates. Since this canonical variate provides the first canonical correlation coefficient, it is called the first canonical variate. Moreover, like $\{a_j\}$ and $\{b_j\}$, coefficients that give canonical variates are called canonical coefficients or canonical weights.

The values of $(\cos(\theta))^2$ and $\hat{\mathbf{a}}$ $(= (\hat{a}_1, \hat{a}_2, \ldots, \hat{a}_p)^t)$ are identical to the maximum value of the eigenvalues and the corresponding eigenvector (the length of the vector is adjusted to satisfy eq(4.146)), respectively, which are the solutions of the eigenvalue problem. For more detailed explanation, refer to the literature on canonical correlation analysis.

$$(\mathbf{Cov}_{AA})^{-1}\mathbf{Cov}_{AB}(\mathbf{Cov}_{BB})^{-1}\mathbf{Cov}_{BA}\mathbf{a} = \gamma\mathbf{a} \qquad (4.148)$$

Assume that $\mathbf{A}$ is a matrix that satisfies $[\mathbf{A}]_{ij} = A_{ij} - \bar{A}_j$ $(1 \leq i \leq n, \ 1 \leq j \leq p)$, and $\mathbf{B}$ is a matrix that satisfies $[\mathbf{B}]_{ij} = B_{ij} - \bar{B}_j$ $(1 \leq i \leq n, \ 1 \leq j \leq q)$. Then, $\mathbf{Cov}_{AA}$, $\mathbf{Cov}_{AB}$, $\mathbf{Cov}_{BA}$, and $\mathbf{Cov}_{BB}$ are

$$\mathbf{Cov}_{AA} = \frac{1}{n-1}\mathbf{A}^t\mathbf{A},$$

$$\mathbf{Cov}_{AB} = \frac{1}{n-1}\mathbf{A}^t\mathbf{B},$$

$$\mathbf{Cov}_{BA} = \frac{1}{n-1}\mathbf{B}^t\mathbf{A} = (\mathbf{Cov}_{AB})^t,$$

$$\mathbf{Cov}_{BB} = \frac{1}{n-1}\mathbf{B}^t\mathbf{B}. \qquad (4.149)$$

These matrices are variance-covariance matrices (or covariance matrices) of the given data. For the definition of a variance-covariance matrix, while division by $(n-1)$ renders an unbiased estimator, division by n yields a maximum likelihood estimator. Although a maximum likelihood estimator should be employed to conform with eq(4.143), an unbiased estimator is adopted here to comply with the definition of a variance-covariance matrix in S-Plus. However, whichever definition is employed, the results of canonical correlation analysis would remain the same; eq(4.148) proves this.

Eq(4.148) is rewritten as

$$\mathbf{A}^t\mathbf{B}(\mathbf{B}^t\mathbf{B})^{-1}\mathbf{B}^t\mathbf{A}\mathbf{a} = (\cos(\theta))^2\mathbf{A}^t\mathbf{A}\mathbf{a}. \qquad (4.150)$$

Both sides of this equation are multiplied by $\mathbf{A}(\mathbf{A}^t\mathbf{A})^{-1}$ from the left, giving

$$\mathbf{A}(\mathbf{A}^t\mathbf{A})^{-1}\mathbf{A}^t\mathbf{B}(\mathbf{B}^t\mathbf{B})^{-1}\mathbf{B}^t\mathbf{A}\mathbf{a} = (\cos(\theta))^2\mathbf{A}(\mathbf{A}^t\mathbf{A})^{-1}\mathbf{A}^t\mathbf{A}\mathbf{a}. \qquad (4.151)$$

Then, the equation below is obtained.

$$\mathbf{A}(\mathbf{A}^t\mathbf{A})^{-1}\mathbf{A}^t\mathbf{B}(\mathbf{B}^t\mathbf{B})^{-1}\mathbf{B}^t\alpha^* = (\cos(\theta))^2\alpha^*. \qquad (4.152)$$

Next, to consider the meaning of eq(4.152), the values of $\{c_1, c_2, \ldots, c_p\}$ are derived by the least squares to obtain the relationship

$$B_i^* = \sum_{j=1}^{p} c_j (A_{ij} - \bar{A}_j) + \epsilon_i, \qquad (4.153)$$

where $\{\epsilon_i\}$ are residuals. Then, the estimates are defined as $\hat{B}_i^*$ when the values of predictors are $\{A_{i1}, A_{i2}, \ldots, A_{ip}\}$, and $\hat{\boldsymbol{\beta}}^* (= (\hat{B}_1^*, \hat{B}_2^*, \ldots, \hat{B}_n^*)^t)$ is defined. Then, we have

$$\hat{\boldsymbol{\beta}}^* = \mathbf{H}_A \boldsymbol{\beta}^*, \qquad (4.154)$$

where $\mathbf{H}_A$, which is given as below, is a hat matrix to derive the estimates $(\hat{\boldsymbol{\beta}}^*)$ using data $(\boldsymbol{\beta}^*)$.

$$\mathbf{H}_A = \mathbf{A}(\mathbf{A}^t \mathbf{A})^{-1} \mathbf{A}^t. \qquad (4.155)$$

Next, let us discuss the following problem: the values of $\{d_1, d_2, \ldots, d_q\}$ are obtained by the least squares to give the relationship

$$A_i^* = \sum_{j=1}^{q} d_j (B_{ij} - \bar{B}_j) + \epsilon_i', \qquad (4.156)$$

where $\{\epsilon_i'\}$ are residuals. Then, estimates are defined as $\hat{A}_i^*$ when the values of predictors are $\{B_{i1}, B_{i2}, \ldots, B_{iq}\}$, and $\hat{\boldsymbol{\alpha}}^* = (\hat{A}_1^*, \hat{A}_2^*, \ldots, \hat{A}_n^*)^t$ (a vector) is defined. As a consequence, we obtain

$$\hat{\boldsymbol{\alpha}}^* = \mathbf{H}_B \boldsymbol{\alpha}^*, \qquad (4.157)$$

where $\mathbf{H}_B$ is a hat matrix used to derive estimates $(\hat{\boldsymbol{\alpha}}^*)$ from data $(\boldsymbol{\alpha}^*)$. This hat matrix is written as

$$\mathbf{H}_B = \mathbf{B}(\mathbf{B}^t \mathbf{B})^{-1} \mathbf{B}^t. \qquad (4.158)$$

By using eq(4.155) and eq(4.158), eq(4.152) is transformed into

$$\mathbf{H}_A \mathbf{H}_B \boldsymbol{\alpha}^* = (\cos(\theta))^2 \boldsymbol{\alpha}^*. \qquad (4.159)$$

This equation is equivalent to eq(4.148); both equations yield the first canonical correlation coefficient (Exercise 7.8 in Chapter 7 of [7]). That is, the maximum eigenvalue given by the matrix $(\mathbf{H}_A \mathbf{H}_B)$ is $(\cos(\theta))^2$, and the corresponding eigenvector is $\boldsymbol{\alpha}^*$. Hence, the eigenvalue problem eq(4.159) is the same as eq(4.148).

Now, let us proceed with the discussion of the problem of obtaining eq(4.153). The values of $\{c_j\}$ are derived by minimizing

$$E_{proportion} = \sum_{i=1}^{n} \left(B_i^* - \sum_{j=1}^{p} c_j (A_{ij} - \bar{A}_j) \right)^2$$

$$= n - 2 \sum_{i=1}^{n} \left(B_i^* \sum_{j=1}^{p} c_j (A_{ij} - \bar{A}_j) \right) + \sum_{i=1}^{n} \left(\sum_{j=1}^{p} c_j (A_{ij} - \bar{A}_j) \right)^2.$$

$$(4.160)$$

Then we set

$$\tau = \sqrt{\sum_{i=1}^{n}\left(\sum_{j=1}^{p} c_j (A_{ij} - \bar{A}_j)\right)^2}, \tag{4.161}$$

$$B_i^+ = \frac{\sqrt{n}\sum_{j=1}^{p} c_j (A_{ij} - \bar{A}_j)}{\tau}. \tag{4.162}$$

As a result, $\boldsymbol{\beta}^+ \ (= (B_1^+, B_2^+, \ldots, B_n^+)^t)$ is a vector with the length of $\sqrt{n}$. Then, $E_{proportion}$ is written as

$$
\begin{aligned}
E_{proportion} &= n - \frac{2\tau}{\sqrt{n}}\sum_{i=1}^{n} B_i^* B_i^+ + \tau^2 \\
&\geq n - 2\sqrt{n}\tau\cos(\theta) + \tau^2 \\
&= (\tau - \sqrt{n}\cos(\theta))^2 - n(\cos(\theta))^2 + n, \tag{4.163}
\end{aligned}
$$

where the inequality between the first line and the second one is derived, since the maximum value of $\sum_{i=1}^{n} B_i^* B_i^+$ is $n \cdot \cos(\theta)$ because both $\boldsymbol{\beta}^* \ (= (B_1^*, B_2^*, \ldots, B_n^*)^t)$ and $\boldsymbol{\beta}^+$ have the length of $\sqrt{n}$, Hence, equality holds in this inequality when $B_i^+ = A_i^*$. Then, we have

$$\hat{c}_j = \frac{\tau \cdot \hat{a}_j}{\sqrt{n}} \quad (1 \leq j \leq p). \tag{4.164}$$

Furthermore, the value of the third line of eq(4.163) takes a minimum value when $\tau = \sqrt{n}\cos(\theta)$. Then, we obtain

$$\hat{c}_j = \cos(\theta) \cdot \hat{a}_j \quad (1 \leq j \leq p). \tag{4.165}$$

To derive $\{c_j\}$ that minimizes $E_{proportion}$, the condition for rendering equal the inequality in the second line of eq(4.163) and the condition required to minimize the third line are derived independently. Then using these two conditions, $\{\hat{c}_j\}$ is derived. The justification of this procedure is that, in minimizing $E_{proportion}$, optimization of $\boldsymbol{\beta}^+ \ (= (B_1^+, B_2^+, \ldots, B_n^+)^t)$ (eq(4.164)) corresponds to the direction of the vector $(\sum_{j=1}^{p} c_j (A_{1j} - \bar{A}_1), \sum_{j=1}^{p} c_j (A_{2j} - \bar{A}_2), \ldots, \sum_{j=1}^{p} c_j (A_{nj} - \bar{A}_n))^t$, and optimization of τ (eq(4.165)) corresponds to the length of the vector.

Therefore, when $\{A_{i1}, A_{i2}, \ldots, A_{ip}\}$ are values of predictors, the estimates $(\hat{B}_i^*)$, described using eq(4.165), are

$$\hat{B}_i^* = \cos(\theta) \sum_{j=1}^{p} \hat{a}_j (A_{ij} - \bar{A}_j). \tag{4.166}$$

$E_{proportion}$ here becomes $n\left(1 - (\cos(\theta))^2\right)$.

On the other hand, if $\tau = \sqrt{n}$ is set in eq(4.164), the following $\hat{B}_i^+$ are estimates.

$$\hat{B}_i^+ = \sum_{j=1}^{p} \hat{a}_j (A_{ij} - \bar{A}_j). \tag{4.167}$$

Then, we have

$$\hat{c}_j = \hat{a}_j \quad (1 \le j \le p). \tag{4.168}$$

$E_{proportion}$ here is $n\left(1 - (\cos(\theta))^2 + (1 - \cos(\theta))^2\right)$. Unless $\cos(\theta) = 1$ holds, the value of $E_{proportion}$ is larger than that given by eq(4.166).

These are confirmed by the relationship shown in figure 4.22: $\beta^*, \hat{\beta}^+ (= (\hat{B}_1^+, \hat{B}_2^+, \ldots, \hat{B}_n^+)^t$, and $\hat{\beta}^*$ are illustrated. Both the value of $Cor(\beta^*, \hat{\beta}^*)$ and the value of $Cor(\beta^*, \hat{\beta}^+)$ are $\cos(\theta)$. However, among vectors whose coefficient of correlation with β^* is $\cos(\theta)$, $\hat{\beta}^*$ is the one that minimizes $E_{proportion}$.

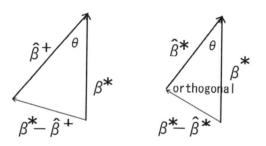

Figure 4.22   Relationship between β^* and $\hat{\beta}^+$. Both vectors have a length of $\sqrt{n}$. The angle between the two vectors is θ. $\cos(\theta)$ is the maximum value of $Cor(\{\hat{A}_i^*, \hat{B}_i^*\})$ (left). Relationship between β^* and $\hat{\beta}^*$. The length of $\hat{\beta}^*$ is $\sqrt{n}\cos(\theta)$ (right). The length of $(\beta^* - \hat{\beta}^*)$ is shorter than that of $(\beta^* - \hat{\beta}^+)$.

To understand the meanings of eq(4.165) and eq(4.168), the simplest multiple regression equation with the form eq(4.166) is considered. That is, $p = q = 1$. This setting corresponds to the following simple regression equation.

$$A_{i1} = X_i, \qquad B_{i1} = Y_i. \tag{4.169}$$

Eq(4.146) (page 271) gives the absolute values of $\hat{a}_1$ and $\hat{b}_1$. Substituting these values into eq(4.143) (page 271) yields two sets of $(\hat{a}_1, \hat{b}_1)$ that maximize $Cor(\{A_i^*, B_i^*\})$. These two sets of $(\hat{a}_1, \hat{b}_1)$ have opposite signs. Both sets give $Cor(\{A_i^*, B_i^*\})$, which takes the value of $\cos(\theta)$. That is, without using canonical correlation analysis, the values of $\hat{a}_1$, $\hat{b}_1$, and $\cos(\theta)$ are obtained. Hence, using eq(4.166), the relationship between X_i and corresponding estimates $(\hat{Y}_i)$ is obtained:

$$\hat{b}_1(\hat{Y}_i - \bar{Y}) = \cos(\theta)\hat{a}_1(X_i - \bar{X}), \tag{4.170}$$

where $\bar{X}$ and $\bar{Y}$ are averages of $\{X_i\}$ and $\{Y_i\}$, respectively. $\cos(\theta)\hat{a}_1$ corresponds to $\hat{c}_1$ in eq(4.165). The two sets of $(\hat{a}_1, \hat{b}_1)$ result in the same eq(4.170). Eq(4.170) holds even when X_i is substituted by an arbitrary value of a predictor (x) and $\hat{Y}_i$ is substituted by an arbitrary value of a target variable (y). This corresponds to simple regression by the least squares. Hence, $\sum_{i=1}^{n}\left(\hat{b}_1(\hat{Y}_i - \bar{Y})\right)^2$ becomes $n(\cos(\theta))^2$.

On the other hand, if $\sum_{1=1}^{n}\left(\hat{b}_1(\hat{Y}_i - \bar{Y})\right)^2$ is made to be n, that is, $\sum_{1=1}^{n}\left(\hat{b}_1(\hat{Y}_i - \bar{Y})\right)^2$ is equal to $\sum_{1=1}^{n}\left(\hat{b}_1(Y_i - \bar{Y})\right)^2$, eq(4.167) results in the following relationship between X_i and corresponding estimates ($\hat{Y}_i$).

$$\hat{b}_1(\hat{Y}_i - \bar{Y}) = \hat{a}_1(X_i - \bar{X}). \tag{4.171}$$

$\hat{a}_1$ corresponds to $\hat{c}_1$ in eq(4.168). This equation also holds even if X_i is replaced with an arbitrary value of a predictor (x) and $\hat{Y}_i$ is substitued by an arbitrary value of a target variable (y). Figure 4.23 illustrates the difference between figure 4.170 and figure 4.171. Since figure 4.23(left) shows the result of eq(4.170), the straight line given by this regression is the same as that of simple regression by the least squares. Since figure 4.23(right) shows the result of eq(4.171), the gradient of the line is larger than that obtained by the least squares. The characterictics that both $\sum_{i=1}^{n}\left(\hat{a}_1(X_i - \bar{X})\right)$ and $\sum_{i=1}^{n}\left(\hat{b}_1(Y_i - \bar{Y})\right)$ are 0 and both $\sum_{i=1}^{n}\left(\hat{a}_1(X_i - \bar{X})\right)^2$ and $\sum_{i=1}^{n}\left(\hat{b}_1(Y_i - \bar{Y})\right)^2$ are n may indicate that estimates ($\hat{Y}_i$) are represented by eq(4.171). However, the regression equation by the least squares (eq(4.160) (page 273)) is eq(4.170), and hence the two lines do not coincide unless $\cos(\theta) = 1$. This reflects the fact that while canonical correlation analysis does not distinguish between predictors and target variables, the least squares method does not deal with predictors and target variables equally.

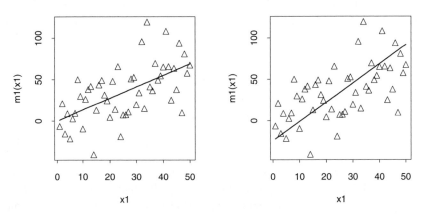

Figure 4.23   Result of regression using eq(4.170) (left). Result of regression using eq(4.171) (right).

Through the use of these features of canonical correlation analysis, a regression equation with the form of eq(4.141) (page 270) is derived. In a simplistic example, the number of predictors is two ($p = 2$) and quadratic equations are utilized as $m_1(x_1)$, $m_2(x_2)$, and $\eta(y)$. That is, we assume

$$A_{i1} = X_{i1}, \qquad A_{i2} = (X_{i1})^2, \tag{4.172}$$

$$A_{i3} = X_{i2}, \qquad A_{i4} = (X_{i2})^2, \tag{4.173}$$

$$B_{i1} = \hat{Y}_i, \qquad B_{i2} = (\hat{Y}_i)^2. \qquad (4.174)$$

Then, eq(4.166) is replaced with

$$
\begin{aligned}
\hat{b}_1(\hat{Y}_i - \bar{Y}) + \hat{b}_2(\hat{Y}_i^2 - \bar{Y}^2) \;=\;& \cos(\theta)\Big(\hat{a}_1(X_{i1} - \bar{X}_1) + \hat{a}_2(X_{i1}^2 - \bar{X}_1^2) \\
&+ \hat{a}_3(X_{i2} - \bar{X}_2) + \hat{a}_4(X_{i2}^2 - \bar{X}_2^2)\Big), \quad (4.175)
\end{aligned}
$$

where $\bar{X}_1$, $\bar{X}_1^2$, $\bar{X}_2$, $\bar{X}_2^2$, $\bar{Y}$, and $\bar{Y}^2$ are averages of $\{X_{i1}\}$, $\{X_{i1}^2\}$, $\{X_{i2}\}$, $\{X_{i2}^2\}$, $\{\hat{Y}_i\}$, and $\{\hat{Y}_i^2\}$, respectively. Then, we obtain

$$\hat{m}_1(X_{i1}) = \cos(\theta)\Big(\hat{a}_1(X_{i1} - \bar{X}_1) + \hat{a}_2(X_{i1}^2 - \bar{X}_1^2)\Big), \qquad (4.176)$$

$$\hat{m}_2(X_{i2}) = \cos(\theta)\Big(\hat{a}_3(X_{i2} - \bar{X}_2) + \hat{a}_4(X_{i2}^2 - \bar{X}_2^2)\Big), \qquad (4.177)$$

$$\hat{\eta}(Y_i) = \hat{b}_1(\hat{Y}_i - \bar{Y}) + \hat{b}_2(\hat{Y}_i^2 - \bar{Y}^2). \qquad (4.178)$$

Only when $m_1(x_1)$, $m_2(x_2)$, and $\eta(y)$ have the functional form to yield a linear estimator, can such equations be constructed. Therefore, various functions can be employed as $m_1(x_1)$, $m_2(x_2)$, and $\eta(y)$, such as the B-spline, along with polynomial equations.

Figure 4.24 illustrates an example of regression given by canonical correlation analysis; a polynomial equation is given as an example of a regression equation. The number of data is 100, and uniform pseudorandom numbers between 0 and 10 are used as values of $\{X_{i1}\}$ and $\{X_{i2}\}$. $\{Y_i\}$ are obtained by

$$Y_i = \big(\sin(0.2\pi X_{i1}) + 0.05X_{i2}^2 - 0.4X_{i2} + 3 + \epsilon_i\big)^2, \qquad (4.179)$$

where $\{\epsilon_i\}$ are realizations of $N(0, 0.1^2)$ (normal distribution with mean 0, standard deviation 0.1^2). $m_1(x_1)$ is a quartic equation, and $m_2(x_2)$ and $\eta(y)$ are quadratic equations. As a result, the value of $\cos(\theta)$ becomes 0.98786.

Figure 4.25 displays the results of directly solving the eigenvalue problem described by eq(4.159) (page 273); the data are the same as that in the case of figure 4.24. Figure 4.25(left top) shows eigenvalues aligned in descending order. The positive square root of the maximum eigenvalue is 0.98786. The value is identical to that obtained by canonical correlation analysis. The resulting regression equation is the same as that in the case of figure 4.24.

There is another way of obtaining a regression equation given by canonical correlation analysis. An iterative calculation, as below, is also possible.
(1) We set

$$Y_i^* = \frac{Y_i - \bar{Y}}{sd(\{Y_i\})} \qquad (1 \le i \le n). \qquad (4.180)$$

(2) Using the data $\{(X_{i1}, X_{i2}, Y_i^*)\}$, an additive model is obtained by the least squares to define $\{\hat{m}_1(X_{i1}) + \hat{m}_2(X_{i2})\}$ as $\{\hat{Y}_i\}$. That is, the relationship below is obtained.

$$Y_i^* = \hat{m}_1(X_{i1}) + \hat{m}_2(X_{i2}) + \epsilon_i = \hat{Y}_i^* + \epsilon_i, \qquad (4.181)$$

where $\hat{m}_1(\cdot)$ and $\hat{m}_2(\cdot)$ are polynomial equations and $\{\epsilon_i\}$ are residuals.

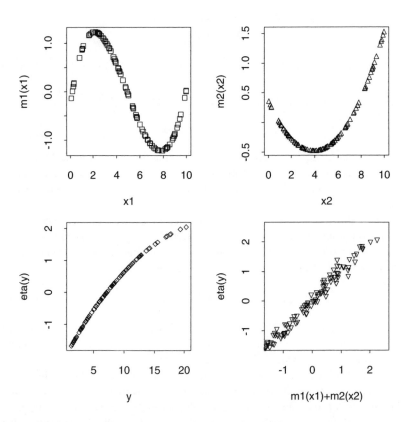

Figure 4.24 Example of regression in which each regression equation is a polynomial and canonical correlation analysis is used to construct regression equations with the form of eq(4.141) (page 270). $\{X_{i1}, \hat{m}_1(X_{i1})\}$ (quartic equation), (top left). $\{X_{i2}, \hat{m}_2(X_{i2})\}$ (quadratic equation), (top right). $\{Y_i, \hat{\eta}(Y_i)\}$ (quadratic equation), (bottom left). $\{(\hat{m}_1(X_{i1}) + \hat{m}_2(X_{i2}), \hat{\eta}(Y_i))\}$, (bottom right).

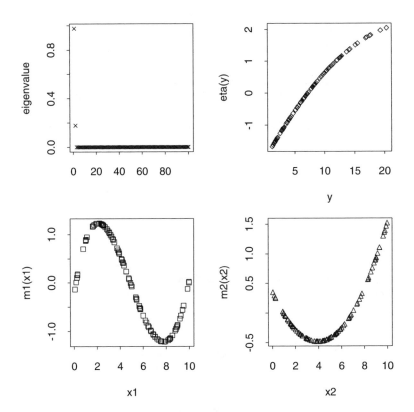

Figure 4.25 Results of the eigenvalue problem shown as eq(4.159) (page 273) for creating a regression equation with the form of eq(4.141) (page 270). Eigenvalues aligned in descending order (top left). $\{Y_i, \hat{\eta}(Y_i)\}$ (quadratic equation), (top right). $\{X_{i1}, \hat{m}_1(X_{i1})\}$(quartic equation), (bottom left). $\{X_{i2}, \hat{m}_2(X_{i2})\}$ (quadratic equation), (bottom right).

(3) A polynomial equation is fitted using the data $\{(Y_i, \hat{Y}_i^*)\}$. That is, we have

$$\hat{Y}_i^* = \tilde{\eta}(Y_i) + \epsilon_i' = \tilde{Y}_i^* + \epsilon_i', \tag{4.182}$$

where $\tilde{\eta}(\cdot)$ is a polynomial equation, and $\{\epsilon_i'\}$ are residuals.

(4) $\left\{ \dfrac{\tilde{Y}_i^*}{sd(\{\tilde{Y}_i^*\})} \right\}$ is defined as $\{Y_i^*\}$.

(5) The calculation from (2) through (4) is repeated. $\{Y_i^*\}$, which is obtained by kth iterative calculation, is defined as $\{Y_i^{*(k)}\}$. When the values of $\{Y_i^{*(k)}\}$ converge sufficiently, $\{Y_i^{*(k)}\}$ is defined as $\{\hat{\eta}(Y_i)\}$. An additive model is created using $\{(X_{i1}, X_{i2}, \hat{\eta}(Y_i))\}$ by the same method as in (2). $\{\hat{m}_1(X_{i1})\}$ and $\{\hat{m}_2(X_{i2})\}$ are calculated.

This procedure is called the ACE (Alternating Conditional Expectation) algorithm. The regression equation derived using the ACE algorithm is considered to be the same as that obtained by canonical correlation analysis. The proof is shown as follows.

First, in eq(4.181), a similar relationship to eq(4.154) (page 273) holds. That is,

$$\hat{\mathbf{y}}^* = \mathbf{H}_X \mathbf{y}^*, \tag{4.183}$$

where $\mathbf{y}^* = (Y_1^*, Y_2^*, \ldots, Y_n^*)^t$ and $\hat{\mathbf{y}}^* = (\hat{Y}_1^*, \hat{Y}_2^*, \ldots, \hat{Y}_n^*)^t$. $\mathbf{H}_X$ defined as below is a hat matrix corresponding to this regression.

$$\mathbf{H}_X = \mathbf{X}(\mathbf{X}^t\mathbf{X})^{-1}\mathbf{X}^t. \tag{4.184}$$

When both $\hat{m}_1(\cdot)$ and $\hat{m}_2(\cdot)$ are quadratic equations, this $\mathbf{X}$ is a matrix in which $[\mathbf{X}]_{i1} = X_{i1}$, $[\mathbf{X}]_{i2} = X_{i1}^2$, $[\mathbf{X}]_{i3} = X_{i2}$, and $[\mathbf{X}]_{i4} = X_{i2}^2$ $(1 \leq i \leq n)$. Furthermore, in eq(4.182), a similar equation to eq(4.157) (page 273) holds. That is,

$$\tilde{\mathbf{y}}^* = \mathbf{H}_Y \hat{\mathbf{y}}^*, \tag{4.185}$$

where $\tilde{\mathbf{y}}^* = (\tilde{Y}_1^*, \tilde{Y}_2^*, \ldots, \tilde{Y}_n^*)^t$. $\mathbf{H}_Y$ below is a hat matrix corresponding to this regression.

$$\mathbf{H}_Y = \mathbf{Y}(\mathbf{Y}^t\mathbf{Y})^{-1}\mathbf{Y}^t. \tag{4.186}$$

When $\tilde{\eta}(\cdot)$ is a quadratic equation, this $\mathbf{Y}$ is a matrix in which $[\mathbf{Y}]_{i1} = Y_i$, and $[\mathbf{Y}]_{i2} = Y_i^2$ $(1 \leq i \leq n)$.

Hence, $\mathbf{y}^{*(k)}$ $(= (Y_1^{*(k)}, Y_2^{*(k)}, \ldots, Y_n^{*(k)})^t)$ is written as

$$\mathbf{y}^{*(k+1)} = \frac{\sqrt{n}\mathbf{H}_Y\mathbf{H}_X\mathbf{y}^{*(k)}}{\| \mathbf{H}_Y\mathbf{H}_X\mathbf{y}^{*(k)} \|}, \tag{4.187}$$

where the equation below is used.

$$sd(\{\tilde{Y}_i^*\}) = \frac{\| \tilde{\mathbf{y}}^* \|}{\sqrt{n}} = \frac{\| \mathbf{H}_Y\mathbf{H}_X\mathbf{y}^* \|}{\sqrt{n}}. \tag{4.188}$$

If the average of $\{Y_i^*\}$ is 0, eq(2.46) (page 41) indicates that the average of $\{\hat{Y}_i^*\}$ becomes 0. In the same manner, if the average of $\{\hat{Y}_i^*\}$ is 0, the average of $\{\tilde{Y}_i^*\}$

becomes 0. Furthermore, if the average of $\{\tilde{Y}_i^*\}$ is 0, the average of $\{Y_i^*\}$ becomes 0. Hence, since the average of $\{\tilde{Y}_i^*\}$ is always 0, eq(4.188) holds.

Since $\mathbf{y}^{*(k)}$ is a vector with the length of $\sqrt{n}$ regardless of the value of k, no element of $\mathbf{y}^{*(k)}$ diverges when $k \to \infty$. By a calculation similar to eq(2.87) (page 58), $\mathbf{y}^{*(k+1)}$ is rewritten as

$$\mathbf{y}^{*(k+1)} = \frac{\sqrt{n} \sum_{i=1}^{n} \gamma_i \mathbf{u}_i < \mathbf{u}_i \cdot \mathbf{y}^{*(k)} >}{\| \sum_{i=1}^{n} \gamma_i \mathbf{u}_i < \mathbf{u}_i \cdot \mathbf{y}^{*(k)} >\|}, \tag{4.189}$$

where $\{\gamma_i\}$ $(1 \le i \le n)$ is an eigenvalue of $(\mathbf{H}_Y \mathbf{H}_X)$; all of the eigenvalues are real numbers. Each of $\{\mathbf{u}_i\}$ $(1 \le i \le n)$ is an eigenvector (length is $\sqrt{n}$) corresponding to γ_i. If $\mathbf{H}_Y \mathbf{H}_X$ is a quasi-definite matrix (positive semidefinite matrix, nonnegative definite matrix), by the process illustrated in figure 4.26, $\mathbf{y}^{*(k)}$ converges to

$$\mathbf{y}^{*(\infty)} = \frac{\sqrt{n} \mathbf{u}_1 < \mathbf{u}_1 \cdot \mathbf{y}^{*(\infty)} >}{\| \mathbf{u}_1 < \mathbf{u}_1 \cdot \mathbf{y}^{*(\infty)} >\|} = \mathbf{u}_1. \tag{4.190}$$

The vector of $\mathbf{u}_1$ is the eigenvector (length of $\sqrt{n}$) corresponding to the maximum eigenvalue of $\mathbf{H}_Y \mathbf{H}_X$. Consequently, since one more execution of $(\mathbf{H}_Y \mathbf{H}_X)$ to $\mathbf{y}^{*(\infty)}$ results in a vector whose direction is the same as that of $\mathbf{u}_1$, we obtain

$$\mathbf{H}_Y \mathbf{H}_X \mathbf{y}^{*(\infty)} = \gamma_{max} \mathbf{y}^{*(\infty)}. \tag{4.191}$$

Since this $\mathbf{y}^{*(\infty)}$ is identical to $\mathbf{u}_1$, γ_{max} in this equation is the maximum eigenvalue of $\mathbf{H}_Y \mathbf{H}_X$. Furthermore, $\mathbf{y}^{*(\infty)}$ is a vector with the length of $\sqrt{n}$. By comparing eq(4.159) (page 273) and eq(4.191), we conclude that $\mathbf{H}_X$ is considered to be $\mathbf{H}_B$, and $\mathbf{H}_Y$ is considered to be $\mathbf{H}_A$. While $\mathbf{H}_X$ is the hat matrix to transform predictors, $\mathbf{H}_B$ is the hat matrix to transform target variables; the same is true of $\mathbf{H}_Y$ and $\mathbf{H}_A$. However, the distinction between predictors and target variables is not essential. Therefore, one of the two first canonical correlation coefficients given by canonical correlation analysis is proved to be $\mathbf{y}^{*(\infty)}$ through this iterative calculation. Note that the similarity of the method above to the power method (e.g., page 357 of [10]) implies that the ACE algorithm provides the maximun eigenvalue and the corresponding eigenvector of a symmetric matrix.

Figure 4.27 shows the results of deriving $\{Y_i^{*(k)}\}$ $(k = 1, 2, 3, 4)$ by iterative calculation using the same form of the function and data as those in the cases of figure 4.24 and figure 4.25. It seems to converge at the step of $k = 1$. That is, $\{\hat{\eta}(Y_i)\}$ is obtained. This is confirmed by comparison with figure 4.25(top right).

Mathematical investigation and many actual examples show that the ACE algorithm generally converges solutions to the optimal ones in a similar manner when polynomial equations or the B-spline is used to obtain $\hat{m}_1(x_1)$, $\hat{m}_2(x_2)$, and $\eta(y)$, even when $\mathbf{H}_X$ and $\mathbf{H}_Y$ are not fixed. In fact, the first program with the name of ACE employs supersmoother (cf. page 191 of Chapter 3) [4]. Hence, the ACE algorithm with supersmoother is usually called ACE. Supersmoother has the advantages of being carried out quickly and of being robust to outliers. Furthermore, a great number of practical examples demonstrate that the ACE algorithm with supersmoother results in beneficial regression equations. Note that while continuous regression equations such as $\hat{m}_1(x_1)$, $\hat{m}_2(x_2)$, and $\hat{\eta}(y)$ are derived as well as $\{\hat{m}_1(X_{i1})\}$, $\{\hat{m}_2(X_{i2})\}$,

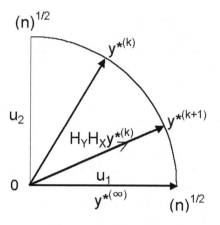

Figure 4.26   Illustration of how the ACE algorithm leads to eq(4.191). Since eq(4.187)(or eq(4.189)) guarantees that all $\mathbf{y}^{*(k)}$ are vectors with the length of $\sqrt{n}$, $\mathbf{y}^{*(k)}$ approaches $\mathbf{y}^{*(\infty)}$ as k becomes larger and $\mathbf{y}^{*(\infty)}$ satisfies eq(4.191).

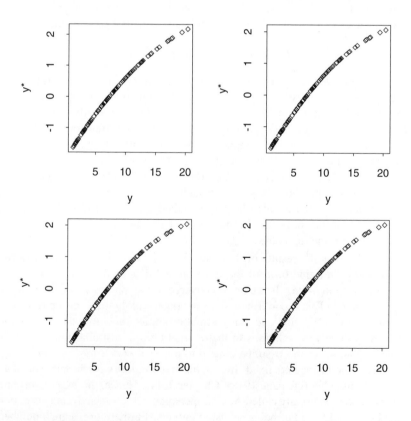

Figure 4.27 $\{(Y_i, Y_i^{*(k)})\}$ $(k = 1, 2, 3, 4)$ given by the ACE algorithm. $\{(Y_i, Y_i^{*(1)})\}$ (top left). $\{(Y_i, Y_i^{*(2)})\}$ (top right). $\{(Y_i, Y_i^{*(3)})\}$ (bottom left). $\{(Y_i, Y_i^{*(4)})\}$ (bottom right).

and $\{\hat{\eta}(Y_i)\}$ when polynomial equations or other regression equations are used as $\hat{m}_1(x_1)$, $\hat{m}_2(x_2)$, and $\hat{\eta}(y)$, the ACE algorithm with supersmoother requires interpolation to obtain continuous regression functions. Figure 4.28 displays the solutions of regression equations obtained by the ACE algorithm using the same data as those in figure 4.24 and figure 4.27. Moreover, though the number of predictors here is two, similar discussion is possible even when the number of predictors is larger.

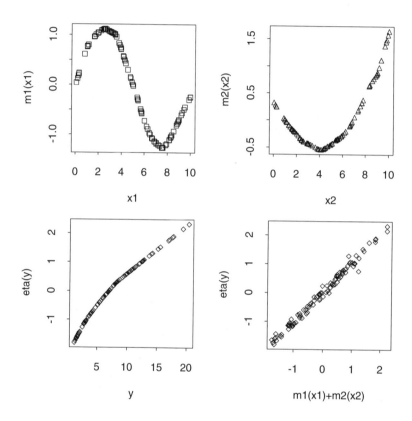

Figure 4.28 Regression equations given by the ACE algorithm. $\{X_{i1}, \hat{m}_1(X_{i1})\}$ (top left). $\{X_{i2}, \hat{m}_2(X_{i2})\}$ (top right). $\{Y_i, \hat{\eta}(Y_i)\}$ (bottom left). $\{(\hat{m}_1(X_{i1}) + \hat{m}_2(X_{i2})), \hat{\eta}(Y_i)\}$ (bottom right).

4.8 PROJECTION PURSUIT REGRESSION

In the additive model, we assume that the value of a target variable approximates the sum of the transformations of each predictor (x_j) by a smooth function. Although this is a flexible model in comparison with a simple multiple regression equation $(y = a_0 + \sum_{j=1}^{r} a_j x_j)$, it can be regarded as a model with somewhat strict limitations.

Let us discuss the use of the regression equation

$$m(x_1, x_2, \ldots, x_r) = \alpha_0 + \sum_{j=1}^{M} m_j \left(\sum_{k=1}^{r} \alpha_{kj} x_k \right), \qquad (4.192)$$

where $\alpha_0 = \frac{1}{n} \sum_{i=1}^{n} Y_i$ and $\{x_1, x_2, \ldots, x_r\}$ are predictors. $\{m_j(\cdot)\}$ $(1 \le j \le M)$ are smooth functions given by nonparametric regression. Furthermore, $\{\alpha_{kj}\}$ are constants that satisfy

$$\sum_{k=1}^{r} \alpha_{kj} \cdot \alpha_{kj} = 1 \qquad (1 \le j \le M). \qquad (4.193)$$

That is, each vector has the length of 1. This method is called projection pursuit regression. Since this method represents predictors on another orthogonal coordinate system, it is considered to be an extension of principal component regression. Principal component regression gives the following regression equation.

$$m(x_1, x_2, \ldots, x_r) = \alpha_0 + \sum_{j=1}^{M} \left(\beta_j \cdot \left(\sum_{k=1}^{r} \alpha_{kj} x_k \right) \right). \qquad (4.194)$$

$\boldsymbol{\alpha}_j \ (= (\alpha_{1j}, \alpha_{2j}, \ldots, \alpha_{rj})^t)$ is a vector that represents a transformation; principal component analysis yields these vectors.

$$\sum_{k=1}^{r} \alpha_{ki} \cdot \alpha_{kj} = \begin{cases} 1 & \text{if } i = j \\ 0 & \text{if } i \ne j. \end{cases} \qquad (4.195)$$

Vector $\boldsymbol{\alpha}_i \ (= (\alpha_{1i}, \alpha_{2i}, \ldots, \alpha_{ri})^t)$ is orthogonal to vector $\boldsymbol{\alpha}_j \ (= (\alpha_{1j}, \alpha_{2j}, \ldots, \alpha_{rj})^t)$ $(i \ne j)$, and the length of each vector is 1. Comparing eq(4.194) with eq(4.192), nonparametric regression functions $(\{m_j(\cdot)\})$ in eq(4.192) perform similarly to constants $(\{\beta_j\})$ in eq(4.194). This change generalizes a regression function.

Eq(4.192) is rewritten as

$$m(x_1, x_2, \ldots, x_r) = \alpha_0 + \sum_{j=1}^{M} \left(\beta_j \cdot \phi_j \left(\sum_{k=1}^{r} \alpha_{kj} x_k \right) \right). \qquad (4.196)$$

These $\{\beta_j\}$ and $\{\phi_j(\cdot)\}$ (smooth functions) satisfy

$$\sum_{i=1}^{n} \phi_j \left(\sum_{k=1}^{r} \alpha_{kj} X_{ik} \right) = 0 \quad (1 \le j \le M), \qquad (4.197)$$

$$\frac{1}{n} \sum_{i=1}^{n} \left(\phi_j \left(\sum_{k=1}^{r} \alpha_{kj} X_{ik} \right) \right)^2 = 1 \quad (1 \le j \le M). \qquad (4.198)$$

By obtaining $\{\beta_j\}$ and $\{\phi_j(x)\}$ to reduce the value below, a regression equation with the form of eq(4.192) is derived.

$$E_{pursuit} = \sum_{i=1}^{n} \left(Y_i - \alpha_0 - \sum_{j=1}^{M} \left(\beta_j \cdot \phi_j \left(\sum_{k=1}^{r} \alpha_{kj} X_{ik} \right) \right) \right)^2. \qquad (4.199)$$

A regression equation of this sort is obtained by the following procedure.
(1) $\{\alpha_{k1}\}$ $(1 \le k \le n)$ is optimized to realize the approximation below.

$$Y_i \approx \hat{\alpha}_0 + \beta_1 \cdot \phi_1\left(\sum_{k=1}^{r} \alpha_{k1} X_{ik}\right). \tag{4.200}$$

Here, $\hat{\alpha}_0$ is $\frac{1}{n}\sum_{i=1}^{n} Y_i$. Supersmoother (Chapter 3 (page 191)) is employed to obtain $\beta_1 \cdot \phi_1(\sum_{k=1}^{r} \alpha_{k1} X_{ik})$. Estimates given by the right-hand side of eq(4.200) are defined as $\{\hat{Y}_i\}$. The values of β_1 are adjusted to satisfy eq(4.198), and the corresponding $\phi_1(\cdot)$ is derived.
(2) Each $\{\alpha_{k2}\}$ $(1 \le k \le n)$ is optimized to realize the approximation below.

$$Y_i \approx \hat{\alpha}_0 + \hat{\beta}_1 \cdot \hat{\phi}_1\left(\sum_{k=1}^{r} \hat{\alpha}_{k1} X_{ik}\right) + \beta_2 \cdot \phi_2\left(\sum_{k=1}^{r} \alpha_{k2} X_{ik}\right). \tag{4.201}$$

Supersmoother is utilized to calculate $\beta_2 \cdot \phi_2(\sum_{k=1}^{r} \alpha_{k2} X_{ki})$. The value of β_2 is adjusted to satisfy eq(4.198), and the corresponding $\phi_2(\cdot)$ is derived.
(3) Repetition of a procedure similar to (2) yields the approximate expression

$$Y_i \approx \hat{\alpha}_0 + \sum_{j=1}^{M_{max}} \left(\hat{\beta}_j \cdot \hat{\phi}_j\left(\sum_{k=1}^{r} \hat{\alpha}_{kj} X_{ik}\right)\right). \tag{4.202}$$

(4) The term in which the absolute value of one $\{\hat{\beta}_j\}$ is the smallest is omitted in the regression equation on the right-hand side of eq(4.202). This procedure reduces the number of terms on the right-hand side of eq(4.202) by one. The values of $\{\hat{\beta}_j\}$, $\{\hat{\alpha}_{kj}\}$, and $\{\hat{\phi}_j(\cdot)\}$ are finely tuned to realize a good approximation in the regression equation obtained here.
(5) Repetition of the procedure similar to that described in (4) leads to the approximate expression

$$Y_i \approx \hat{\alpha}_0 + \sum_{j=1}^{M} \left(\hat{\beta}_j \cdot \hat{\phi}_j\left(\sum_{k=1}^{r} \hat{\alpha}_{kj} X_{ik}\right)\right), \tag{4.203}$$

where $M = M_{min}, M_{min} + 1, \ldots, M_{max}$ $(M_{min} < M_{max})$ is assumed.
(6) The optimal regression equation is chosen among $(M_{max} - M_{min} + 1)$ regression equations and used as a projection pursuit regression equation. To determine the optimal number of M, the residual sum of squares given by the difference between the values of a target variable of data and the estimates rendered by eq(4.196) is computed; the residual sum of squares is calculated for each value of M. Then, the value of M_0 is usually found: the residual sum of squares plunges when the value of M becomes M_0, and a slightly larger value of M does not greatly increase this value. This value of M_0 is considered to be the optimal M.

Figure 4.29 exemplifies the performance of projection pursuit regression using simple simulation data. The number of data here is 100, the values of predictors $(\{(X_{i1}, X_{i2})\}$ $(1 \le i \le 100))$ are uniform pseudorandom numbers that take values between 0 and 10. The values of a target value $(\{Y_i\})$ are

$$Y_i = \cos(0.6X_{i1} + 0.4X_{i2}) + 2 + \epsilon_i, \tag{4.204}$$

where $\{\epsilon_i\}$ are realizations of $N(0, 0.1^2)$ (normal distribution with mean 0 and variance 0.1^2). Calculation based on the algorithm above is carried out setting $M_{max} = 7$ and $M_{min} = 1$. For each value of M, the residual sum of squares is divided by the number of data, and the result is divided by the variance of a target variable of data; that is, the ratio of the variance which the regression equation does not represent to the variance of a target variable of data is calculated. Since the values that this example gives are 0.01853622, 0.01791926, and 0.01648599 for $M = 1, 2, 3$, respectively, $M = 1$ is determined to be optimal. The regression equation with $M = 1$ is

$$\hat{m}_1(z) = \hat{\alpha}_0 + \hat{\beta}_1 \cdot \hat{\phi}_1(z). \tag{4.205}$$

The corresponding z is

$$z = 0.8233234x_1 + 0.5675725x_2. \tag{4.206}$$

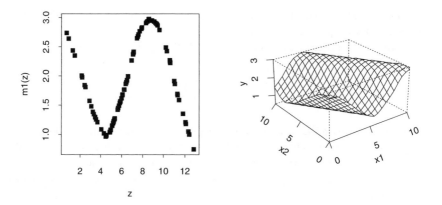

Figure 4.29   Example of a regression equation for projection pursuit regression ($\hat{m}(Z_i)$) ($Z_i = \sum_{k=1}^{2} \alpha_{k1} X_{ik}$, $\hat{m}_1(Z_i) = \hat{\alpha}_0 + \hat{\beta}_1 \cdot \hat{\phi}_1(Z_i)$) ($1 \le i \le n$) (left). Regression surface obtained by projection pursuit regression (right).

$\hat{m}_1(Z_i)$ is exhibited in figure 4.29(left). Z_i is represented as

$$Z_i = 0.8233234X_{i1} + 0.5675725X_{i2}. \tag{4.207}$$

Furthermore, figure 4.29(right) shows eq(4.205) as a function of x_1, x_2.

Projection pursuit in a more general sense means that by transforming the data ($\{(X_{i1}, X_{i2}, \ldots, X_{ir})\}$ ($1 \le i \le n$)) on the basis of $X'_{ij} = \sum_{k=1}^{r} \alpha_{kj} X_{ik}$ ($1 \le j \le s$, $1 \le i \le n$), the data ($\{(X'_{i1}, X'_{i2}, \ldots, X'_{is})\}$ ($1 \le i \le n$)) is obtained. Here, the transformed data should sufficiently represent the characteristics of the original data in the space with a smaller number of dimensions than that of the original data. That is, projection is conducted in the direction that enhances the interestingness (also called figure of merit). Projection with high interestingness in projection pursuit regression realizes nonparametric regression, which gives a small value of residuals. Projection pursuit without a target variable makes use of the discrepancy between the distribution obtained by the projection and the normal distribution as an index representing interestingness (projection index) [8]; other indices are also possible. Coordinate transformation using principal component analysis is considered to be a kind of projection pursuit without target variables.

4.9 EXAMPLES OF S-PLUS OBJECT

(A) Calculation of CV and GCV for local linear regression with two predictors
Object: llin2dcv()

```
function(h12, nd, xx1, xx2, yy)
{
#   (1)
    influence1 <- function(ii, xx1, xx2, yy, h1, h2)
    {
#   (2)
        cpx1 <- xx1[ii]
        cpx2 <- xx2[ii]
#   (3)
        wts <- exp(-0.5 * ((xx1 - cpx1)/h1)^2. - 0.5 *
        ((xx2 - cpx2)/h2)^2.)
#   (4)
        data1 <- data.frame(x1 = xx1 - cpx1, x2 = xx2 - cpx2,
        y = yy, www = wts)
        fit.lm <- lm(y ~ x1 + x2, data = data1,
        weights = www, x = T)
#   (5)
        res <- yy[ii] - fit.lm$coef[1.]
        names(res) <- NULL
        hat1 <- lm.influence(fit.lm)$hat[ii]
        return(res, hat1)
    }
#   (6)
    influence1.out <- lapply(as.list(c(1.:nd)),
    influence1, xx1 = xx1, xx2 = xx2, yy = yy, h1 =
    h12[1.], h2 = h12[2.])
#   (7)
    resdhat <- unlist(influence1.out)
    res <- resdhat[attr(resdhat, "names") == "res"]
    dhat <- resdhat[attr(resdhat, "names") == "hat1"]
#   (8)
    trhat <- sum(dhat)
    sse <- sum(res^2.)
    cv <- sum((res/(1. - dhat))^2.)/nd
    gcv <- sse/(nd * (1. - (trhat/nd))^2.)
#   (9)
    return(cv, gcv)
}
```

(1) The function `influence1()` is defined; this function yields the estimate at the position where `ii`-th data are located, and outputs the `ii`-th values of diagonal elements of a hat matrix and a corresponding residual. The objects of `xx1`, `xx2`, and `yy` are data. The objects of `h1` and `h2` are bandwidths for each direction.

(2) The values of predictors at the position where estimates are obtained are stored as `cpx1` and `cpx2`.

(3) The values of x_1^* (`cpx1`) and x_2^* (`cpx2`) are substituted into a kernel function (eq(4.5) (page 233)); the resulting kernel function is saved as `wts`.

(4) The data used to obtain estimates at the point defined by `cpx1` and `cpx2` are accumulated in a data frame called `data1`. Regression is carried out using `lm()` and the result is stored in `fit.lm` (a list object).

(5) The residual at the point defined by `cpx1` and `cpx2` is saved as `res`. The `ii`-th values of diagonal elements of a hat matrix are stored as `hat1`. The procedure `names(res) <- NULL` facilitates the handling of the list object which is outputted by `lapply()` ((6)). The objects `res` and `hat1` are output arguments of `influence1()`.

(6) The residuals at the positions where data are located and the values of diagonal elements of a hat matrix are obtained using `influence1()`.

(7) A vector of residuals and that of diagonal elements of a hat matrix are extracted from `influence1.out`; they are saved as `res` and `dhat`, respectively.

(8) The sum of the elements of `dhat` gives the sum of diagonal elements of a hat matrix (trace); it is stored as `trhat`. The residual sum of squares is saved as `sse`. The values of CV and GCV are calculated and defined as `cv` and `gcv`, respectively.

(9) The objects `cv` and `gcv` are outputted.

An example of an object to use `llin2dcv()` (construction of figure 4.1 (page 233))

```
function()
{
#   (1)
    nd <- 100
    set.seed(2525)
    xx1 <- runif(nd, min = 0, max = 10)
    xx2 <- runif(nd, min = 0, max = 10)
    yy <- (xx1 - 4)^2 + (xx2 - 6)^2 + rnorm(nd,
      mean = 0, sd = 3)
#   (2)
    ntrialx1 <- 10
    ntrialx2 <- 10
    hx1 <- seq(from = 0.5, to = 1.4, length = ntrialx1)
    hx2 <- seq(from = 0.5, to = 1.4, length = ntrialx2)
    hx12 <- expand.grid(hx1, hx2)
    listhx12 <- list(unlist(hx12[1, ]))
    for(ii in 2:(ntrialx1 * ntrialx2)) {
        listhx12[[ii]] <- unlist(hx12[ii, ])
    }
#   (3)
    cvgcv <- lapply(listhx12, llin2dcv, nd = nd,
```

```
    xx1 = xx1, xx2 = xx2, yy = yy)
    cvgcv <- unlist(cvgcv)
    cvt <- cvgcv[attr(cvgcv, "names") == "cv"]
    gcvt <- cvgcv[attr(cvgcv, "names") == "gcv"]
#   (4)
    gcvmin <- min(gcvt)
    igcvmin <- (1:(ntrialx1 * ntrialx2))[gcvt == gcvmin]
    print("optimal bandwidth")
    print(hx12[igcvmin, 1])
    print(hx12[igcvmin, 2])
#   (5)
    par(mfrow = c(1, 2), mai = c(2.5, 1.5, 2.5, 0.5),
      oma = c(1, 1, 1, 1))
    contour(hx1, hx2, matrix(cvt, nrow = ntrialx1), v =
      seq(from = 0, to = 30, by = 0.5), xlab = "h1",
      ylab = "h2", main = "CV")
    contour(hx1, hx2, matrix(gcvt, nrow = ntrialx1), v =
      seq(from = 0, to = 30, by = 0.5), xlab = "h1",
      ylab = "h2", main = "GCV")
}
```

(1) Simulation data are generated. The object nd is the number of data. The objects xx1 and xx2 are the values of predictors. The object yy denotes values of a target variable.

(2) The values of ntrialx1 (number of bandwidths in the direction x_1) and ntrialx2 (number of bandwidths in the direction x_2) are given. Vectors (hx1 and hx2) that store the bandwidth of each direction are created. All combinations of the elements of hx1 and those of hx2 are defined as hx12 (a data frame that consists of two vectors). A list object (listhx12) is created to store all combinations of bandwidths (saved in hx12) as components.

(3) Using llin2dcv(), CV and GCV for each value of bandwidth are calculated and named cvgcv. CV stored in cvgcv is saved as cvt, and GCV stored in the same object is saved as gcvt.

(4) Bandwidths (hx12[igcvmin, 1], hx12[igcvmin, 2]) that minimize GCV (gcvmin) are derived and desplayed on a screen.

(5) A contour is drawn to exhibit the relationship between the bandwidth and CV. The contour for displaying the relationship between the bandwidth and GCV is illustrated.

(B) Local linear regression with two predictors
Object: llin2d()

```
function(ex, nd, xx1, xx2, yy, hx1, hx2)
{
#   (1)
    wts <- exp(-0.5 * ((xx1 - ex[1.])/hx1)^2. - 0.5 *
      ((xx2 - ex[2.])/hx2)^2.)
#   (2)
    data1 <- data.frame(x1 = xx1 - ex[1.], x2 = xx2 - ex[2.],
```

```
      y = yy, www = wts)
#   (3)
      fit.lm <- lm(y ~ x1 + x2, data = data1, weights = www)
#   (4)
      est <- fit.lm$coef[1.]
#   (5)
      return(est)
}
```

(1) Gaussian weights for local linear equations are saved as wts. The objects ex[1] and ex[2] indicate the position of the estimation point; ex[1] is the value on the x_1-axis and ex[2] is the value on the x_2-axis.

(2) Data for local linear regression are accumulated in a data frame (data1).

(3) Local linear regression is carried out and the result is stored as fit.lm.

(4) Resultant estimates (constant terms of linear equations) are saved as est.

(5) The object est is outputted.

An example of an object to use llin2d() (construction of figure 4.2 (page 234))

```
function()
{
#   (1)
      nd <- 100.
      set.seed(2525.)
      xx1 <- runif(nd, min = 0., max = 10.)
      xx2 <- runif(nd, min = 0., max = 10.)
      yy <- (xx1 - 4.)^2. + (xx2 - 6.)^2. + rnorm(nd,
        mean = 0., sd = 3.)
#   (2)
      hx1 <- 0.7
      hx2 <- 0.8
#   (3)
      nex1 <- 11.
      gridx1 <- seq(from = 0., to = 10., length = nex1)
      nex2 <- 21.
      gridx2 <- seq(from = 0., to = 10., length = nex2)
      grid <- expand.grid(gridx1, gridx2)
#   (4)
      ey <- apply(grid, 1., llin2d, nd = nd, xx1 = xx1,
        xx2 = xx2, yy = yy, hx1 = hx1, hx2 = hx2)
#   (5)
      par(mfrow = c(1., 2.), mai = c(2.5, 1.5, 2.5, 0.5),
        oma = c(1., 1., 1., 1.))
      contour(gridx1, gridx2, matrix(ey, nrow = nex1), v =
        seq(from = 0., to = 100., by = 5.), xlab = "x1",
        ylab = "x2")
#   (6)
```

```
    xyz.in <- interp(xx1, xx2, yy)
    contour(xyz.in, v = seq(from = 0., to = 100., by = 5.),
      xlab = "x1", ylab = "x2")
}
```

(1) Simulation data are generated. The object nd is the number of data. The objects xx1 and xx2 are values of predictors. The object yy denotes values of a target variable.
(2) The bandwidth of the x_1-direction is named hx1, and the bandwidth of the x_2-direction is named hx2.
(3) The coordinates of estimation points are accumulated in a data frame (grid) (the first object of the data frame is the vector representing the values on the x_1-axis, the second object of the data frame is the vector representing the values on the x_2-axis). The number of grid points on the x_1 axis is defined as nex1, and that on the x_2-axis is defined as nex2; the coordinates of the grid points are named gridx1 and gridx2, respectively.
(4) Using llin2d(), estimates at each grid point are calculated and stored as ey.
(5) A contour is drawn using ey.
(6) Interpolation is carried out by Akima's method to illustrate a contour of estimates.

(C) Smoothing using the thin plate smoothing splines for data with two predictors (dtpss() ([1]; also refer to Chapter 3 (page 211)) is utilized. Construction of figure 4.3 (page 236))

```
function()
{
#    (1)
    nobs <- 100.
    set.seed(2525.)
    xx1 <- runif(nobs, min = 0., max = 10.)
    xx2 <- runif(nobs, min = 0., max = 10.)
    yy <- (xx1 - 4.)^2. + (xx2 - 6.)^2. + rnorm(nobs, mean = 0.,
      sd = 3.)
#    (2)
    des <- cbind(xx1, xx2)
    y <- yy
#    (3)
    nex1 <- 11.
    gridx1 <- seq(from = 0., to = 10., length = nex1)
    nex2 <- 21.
    gridx2 <- seq(from = 0., to = 10., length = nex2)
    grid <- expand.grid(gridx1, gridx2)
#    (4)
    ntbl <- 10.
    lamlim <- c(-2., -0.2)
#    (5)
    m <- 2.
    job <- 1110.
    dim <- 2.
```

```
    maxobs <- nobs
    maxtbl <- ntbl
    mxncts <- m + dim - 1.
    maxuni <- nobs
    maxpar <- maxuni + mxncts
    lwa <- maxuni * (2. + mxncts + maxuni) + maxobs
    liwa <- 2. * maxobs + maxuni
    info <- 0.
    iwork <- rep(0., liwa)
    iout <- rep(0., 4.)
    adiag <- rep(0., maxobs)
    tbl2 <- rep(0., (maxtbl * 3.))
    tbl <- matrix(tbl2, ncol = 3.)
    coef <- rep(0., maxpar)
    auxtbl <- matrix(rep(0., length = 9.), ncol = 3.)
    svals <- rep(0., maxpar)
    dout <- rep(0., 5.)
    work <- rep(0., lwa)
    lddes <- maxobs
    lds <- maxobs
    ncov <- 0.
    ldtbl <- maxtbl
    s <- rep(0., nobs)
#   (6)
    fit <- dtpss(des, lddes, nobs, dim, m, s, lds, ncov,
      y, ntbl, adiag, lamlim, dout, iout, coef, svals, tbl,
      ldtbl, auxtbl, work, lwa, iwork, liwa, job, info)
#   (7)
    auxtbl <- fit$auxtbl
#   (8)
    print("info=", quote = F)
    print(fit$info)
#   (9)
    estimation1 <- function(ex, coef, m1, nd, xd1, xd2)
    {
        ex1 <- ex[1.]
        ex2 <- ex[2.]
        const <- ((-1.)^(m1) * 2.^(1. - 2. * m1))/pi
          /gamma(m1)^2.
        nobsm <- nd + m1 + 1.
        ti <- sqrt((ex1 - xd1)^2. + (ex2 - xd2)^2.)
        green <- const * ti^(2. * m1 - 2.) * logb(ti)
        green[is.na(green) == T] <- 0.
        est <- sum(coef[(m1 + 2.):nobsm] * green)
        est <- est + coef[1.]
        coefp <- 1.
        for(j in 1.:(m1 - 1.)) {
```

```
         for(k in 0.:j) {
             coefp <- coefp + 1.
             est <- est + ex1^(j - k) * ex2^(k)
             * coef[coefp]
         }
     }
     return(est)
  }
# (10)
  ey <- apply(grid, 1., estimation1, coef = fit$coef,
    m1 = m, nd = nobs, xd1 = xx1, xd2 = xx2)
# (11)
  par(mfrow = c(1., 2.), mai = c(2.5, 1.5, 2.5, 0.5),
    oma = c(1., 1., 1., 1.))
  tblm <- matrix(fit$tbl, ncol = 3.)
  spara <- tblm[, 1.]
  gcv <- tblm[, 2.]
  plot(spara, gcv, type = "n", xlab =
    "log10(smoothing parameter)", ylab = "GCV")
  points(spara, gcv, pch = 3.)
  lines(spara, gcv, lwd = 2.)
  sparagcv <- auxtbl[1., 1.]
  gcvmin <- auxtbl[1., 2.]
  points(sparagcv, gcvmin, pch = 15., cex = 1.2)
# (12)
  grid.es <- NULL
  grid.es$x <- gridx1
  grid.es$y <- gridx2
  grid.es$z <- matrix(ey, ncol = nex2)
  contour(grid.es, xlab = "x1", ylab = "x2",
    v = seq(from = 0., to = 100., by = 5.))
  invisible()
}
```

(1) Simulation data are generated. The object nobs is the number of data.

(2) The form of data is adjusted for the use of dtpss(). That is, the two vectors (xx1 and xx2) are combined to create a matrix (des). Furthermore, the vector (yy) is copied to y.

(3) The coordinates of the estimation point are given. The number nex1 is the number of grid points on the x_1-direction. The object nex2 is the number of grid points in the x_2-direction. The object gridx1 is the coordinate along the x_1-direction. The object gridx2 is the coordinate along the x_2-direction. All combinations of these two coordinates are accumulated in a data frame (grid). The variable name of the first variable in grid becomes Var1 (a vector indicating the coordinates along the x_1-direction), and that of the second variable becomes Var2 (a vector indicating the coordinates along the x_2-direction).

(4) We attempt all `ntbl` (10 in this example) values of a smoothing parameter. When `c(-2, -0.2)` is used as `lamlim`, the values of a smoothing parameter become $\{10^{-2}, 10^{-1.8}, \ldots, 10^{-0.2}\}$.

(5) The values of argument for `dtpss()` are given. When `dim` is set at 2, smoothing data with two predictors is performed. For the other arguments, refer to Chapter 3 (page 212).

(6) Using `dtpss()`, calculation of the thin plate smoothing splines using dtpss in GCVPACK is carried out and the results are saved in `fit`.

(7) The object `fit$auxtbl` is extracted from `fit` and defined as `auxtbl`.

(8) The object `fit$info` stored in `fit` is displayed. For the meanings of `fit$info`, refer to Chapter 3 (page 215).

(9) The function `estimation1()` is defined; this function yields estimates using eq(4.11) (page 236) and eq(4.12) (page 236). The object `ex`, which is an argument of `estimation1()` and consists of two elements, saves the coordinates of an estimation point $((x_1, x_2))$.

(10) The function `estimation1()` yields estimates corresponding to all values of coordinates stored in `grid`. The regression coefficients (`fit$coef`) used here are given by the smoothing parameter that is optimized by the golden ratio search.

(11) The values of GCV corresponding to each value of a smoothing parameter (given in (4)) are illustrated. The value of a smoothing parameter is optimized by the golden ratio search based on GCV. The resulting value of the smoothing parameter is transformed by logarithmic transformation base 10 to be stored as `auxtbl[1, 1]`; the corresponding value of GCV is saved as `auxtbl[1, 2]`. These two values are named `sparagcv` and `gcvmin`, respectively. Furthermore, the point indicated by `sparagcv` and `gcvmin` is displayed on a graph.

(12) A contour is charted to illustrate estimates (obtained in (9)) given using the optimal value of a smoothing parameter.

Note that the following three points should be taken into consideration when both values of predictors are tied in some data.

(i) The values of the first predictor of data are aligned in ascending order. For example, the data are preprocessed as below.

```
ord1 <- order(xx1)
xx1 <- xx1[ord1]
xx2 <- xx2[ord1]
yy <- yy[ord1]
```

Even when the two values of predictors are never tied in any data, this preprocessing does no harm.

(ii) The value of `job`, which is one of the arguments of `dptss()`, is set at 1111 instead of 1110. Even if the two values of predictors are never tied in any data, this value can be 1111.

(iii) Part (9) in the S-Plus object above is replaced with

```
#   (9)'
    estimation1 <- function(ex, coef, m1, nd, xd1, xd2)
    {
        ex1 <- ex[1.]
```

```
    ex2 <- ex[2.]
    dataxx <- data.frame(xx1 = xd1, xx2 = xd2)
    u1 <- unique.data.frame(dataxx)
    xd1u <- u1[[1]]
    xd2u <- u1[[2]]
    ndu <- length(xd1u)
    ex1 <- ex[1.]
    ex2 <- ex[2.]
    const <- ((-1.)^(m1) * 2.^(1. - 2.
     * m1))/pi/gamma(m1)^2.
    nobsm <- ndu + m1 + 1.
    ti <- sqrt((ex1 - xd1u)^2. + (ex2 - xd2u)^2.)
    green <- const * ti^(2. * m1 - 2.) * logb(ti)
    green[is.na(green) == T] <- 0.
    est <- sum(coef[(m1 + 2):nobsm] * green)
    est <- est + coef[1.]
    coefp <- 1.
    for(j in 1.:(m1 - 1.)) {
        for(k in 0.:j) {
            coefp <- coefp + 1.
            est <- est + ex1^(j - k) * ex2^(k) *
            coef[coefp]
        }
    }
    return(est)
}
```

The objects xd1 and xd2 give the coordinates in a two-dimensional space. If both values of predictors are tied at these points, a procedure is performed to leave one data among tied data and discard the others; unique.data.frame() is used for this procedure. Part (9)' can be used even when the data do not contain tied data.

(D) Smoothing by LOESS with two predictors (usage of loess(). Construction of figure 4.4 (page 238))

```
function()
{
#   (1)
    nd <- 100.
    set.seed(2525.)
    xx1 <- runif(nd, min = 0., max = 10.)
    xx2 <- runif(nd, min = 0., max = 10.)
    yy <- (xx1 - 4.)^2. + (xx2 - 6.)^2. + rnorm(nd,
      mean = 0., sd = 3.)
#   (2)
    nex1 <- 11.
    nex2 <- 11.
    gridx1 <- seq(from = 0., by = 1., length = nex1)
```

```
      gridx2 <- seq(from = 0., by = 1., length = nex2)
      grid <- expand.grid(gridx1, gridx2)
      assign("data1", data.frame(x1 = xx1, x2 = xx2, y1 = yy),
        frame = 1.)
#   (3)
      par(mfrow = c(1., 2.), mai = c(2.5, 1.5, 2.5, 0.5),
        oma = c(1., 1., 1., 1.))
#   (4)
      sp1 <- 0.1
      fit.lo <- loess(y1 ~ x1 + x2, data = data1,
        span = sp1, family = "gaussian", degree = 1.,
        surface = "direct")
      data2 <- data.frame(x1 = grid[, 1.], x2 = grid[, 2.])
      ey <- predict.loess(fit.lo, newdata = data2)
#   (5)
      grid.es <- NULL
      grid.es$x <- gridx1
      grid.es$y <- gridx2
      grid.es$z <- matrix(ey, nrow = nex1)
      contour(grid.es, v = seq(from = 0., to = 100., by = 5.),
        xlab = "x1", ylab = "x2")
#   (6)
      sp1 <- 0.2
      fit.lo <- loess(y1 ~ x1 + x2, data = data1,
        span = sp1, family = "gaussian", degree = 1.,
        surface = "direct")
      data2 <- data.frame(x1 = grid[, 1.], x2 = grid[, 2.])
      ey <- predict.loess(fit.lo, newdata = data2)
      print(ey)
      grid.es <- NULL
      grid.es$x <- gridx1
      grid.es$y <- gridx2
      grid.es$z <- matrix(ey, nrow = nex1)
      contour(grid.es, v = seq(from = 0., to = 100., by = 5.),
        xlab = "x1", ylab = "x2")
      invisible()
}
```

(1) Data are simulated.

(2) The coordinates of estimation points are stored as grid. Data are stored in the data frame (data1).

(3) The area in which to draw graphs is set.

(4) The value of span (sp1) is given and loess() is performed; the results are stored in fit.lo. If surface = "direct" is assigned, the estimates by LOESS are carried out without any modification. If surface = "direct" is not assigned or surface = "interpolate" is assigned, LOESS produces a small number of estimates and the remaining estimates are obtained by interpolation. In this case,

estimates at points where extrapolation is required are not derived. Using the resulting `fit.lo`, estimates at the coordinates supplied by `data2` (grid points in this example) are obtained and stored as `ey`.

(5) The coordinates of estimation points and corresponding estimates are saved as `grid.es` and they chart a contour of estimates.

(6) The object `loess()` is executed using a different value of span in a manner similar to that of (4)-(5), and the contour of estimates is illustrated.

(E) Ordinary kriging with one predictor
Object: `mykrig1()`

```
function(nd, ne, xx, yy, ex, bw)
{
#   (1)
    full <- matrix(0., nd, nd)
    full[lower.tri(full)] <- dist(xx)
    dist1 <- full + t(full)
    cc <- exp( - dist1^2./bw^2.)
#   (2)
    dvector <- function(exs, xx, bw)
    {
        dv <- exp( - (exs - xx)^2./bw^2.)
        return(dv)
    }
    dmat <- apply(matrix(ex, nrow = 1.), 2., dvector, xx = xx,
      bw = bw)
#   (3)
    ccr <- solve(cc)
    ccs <- sum(apply(ccr, 2., sum))
    chi <- (-2. * (1. - apply((ccr %*% dmat), 2., sum)))/ccs
    chimat <- matrix(rep(chi, times = nd), ncol = ne,
      byrow = T)
#   (4)
    ww <- ccr %*% (dmat - 0.5 * chimat)
    ey <- t(ww) %*% yy
#   (5)
    return(ey)
}
```

(1) C defined by eq(4.19) (page 239) is derived and saved as `cc`. The object `dist()` gives a vector whose elements are distances between all combinations of two elements (except the identical elements) of `xx`. When `xx` is a matrix, `dist()` yields distances between column vectors constituting `xx`. If a specific assignment is not set out, euclidean distance is supplied. Other definitions of distance can be used. Furthermore, the matrix that `lower.tri(full)` gives has the same size as that of `full`; the elements that are located under diagonal elements (diagonal elements are not included) are T, and other elements (i.e., elements over diagonal elements; diagonal elements are included) are F. If T is to be included as the diagonal elements,

`lower.tri(full, diag=T)` is designated. Then, eq(4.54) (page 245) is employed for the theoretical correlogram, and `bw` is used as δ and `cc` is derived.

(2) $\mathbf{d}_0 \ (= (d_1, d_2, \ldots, d_n)^t)$ (eq(4.21) (page 239)) is calculated at `ex` (values of coordinates of estimation points), and the resultant column vectors are combined to construct `dmat`.

(3) Using `cc` obtained in (1), χ defined by eq(4.45) (page 243) are derived as `chi`. The object `chi` (a column vector), whose elements correspond to each elment of `ex`, is combined to create a matrix (`chimat`).

(4) The object **w** defined in eq(4.44) (page 243) is calculated and a matrix (`ww`), whose column vector is **w**, is constructed. Using this `ww`, estimates (`ey`) corresponding to `ex` are computed. This `ww <- ccr %*% (dmat - 0.5 * chimat)` is replaced with `ww <- ccr %*% dmat`; simple kriging is realized. Hence, simple kriging does not need the derivation of `chimat`. Furthermore, `ey <- t(ww) %*% yy` can be substituted by `ey <- crossprod(ww, yy)`. This substitution saves computational cost.

(5) The object `ey` is outputted.

An example of an object to use `mykrig1()` (construction of figure 4.5 (page 246))

```
function()
{
#   (1)
    nd <- 20.
    set.seed(100.)
    xx <- seq(from = 1., by = 1., length = nd)
    yy <- sin(0.1 * pi * xx) + 1. +
     rnorm(nd, mean = 0., sd = 0.5)
#   (2)
    ne <- 96.
    ex <- seq(from = 1., by = 0.2, length = ne)
    band <- c(0.3, 1., 3.)
#   (3)
    par(mfrow = c(2., 2.), mai = c(1., 1.5, 0.1, 0.5),
     oma = c(1.5, 1.5, 1.5, 1.5))
#   (4)
    for(jj in 1.:3.) {
#   (5)
        bw <- band[jj]
        ey <- mykrig1(nd, ne, xx, yy, ex, bw)
#   (6)
        plot(xx, yy, type = "n", xlab = "x",
         ylab = "y", ylim = c(-1., 3.))
        points(xx, yy, cex = 1.2, pch = 2.)
        lines(ex, ey, lwd = 2.)
    }
#   (7)
    exy <- spline(xx, yy, n = ne, boundary = 0.)
```

```
      ey <- exy$y
#    (8)
      plot(xx, yy, type = "n", xlab = "x",
       ylab = "y", ylim = c(-1., 3.))
      points(xx, yy, cex = 1.2, pch = 2.)
      lines(ex, ey, lwd = 2.)
}
```

(1) Simulation data are generated.

(2) The coordinates of estimation points are saved as ex. The bandwidths to be examined (band) are set at three values $(0.3, 1, 3)$.

(3) The area in which to draw graphs is set.

(4) Estimates are obtained using three bandwidths defined by band and are exhibited in a graph.

(5) One element of band is defined as bw, and estimates are derived by mykrig1(); the estimates are stored as ey.

(6) Data and estimates are illustrated in a graph.

(7) Interpolation by cubic spline is carrid out using spline(); the result is saved in exy. If boundary = 0 is designated in spline(), the natural spline is materialized. Estimates are extracted from exy and stored as ey.

(8) Estimates derived in (7) and data are displayed on a graph.

(F) Smoothing by simple kriging with one predictor
Object: mykrigs()

```
function(nd, ne, xx, yy, ex, bw, sig2)
{
#    (1)
      full <- matrix(0., nd, nd)
      full[lower.tri(full)] <- dist(xx)
      dist1 <- full + t(full)
      cc <- exp( - dist1^2./bw^2.)
#    (2)
      dvector <- function(exs, xx, bw)
      {
           dv <- exp( - (exs - xx)^2./bw^2.)
           return(dv)
      }
      dmat <- apply(matrix(ex, nrow = 1.),
       2., dvector, xx = xx, bw = bw)
#    (3)
      ccr <- solve(cc + sig2 * diag(nd))
      ccs <- sum(apply(ccr, 2., sum))
#    (4)
      ww <- ccr %*% dmat
      ey <- t(ww) %*% yy
#    (5)
      return(ey)
```

```
}
```

The object sig2 (σ^2 in eq(4.63) (page 248); it is 0.25 in figure 4.7 (page 249)) is included in input arguments and calculation based on eq(4.63) (page 248) is implemented in (3), thereby, smoothing instead of interpolation is realized. The remaining procedures are the same as those of mykrig1() in which the parts concerning chimat are omitted.

(G) Universal kriging with one predictor (a rough trend is represented by a polynomial equation)
Object: mykrig2()

```
function(nd, ne, xx, yy, ex, bw, np)
{
#   (1)
    full <- matrix(0., nd, nd)
    full[lower.tri(full)] <- dist(xx)
    dist1 <- full + t(full)
    cc <- exp( - dist1^2./bw^2.)
#   (2)
    ll <- t(chol(cc))
    llr <- solve(ll)
#   (3)
    powerf <- function(jj, x1)
    {
        pw <- x1^jj
        return(pw)
    }
    gg <- apply(matrix(c(0.:(np - 1.)), nrow = 1.), 2., powerf,
      x1 = xx)
    aa <- qr(llr %*% gg)
    qq <- qr.Q(aa)
    rr <- qr.R(aa)
#   (4)
    bb <- t(qq) %*% llr %*% yy
    beta1 <- solve(rr[1.:np,   ], bb[1.:np])
#   (5)
    dvector <- function(exs, xx, bw)
    {
        dv <- exp( - (exs - xx)^2./bw^2.)
        return(dv)
    }
    dmat <- apply(matrix(ex, nrow = 1.), 2., dvector, xx = xx,
      bw = bw)
#   (6)
    ccr <- solve(cc)
    almat <- ccr %*% dmat
#   (7)
```

```
    mm <- as.vector(beta1 %*% t(gg))
    exmat <- t(apply(matrix(c(0.:(np - 1.)), nrow = 1.), 2.,
      powerf, x1 = ex))
    ey <- beta1 %*% exmat + as.vector(t(almat) %*% (yy - mm))
#   (8)
    return(ey)
}
```

(1) **C**, which is defined by eq(4.19) (page 239), is calculated and saved as cc. Eq(4.54) (page 245) is employed as a theoretical correlogram (correlogram model), and bw is used as δ.

(2) **L** defined in eq(4.77) (page 251) is obtained and is set as 11. Its inverse matrix is named 11r. Since a matrix given by chol() is $\mathbf{L}^t$ in eq(4.77), t() must be executed to obtain **L**.

(3) **G** defined by eq(4.80) (page 251) is named gg, and $\mathbf{L}^{-1}\mathbf{G}$ is set as aa. The object aa is decomposed by QR decomposition. The resultant **Q** is defined as qq, and **R** is named rr. Since this np indicates the number of coefficients of a polynomial equation, it is 4 when a cubic equation is adopted.

(4) $\mathbf{Q}^t\mathbf{L}^{-1}$ in eq(4.82) (page 251) is defined as bb. $[\beta]_{(1:P)}$ obtained by solving eq(4.82) is saved as beta1.

(5) The vector of $\mathbf{d}_0$ $(= (d_1, d_2, \ldots, d_n)^t)$ corresponding to each element of ex (values of coordinates of estimation points) is derived. These column vectors are combined to construct a matrix (dmat).

(6) α $(= (\alpha_1, \alpha_2, \ldots, \alpha_n)^t)$ (eq(4.28) (page 240)) to calculate each estimate are used as column vectors to create a matrix (almat).

(7) Estimates are derived using eq(4.83) (page 251). The object mm indicates $\{m_i\}$ defined in eq(4.70) (page 250).

(8) The object ey is outputted.

The object mykrig2() is used in the same manner as that of mykrig1(); however, the input arguments contain np (degree of polynomial equation plus 1). Hence, an object to utilize mykrig2() is identical to that for mykrig1() except that mykrig2() requires a specific value (4 is adopted in figure 4.8 (page 253)) for np.

(H) Universal kriging with two predictors (a rough trend is represented by a polynomial equation) (usage of fit.gls() and prmat(); construction of figure 4.10 (page 255))

```
function()
{
#   (1)
    library(spatial)
#   (2)
    xx1 <- c(1., 2., 3., 4., 5., 6., 7., 8., 9., 10.)
    xx2 <- c(7., 5., 4., 2., 9., 5., 3., 4., 2., 9.)
    yy <- c(2., 7., 4., 3., 5., 2., 9., 4., 1., 8.)
#   (3)
    par(mfrow = c(2., 2.), mai = c(1., 1., 0.5, 0.5),
      oma = c(1., 1., 1., 1.))
```

```
#  (4)
    fit.gls <- surf.gls(np = 2., gaucov, x = xx1, y = xx2,
    z = yy, d = 0.2)
#  (5)
    krig <- prmat(fit.gls, 1., 10., 1., 10., 19.)
#  (6)
    persp(krig, zlim = c(0., 12.), xlab = "x1", ylab = "x2",
    zlab = "y", lab = c(3., 3., 3.))
#  (7)
    fit.gls <- surf.gls(np = 2., gaucov, x = xx1, y = xx2,
    z = yy, d = 0.5)
    krig <- prmat(fit.gls, 1., 10., 1., 10., 19.)
    persp(krig, zlim = c(0., 12.), xlab = "x1", ylab = "x2",
    zlab = "y", lab = c(3., 3., 3.))
    fit.gls <- surf.gls(np = 2., gaucov, x = xx1, y = xx2,
    z = yy, d = 1.)
    krig <- prmat(fit.gls, 1., 10., 1., 10., 19.)
    persp(krig, zlim = c(0., 12.), xlab = "x1", ylab = "x2",
    zlab = "y", lab = c(3., 3., 3.))
    fit.gls <- surf.gls(np = 2., gaucov, x = xx1, y = xx2,
    z = yy, d = 2.)
    krig <- prmat(fit.gls, 1., 10., 1., 10., 19.)
    persp(krig, zlim = c(0., 12.), xlab = "x1", ylab = "x2",
    zlab = "y", lab = c(3., 3., 3.))
}
```

(1) To materialize universal kriging with two predictors (a rough trend is depicted by a polynomial equation), the library "spatial," which is treated in [13], [14], and [15], is used. The library "spatial" is included in S-Plus 6.0J by default.

(2) Data are given.

(3) The area in which to draw graphs is set out.

(4) Eq(4.54) (page 245) is employed as a theoretical correlogram. The value of δ is set at 0.2 and calculation of universal kriging is carried out; the result is saved in fit.gls. The object np in surf.gls() indicates the degree of the polynomial equation; its meaning is different from np (the number of coefficients in a regression equation with one predictor) in mykrig2().

(5) Using fit.gls, estimates at grid points (20×20), obtained by dividing the regions equally into 19 parts between 1 and 10 along the x_1-axis and x_2-axis, are determined; the results are saved as krig.

(6) The values in krig are charted as a perspective plot.

(7) In a similar fashion to (4), (5), and (6) above, δ for a theoretical correlogram (eq(4.54)) is altered among 0.5, 1, and 2 to calculate estimates; a perspective plot of the estimates is illustrated.

(I) Universal kriging with two predictors (a rough trend is represented by a polynomial equation; the broadness of the theoretical correlogram is determined by reference to data) (use of fit.gls() and prmat(); construction of figure 4.12(top left) (page 257), figure 4.12 (top right), and figure 4.12 (bottom left))

```
function()
{
#   (1)
    library(spatial)
#   (2)
    dd <- 0.8
#   (3)
    nd <- 100.
    set.seed(199.)
    xx1 <- runif(nd, min = 0., max = 10.)
    xx2 <- runif(nd, min = 0., max = 10.)
    yy <- 8. + 5. * sin(xx1^2./10.) + 3. * cos(xx2 * 2.) +
      3. * sin(xx2 * 2.5) + xx2 * 0.6
#   (4)
    par(mfrow = c(2., 2.), mai = c(1, 1, 1, 1),
      oma = c(1., 1., 1., 1.))
#   (5)
    fit.gls <- surf.gls(1., gaucov, x = xx1, y = xx2,
      z = yy, d = dd)
#   (6)
    cor1 <- correlogram(fit.gls, nint = 40., plotit = F)
    plot(cor1$x, cor1$y, type = "n", xlab = "distance",
      ylab = "cor")
    points(cor1$x, cor1$y, pch = 0., cex = 0.8)
#   (7)
    rr <- seq(from = 0., to = 12., by = 0.2)
    lines(rr, gaucov(r = rr, d = dd), lwd = 3.)
#   (8)
    var1 <- variogram(fit.gls, nint = 40., plotit = F)
    plot(var1$x, var1$y, type = "n", xlab = "distance",
      ylab = "var")
    points(var1$x, var1$y, pch = 5., cex = 0.8)
#   (9)
    fit.gls <- surf.gls(1., expcov, x = xx1, y = xx2,
      z = yy, d = dd)
    krig <- prmat(fit.gls, 1., 10., 1., 10., 19.)
    persp(krig, zlim = c(0., 18.), xlab = "x1", ylab = "x2",
      zlab = "y", lab = c(3., 3., 3.))
}
```

(1) The library "spatial" [13] is used to carry out universal kriging with two predictors (a polynomial equation is used to represent a rough trend).
(2) The value of δ is set at 0.8 in order to use eq(4.54) (page 245) as a theoretical correlogram. In fact, the empirical correlogram is compared with a theoretical cor-

relogram to select the form of the function for the theoretical correlogram and to optimize the broadness of the selected function.

(3) Simulation data are generated.

(4) The area in which to draw graphs is set.

(5) Calculation of universal kriging based on a linear equation as a rough trend is executed; the result is saved in `fit.gls`.

(6) An empirical correlogram is computed using the values given by subtracting values supplied by the linear equation ((5)) from data; the results are stored as `cor1` and illustrated in a graph.

(7) A curve of the theoretical correlogram (eq(4.54)) is superimposed in the graph drawn in (6). If the theoretical correlogram does not appear to approximate the empirical correlogram, the values of dd are adjusted. Otherwise, another form of a function is adopted.

(8) An empirical variogram is derived using the values given by subtracting values supplied by the linear equation ((5)) from data; the results are saved as `var1` and displayed on a graph.

(9) The theoretical variogram drawn in (7) is used to compute estimates, and the estimates are exhibited as a perspective plot.

(J) Derivation of an additive model by solving a normal equation by Gauss-Seidel method (polynomial equations are used as equations ($m_1(x_1)$, $m_2(x_2)$) to transform predictors)

Object: `myadd1()`

```
function(nd, it, xx1, xx2, yy, deg1, deg2)
{
#   (1)
    diag1 <- diag(nd)
#   (2)
    polyreg <- function(yy, xx, degree)
    {
        data1 <- data.frame(x1 = xx, y1 = yy)
        d1 <- degree
        formula.name <- substitute(y1 ~ poly(x1, degree = d2),
          list(d2 = d1))
        fit.lm <- lm(formula.name, data = data1)
        return(fit.lm$fitted.values)
    }
    hh1 <- apply(diag1, 2., polyreg, xx = xx1, degree = deg1)
#   (3)
    hh2 <- apply(diag1, 2., polyreg, xx = xx2, degree = deg2)
#   (4)
    av1 <- 0.
    mm2 <- rep(av1, length = nd)
    for(ii in 1.:it) {
        mm1 <- hh1 %*% (yy - mm2)
        mm2 <- hh2 %*% (yy - mm1)
    }
```

```
#   (5)
    return(mm1, mm2)
}
```

(1) An identity matrix (`diag1`) (size of nd × nd) is created.

(2) $\mathbf{H}^{(1)}$ (hat matrix to obtain a deg1-degreee polynomial equation) in eq(4.102) (page 260) is computed and is defined as `hh1`. By substituting values of predictors into `xx` (a vector) and those of a target variable into `yy` (a vector), `polyreg()` fits a degree-degree polynomial equation and outputs estimates corresponding to each data. For `substitute()`, refer to Chapter 6. The command `d1 <- degree` in the second line of `polyreg()` intends a clear description; if `degree` is used as is, the third line becomes confusing as follows (since `substitute()` conducts syntax analysis, the following expression is also effective).

```
formula.name <- substitute(y1 ~ poly(x1, degree = degree),
  list(degree = degree))
```

For the sum of the elements of `mm1` to be 0, the line below is added after the above part.

```
hh1 <- hh1 - matrix(rep(1, length = nd * nd), ncol = nd)/nd
```

(3) $\mathbf{H}^{(2)}$ (a hat matrix to obtain a deg2-degree polynomial equation) defined in eq(4.102) is obtained and saved as `hh2`. If, in place of polynomial equations, a one-degree B-spline (a broken line) that uses the maximum and minimum values of `xx1` as boundary knots (boundary breakpoints) and 2, 4, 6, 8 as internal knots (internal breakpoints) is employed, (2) and (3) are replaced by the following procedure. Furthermore, if `deg1` and `deg2` are set at 1, figure 4.17 (page 265), figure 4.18 (page 266), and figure 4.19 (page 267) can be obtained.

```
#   (2)'
    bsreg <- function(yy, xx, knots, degree)
    {
        data1 <- data.frame(x1 = xx, y1 = yy)
        kn1 <- knots
        d1 <- degree
        formula.name <- substitute(y1 ~ bs(x1, knots = kn2,
          degree = d2), list(kn2 = kn1, d2 = d1))
        fit.lm <- lm(formula.name, data = data1)
        return(fit.lm$fitted.values)
    }
    diag1 <- diag(nd)
    hh1 <- apply(diag1, 2, bsreg, xx = xx1,
      knots = c(2, 4, 6, 8), degree = deg1)
#   (3)'
    hh2 <- apply(diag1, 2, bsreg, xx = xx2,
      knots = c(3, 7), degree = deg2)
```

(4) The iterative calculation based on eq(4.117) (page 264) and eq(4.118) (page 264) is carried out `it` times.

(5) The objects mm1 and mm2 are outputted.

An example of an object to use myadd1() (construction of figure 4.14 (page 263))

```
function()
{
#   (1)
    nd <- 40.
    it <- 1.
#   (2)
    set.seed(100.)
    xx1 <- runif(nd, min = 0., max = 10.)
    xx2 <- runif(nd, min = 0., max = 10.)
    yy <- sin(0.2 * pi * xx1) + xx2^2. * 0.05 - xx2 * 0.4 +
      3. + rnorm(nd, mean = 0., sd = 0.1)
#   (3)
    deg1 <- 4.
    deg2 <- 2.
    fit.add <- myadd1(nd, it, xx1, xx2, yy, deg1, deg2)
    mm1 <- fit.add$mm1
    mm2 <- fit.add$mm2
#   (4)
    par(mfrow = c(1., 2.), mai = c(1., 1.5, 0.1, 0.5),
      oma = c(10., 1., 10., 1.))
    plot(xx1, mm1, type = "n", xlab = "x1", ylab = "m1(x1)",
      ylim = c(1., 4.))
    points(xx1, mm1, cex = 1.2, pch = 2.)
    plot(xx2, mm2, type = "n", xlab = "x2", ylab = "m2(x2)",
      ylim = c(-0.5, 1.5))
    points(xx2, mm2, cex = 1.2, pch = 5.)
}
```

(1) The number of data is saved as nd and the number of repetitions of iterative calculation is saved as it.
(2) Simulation data are generated.
(3) The degrees of polynomial equations ($m_1(\cdot)$ and $m_2(\cdot)$) for transforming the two variables are defined as deg1 and deg2. An additive model is created using myadd1(); the result is saved in fit.add. The objects mm1 (result of transforming xx1) and mm2 (result of transforming xx2) are extracted from fit.add and they are saved as mm1 and mm2, respectively.
(4) The resultant additive model is displayed.

(K) Additive model by solving a normal equation by Gauss-Seidel method (smoothing spline is employed for transforming each predictor; gam() in S-Plus is used as is)
Object: sadd1()

```
function(nd, ne, xx1, xx2, yy, ex1, ex2, ey1)
```

```
{
#  (1)
    assign("data1", data.frame(x1 = xx1, x2 = xx2, y = yy),
    frame = 1.)
#  (2)
    fit.gam <- gam(y ~ s(x1) + s(x2), data = data1)
#  (3)
    data2 <- data.frame(x1 = xx1, x2 = xx2)
    fit.tmd <- predict.gam(fit.gam, newdata = data2,
    type = "terms")
#  (4)
    eyd1 <- fit.tmd[, 1.]
    eyd2 <- fit.tmd[, 2.] + attr(fit.tmd, "constant")
#  (5)
    data3 <- data.frame(x1 = ex1, x2 = ex2)
    fit.tme <- predict.gam(fit.gam, newdata = data3,
    type = "terms")
#  (6)
    ey1 <- fit.tme[, 1.]
    ey2 <- fit.tme[, 2.] + attr(fit.tme, "constant")
#  (7)
    return(eyd1, eyd2, ey1, ey2)
}
```

(1) A data frame (data1) that consists of xx1, xx2, and yy is constructed, and it is saved in the expression frame (top-level frame, frame 1) (frame = 1 indicates expression frame). Assume that a function is executed while the cursor is on a command line of S-Plus; if the function includes a description to save a specific object in an expression frame, the object is preserved during the execution of the function so that it is refered to or is modified. Function is a form of an object; it begins with function and is used to carry out a series of procedures. The order of searching for an object is as follows: local frame (the frame where objects created by <- in a function are put; it corresponds to a number greater than or equal to 2 in frame =), expression frame, working directory (corresponds to where = 1), database (corresponds to a number greater than or equal to 2 in where =. Since a larger number in where = means a lower priority, the priority of database is lower than that of working directory.). If objects of the same name are located in different frames, the objects positioned in frames with lower priorities cannot be referred to (this situation is called "mask"); attention is required to avoid this problem.

(2) An additive model based on a smoothing spline is created using data1; the result is saved in fit.gam. In this object, the degree of smoothing is not specified and hence the value of a smoothing parameter is tuned to set equivalent degrees of freedom (= trace($\mathbf{H}$) − 1) (page 269) at 4. By s(x1, df = 4.5), for example, the equivalent degrees of freedom is specified as df =. Furthermore, an assignment such as s(x1, spar = 0.1) determines the value of a smoothing parameter directly. On the other hand, the assignment of df = in smooth.spline() (page 504) is used to designate the degrees of freedom (= trace($\mathbf{H}$)). Therefore, to use the same hat

matrix in `smooth.spline()` and `s()`, the value of `df` = in the former minus 1 must be assigned in `df` = in the latter (when `spar` = is used, the same values can be shared).

When LOESS instead of a smoothing spline is employed, `fit.gam <- gam(y ~ s(x1) + s(x2), data = data1)` is substituted by the following line.

```
fit.gam <- gam(y ~ lo(x1) + lo(x2), data = data1)
```

(3) The objects xx1 and xx2 are put together in a data frame (data2). Using `predict.gam()`, the values of $\hat{m}_1()$ in which xx1 is used and those of $\hat{m}_2()$ in which xx2 is used are obtained; the results are saved as `fit.tmd`. The command to calculate values of each term of an additive model, `type = "terms"`, is assigned.
(4) The values obtained when xx1 is substituted into $\hat{m}_1()$ and those obtained when xx2 is substituted into $\hat{m}_2()$ are stored as eyd1 and eyd2, respectively. The average of `fit.tmd[, 1]` becomes 0, and that of `fit.tmd[, 2]` is also 0. The value of a constant term is extracted by `attr(fit.tme, "constant")`.
(5) The objects ex1 and ex2 are put together in a data frame (data3). Then, a calculation similar to (3) is carried out.
(6) Using `fit.tme`, a similar calculation to (4) is executed.
(7) The objects eyd1, eyd2, ey1, and ey2 are inputted.

An example of an object to use `sadd1()` (construction of figure 4.20 (page 268))

```
function()
{
#    (1)
     nd <- 40.
     set.seed(100.)
     xx1 <- runif(nd, min = 0., max = 10.)
     xx2 <- runif(nd, min = 0., max = 10.)
     yy <- sin(0.2 * pi * xx1) + xx2^2. * 0.05 - xx2 * 0.4
      + 3. + rnorm(nd, mean = 0., sd = 0.1)
#    (2)
     ne <- 100.
     ex1 <- seq(from = min(xx1), to = max(xx1), length = ne)
     ex2 <- seq(from = min(xx2), to = max(xx2), length = ne)
#    (3)
     fit.add <- sadd1(nd, ne, xx1, xx2, yy, ex1, ex2, ey1)
     eyd1 <- fit.add$eyd1
     eyd2 <- fit.add$eyd2
     ey1 <- fit.add$ey1
     ey2 <- fit.add$ey2
#    (4)
     par(mfrow = c(1., 2.), mai = c(1., 1.5, 0.1, 0.5),
       oma = c(10., 1., 10., 1.))
     plot(ex1, ey1, type = "n", xlab = "x1", ylab = "m1(x1)")
     lines(ex1, ey1, lwd = 2.)
     points(xx1, eyd1, pch = 2.)
```

```
    plot(ex2, ey2, type = "n", xlab = "x2", ylab = "m2(x2)")
    lines(ex2, ey2, lwd = 2.)
    points(xx2, eyd2, pch = 2.)
}
```

(1) Simulation data are generated.

(2) The values of coordinates of estimation points are supplied. When LOESS is used as a smoothing method to create an additive model, all values of elements of ex1 are located between the maximum and the minimum of xx1, and all values of elements of ex2 are positioned between the maximun and the minimum of xx2. If extrapolation is carried out when a smoothing spline is employed, a natural spline is used for extrapolation.

(3) An additive model is constructed using sadd1(); the results are saved in fit.add. Furthermore, fit.add$eyd1 ($m_1$), fit.add$eyd2 (m_2), fit.add$ey1 ($\hat{m}_1(x_1)$), and fit.add$ey2 ($\hat{m}_2(x_2)$) are stored as eyd1, eyd2, ey1, and ey2.

(4) The objects m_1 and m_2, and the values of $\hat{m}_1(x_1)$ and $\hat{m}_2(x_2)$ are illustrated.

(L) Derivation of regression equation with the form of ACE (each regression equation is a polynomial equation) by canonical correlation analysis
Object: acecan1()

```
function(nd, npx1, npx2, npy, xx1, xx2, yy)
{
#   (1)
    powerf <- function(jj, x1)
    {
        pw <- x1^jj
        return(pw)
    }
    xxm1 <- apply(matrix(c(1.:npx1), nrow = 1.), 2.,
      powerf, x1 = xx1)
    xxm2 <- apply(matrix(c(1.:npx2), nrow = 1.), 2.,
      powerf, x1 = xx2)
    xxmat <- cbind(xxm1, xxm2)
    yymat <- apply(matrix(c(1.:npy), nrow = 1.), 2.,
      powerf, x1 = yy)
#   (2)
    can1 <- cancor(xxmat, yymat, xcenter = T, ycenter = T)
    cor1 <- can1$cor[1.]
#   (3)
    ex1 <- sqrt(nd) * can1$xcoef[1.:npx1, 1.] %*%
      sweep(t(xxm1), 1., can1$xcenter[1.:npx1])
    ex2 <- sqrt(nd) * can1$xcoef[(npx1 + 1.):(npx1 + npx2),
      1.] %*% sweep(t(xxm2), 1., can1$xcenter[1.:npx2])
    ey <- sqrt(nd) * can1$ycoef[1.:npy, 1.] %*%
      sweep(t(yymat), 1., can1$ycenter[1.:npy])
    ex1 <- ex1 * cor1
    ex2 <- ex2 * cor1
```

```
    ex12 <- ex1 + ex2
#  (4)
    return(ex1, ex2, ex12, ey, cor1)
}
```

(1) The object xxmat is defined; this matrix is constructed using $(A_{11}, A_{21}, \ldots,$ $A_{n1})^t, (A_{12}, A_{22}, \ldots, A_{n2})^t, (A_{13}, A_{23}, \ldots, A_{n3})^t, \ldots$ (eq(4.172) (page 276) and eq(4.173) (page 276)). The object yymat is defined; this matrix is created using $(B_{11}, B_{21}, \ldots, B_{n1})^t, (B_{12}, B_{22}, \ldots, B_{n2})^t, (B_{13}, B_{23}, \ldots, B_{n3})^t, \ldots$ (eq(4.174) (page 277)). The values of npx1, npx2, and npy are degrees of polynomials used to transform xx1, xx2, and yy, respectively.

(2) Canonical correlation analysis is conducted by cancor(). By designating xcenter = T and ycenter = T, canonical correlation analysis is performed after the averages of the column vectors of xxmat are subtracted from the corresponding column vectors constituting xxmat and the averages of the column vectors of yymat are subtracted from the corresponding column vectors constituting yymat. The object can1$cor[1] is the positive square root of the maximum value of the eigenvalues of $A(A^tA)^{-1}A^tB(B^tB)^{-1}B^t$ (eq(4.152) (page 272)); can1$cor[1] is identical to $\cos(\theta)$.

(3) By the same methods as those with eq(4.176) (page 277), eq(4.177), and eq(4.178), the transformed values of xx1 (ex1(= $\{\hat{m}_1(X_{i1})\}$)), the transformed values of xx2 (ex2(= $\{\hat{m}_2(X_{i2})\}$)), and the transformed values of yy (ey(= $\{\hat{\eta}(Y_i)\}$))) are derived. Furthermore, the sum of ex1 and ex2 (= $\{\hat{m}_1(X_{i1}) + \hat{m}_2(X_{i2})\}$) is saved as ex12. The object sqrt(nd) is multiplied when ex1, ex2, and ey are obtained because cancor() uses the condition below in place of eq(4.146) (page 271).

$$\sum_{i=1}^{n}(A_i^*)^2 = 1, \qquad \sum_{i=1}^{n}(B_i^*)^2 = 1. \tag{4.208}$$

(4) The objects ex1, ex2, ex12, ey, and cor1 are outputted.

An example of an object to use acecan1() (construction of figure 4.24 (page 278))

```
function()
{
#  (1)
    nd <- 100.
    set.seed(100.)
    xx1 <- runif(nd, min = 0., max = 10.)
    xx2 <- runif(nd, min = 0., max = 10.)
    yy <- (sin(0.2 * pi * xx1) + xx2^2. * 0.05 - xx2 * 0.4
     + 3. + rnorm(nd, mean = 0., sd = 0.1))^2.
#  (2)
    npx1 <- 4.
    npx2 <- 2.
    npy <- 2.
#  (3)
```

```
    fit.ace <- acecan1(nd, npx1, npx2, npy, xx1, xx2, yy)
    print(fit.ace$cor1)
    ex1 <- fit.ace$ex1
    ex2 <- fit.ace$ex2
    ex12 <- fit.ace$ex12
    ey <- fit.ace$ey
#   (4)
    par(mfrow = c(2, 2), mai = c(1, 1.5, 0.1, 0.5),
      oma = c(1, 1, 1, 1))
#   (5)
    plot(xx1, ex1, type = "n", xlab = "x1", ylab = "m1(x1)")
    points(xx1, ex1, cex = 0.7, pch = 0.)
#   (6)
    plot(xx2, ex2, type = "n", xlab = "x2", ylab = "m2(x2)")
    points(xx2, ex2, cex = 0.7, pch = 2.)
#   (7)
    plot(yy, ey, type = "n", xlab = "y", ylab = "eta(y)")
    points(yy, ey, cex = 0.7, pch = 5.)
#   (8)
    plot(ex12, ey, type = "n", xlab = "m1(x1)+m2(x2)",
      ylab = "eta(y)", xlim = c(-1.7, 2.5), ylim = c(-1.7, 2.5))
    points(ex12, ey, cex = 0.7, pch = 6.)
}
```

(1) Simulation data are generated.
(2) The degree of a polynomial equation used to transform xx1 is defined as npx1. The degree of a polynomial equation for transforming xx2 is defined as npx2. The degree of a polynomial equation that transforms yy is defined as npy.
(3) A regression equation with the form of ACE is gained by the use of acecan1(), and the result is saved in fit.ace. The value of $\cos(\theta)$ is displayed on a screen. The object fit.ace$ex1 is saved as ex1, fit.ace$ex2 is stored as ex2, fit.ace$ex12 is named ex12, and fit.ace$ey is defined as ey.
(4) The area in which to draw graphs is set.
(5) A graph is displayed to represent $\{(X_{i1}, \hat{m}_1(X_{i1}))\}$.
(6) A graph for $\{(X_{i2}, \hat{m}_2(X_{i2}))\}$ is drawn.
(7) $\{(Y_i, \hat{\eta}(Y_i))\}$ are shown in the graph.
(8) $\{(\hat{m}_1(X_{i1}) + \hat{m}_2(X_{i2}), \hat{\eta}(Y_i))\}$ are plotted in the graph.

(M) Derivation of a regression equation with the form of ACE by solving an eigenvalue problem using iterative calculation (each regression equation is a polynomial equation)
Object: aceit1()

```
function(nd, it, npx1, npx2, npy, xx1, xx2, yy)
{
#   (1)
    yyst <- yy
#   (2)
```

```
    powerf <- function(jj, x1)
    {
        pw <- x1^jj
        return(pw)
    }
    xxm1 <- apply(matrix(c(1.:npx1), nrow = 1.),
     2., powerf, x1 = xx1)
    xxm2 <- apply(matrix(c(1.:npx2), nrow = 1.),
     2., powerf, x1 = xx2)
    xxmat <- cbind(xxm1, xxm2)
    xxmean <- apply(xxmat, 2., mean)
    xxmat <- sweep(xxmat, 2., xxmean)
    yymat <- apply(matrix(c(1.:npy), nrow = 1.),
     2., powerf, x1 = yy)
    yymean <- apply(yymat, 2., mean)
    yymat <- sweep(yymat, 2., yymean)
    hatx <- xxmat %*% solve(crossprod(xxmat)) %*% t(xxmat)
    haty <- yymat %*% solve(crossprod(yymat)) %*% t(yymat)
    hatyx <- haty %*% hatx
#   (3)
    for(ii in 1.:it) {
        yyst <- hatyx %*% yyst
        yyst <- (sqrt(nd) * yyst)/sqrt(sum(yyst^2.))
    }
#   (4)
    return(yyst)
}
```

(1) The value of yy ($= \{Y_i\}$) is employed as an initial value of yyst ($= \{Y_i^*\}$).
(2) $\mathbf{H}_X$ in eq(4.183) (page 280) is defined as hatx, and $\mathbf{H}_Y$ in eq(4.185) (page 280) is defined as haty. Furthermore, $\mathbf{H}_Y \mathbf{H}_X$ is named hatyx. t(xxmat) %*% xxmat is identical to crossprod(xxmat).
(3) The iterative calculation is carried out it times.
(4) The object yyst is outputted.

An example of an object to use aceit1() (construction of figure 4.27 (page 282))

```
function()
{
#   (1)
    nd <- 100.
    set.seed(100.)
    xx1 <- runif(nd, min = 0., max = 10.)
    xx2 <- runif(nd, min = 0., max = 10.)
    yy <- (sin(0.2 * pi * xx1) + xx2^2. * 0.05 - xx2 * 0.4
     + 3. + rnorm(nd, mean = 0., sd = 0.1))^2.
#   (2)
```

```
    npx1 <- 4.
    npx2 <- 2.
    npy <- 2.
#   (3)
    par(mfrow = c(2., 2.), mai = c(1., 1.5, 0.1, 0.5),
      oma = c(1., 1., 1., 1.))
#   (4)
    for(it in 1.:4.) {
        yyst <- aceit1(nd, it, npx1, npx2, npy, xx1, xx2, yy)
        plot(yy, yyst, type = "n", xlab = "y", ylab = "y*")
        points(yy, yyst, cex = 0.7, pch = 5.)
    }
}
```

(1) Simulation data are generated.
(2) The degree of the polynomial equation that transforms xx1 is defined as npx1. The degree of the polynomial equation that transforms xx2 is defined as npx2. The degree of the polynomial equation that transforms yy is defined as npy.
(3) The area in which to draw graphs is set.
(4) By giving the values $1, 2, 3$, and 4 for it (number of repetitions of iterative calculation), yyst ($= \{Y_i^*\}$) is calculated using aceit1() and is illustrated in a graph.

Using resultant yyst ($= \{Y_i^*\}$), $\{X_{i1}, \hat{m}_1(X_{i1})\}$ and $\{X_{i2}, \hat{m}_2(X_{i2})\}$ are derived. To illustrate the values in a graph, the following is added. The object eyd1 indicates $\{\hat{m}_1(X_{i1})\}$, and eyd2 denotes $\{\hat{m}_2(X_{i2})\}$.

```
    assign("data1", data.frame(x1 = xx1, x2 = xx2, y = yyst),
      frame = 1.)
    formula1 <- substitute(y ~ poly(x1, degree = deg1)
      + poly(x2, degree = deg2), list(deg1 = npx1, deg2 = npx2))
    fit.lm <- lm(formula1, data = data1)
    assign("data2", data.frame(x1 = xx1, x2 = xx2), frame = 1.)
    fit.tmd <- predict.gam(fit.lm, newdata = data2,
      type = "terms")
    eyd1 <- fit.tmd[, 1.]
    eyd2 <- fit.tmd[, 2.] + attr(fit.tmd, "constant")
    yhat <- eyd1 + eyd2
    plot(xx1, eyd1, type = "n", xlab = "x1", ylab = "m1(x1)")
    points(xx1, eyd1, cex = 1.2, pch = 5.)
    plot(xx2, eyd2, type = "n", xlab = "x2", ylab = "m2(x2)")
    points(xx2, eyd2, cex = 1.2, pch = 5.)
```

(N) Derivation of a regression equation with the form of ACE by ACE algorithm (each regression equation is a polynomial equation)
Object: aceit2()

```
function(nd, it, npx1, npx2, npy, xx1, xx2, yy)
{
```

```
#   (1)
    yyst <- (yy - mean(yy))/sqrt(sum((yy - mean(yy))^2.)/nd)
#   (2)
    for(ii in 1.:it) {
#   (3)
        data1 <- data.frame(x1 = xx1, x2 = xx2, y = yyst)
        formula1 <- substitute(y ~ poly(x1, degree = deg1) +
         poly(x2, degree = deg2), list(deg1 = npx1, deg2 = npx2))
        fit.lm <- lm(formula1, data = data1)
        yhat <- fitted.values(fit.lm)
#   (4)
        data2 <- data.frame(x = yy, y = yhat)
        formula2 <- substitute(y ~ poly(x, degree = deg3),
         list(deg3 = npy))
        fit.lm <- lm(formula2, data = data2)
        eta <- fitted.values(fit.lm)
#   (5)
        yyst <- eta/sqrt(sum((eta - mean(eta))^2.)/nd)
        print(sqrt(sum((yyst - mean(yyst))^2.)/nd))
    }
#   (6)
    return(yyst)
}
```

(1) The object yyst $(= \{Y_i^*\})$ is computed using eq(4.180) (page 277).
(2) Iterative calculation based on the ACE algorithm is executed it times.
(3) Regression is carried out to derive eq(4.181).
(4) Regression is performed to obtain eq(4.182).
(5) The object yyst $(= \{Y_i^*\})$ is updated.
(6) The object yyst is outputted.
  The usage of aceit2() is the same as that of aceit1().

(O) Derivation of regression equation by the use of ACE based on super-smoother (use of ace(); construction of figure 4.28 (page 283))

```
function()
{
#   (1)
    nd <- 100.
    set.seed(100.)
    xx1 <- runif(nd, min = 0., max = 10.)
    xx2 <- runif(nd, min = 0., max = 10.)
    yy <- (sin(0.2 * pi * xx1) + xx2^2. * 0.05 - xx2 * 0.4
     + 3. + rnorm(nd, mean = 0., sd = 0.1))^2.
#   (2)
    xxmat <- cbind(xx1, xx2)
    fit.ace <- ace(xxmat, yy)
    eys1 <- fit.ace$tx[, 1.]
```

```
      eys2 <- fit.ace$tx[, 2.]
      yy2 <- fit.ace$ty
#    (3)
      par(mfrow = c(2, 2), mai = c(1, 1.5, 0.1, 0.5),
        oma = c(1, 1, 1, 1))
      plot(xx1, eys1, type = "n", xlab = "x1", ylab =
        "m1(x1)")
      points(xx1, eys1, cex = 0.7, pch = 0.)
      plot(xx2, eys2, type = "n", xlab = "x2", ylab =
        "m2(x2)")
      points(xx2, eys2, cex = 0.7, pch = 2.)
      plot(yy, yy2, type = "n", xlab = "y", ylab =
        "eta(y)")
      points(yy, yy2, cex = 0.7, pch = 5.)
      plot(eys1 + eys2, yy2, type = "n", xlab =
        "m1(x1)+m2(x2)", ylab = "eta(y)")
      points(eys1 + eys2, yy2, cex = 0.7, pch = 5.)
}
```

(1) Simulation data are generated.
(2) Values of predictors to be used in ace() must be represented as a matrix, and hence xxmat, which consists of xx1 (a column vector) and xx2 (a column vector), is created to perform ace(). Assignment of monotone= renders a specific transformation function monotonous. For example, monotone=c(2) results in a monotonous transformation function for the second predictor, and monotone=c(0, 1) provides monotonous functions for the target variable and the first predictor. Furthermore, linear= renders specific transformation functions linear. The usage is the same as that of monotone=.
(3) The result obtained by ace() is exhibited in a graph.

(P) Derivation of a regression equation by projection pursuit regression (use of ppreg(); construction of figure 4.29 (page 286))

```
function()
{
#    (1)
      nd <- 100.
      set.seed(100.)
      xx1 <- runif(nd, min = 0., max = 10.)
      xx2 <- runif(nd, min = 0., max = 10.)
      yy <- 2. + cos(xx1 * 0.6 + xx2 * 0.4) +
        rnorm(nd, mean = 0., sd = 0.1)
#    (2)
      xmat <- cbind(xx1, xx2)
      fit.pro <- ppreg(xmat, yy, min.term = 1., max.term = 7.)
      print(fit.pro$esqrsp)
#    (3)
      nex1 <- 21.
```

```
    nex2 <- 21.
    ex1 <- seq(from = 0., by = 0.5, length = nex1)
    ex2 <- seq(from = 0., by = 0.5, length = nex2)
    grid <- expand.grid(ex1, ex2)
    grid12 <- cbind(grid[, 1.], grid[, 2.])
#   (4)
    fit.pro <- ppreg(xmat, yy, min.term = 1., max.term = 7.,
      xpred = grid12)
    print(fit.pro$alpha[1.,   ])
#   (5)
    ff1x <- fit.pro$z[, 1.]
    beta1 <- fit.pro$beta[1.]
    ypred1 <- fit.pro$ypred
    ff1y <- fit.pro$zhat[, 1.] * beta1 + mean(yy)
#   (6)
    exy <- NULL
    exy$x <- ex1
    exy$y <- ex2
    exy$z <- matrix(ypred1, ncol = nex2)
#   (7)
    par(mfrow = c(1., 2.), mai = c(1., 1., 1., 1.),
      oma = c(8., 1., 8., 1.))
    plot(ff1x, ff1y, type = "n", xlab = "z", ylab = "m1(z)")
    points(ff1x, ff1y, pch = 15.)
#   (8)
    persp(exy, xlab = "x1", ylab = "x2", zlab = "y",
      lab = c(3., 3., 1.))
    invisible()
}
```

(1) Simulation data are generated.

(2) The values of predictors for ppreg() must be depicted as a matrix, and hence xmat, which consists of xx1 (a column vector) and xx2 (a column vector), is created. M_{max} is set at 7 and M_{min} is set at 1. Then, ppreg() is carried out. The residual sum of squares is divided by the number of data, and the resulting value is divided by the variance of a target variable of data; this calculation is performed for each value of M. These values are outputted as fit.pro$esqrsp and are illustrated in a graph.

(3) The values of coordinates at estimation points are provided. The object grid12 (data frame) consists of two objects. Each object is a vector, and the corresponding elements of the two vectors denote the coordinates of an estimation point.

(4) M_{max} is fixed at 7 and M_{min} is set at 1. Then, estimates at estimation points given by grid12 are calculated by performing ppreg(). M_{min} is set at 1; it is due to the result of (2). Since the values of $(\hat{\alpha}_{11}, \hat{\alpha}_{21})$ when M is 1 are obtained as fit.pro$alpha[1,   ], these values are displayed on a screen.

(5) $\{\sum_{k=1}^{r} \alpha_{k1} X_{ik}\}$ $(1 \leq i \leq n)$ in eq(4.196) (page 284) is defined as ff1x, and $\{\alpha_0 + \beta_1 \cdot \phi_1(\sum_{k=1}^{r} \alpha_{k1} X_{ik})\}$ $(1 \leq i \leq n)$ in the same equation is named ff1y.

$r = 2$ is adopted in this example. The estimates at estimation points denoted by which `grid12` are saved as `ypred1`.

(6) The object `exy` is created so that `ypred1` is illustrated using `persp()`.

(7) The relationship between `ff1x` and `ff1y` is displayed.

(8) A perspective plot is charted to show `exy`.

REFERENCES

1. D.M. Bates, M.J. Lindstorm, G. Wahba, and B.S. Yandell (1987). GCVPACK - routines for generalized cross validation. *Communication in Statistics, Series B*, Vol. 16, No. 1, pp. 263-297.

2. D. Bates, F. Reames, and G. Wahba (1993). Getting better contour plots with S and GCVPACK. *Computational Statistics & Data Analysis*, Vol. 15, pp. 329-342.

3. A. Buja, T. Hastie, and R. Tibshirani (1989). Linear smoothers and additive models (with discussion). *Annals of Statistics*, Vol. 17, pp. 453-555.

4. L. Breiman and J.H. Friedman (1985). Estimating optimal transformations for multiple regression and correlation (with discussion). *Journal of the American Statistical Association*, Vol. 80, pp. 580-619.

5. P.J. Green and B.W. Silverman (1994). *Nonparametric Regression and Generalized Linear Models, a Roughness Penalty Approach*, Chapman & Hall/CRC.

6. J. Duchon (1976). Interpolation des functions de deux variables suivant le principe de la flexion des plaques minces. *R.A.I.R.O. Analyse numérique*, Vol. 16, No. 3, pp. 201-209.

7. T.J. Hastie and R.J. Tibshirani (1990). *Generalized Additive Model*, Chapman & Hall/CRC.

8. P.J. Huber (1985). Projection pursuit (with discussion). *Annals of Statistics*, Vol. 13, pp. 435-525.

9. D.G. Kringe (1951). A statistical approach to some basic mine valuation problems on the Witwatersrand. *Journal of the Chemical, Metallurgical and Mining Society of South Africa*, Vol. 52, pp. 119-139.

10. D.C. Lay (1996). *Linear Algebra and Its Applications*, 2nd edition, Addison Wesley Longman.

11. K. Takezawa. Three-dimensional Free Curved Surface Design Device, *Japanese Patent (JP6139315)*, Publication date: May 20, 1994. Method for Designing Three-Dimensional Free-Form Surface and Apparatus for Executing the Same, *American Patent (US6590575)*, Publication date: July 8, 2003. "Apparatus and Method for Designing a Three-Dimensional Free-Form Surface," *European Patent* (EP0594276 (except Germany), 69326868.9-08 (Germany)). Designated states: Belgium, Switzerland, Germany, France, Italy, The Netherlands, Sweden, Great Britain, Lichtenstein. Inventor: Kunio Takezawa. Publication date: April 27, 1994. Method for Designing Three-Dimensional Free-Form Surface and Apparatus for Executing the Same, *Canadian Patent* (CA2092217). Publication date: April 23, 1994.

12. J. S. Simonoff (1996). *Smoothing Methods in Statistics*, Springer-Verlag.

13. W.N. Venables and B.D. Ripley (1997). *Modern Applied Statistics with S-PLUS, 2nd ed.*, Springer-Verlag.

14. W.N. Venables and B.D. Ripley (1999). *Modern Applied Statistics with S-PLUS, 3rd ed.*, Springer-Verlag.

15. W.N. Venables and B.D. Ripley (2002). *Modern Applied Statistics with S-PLUS, 4th ed.*, Springer-Verlag.

Problems

4.1   Modify the object `llin2d()` presented in (B) to carry out local quadratic regression for two predictors. Compare the estimates given by this object with those given by the original object using the same simulation data.

4.2   Revise the object presented in (C) for a cubic smoothing spline (cubic smoothing splines) with two different smoothing parameters.

4.3   Consider the use of cross-validation (CV) to optimize the broadness of a theoretical correlogram (e.g., the value of δ in eq(4.54) (page 245)) for ordinary kriging.
(a) Create an S-Plus or R object for optimizing δ in eq(4.54). The cross-validation is carried out by deleting data one by one, except those located at the two ends of the data region, and estimating the prediction error by comparing the estimated values with the deleted data.
(b) Using the data shown in Problem **2.3** in Chapter 2, optimize δ by the use of the object created in (a).

4.4   Use simple kriging with one predictor to carry out the same tasks performed in Problem **4.3**.

4.5   Construct an S-Plus or R object for calculating and charting the elements of a hat matrix for smoothing based on simple kriging (eq(4.59) (page 248) and eq(4.60)). Eq(4.65) (page 248) gives the definition of the hat matrix.

4.6   Consider the smoothing technique based on simple kriging (eq(4.59) (page 248) and eq(4.60)). Construct an S-Plus or R object for optimizing σ^2 (eq(4.54) is used, and δ is fixed) by CV or GCV. Check the validity of the optimized σ^2 using some simulation data.

4.7   Use universal kriging with one predictor to carry out tasks similar to those performed in Problem **4.3**; the degree of the polynomial equation is also optimized.

4.8   Modify the object presented in (I) of this chapter to be able to use various degrees of polynomial equation and compare the results.

4.9   Consider methods for obtaining the regression equation defined in eq(4.116) (page 263) without using an iterative procedure; solve a normal equation by a noniterative method.
(a) Derive a normal equation to obtain eq(4.116); use the matrix below as a design matrix.

$$
\begin{pmatrix}
1 & X_{11} & X_{11}^2 & \cdots & X_{11}^p & X_{12} & X_{12}^2 & \cdots & X_{12}^q \\
1 & X_{21} & X_{21}^2 & \cdots & X_{21}^p & X_{22} & X_{22}^2 & \cdots & X_{22}^q \\
\vdots & \vdots & \vdots & \ddots & \vdots & \vdots & \vdots & \ddots & \vdots \\
1 & X_{n1} & X_{n1}^2 & \cdots & X_{n1}^p & X_{n2} & X_{n2}^2 & \cdots & X_{n2}^q
\end{pmatrix}
\tag{4.209}
$$

(b) Create an S-Plus or R object for computing regression coefficients using the normal equation obtained in (a). Moreover, produce an S-Plus or R object for calculating a corresponding hat matrix.

(c) Construct an S-Plus or R object for deriving a regression equation in the form of eq(4.116) using the design matrix below.

$$
\begin{pmatrix}
r_0(X_{11}) & r_1(X_{11}) & r_2(X_{11}) & \cdots & r_p(X_{11}) & s_1(X_{12}) & \cdots & s_q(X_{12}) \\
r_0(X_{21}) & r_1(X_{21}) & r_2(X_{21}) & \cdots & r_p(X_{21}) & s_1(X_{22}) & \cdots & s_q(X_{22}) \\
\vdots & \vdots & \vdots & \ddots & \vdots & \vdots & \ddots & \vdots \\
r_0(X_{n1}) & r_1(X_{n1}) & r_2(X_{n1}) & \cdots & r_p(X_{n1}) & s_1(X_{n2}) & \cdots & s_q(X_{n2})
\end{pmatrix},
$$
$$(4.210)$$

where $\{r_1(\cdot), r_2(\cdot), \ldots, r_p(\cdot)\}$ and $\{s_1(\cdot), s_2(\cdot), \ldots, s_q(\cdot)\}$ are orthonormal polynomials. Therefore, the equations below hold.

$$
\sum_{k=1}^{n} r_i(X_{k1}) r_j(X_{k1}) = \begin{cases} 1 & \text{if } i = j \\ 0 & \text{if } i \neq j, \end{cases} \qquad (4.211)
$$

$$
\sum_{k=1}^{n} s_i(X_{k2}) s_j(X_{k2}) = \begin{cases} 1 & \text{if } i = j \\ 0 & \text{if } i \neq j. \end{cases} \qquad (4.212)
$$

Furthermore, $r_0(\cdot) \equiv \frac{1}{\sqrt{n}}$ (refer to Problem **2.16**).

(d) Confirm that the hat matrix given by the object created in **(b)** is identical to that given by the object constructed in **(c)** using the simulation data.

4.10   Consider the derivation of regression coefficients in the regression equation defined in eq(4.119) (page 266) without using an iterative calculation. That is, solve the normal equation by a noniterative procedure.

(a) Construct an S-Plus or R object to solve the normal equation directly (using the Gauss-Jordan method, for example). Use the data
$\{X_{1i}\} = \{1.0, 4.4, 4.8, 2.1, 4.6, 4.3, 4.0, 1.2, 4.7, 6.8, 4.0, 4.5, 1.9, 4.3, 1.9, 6.7, 4.3, 5.5, 2.7, 7.0\}$,
$\{X_{2i}\} = \{1.0, 4.5, 1.1, 2.0, 2.9, 2.2, 4.8, 2.4, 4.2, 1.9, 4.2, 3.3, 1.2, 2.0, 3.7, 2.0, 2.7, 4.1, 4.2, 5.0\}$,
$\{Y_i\} = \{3.2, 20.8, 18.7, 5.7, 18.3, 16.1, 17.1, 5.1, 23.7, 43.3, 15.8, 20.5, 3.5, 15.9, 8.3, 40.0, 14.7, 29.9, 12.0, 50.7\}$.
Adopt linear B-splines as bases. Fix the internal knots of the linear B-splines at $(3, 5.5)$ (x_1-axis) and $(2.4, 3.3)$ (x_2-axis). Use the following design matrix.

$$
\begin{pmatrix}
g_{11}(X_{11}) & \cdots & g_{14}(X_{11}) & g_{21}(X_{12}) & \cdots & g_{24}(X_{12}) \\
g_{11}(X_{21}) & \cdots & g_{14}(X_{21}) & g_{21}(X_{22}) & \cdots & g_{24}(X_{22}) \\
\vdots & \ddots & \vdots & \vdots & \ddots & \vdots \\
g_{11}(X_{n1}) & \cdots & g_{14}(X_{n1}) & g_{21}(X_{n2}) & \cdots & g_{24}(X_{n2})
\end{pmatrix}. \qquad (4.213)
$$

Note that, when bs() is used, internal knots are assigned as knots = and intercept = T must be designated. Confirm numerically that, when this design matrix is employed, a conventional method cannot yield a unique solution for the simultaneous equations because the solution is not uniquely determined.

(b) A possible method for overcoming this problem is the use of a Moore-Penrose generalized inverse matrix. When the simultaneous equation $St = u$ (S is a matrix and t and u are column vectors; t is unknown) has one or more than one solutions, t is rewritten as (for proof, refer to textbooks on generalized inverse matrices)

$$t = S^- u + (I - S^- S)z, \qquad (4.214)$$

where I is an identity matrix, S^- is a Moore-Penrose generalized inverse matrix corresponding to S, and z is an arbitrary column vector.

Use the object ginverse() (included in S-Plus by default) or the like, which provides a Moore-Penrose generalized inverse matrix, to solve the normal equation using the design matrix defined in eq(4.213). Check that when various values are used as elements of z defined in eq(4.214), the regression coefficients vary but the estimates change only slightly unless the values of the elements of z are extraordinarily high, causing problems concerning numerical calculation.

(c) Construct an S-Plus or R object to yield a hat matrix for this regression by extending the object created in (b).

(d) In addition to the method described in (b), there is another possible technique for solving the normal equation based on eq(4.213). That is, after one of the regression coefficients (e.g., β_{11}) is fixed at 0, a simultaneous equation is solved to derive the values of the other regression coefficients. Then, while a simultaneous equation consisting of eight equations is obtained, seven regression coefficients are unknown. Hence, a Moore-Penrose generalized inverse matrix is employed. The seven regression coefficients are derived using eq(4.214). However, the resultant regression coefficients are unique. Create an S-Plus or R object to realize this method. Check that the estimates given by this method are identical to those given by the method described in (b).

4.11   Consider techniques for solving directly the following equation (eq(4.131) (page 268)) without using an iterative procedure.

$$E_{ss-add} = (y - G_1\beta_1 - G_2\beta_2)^t(y - G_1\beta_1 - G_2\beta_2) + \lambda_1\beta_1^t K_1\beta_1$$
$$+ \lambda_2\beta_2^t K_2\beta_2. \qquad (4.215)$$

(a) Create an S-Plus or R object for deriving the regression coefficients (β_1 and β_2) of an additive model using the technique described in (b) in Problem **4.10**. (Note: To adjust the value of tol assigned in ginverse() (included in S-Plus by default), refer to the value of rank outputted from ginverse().)

(b) By developing the object created in (a), construct an S-Plus or R object for deriving a hat matrix.

(c) By extending the object produced in (b), construct an S-Plus or R object for optimizing the smoothing parameters on the basis of GCV.

(d) Confirm that in some cases this method does not perform well. In what situations does it occur?

4.12   Modify the object myadd1() presented in (J) of this chapter to obtain $\hat{m}_2(x_2)$, which is a one-degree B-spline.

4.13   Create an S-Plus or R object for obtaining a partial spline using dtpss (or dptpss) included in GCVPACK. Revise the object myadd1() presented in (J) of this chapter to realize this partial spline. Compare the results of the two objects for realizing the partial spline.

4.14   Consider the value of the degrees of freedom of the additive model given by using eq(4.209) (page 319) or eq(4.213) (page 320) as a design matrix.
(a) Show that the degrees of freedom (df) of the additive model given by using eq(4.209) is

$$
\begin{aligned}
df &= \mathrm{trace}(\mathbf{X}_3(\mathbf{X}_3^t\mathbf{X}_3)^{-1}\mathbf{X}_3^t) \\
&= \mathrm{trace}(\mathbf{X}_1(\mathbf{X}_1^t\mathbf{X}_1)^{-1}\mathbf{X}_1^t) + \mathrm{trace}(\mathbf{X}_2(\mathbf{X}_2^t\mathbf{X}_2)^{-1}\mathbf{X}_2^t). \quad (4.216)
\end{aligned}
$$

$\mathbf{X}_1, \mathbf{X}_2$, and $\mathbf{X}_3$ are

$$
\mathbf{X}_1 = \begin{pmatrix}
1 & X_{11} & X_{11}^2 & \cdots & X_{11}^p \\
1 & X_{21} & X_{21}^2 & \cdots & X_{21}^p \\
\vdots & \vdots & \vdots & \ddots & \vdots \\
1 & X_{n1} & X_{n1}^2 & \cdots & X_{n1}^p
\end{pmatrix}, \quad (4.217)
$$

$$
\mathbf{X}_2 = \begin{pmatrix}
X_{12} & X_{12}^2 & \cdots & X_{12}^q \\
X_{22} & X_{22}^2 & \cdots & X_{22}^q \\
\vdots & \vdots & \ddots & \vdots \\
X_{n2} & X_{n2}^2 & \cdots & X_{n2}^q
\end{pmatrix}, \quad (4.218)
$$

$$
\mathbf{X}_3 = \begin{pmatrix}
1 & X_{11} & X_{11}^2 & \cdots & X_{11}^p & X_{12} & X_{12}^2 & \cdots & X_{12}^q \\
1 & X_{21} & X_{21}^2 & \cdots & X_{21}^p & X_{22} & X_{22}^2 & \cdots & X_{22}^q \\
\vdots & \vdots & \vdots & \ddots & \vdots & \vdots & \vdots & \ddots & \vdots \\
1 & X_{n1} & X_{n1}^2 & \cdots & X_{n1}^p & X_{n2} & X_{n2}^2 & \cdots & X_{n2}^q
\end{pmatrix}. \quad (4.219)
$$

(Hint: trace($\mathbf{AB}$) = trace($\mathbf{BA}$).)
(b) Show that the degrees of freedom (df) of the additive model given by using eq(4.213) is

$$
df = \mathrm{trace}(\mathbf{G}_1(\mathbf{G}_1^t\mathbf{G}_1)^{-1}\mathbf{G}_1^t) + \mathrm{trace}(\mathbf{G}_2(\mathbf{G}_2^t\mathbf{G}_2)^{-1}\mathbf{G}_2^t) - 1. \quad (4.220)
$$

$\mathbf{G}_1$ and $\mathbf{G}_2$ are

$$
\mathbf{G}_1 = \begin{pmatrix}
g_{11}(X_{11}) & g_{12}(X_{11}) & g_{13}(X_{11}) & g_{14}(X_{11}) \\
g_{11}(X_{21}) & g_{12}(X_{21}) & g_{13}(X_{21}) & g_{14}(X_{21}) \\
\vdots & \vdots & \vdots & \vdots \\
g_{11}(X_{n1}) & g_{12}(X_{n1}) & g_{13}(X_{n1}) & g_{14}(X_{n1})
\end{pmatrix}, \quad (4.221)
$$

$$
\mathbf{G}_2 = \begin{pmatrix}
g_{21}(X_{12}) & g_{22}(X_{12}) & g_{23}(X_{12}) & g_{24}(X_{12}) \\
g_{21}(X_{22}) & g_{22}(X_{22}) & g_{23}(X_{22}) & g_{24}(X_{22}) \\
\vdots & \vdots & \vdots & \vdots \\
g_{21}(X_{n2}) & g_{22}(X_{n2}) & g_{23}(X_{n2}) & g_{24}(X_{n2})
\end{pmatrix}. \quad (4.222)
$$

(Hint: transform G_1 or G_2 into a centered smoother.)

(c) Confirm numerically that the value of the trace of a hat matrix obtained by in Problem **4.10(c)** is identical to that calculated by eq(4.220).

4.15   Produce an S-Plus or R object for carrying out the power method to obtain all of the eigenvalues of a symmetric matrix. Note that the power method enables the calculation of all the eigenvalues of a symmetric matrix, while the ACE algorithm enables the computation of the largest eigenvalues. The object aceit1() in (M) is useful.

4.16   Modify the object aceit1() presented in (M) to be able to use the smoothing spline as a smoother for each variable.

CHAPTER 5

NONPARAMETRIC REGRESSION WITH PREDICTORS REPRESENTED AS DISTRIBUTIONS

5.1 INTRODUCTION

Ordinary regression analysis represents predictors as scalars or vectors. When a predictor is a vector, the order of the element does not make any sense. However, if the predictors are represented as distributions, by regarding frequencies in a histogram as predictors, the order of the predictors has important implications, because the influences of neighboring predictors on the target variable (object variable) are considered to be similar.

For example, when the predictor is a histogram of daily average air temperatures during the growth of a plant, the effect of the frequency between $15°$ C and $16°$ C on the target variable is not supposed to be widely different from that of the frequency between $16°$ C and $17°$ C. Hence, nonparametric regression with predictors represented as distributions is an application of nonparametric regression in which the effects of neighboring predictors on the target variable are similar [2]. The similarity of the effects of neighboring predictors on the target variable means that the value of $m(\Xi_1, \Xi_2, \ldots, \Xi_k, \Xi_{k+1}, \ldots, \Xi_r)$ is close to those of $m(\Xi_1, \Xi_2, \ldots, \Xi_k, \Xi_k, \ldots, \Xi_r)$ and $m(\Xi_1, \Xi_2, \ldots, \Xi_{k+1}, \Xi_k, \ldots, \Xi_r)$.

Introduction to Nonparametric Regression, Kunio Takezawa.
Copyright © 2006 John Wiley & Sons, Inc.

Ordinary nonparametric regression assumes that, when the regression equation (regression function) is dipicted as $m(x_1, x_2, \ldots, x_r)$, the value of $m(X_1, X_2, \ldots, X_k, \ldots, X_r)$ ($\{X_1, X_2, \ldots, X_k, \ldots, X_r\}$ is the value given to $\{x_1, x_2, \ldots, x_k, \ldots, x_r\}$), which is not widely different from that of $m(X_1, X_2, \ldots, X_k + \Delta X_k, \ldots, X_r)$ (ΔX_k is a small value compared with the width of the regions where X_k has a value). The nonparametric regression treated here overcomes the problem caused by the large number of predictors when the influences of neighboring predictors are similar.

A nonparametric regression of this kind attributes great importance to the order relation of the predictor ($\{\xi_1, \xi_2, \ldots, \xi_r\}$). Furthermore, the dimension of a predictor (r) is considerably large in most situations. Therefore, $\{\xi_1, \xi_2, \ldots, \xi_r\}$ is considered to be an approximate expression of the continuous function $\xi(x)$, which satisfies $\xi(j) = \xi_j$. Then, the resulting regression equation accompanies a predictor represented as the distribution ($m'(\xi(x))$). This explains the background of classification of a regression in which the effects of neighboring predictors on the target variable are similar: it is sometimes identified as a functional data analysis [5], [6].

This chapter first describes the importance of and the calculations for this method through the investigation of the problem of predicting the rice yield (yield (kg) per 10 a, 10 a $= 1000 \ m^2$) using temperature data as predictors. Next, the method, which is similar to that described, of predicting the day of a developmental stage of a plant (e.g., the heading day of rice or the flowering day of a flower) using environmental conditions is mentioned. Finally, the contents of this method are summarized and simple simulations are illustrated. Although these methods assume that the predictors represent distributions, similar calculations result in beneficial regressions if the predictors share similar characteristics. Many examples that yielded results for practical purposes using such techniques are known (e.g., [3], [8], [4]).

5.2   USE OF DISTRIBUTIONS AS PREDICTORS

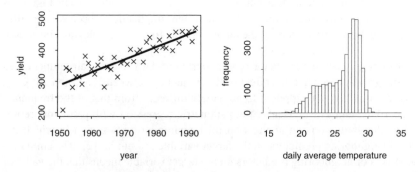

Figure 5.1   Rice yield (kg / 10 a) from 1951 through 1992 in Kagoshima Prefecture in Japan, and the linear curve fitted by the least squares (left). The accumulation of histograms of daily average air temperature ($^\circ$ C) in June, July, and August in Kagoshima Prefecture from 1951 through 1992 (right).

Figure 5.1(left) displays the rice yield (kg / 10 a) from 1951 through 1992 in Kagoshima Prefecture in Japan. These 42 data points are used as data for a target variable; they are denoted by $\{Y_i\}$ $(1 \leq i \leq 42)$. The linear line in this graph is obtained by fitting a linear line to these data by the least squares (least square method). The linear line is represented as

$$y = 4.086865x + 288.0848, \tag{5.1}$$

where x stands for the year (from 1951 through 1992) minus 1950. That is, $x = 1, 2, \ldots, 42$. The term y means the value of the estimate (estimated value) (kg / 10 a) of rice yield corresponding to the value of x. This equation is considered to depict the increase in rice yield realized by enhancing rice cultivation technology and improving the environment surrounding the rice plants.

Figure 5.1(right) displays a histogram of daily average air temperature ($^\circ$ C) in June, July, and August in Kagoshima Prefecture from 1951 through 1992. The frequency of the kth bin of this histogram shows the number of data points that satisfy the following conditions among the daily average air temperature data (t) observed over 42 years.

$$T_0 \leq t \leq T_0 + \Delta T \quad (k = 1),$$
$$T_0 + (k - 1)\Delta r < t \leq T_0 + k \cdot \Delta T \quad (2 \leq k \leq K), \tag{5.2}$$

where T_0 is the minimum value of the region used by this histogram (in this example, 15) and ΔT is the binwidth (in this example, 0.5). The value of K is the number of bins. In this example, $K = 40$ is set so that the region of this histogram covers the entire area where the data are located. The definition of a histogram (eq(5.2)) is along the line of the object of `hist()`, which S-Plus includes by default.

The estimates given by eq(5.1) are identified as $\{\tilde{Y}_i\}$ $(1 \leq i \leq 42)$. Hence, the examination of the relationship between $\{Y_i'\}$ $(Y_i' = Y_i - \tilde{Y}_i, \; 1 \leq i \leq 42)$ and air temperature in these years is expected to enable prediction of rice yield using air temperature. Then, the averages of daily average air temperatures ($^\circ$ C) from June through August (the number of data points is 92) are computed to derive the averages of daily average air temperatures during the period. The relationship between the averages and $\{Y_i'\}$ is exhibited in figure 5.2. Since the data positioned at the left end correspond to that observed in 1951 (cool summer damage occurred that year), the cool summer damage explains the considerable decrease in rice yield that year. However, we cannot find any specific relationship between the air temperature in this period and the rice yield in the same period concerning other data. These data, therefore, dissuade us from predicting rice yield using air temperature. This means that the association shown in eq(5.1) between the year and the rice yield is the only functional relationship we obtained. Since the value of GCV given by eq(5.1) is 888.9416, the amount of error in eq(5.1), which is the most beneficial functional relationship available, is approximately the square root of this value (29.8 kg / 10 a).

This result is obtained by summarizing the daily average air temperatures of June, July, and August as the average of the 92-day data and using them as predictor values. The influence of environment on plants, however, should not be depicted by their average during a certain period, but should be regarded as the synthesis of the impacts of the daily environment on plants. From this point of view, the following model is considered to be a yield prediction model.

$$y = \sum_{j=1}^{J} m(t_j), \tag{5.3}$$

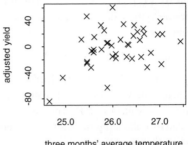

three months' average temperature

Figure 5.2   Rice yield (kg/ 10 a) in Kagoshima Prefecture from 1951 through 1992 minus the linear trend brought about by the year versus the averages of daily average air temperatures in June, July, and August.

where the period which is important for plant growth is set at $1 \leq j \leq J$, and t_j is defined as the $j$the day average air temperature. The term $m(t_j)$ is a function that represents the effect of the jth day average air temperature on plants. y denotes the rice yield, from which the influence of the year is subtracted. Since the influence of environment on plants is not limited to daily air temperature, $m(t_j)$ can be a function that also represents the effects of other environmental factors.

The simplest function considered as $m(t_j)$ is $m(t_j) = a \cdot t_j$ (a is a positive constant). Then, eq(5.3) becomes

$$y = a \cdot \sum_{j=1}^{J} t_j, \tag{5.4}$$

where $\sum_{j=1}^{J} t_j$ is called the accumulated temperature in the field of agricultural science. Furthermore, when we assume that plants are influenced by air temperature above a specific value, the summation of $(t_j - t_0)$ is carried out only if the value of t_j is larger than a given value (t_0, base temperature); a sum of this kind is called the effective accumulated temperature. This is equivalent to using the following equation as $m(t_j)$ in eq(5.3).

$$m(t_j) = \begin{cases} t_j - t_0 & \text{if} \quad t_j \geq t_0 \\ 0 & \text{if} \quad t_j < t_0. \end{cases} \tag{5.5}$$

Note that the effective accumulated temperature is also called the accumulated temperature in some situations.

Generalization of the concept of accumulated temperature by obtaining $m(t_j)$ in eq(5.3) on the basis of nonparametric regression leads to more general expressions for $m(t_j)$. Then, the data are assumed to consist of n sets, and the n sets of temperature data and the n rice yield data are represented as temperature data ($\{t_{ij}\}$ ($1 \leq i \leq n$, $1 \leq j \leq J$)) and rice yield data ($\{Y_i\}$ ($1 \leq i \leq n$) ($\{Y_i'\}$ plotted in figure 5.2)), respectively. Using this form of data, a regression equation based on this philosophy is derived.

First, daily average air temperatures are discretized. Among the ith set of temperature data, the number of data points that satisfy the following equation is defined as f_{ik}.

$$T_0 \le t_{ij} \le T_0 + \Delta T \quad (k = 1, \ 1 \le i \le n),$$
$$T_0 + (k - 1)\Delta r < t_{ij} \le T_0 + k \cdot \Delta T \quad (2 \le k \le K, \ 1 \le i \le n). \quad (5.6)$$

T_0 is configured to be smaller than the minimum daily average air temperature ($\{t_{ij}\}$). $(T_0 + K\Delta T)$ is set to be larger than the maximum value of $\{t_{ij}\}$. Then, f_{ik} becomes a distribution of data (a histogram) of the i-th set of temperature data. Therefore, the summation of f_{ik} with respect to k ends up with J for any value of i.

Next, $m(t_j)$ is also dicretized. That is, we set

$$m(t_j) = \bar{m}_k \quad \text{if} \quad T_0 \le t_j \le T_0 + \Delta T \quad (k = 1),$$
$$m(t_j) = \bar{m}_k \quad \text{if} \quad T_0 + (k - 1)\Delta T < t_j \le T_0 + k\Delta T. \quad (5.7)$$
$$(2 \le k \le K).$$

Then, if the transition of daily average air temperature is $\{t_1^*, t_2^*, \ldots, t_J^*\}$, and the distribution given by discretized temperature data is represented as $\{f_k^*\}$ ($1 \le k \le K$), the resultant estimate of rice yield (Y^*) is written as

$$Y^* = \sum_{j=1, k=1}^{J,K} \bar{m}_k \quad (5.8)$$

$$= \sum_{k=1}^{K} f_k^* \bar{m}_k, \quad (5.9)$$

where $\sum_{j=1,k=1}^{J,K} \bar{m}_k$ in the first line indicates that the summation takes place if the jth value of the ith temperature data satisfies $T_0 + (k - 1)\Delta T \le t_{ij} < T_0 + k\Delta T$.

The value below is minimized to derive $\{\bar{m}_k\}$ based on data by the least squares.

$$E_{distribution}(\{\bar{m}_k\}) = \sum_{i=1}^{n} (\sum_{k=1}^{K} f_{ik}\bar{m}_k - Y_i)^2 + \lambda \sum_{k=2}^{K-1} (\bar{m}_{k-1} - 2\bar{m}_k + \bar{m}_{k+1})^2.$$
$$(5.10)$$

As in the case with a smoothing spline (smoothing splines) (cf. Chapter 3 and Chapter 4), λ (> 0) is the smoothing parameter. This equation is represented as

$$E_{distribution}(\bar{\mathbf{m}}_1) = \| \mathbf{F}_1 \bar{\mathbf{m}}_1 - \mathbf{y} \|^2 + \lambda(\bar{\mathbf{m}}_1)^t \mathbf{S}_1 \bar{\mathbf{m}}_1, \quad (5.11)$$

where superscript t denotes transpose, and $\| \ \|$ indicates the length of a vector. Furthermore, $\mathbf{y}$ (a column vector with n elements), $\mathbf{F}_1$ (a matrix with a size of $n \times K$), $\bar{\mathbf{m}}_1$ (a column vector of K elements), and $\mathbf{S}_1$ (a matrix with a size of $K \times K$) are

$$\mathbf{y} = (Y_1, Y_2, Y_3, \ldots, Y_n)^t, \quad (5.12)$$

$$
\mathbf{F}_1 = \begin{pmatrix}
f_{11} & f_{12} & f_{13} & f_{14} & \cdots & \cdots & \cdots & f_{1K} \\
f_{21} & f_{22} & f_{23} & f_{24} & \cdots & \cdots & \cdots & f_{2K} \\
f_{31} & f_{32} & f_{33} & f_{34} & \cdots & \cdots & \cdots & f_{3K} \\
\vdots & \vdots & \vdots & \vdots & \vdots & \vdots & \vdots & \vdots \\
f_{(n-1)1} & f_{(n-1)2} & f_{(n-1)3} & f_{(n-1)4} & \cdots & \cdots & \cdots & f_{(n-1)K} \\
f_{n1} & f_{n2} & f_{n3} & f_{n3} & \cdots & \cdots & \cdots & f_{nK}
\end{pmatrix},
\tag{5.13}
$$

$$
\mathbf{m}_1 = (\bar{m}_1, \bar{m}_2, \bar{m}_3, \ldots, \bar{m}_K)^t,
\tag{5.14}
$$

$$
\mathbf{S}_1 = \begin{pmatrix}
1 & -2 & 1 & 0 & \cdots & \cdots & \cdots & \cdots & \cdots & \cdots & \cdots & 0 \\
-2 & 5 & -4 & 1 & 0 & \cdots & \cdots & \cdots & \cdots & \cdots & \cdots & 0 \\
1 & -4 & 6 & -4 & 1 & 0 & \cdots & \cdots & \cdots & \cdots & \cdots & 0 \\
0 & 1 & -4 & 6 & -4 & 1 & 0 & \cdots & \cdots & \cdots & \cdots & 0 \\
\vdots & \vdots & \vdots & \vdots & \vdots & \vdots & \vdots & \vdots & \vdots & \vdots & \vdots & \vdots \\
0 & \cdots & \cdots & \cdots & \cdots & 0 & 1 & -4 & 6 & -4 & 1 & 0 \\
0 & \cdots & \cdots & \cdots & \cdots & 0 & 1 & -4 & 6 & -4 & 1 \\
0 & \cdots & \cdots & \cdots & \cdots & \cdots & 0 & 1 & -4 & 5 & -2 \\
0 & \cdots & \cdots & \cdots & \cdots & \cdots & \cdots & 0 & 1 & -2 & 1
\end{pmatrix}.
\tag{5.15}
$$

The term $\{\bar{m}_k\}$ is differentiated with respect to each element to be 0 to minimize $E_{distribution}(\bar{\mathbf{m}}_1)$. Then, a simultaneous equation is derived. Hence, a hat matrix and GCV and the like are obtainable. This calculation, however, is ill-conditioned in many cases. GCVPACK [1] is a program that takes this point into account sufficiently. The program is written in standard FORTRAN77 and is available on the Internet free of charge.

The subroutine named dsnsm in GCVPACK and subordinate subroutines included in GCVPACK and LINPACK provide the values of $\{a_j\}$ $(1 \le j \le K)$ as output by minimizing

$$
E_{dsnsm} = \sum_{i=1}^{n} \left(\sum_{k=1}^{K} \zeta_{ik} a_k - Y_i \right)^2 + \lambda \sum_{j=1}^{K} \sum_{k=1}^{K} a_j \Psi_{jk} a_k,
\tag{5.16}
$$

where, since $\{\zeta_{ik}\}$ is the kth element of the ith data set, the matrix in which $\{\zeta_{ik}\}$ is the ikth element is regarded as a design matrix. Ψ_{jk} is the jkth element of a quasi-definite matrix (positive semidefinite matrix, nonnegative definite matrix) (Ψ) with a size of $K \times K$. If $\lambda = 0$ is substituted in this equation, it becomes a simultaneous equation (the constant term equals 0) for deriving the values of the regression coefficient ($\{a_j\}$). Furthermore, when Ψ is a diagonal matrix, it becomes an equation for carrying out a ridge regression to obtain a multiple regression equation (the constant term is 0). To delete the condition under which the constant term is 0, $\zeta_{i(K+1)} = 1$ $(1 \le i \le n)$ are set, and K in eq(5.16) is replaced by $(K+1)$.

In this manner, the form of eq(5.16) contains an ordinary multiple regression equation and one based on ridge regression as special cases. Furthermore, if Ψ is used as $\mathbf{S}_1$ in eq(5.11), it is used to derive each value of $\{\bar{m}_k\}$, which is a discretized representation of $m(t_j)$ in eq(5.3).

Figure 5.3 shows the results of calculations in which rice yields (kg/ 10 a) in Kagoshima Prefecture in Japan from 1951 to 1992 minus a yearly linear trend (Y_i'

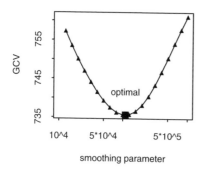 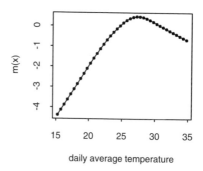

Figure 5.3   Rice yields (kg / 10 a) in Kagoshima Prefecture in Japan from 1951 to 1992 minus a yearly linear trend are used as $\{Y_i\}$ in eq(5.10) to carry out calculations using dsnsm included in GCVPACK. Optimization of the value of the smoothing parameter (eq(5.10)) to obtain $m(t_j)$ (■ is the optimal value) (left). $\{\bar{m}_k\}$ is given by the optimal smoothing parameter (right).

$(= Y_i - \tilde{Y}_i))$ are used as $\{Y_i\}$ in eq(5.10). After the functional relationship between the value of a smoothing parameter and that of GCV is outlined by calculating the values of GCV using 20 values $(10^{4.1}, 10^{4.2}, 10^{4.3}, \ldots, 10^6)$ as the smoothing parameter, the golden ratio searches are conducted; the results are shown in figure 5.3(left). Figure 5.3(right) demonstrates $\{\bar{m}_k\}$ given by the optimal smoothing parameter ($\lambda = 10^{5.041624}$). These calculations are based on the configurations of $T_0 = 15$, $\Delta T = 0.5$, and $K = 40$, and the subroutine dsnsm is used. The minimum value of GCV is 735.634. This indicates substantial progress from the GCV value of 888.9416 provided by fitting a linear equation in which the year is the predictor (figure 5.1(left) (page 326)). However, this reduction in the value of GCV can be regarded as no mystery because this calculation consists of two sequential steps: (1) fitting of a linear equation to the data in which the year is the predictor, and (2) estimation of $\{\bar{m}_k\}$. Therefore, this result cannot be firmly asserted to be beneficial even if the shape of $\{\bar{m}_k\}$ drawn in figure 5.3(right), which shows that the effect of daily average air temperature is saturated and maximized in a high temperature region, is persuasive.

Hence, the influences of the year on the rice yield and the distribution of air temperatures must be derived simultaneously. Furthermore, if the influences are represented as nonparametric functions, it is a step forward from the rough approximations of the relationship between the year and the yield using a linear curve. This calculation results in an exact comparison of models using GCV. Then, the model predicting rice yield as follows is used in place of eq(5.3).

$$y = \sum_{j=1}^{J} m_1(t_j) + \rho m_2(i), \qquad (5.17)$$

where $m_1(t_j)$ plays the same role as $m(t_j)$ in eq(5.3); it is a nonparametric regression that represents the influence of daily average air temperature on rice plants. In addition, $m_2(i)$ is another nonparametric function which represents the effect that the observation is carried out in the ith year. ρ of the second term on the right-hand side is a positive constant. It plays a role in adjusting the difference between the value of the smoothing parameter for $m_1(t_j)$ and that for $m_2(i)$. When regression is conducted using a small value of ρ, the roughness of $m_2(\cdot)$ is restrained even if the smoothing parameter for $m_1(\cdot)$ is the same as that for $m_2(\cdot)$, because $m_2(\cdot)$ becomes a violently oscillating function.

$m_1(t_j)$ is discretized in the same manner as $m(t_j)$. Then, $\{\bar{m}_{1,1}, \bar{m}_{1,2}, \ldots, \bar{m}_{1,K}\}$ is set, and $m_2(i)$ is defined as $m_2(1) = \bar{m}_{2,1}, m_2(2) = \bar{m}_{2,2}, \ldots, m_2(n) = \bar{m}_{2,n}$. Then, to obtain eq(5.17), $\{\bar{m}_{1,k}\}$ and $\{\bar{m}_{2,i}\}$ are derived by minimizing

$$
\begin{aligned}
E_{two-dim}(\{\bar{m}_{1,k}, \bar{m}_{2,i}\}) = & \sum_{i=1}^{n}\left(\sum_{k=1}^{K} f_{ik}\bar{m}_{1,k} + \rho \cdot \bar{m}_{2,i} - Y_i\right)^2 \\
& + \lambda\left(\sum_{k=2}^{K-1}(\bar{m}_{1,k-1} - 2\bar{m}_{1,k} + \bar{m}_{1,k+1})^2\right. \\
& \left. + \sum_{i=2}^{n-1}(\bar{m}_{2,i-1} - 2\bar{m}_{2,i} + \bar{m}_{2,i+1})^2\right) \\
& + \chi\left(\sum_{k=1}^{K}\bar{m}_{1,k}^2 + \sum_{i=1}^{n}\bar{m}_{2,i}^2\right).
\end{aligned}
\tag{5.18}
$$

If the fourth line (χ is a positive constant) in eq(5.18) does not exist, the values of $\{\bar{m}_{1,k}\}$ and $\{\bar{m}_{2,i}\}$ are not determined uniquely, because in eq(5.17), even if $m_1(t_j)$ is defined as $m_1(t_j)$ plus $(1/J)$ and $m_2(i)$ is defined as $m_2(i)$ minus $(1/\rho)$, y remains the same. To overcome this problem, a term, which is analogous to a term in ridge regression for deriving a multiple regression equation, is added. This term functions not just to yield a unique solution, but to stabilize the solution. Since this term in this example mainly aims at providing a unique solution, the value of χ is not required to be large; $\chi = 10^{-10}$ is adopted. However, the characteristics of data can require stabilization of regression coefficients ($\{\bar{m}_{1,k}\}$ and $\{\bar{m}_{2,i}\}$) using a somewhat larger value as χ for beneficial regression.

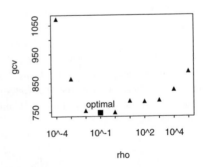

Figure 5.4   Optimization of ρ in eq(5.18).

Eq(5.18) is rewritten as

$$
E_{two-dim}(\bar{\mathbf{m}}_2) = \|\mathbf{F}_2\bar{\mathbf{m}}_2 - \mathbf{y}\|^2 + \lambda\bar{\mathbf{m}}_2^t\mathbf{S}_2\bar{\mathbf{m}}_2 + \chi\bar{\mathbf{m}}_2^t\mathbf{I}\bar{\mathbf{m}}_2,
\tag{5.19}
$$

where $\mathbf{I}$ is an identity matrix with a size of $(K+n) \times (K+n)$. Furthermore, $\mathbf{F}_2$, $\mathbf{m}_2$, and $\mathbf{S}_2$ are

$$
\mathbf{F}_2 = \begin{pmatrix}
f_{11} & f_{12} & f_{13} & \cdots & f_{1K} & \rho & 0 & \cdots & \cdots & \cdots & 0 & 0 \\
f_{21} & f_{22} & f_{23} & \cdots & f_{2K} & 0 & \rho & \cdots & \cdots & \cdots & 0 & 0 \\
\vdots & \vdots & \vdots & \vdots & \vdots & \vdots & \vdots & \vdots & \vdots & \vdots & \vdots & \vdots \\
f_{n1} & f_{n2} & f_{n3} & \cdots & f_{nK} & 0 & 0 & \cdots & \cdots & \cdots & 0 & \rho
\end{pmatrix}, \quad (5.20)
$$

$$
\mathbf{m}_2 = (\bar{m}_{1,1}, \bar{m}_{1,2}, \ldots, \ldots, \bar{m}_{1,K}, \bar{m}_{2,1}, \bar{m}_{2,2}, \ldots, \ldots, \bar{m}_{2,n})^t, \quad (5.21)
$$

$$
\mathbf{S}_2 = \begin{pmatrix}
1 & -2 & 1 & 0 & \cdots & \cdots & \cdots & \cdots & \cdots & \cdots & \cdots & \cdots & 0 \\
-2 & 5 & -4 & 1 & 0 & \cdots & \cdots & \cdots & \cdots & \cdots & \cdots & \cdots & 0 \\
1 & -4 & 6 & -4 & 1 & 0 & \cdots & \cdots & \cdots & \cdots & \cdots & \cdots & 0 \\
0 & 1 & -4 & 6 & -4 & 1 & 0 & \cdots & \cdots & \cdots & \cdots & \cdots & 0 \\
\vdots & \vdots & \vdots & \vdots & \vdots & \vdots & \vdots & \vdots & \vdots & \vdots & \vdots & \vdots & \vdots \\
0 & \cdots & 0 & 1 & -4 & 6 & -4 & 1 & 0 & \cdots & \cdots & \cdots & 0 \\
0 & \cdots & \cdots & 0 & 1 & -4 & 5 & -2 & 0 & \cdots & \cdots & \cdots & 0 \\
0 & \cdots & \cdots & \cdots & 0 & 1 & -2 & 1 & 0 & \cdots & \cdots & \cdots & 0 \\
\ddots & \ddots & \ddots & \ddots & \ddots & \ddots & \ddots & \ddots & \ddots & \ddots & \ddots & \ddots & \ddots \\
0 & \cdots & \cdots & 0 & 0 & 1 & -2 & 1 & 0 & 0 & 0 & \cdots & 0 \\
0 & \cdots & \cdots & 0 & 0 & -2 & 5 & -4 & 1 & 0 & 0 & \cdots & 0 \\
0 & \cdots & \cdots & 0 & 0 & 1 & -4 & 6 & -4 & 1 & 0 & \cdots & 0 \\
\vdots & \vdots & \vdots & \vdots & \vdots & \vdots & \vdots & \vdots & \vdots & \vdots & \vdots & \vdots & \vdots \\
0 & \cdots & \cdots & 0 & 0 & 0 & 0 & \cdots & 0 & 1 & -4 & 5 & -2 \\
0 & \cdots & \cdots & 0 & 0 & 0 & 0 & \cdots & 0 & 0 & 1 & -2 & 1
\end{pmatrix}. \quad (5.22)
$$

Figure 5.4 demonstrates the values of GCV for optimizing the value of λ using GCV when the value of ρ in eq(5.18) varies as $\rho = 10^{-4}, 10^{-3}, \ldots, 10^5$; $T_0 = 15$, $\Delta T = 0.5$, $K = 40$, and $n = 42$ are fixed. The value $\rho = 0.1$ is obtained as the optimal one. Then, ρ in eq(5.18) is set as described and the optimal value of λ is used to obtain $\{\bar{m}_{1,1}, \bar{m}_{1,2}, \ldots, \bar{m}_{1,K}\}$. Figure 5.5 shows the line yielded by connecting the estimates (approximations of $m_1(t_j)$) and the linear interpolation of $\{\bar{m}_{2,1}, \bar{m}_{2,2}, \ldots, \bar{m}_{2,n}\}$ (approximations of $m_2(i)$). The resulting value of GCV is 771.725. Substantial progress is observed compared with the regression rendered by fitting a linear curve to the relationship between the year and the rice yield. Since regression is carried out for data only once in both cases, the comparison of regression equations using GCV is justified.

This result is rewritten as

$$
y = \sum_{j=1}^{J} m_1'(t_j) + m_2'(i), \quad (5.23)
$$

where $m_1'(t_j)$ and $m_2'(i)$ are given, respectively, as

$$
m_1'(t_j) = m_1(t_j) - 3.72, \quad (5.24)
$$

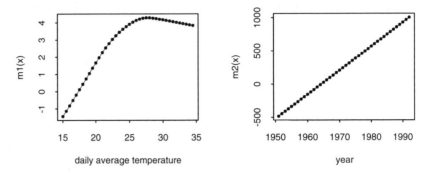

Figure 5.5 $m_1(t_j)$ (in eq(5.17)) obtained using eq(5.18) (left); $m_2(i)$ (in eq(5.17)) obtained using eq(5.18) (right).

$$m_2'(i) = 0.1 \cdot m_2(i) + 342.2. \tag{5.25}$$

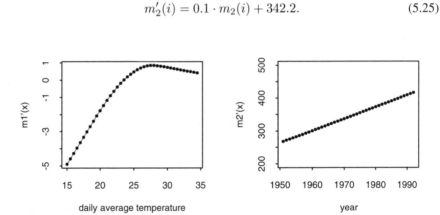

Figure 5.6 $m_1'(t_j)$ given by eq(5.24) (left); $m_2'(i)$ given by eq(5.25) (right).

Figure 5.6 shows $m_1'(t_j)$ and $m_2'(i)$ defined above. A comparison of this graph with figure 5.1(left) (page 326) and figure 5.3(right) (page 331) implies that these results are fairly similar to those from fitting a linear curve to the yearly change in rice yield by the least squares and by estimating $m(t_j)$ in eq(5.3).

The method described in this section derives regression equations using nonparametric regression with predictors represented as distributions. Furthermore, this method is easily extended to the form of an additive model or the form with some predictors without distributions (e.g., year). Regression and modelling given by fuzzy inference assume that the values of data are accompanied by broadness in some cases. Diverse methods have been developed thus far for carrying out regressions using such data. However, since the method based on functional data analysis has nonparametric regression as its background, this methodology characteristically realizes regression accompanied by the sense of security supported by concepts such as a hat matrix.

5.3 NONPARAMETRIC DVR METHOD

If prediction of the advancement of developmental stages of a plant, or forecasting events such as flowering, heading, and ripening, is realized, it is expected to contribute to efficient farm management, prevention of meteorological disasters, and so forth. Therefore, diverse methods have been attempted for this purpose. The value of the application of nonparametric regression to this problem has been realized [7]. This method is called the nonparametric DVR method (nonparametric DeVelopmental Rate method). The essence of this method is the same as the method mentioned in the previous section in which $Y_i = 1$ $(1 \leq i \leq n)$ is configured. The outline of this method is described next.

This method may be explained using the example of paddy rice; the number of days from the planting day through the heading day is predicted, because this method was originally motivated by the demand to predict the heading day to realize appropriate cultivation management and is utilized commonly in this field. However, prediction of the day of a developmental stage of other plants such as the silking day of wheat, the harvesting day of cabbage, or the flowering day of a flowering plant has verified the effectiveness of this method, and hence the practical use of this method has spread steadily.

First, a simple equation for representing the relationship between the environmental factors and developmental stage is set as follows (e.g., [9]); it is based on the assumption that the advancement of a developmental stage is determined by the daily environment.

$$DVI_J = \sum_{j=1}^{J} DVR(Env_j), \qquad (5.26)$$

where Env_j stands for a vector that depicts the environmental factors (daily average air temperature, amount of solar radiation, precipitation, and so on) on the jth day. $DVR(Env_j)$ is a function with environmental factors as predictors, and the value given by this function is the DVR (DeVelopmental Rate). DVI_J denotes DVI (DeVelopmental Index) on the Jth day. The previous day of the planting data, for example, is set as $j = 0$ and $DVI_0 = 0$. Then, if $DVR(\cdot)$ gives about 1.0 as the value of DVI on the heading day, the heading day is predicted using the results of the prediction of environmental factors (i.e., weather forecasting).

To obtain $DVR(\cdot)$, we face two problems: (1) the selection of predictors among intricate environmental factors surrounding plants, and (2) the choice of the form of the function to be used as $DVR(\cdot)$. The former is not a serious problem because qualitative knowledge concerning effects of environmental factors on the developmental stage of plants is usually at hand. As for the heading day of rice, the chief environmental factors are known to be daily average air temperature and daytime. The daytime is the amount of time from sunrise to sunset, and the possible sunshine duration is the length of time that the center of the sun appears; the daytime is slightly longer than the possible sunshine duration. Hence, the problem is the second item, that is, the choice of the functional form of $DVR(\cdot)$. The assumed procedure for this purpose consists of deriving the values of parameters in possible parametric functions and examining the fitness; this is similar to ordinary smoothing. When daily average air temperature is used as the only environmental factor, we use a linear equation, a quadratic equation, a cubic equation, and a logistic function with daily average air temperature as a predictor. This process, however, is essentially different from that of ordinary smoothing and interpolation. It is impossible to plot the relationship between the value of the predictor and the value of the target variable on a graph to discuss the selection of a function to be used, even if the number of predictor is one. The reason is that the value of the predictor for each data point for estimation of $DVR(\cdot)$ is not a scalar but a distribution constructed by the daily environmental factor, and the value of the target variable is fixed at 1. This fact enhances the difficulty in judging intuitively if there is room for improvement in the regression equations given by parametric functions. On the other hand, use of nonparametric regression realizes highly practical regression equations without determining the relationship between data and a regression equation in a graph. Various techniques for obtaining curves which have similar appearance to those drawn by human beings articulate the significance of existence in terms of coping with a problem in which the functional relationship is difficult to grasp in a graph. Furthermore, this fact straightforwardly indicates that nonparametric regression with predictors represented by distributions hold considerable value, which is not limited as simply an automatic procedure of a human task.

Then, let us introduce the simplest case among methods to handle this problem in the context of nonparametric regression. When air temperature is the only environmental factor available, the discretization using eq(5.6) (page 329) summarizes the meteorological data as $\{f_{ik}\}$ $(1 \leq i \leq n,\ 1 \leq k \leq K)$. Note that the method of obtaining a histogram here is the ordinary one. The following equation is used in place of eq(5.6) to obtain $\{f_{ik}\}$.

$$T_0 + (k-1)\Delta r \leq t_{ij} < T_0 + k \cdot \Delta T \quad (1 \leq k \leq K,\ 1 \leq i \leq n). \quad (5.27)$$

The values of $\sum_{k=1}^{K} f_{ik}$ corresponding to the value of i depend on the value of i because they are the days from the planting day through the heading day in the year.

Furthermore, $DVR(t_j)$ is discretized as

$$DVR(t_j) = \bar{r}_k \quad \text{if} \quad T_0 + (k-1)\Delta T \leq t_j < T_0 + k\Delta T \quad (1 \leq k \leq K). \quad (5.28)$$

Hence, DVI $(DVI_{heading}(i))$ on the heading day of the ith year is written as

$$DVI_{heading}(i) = \sum_{k=1}^{K} f_{ik}\bar{r}_k. \quad (5.29)$$

Therefore, the following value is minimized to estimate $\{\bar{r}_k\}$.

$$E_{DVR1}(\{\bar{r}_k\}) = \sum_{i=1}^{n}\left(\sum_{k=1}^{K} f_{ik}\bar{r}_k - 1.0\right)^2 + \lambda \sum_{k=2}^{K-1}(\bar{r}_{k-1} - 2\bar{r}_k + \bar{r}_{k+1})^2. \quad (5.30)$$

Next, let us consider the problem in which the number of environmental factors is two; that is, the value of DVR on the day j is, for example, represented as a function of air temperature (t_j) and daytime (l_j). This DVR is assumed to be depicted as $DVR^t(t_j) + DVR^l(l_j)$ (t_j is the air temperature on the jth day and l_j is the daytime on the jth day). Both $DVR^t(t_j)$ and $DVR^l(l_j)$ are smooth nonparametric functions. Then, $\{\bar{r}_k\}$ and $\{\bar{s}_k\}$ are obtained to minimize

$$
\begin{aligned}
E_{DVR2}(\{\bar{r}_k\}, \{\bar{s}_k\}) \;=\; & \sum_{i=1}^{n} \left(\sum_{k=1}^{K^t} f_{ik}^t \bar{r}_k + \sum_{k=1}^{K^l} f_{ik}^l \bar{s}_k - 1.0 \right)^2 \\
& + \lambda^t \sum_{k=2}^{K^t-1} (\bar{r}_{k-1} - 2\bar{r}_k + \bar{r}_{k+1})^2 \\
& + \lambda^l \sum_{k=2}^{K^l-1} (\bar{s}_{k-1} - 2\bar{s}_k + \bar{s}_{k+1})^2 \\
& + \chi \left(\sum_{k=1}^{K^t} \bar{r}_k^2 + \sum_{k=1}^{K^l} \bar{s}_k^2 \right),
\end{aligned}
\tag{5.31}
$$

where K in eq(5.30) is defined as K^t, and $\{f_{ik}\}$ in the same equation is denoted by $\{f_{ik}^t\}$. Furthermore, the data on daytime are discretized into K^l bins to obtain $\{f_{ik}^l\}$ $(1 \le i \le n, 1 \le k \le K^l)$, and $DVR^l(\cdot)$ is discretized into K^l bins to become $\{\bar{s}_k\}$ $(1 \le k \le K^l)$. The smoothing parameters for $\{\bar{r}_k\}$ and $\{\bar{s}_k\}$ are λ^t and λ^l, respectively. The meaning of the term on the third line of the equation is the same as that of the fourth line in eq(5.18) (page 332).

5.4 FORM OF NONPARAMETRIC REGRESSION WITH PREDICTORS REPRESENTED AS DISTRIBUTIONS

The methods of calculating nonparametric regression with predictors represented as distributions can take a variety of forms. The methods presented in the previous sections are simple but highly valuable from a practical point of view. However, this method should be understood in more general forms to realize the extensive use of nonparametric regression with predictors represented as distributions. This section develops the methods treated in the previous sections into slightly genenal forms and indicates their significance by showing a simple example.

This section assumes that the values of predictors of data are represented as the functions $\{\xi_i(x)\}$ $(1 \le i \le n)$, and the number of predictor values for the ith data, which take a value between x_1 and x_2 $(x_1 < x_2)$, is defined as $\int_{x_1}^{x_2} \xi_i(x)dx$ (not necessarily an integer). The values of the target variables for this data are depicted as $\{Y_i\}$ $(1 \le i \le n)$. Furthermore, if the value of $\{\xi_i(x)\}$ is 0 outside $x_{min} \le x \le x_{max}$, $a(x)$, which minimizes $E_{function}$ defined as follows, realizes a plain regression which uses a function representing distributions and the like as a predictor.

$$
E_{function} = \sum_{i=1}^{n} \left(\int_{x_{min}}^{x_{max}} a(x)\xi_i(x)dx - Y_i \right)^2.
\tag{5.32}
$$

We derive $a(x)$, by which Y_i closely approximates the inner product between the function of $a(x)$ and that of $\xi_i(x)$. When the predictor is not a distribution but a scalar, $\{X_i\}$ $(1 \leq i \leq n)$ are rewritten as $\xi_i(x) = \delta(x - X_i)$. As for these data, eq(5.32) is transformed into

$$
\begin{aligned}
E_{delta} &= \sum_{i=1}^{n} \left(\int_{x_{min}}^{x_{max}} a(x)\delta(x - X_i)dx - Y_i \right)^2 \\
&= \sum_{i=1}^{n} (a(X_i) - Y_i)^2.
\end{aligned}
\tag{5.33}
$$

We arrive at an ordinary regression: $a(x)$, by which $a(X_i)$ is close to Y_i, is derived on the basis of data $((X_i, Y_i)$ $(1 \leq i \leq n))$. In the context of the regression in which the function of $\xi_i(x)$ is used as the ith data, an ordinary regression corresponds to the use of a delta function as $\xi_i(x)$.

To derive $a(x)$ based on the general $\{\xi_i(x)\}$, one method is the expansion of $a(x)$ using $\{b_j(x)\}$ $(1 \leq j \leq M)$ as a basis as follows.

$$
a(x) = \sum_{j=1}^{M} a_j b_j(x).
\tag{5.34}
$$

Then, eq(5.32) is rewritten as

$$
E_{expansion} = \sum_{i=1}^{n} \left(\int_{x_{min}}^{x_{max}} \sum_{j=1}^{M} a_j b_j(x)\xi_i(x)dx - Y_i \right)^2.
\tag{5.35}
$$

Next, $\{b_j(x)\}$ $(1 \leq j \leq M)$ as below are used.

$$
b_j(x) = \begin{cases} 1 & \text{if } x_{min} + (j-1)\Delta x \leq x < x_{min} + j\Delta x \\ 0 & \text{otherwise}, \end{cases}
\tag{5.36}
$$

where Δx is

$$
\Delta x = \frac{x_{max} - x_{min}}{M}.
\tag{5.37}
$$

Furthermore, $\xi_i(x)$ is also expanded using $\{\xi_{ij}\}$ $(1 \leq j \leq M)$ as

$$
\xi_i(x) = \sum_{j=1}^{M} \xi_{ij} c_j(x),
\tag{5.38}
$$

where $\{c_k(x)\}$ $(1 \leq k \leq M)$ are as below.

$$
c_j(x) = \begin{cases} \frac{1}{\Delta x} & \text{if } x_{min} + (j-1)\Delta x \leq x < x_{min} + j\Delta x \\ 0 & \text{otherwise}. \end{cases}
\tag{5.39}
$$

Then, eq(5.35) is transformed into

$$
E_{expansion} = \sum_{i=1}^{n} \left(\int_{x_{min}}^{x_{max}} \sum_{j=1}^{M} (a_j b_j(x)) \sum_{k=1}^{M} (\xi_{ik} c_k(x))dx - Y_i \right)^2
$$

$$
\begin{aligned}
&= \sum_{i=1}^{n} \left(\int_{x_{min}}^{x_{max}} \sum_{j=1}^{M} \sum_{k=1}^{M} (a_j \xi_{ik} b_j(x) c_k(x)) dx - Y_i \right)^2 \\
&= \sum_{i=1}^{n} \left(\sum_{j=1}^{M} \xi_{ij} a_j - Y_i \right)^2.
\end{aligned}
\tag{5.40}
$$

To add a condition that $\{a_j\}$ must vary smoothly with respect to j, the equation as follows accompanied by λ as a smoothing parameter is employed.

$$
E_{expansion-smooth}(\{a_k\}) = \sum_{i=1}^{n} \left(\sum_{j=1}^{M} \xi_{ij} a_j - Y_i \right)^2 + \lambda \sum_{k=2}^{M-1} (a_{k-1} - 2a_k + a_{k+1})^2
\tag{5.41}
$$

This equation has the same form as eq(5.10) (page 329).

If the integration of $\xi_i(x)$ ($\approx \sum_{j=1}^{M} \xi_{ij} c_j(x)$) in the entire region results in J for all the values of i, the following equation is satisfied.

$$
\begin{aligned}
J &= \int_{x_{min}}^{x_{max}} \sum_{j=1}^{M} \xi_{ij} c_j(x) dx \\
&= \sum_{j=1}^{M} \xi_{ij} \qquad (1 \le i \le n).
\end{aligned}
\tag{5.42}
$$

Hence, the corresponding $\{\xi_{ij}\}$ ($1 \le k \le M$) are regarded as frequencies of histograms when $\sum_{j=1}^{M} \xi_{ij} = J$ (J is the sum of frequencies of a histogram) is satisfied for all data (i.e., for all values of i). Then, eq(5.41) is the equation for deriving a nonparametric regression equation ($\sum_{j=1}^{M} a_j b_j(x)$) when the sum of frequencies is constant; the frequencies are obtained by representing distributions as histograms. We obtain the equation as

$$
\sum_{j=1}^{M} a_j b_j(x) = a_j \quad \text{if } x_{min} + (j-1)\Delta x \le x < x_{min} + j\Delta x,
\tag{5.43}
$$

for which a_j is the local value of $a(x)$.

Furthermore, if $\{\xi_i(x)\}$ is a probability density function, we have

$$
\begin{aligned}
1 &= \int_{x_{min}}^{x_{max}} \sum_{j=1}^{M} \xi_{ij} c_j(x) dx \\
&= \sum_{j=1}^{M} \xi_{ij} \qquad (1 \le i \le n).
\end{aligned}
\tag{5.44}
$$

Consequently, if the setting satisfies $\sum_{j=1}^{M} \xi_{ij} = 1$, nonparametric regression equations based on predictors represented as probability density functions are realized.

Hence, $\sum_{j=1}^{M} a_j b_j(x)$ becomes

$$\sum_{j=1}^{M} a_j b_j(x) = a_j \quad \text{if } x_{min} + (j-1)\Delta x \le x < x_{min} + j\Delta x. \tag{5.45}$$

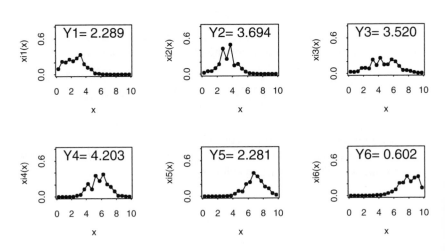

Figure 5.7   Values of predictors of six simulation data ($\{\xi_i(x_j)\}$). The values of a target variable are also shown.

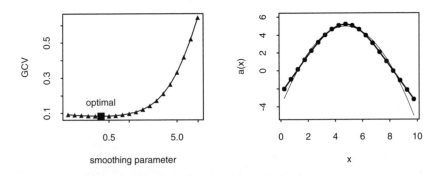

Figure 5.8   Optimization of the smoothing parameter using GCV (the result of the optimization using the golden ratio search is indicated by ■) to regress the data shown in figure 5.7 using eq(5.41). The thick solid line shows estimates given by $\sum_{j=1}^{M}(a_j \cdot b_j(x))$ ($\lambda = 10^{-0.4165415} = 0.3832291$), and the thin line shows the values from the equation that generates the data ($y = 5 - 0.4(x-5)^2$) (right).

Figure 5.7 shows the six simulated data sets ($\{\xi_i(x_j)\}$ ($1 \le i \le 6$, $\{x_j\} = \{0.5, 1, 1.5, \ldots, 10\}$)) for examining the role of the calculation given by eq(5.41). We set $\Delta x = 0.5$, $M = 20$, $x_{min} = 0.25$, and $x_{max} = 9.75$. Since $\{\xi_i(x_j)\}$

is assumed to be supplied by discretization of the probability density function, the following equations hold:

$$\sum_{j=1}^{20} \xi_i(x_j) = 2 \qquad (1 \le i \le 6), \qquad (5.46)$$

$$\sum_{j=1}^{20} \xi_{ij} = \sum_{j=1}^{20} \xi_i(x_j)\Delta x = 1 \qquad (1 \le i \le 6). \qquad (5.47)$$

That is, $\{\xi_{ij}\}$ $(1 \le i \le 6, \ 1 \le j \le 20)$ are the values in figure 5.7 multiplied by 0.5. "$Y_i =$" in the six graphs represents the value of the target variable for these data ($\{Y_i\}$). These values of $\{Y_i\}$ are calculated by

$$Y_i = \sum_{j=1}^{20}\Big(5 - 0.4(x_j - 5)^2\Big)\xi_{ij} + \epsilon_i \qquad (1 \le i \le 6), \qquad (5.48)$$

where $\{\epsilon_i\}$ are realizations of $N(0.0, 0.2^2)$ (a normal distribution; the mean is 0.0 and the standard deviation is 0.2). The results of the calculation based on eq(5.41) using these data are shown in figure 5.8. Figure 5.8(left) demonstrates the process of optimization of the value of the smoothing parameter using GCV. The values of $\{a_j\}$ are obtained using this smoothing parameter; the resultant $\sum_{j=1}^{M} a_j b_j(x)$ are shown in figure 5.8(right). The equation ($y = 5 - 0.4(x - 5)^2$), that generates the data, is reproduced almost correctly.

Next, the means of the six distributions are used as the values of a predictor, and $\{Y_i\}$ are utilized as values of a target variable to be regressed; figure 5.9 shows the result. When nonparametric regression with predictors represented as distributions is not employed and regression with a scalar as a predictor is adopted, these data are used. The result of regression using a smoothing spline with these data is shown in figure 5.10. Figure 5.10(left) shows the process of optimization of the smoothing parameter using GCV. Figure 5.10(right) indicates the resulting estimates. A comparison with figure 5.8(right) reveals that the estimates are substantially far from the equation ($y = 5 - 0.4(x - 5)^2$) that generates the data. Even if $\{\epsilon_i\}$ in eq(5.48) are given by random numbers generated by another initial value, the results are similar in most cases. This implies that the wrap-up of distributions by substituting their means for them results in appreciable loss of information contained in data.

Next, let us consider the probability density function, which takes the value of 0 except in a narrow region. That is, the value of $(\xi_{i1}, \ldots, \xi_{ij}, \ldots, \xi_{iM})^t$ takes the value of 1 only if $j = D(i)$ is satisfied; otherwise, it is 0. Then, eq(5.41) is transformed into

$$E_{density}(\{a_k\}) = \sum_{i=1}^{n}(a_{D(i)} - Y_i)^2 + \lambda \sum_{k=2}^{M-1}(a_{k-1} - 2a_k + a_{k+1})^2. \qquad (5.49)$$

$\{\hat{a}_k\}$, which are obtained by minimizing this value, are equivalent to the result of smoothing $\{a_{D(i)}, Y_i\}$ $(1 \le i \le n)$ using a smoothing spline (the second difference, instead of the second derivatives, is employed as the roughness penalty). This is

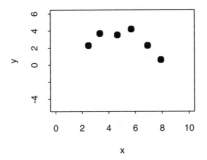

Figure 5.9   The means of the distributions of the six simulated data sets shown in figure 5.7 are calculated, and the relationship between the means and the corresponding values of a target variable is illustrated.

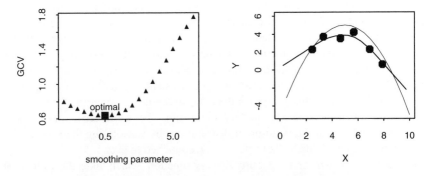

Figure 5.10   Optimization of the smoothing parameter using GCV (the result of the optimization using the golden ratio search is indicated by ■) to smooth the six data sets in figure 5.9 using a smoothing spline. The resulting regression curve is shown by a thick solid line. The equation $(y = 5 - 0.4(x - 5)^2)$, which generates the data, is drawn using the thin solid line (right).

identical to the derivation of $a(x)$ in eq(5.33) (page 338) using a smoothing spline, for which $X_i = D(i)$ in eq(5.33) holds approximately. By augmenting the value of M, the roughness penalty in eq(5.49) can be closely approximated by derivatives. Furthermore, eq(5.43) implies that a_j is a local value of $a(x)$. Therefore, when the distributions (probability density functions), which correspond to the values of predictors of data, take positive values only in the narrow regions in eq(5.41), the problem ends up with ordinary smoothing by the smoothing spline. This shows that eq(5.41) is the form given by generalizing the ordinary smoothing spline.

Another method that does not discretize $\{\xi_i(x)\}$ is possible. It aims at allowing for $\{\xi_i(x)\}$, which are supplied by actual experiments and observations; they are not smooth.

$$\xi_i(x) = \sum_{l=1}^{L} \delta(x - X_{il}).\tag{5.50}$$

The predictor values of the ith data are given as $\{X_{i1}, X_{i2}, \ldots, X_{iL}\}$. In an example of an experiment with plants, this corresponds to the experiment in which the growing period is L days and average air temperature is X_{il} on the lth day of the ith experiment; the yield of the ith experiment is Y_i. The initial part of this chapter introduced a method that transforms $\{X_{i1}, X_{i2}, \ldots, X_{iL}\}$ into a histogram. However, by expressing a distribution (constituted by a finite number of data points) in the form of a histogram, a number of similar data points falling in the same bin are treated as data with the same value. This means that a partial loss of information which data possess occurs. Consequently, it is desirable to use the form of eq(5.50) in order to depict data that have the form of distributions. Then, eq(5.35) (page 338) becomes

$$E'_{expansion-L} = \sum_{i=1}^{n} \left(\sum_{j=1}^{M} \sum_{l=1}^{L} a_j b_j(X_{il}) - Y_i \right)^2.\tag{5.51}$$

In particular, when $L = 1$ is set; each distribution consists of one data point, and eq(5.50) is rewritten as

$$E'_{expansion-1} = \sum_{i=1}^{n} \left(\sum_{j=1}^{M} a_j b_j(X_{i1}) - Y_i \right)^2.\tag{5.52}$$

When bases of the B-spline (B-splines) are used as $\{b_j(x)\}$, and the roughness penalty is employed to represent that $\sum_{j=1}^{M} a_j b_j(x)$ as a smooth function, we reach the equation for the smoothing spline (cf. eq(3.240) (page 180)). Hence, if the smoothing spline is generalized to the form of eq(5.51), regression with predictors represented as distributions using the B-spline as a base is realized. Application of this idea raises the sophistication of the nonparametric DVR method much further.

Moreover, Gaussian functions can be used as bases as follows:

$$b_j(x) = \frac{1}{\sqrt{2\pi\sigma_j^2}} \exp\left(-\frac{(x - \mu_j)^2}{2\sigma_j^2}\right).\tag{5.53}$$

This is the same as the equation used in a RBF (Radial Basis Function) network. This indicates that since the method based on eq(5.51) uses distributions as predictors and uses generalized bases, it is a generalized version of a RBF network.

A RBF network is understood as a metaphor of neurons (bases) and their stimuli (predictors). In line with this schema, distributional stimuli to neurons correspond to the distribution formed by the strength of stimuli over a certain time span. If only the mean of the strengths of stimuli in a certain time span is treated, much information is missed. Hence, distributional stimuli are preferable to represent the influence of stimuli on neurons. Furthermore, distributional stimuli are regarded as those with uncertainty. Along the lines of this methodology, regression reflected by the difference in uncertainty in data is realized; this regression does not use data that are means of distributions representing uncertainty, but uses the distributions as stimuli as they are.

5.5   EXAMPLES OF S-PLUS OBJECT

(A) An object for carrying out the same calculation as dsnsm (a routine included in GCVPACK) in S-Plus
Object: dsnsm()

```
function(x, ldx, y, sigma, ldsigm, nobs, npar, nnull,
     adiag, tau, lamlim, ntbl, dout, iout, coef, svals, tbl,
     ldtbl, auxtbl, iwork, liwa, work, lwa, job, info)
{
#   (1)
     dll.load("c:\\datasets\\gcvpack.dll", symbols = "dsnsm_")
#   (2)
     storage.mode(tbl) <- "double"
     storage.mode(auxtbl) <- "double"
#   (3)
     fit <- .C("dsnsm_",
          x = as.double(x),
          ldx = as.integer(ldx),
          y = as.double(y),
          sigma = as.double(sigma),
          ldsigm = as.integer(ldsigm),
          nobs = as.integer(nobs),
          npar = as.integer(npar),
          nnull = as.integer(nnull),
          adiag = as.double(adiag),
          tau = as.double(tau),
          lamlim = as.double(lamlim),
          ntbl = as.integer(ntbl),
          dout = as.double(dout),
          iout = as.integer(iout),
          coef = as.double(coef),
          svals = as.double(svals),
```

```
      tbl = tbl,
      ldtbl = as.integer(ldtbl),
      auxtbl = auxtbl,
      iwork = as.integer(iwork),
      liwa = as.integer(liwa),
      work = as.double(work),
      lwa = as.integer(lwa),
      job = as.integer(job),
      info = as.integer(info))
#   (4)
    return(fit)
}
```

(1) The object file ""c:\\datasets\\gcvpack.dll"" (cf. Chapter 3) is made available in S-Plus.

(2) If `tbl` and `auxtbl` are replaced with `tbl = as.double (tbl)`, and `auxtbl = as.double(auxtbl)` in `.C()`, `tbl` and `auxtbl` are the outputs from `.C()` not as matrices but as vectors. To avoid this problem, prior to conducting `.C()`, `storage.mode(tbl) <- "double"` and `storage.mode(auxtbl) <- "double"` are carried out, and `tbl = tbl` and `auxtbl = auxtbl` are assigned in `.C()`.

(3) Calculations yielded by dsnsm included in gcvpack.f are carried out. For this purpose, all the arguments which dsnsm requires (input arguments, output arguments, and input output arguments) are transformed into the forms used in dsnsm; if it requires a double precision form, `as.double()` is used for transformation; if an integer form is required, `as.integer()` is used for transformation. No transformations and declarations are required for distinguishing among scalars, vectors, and matrices, or for specifying the number of elements of vectors and the size of matrices, since the forms used in S-Plus objects are taken over to dsnsm. As for contents and forms of arguments for dsnsm, refer to [1] or comment lines in the GCVPACK program.

(4) All of the arguments of dsnsm are outputted from `dsnsm()`.

(B) Object for conducting nonparametric regression with predictors represented as distributions using the object "dsnsm"
Object: distnon()

```
function(nobs, npar, hist1, y, lamlim, ntbl)
{
#   (1)
    ldx <- max(nobs, npar)
    dimhist <- dim(hist1)
    if(dimhist[1] > ldx) {
        stop("The value of nobs must be the number of elements of
        the row vector which constitutes hist1. dsnsm as it is
        does not function. S-Plus possibly stops.")
    }
    if(dimhist[2] < npar) {
        stop("The value of npar must be the number of elements of
        the column vector which constitutes hist1. dsnsm as it
```

```
            is does not function. S-Plus possibly stops.")
    }
    if(dimhist[1] < ldx) {
        hist1 <- rbind(hist1, matrix(rep(0, length =
        (ldx - dimhist[1]) * dimhist[2]), ncol = dimhist[2]))
    }
#   (2)
    ldsigm <- npar
    nnull <- 2
    adiag <- rep(0, length = nobs)
    tau <- 1
    dout <- rep(0, length = 5)
    iout <- rep(0, length = 3)
    coef <- rep(0, length = npar)
    svals <- rep(0, length = npar - nnull)
    tbl <- rep(0, length = (ntbl * 3))
    tbl <- matrix(tbl, ncol = 3)
    ldtbl <- ntbl
    auxtbl <- rep(0, length = 9)
    auxtbl <- matrix(auxtbl, ncol = 3)
    liwa <- 2 * npar - nnull
    iwork <- rep(0, length = liwa)
    lwa <- (npar - nnull) * (npar - 2 * nnull + 2 + nobs)
    + npar + nobs
    lwa1 <- (npar - nnull) * (npar - 2 * nnull + 2 + nobs)
    + npar + nobs
    lwa2 <- (npar - nnull) * (npar - nnull) * (nobs - nnull)
    + npar - nnull + nobs
    lwa <- max(lwa1, lwa2)
    work <- rep(0, length = lwa)
    job <- 111
    info <- 0
#   (3)
    sigma <- matrix(rep(0, times = npar * npar), nrow = npar)
    sigma[1, 1:3] <- c(1, -2, 1)
    sigma[2, 1:4] <- c(-2, 5, -4, 1)
    for(ii in 3:(npar - 2)) {
        sigma[ii, (ii - 2):(ii + 2)] <- c(1, -4, 6, -4, 1)
    }
    sigma[npar - 1, (npar - 3):npar] <- c(1, -4, 5, -2)
    sigma[npar, (npar - 2):npar] <- c(1, -2, 1)
#   (4)
    result <- dsnsm(hist1, ldx, y, sigma, ldsigm, nobs,
     npar, nnull, adiag, tau, lamlim, ntbl, dout,
     iout, coef, svals, tbl, ldtbl, auxtbl, iwork, liwa,
     work, lwa, job, info)
#   (5)
```

```
      tbl <- result$tbl
      auxtbl <- result$auxtbl
      coef <- result$coef
      info <- result$info
#    (6)
      return(tbl, auxtbl, coef, info)
}
```

(1) The value of ldx is required to be an integer, which has a value equal or greater than the value of max(nobs, npar) and to be the same as the number of row vectors that constitute hist1 (the design matrix in this context). This section examines whether the conditions are satisfied or not. If they are not satisfied, the execution of the object stops and the reason is shown. If the value of ldx does not fit that of hist1, distnon() does not function and, additionally, S-Plus itself may stop; the measure here prevents this from occurring. Furthermore, if the number of values of ldx is larger than the number of row vectors constituting hist1, a matrix with 0 elements is combined with hist1.

(2) Arguments whose values scarcely require alteration among those of dsnsm() are provided. Arguments whose values often require alteration are adopted as those of distnon() (nobs, npar, hist1, y, lamlim, and ntbl).

The object ntbl is the number of $\log_{10}$(smoothing parameter) to be attempted to minimize the value of GCV. The object tau is the constant value for supplying the criterion for censoring in singular value decomposition. It is recommended that it be configured at 1, and if it does not function well, 10 and 100 are used sequentially. In this example, the second digit of the value of job in distnon() is set at 0; hence, censoring in singular value decomposition does not take place, and the value of tau does not affect the result. To implement censoring in singular value decomposition, the second digit of job must have a value other than 0 (e.g., job=1111).

Moreover, the form and size of the arguments to be used as workplaces are designated. As for the value of lwa (the number of elements of work), explanation in the comment lines in GCVPACK indicates that it must be equal to or greater than ((npar-nnull)*(npar-2* nnull+2+nobs)+npar+nobs). Additionally, it says that if it is less than ((npar-nnull)* (nobs-nnull+1)+nobs), error occurs and the value of info ends up at 131 (in fact, if the values of nobs and npar are certain values, calculation stops along the way and the value of info becomes 131). To prevent this from occurring, the larger of these two values is set as lwa.

(3) The object sigma is configured at the same values as S_1 in eq(5.11) (page 329).

(4) The object dsnsm() conducts a calculation which is equivalent to that performed by dsnsm; the result is stored in the object result in the form of a list.

(5) The important objects (tbl, auxtbl, coef, and info) among the output arguments of dsnsm() are extracted and treated as sole objects.

(6) The objects tbl, auxtbl, coef, and info are outputted from distnon(). As for more details of the calculation of dsnsm(), the objects result$dout, result$iout, result$svals, and result$adiag are designated as output arguments of distnon().

An example of an object to use distnon() (construction of figure 5.3 (page 331))

```
function()
{
#   (1)
    d1 <- read.table("c:\\datasets\\kahist1.txt")
    hist1 <- t(d1)
    d2 <- read.table("c:\\datasets\\karice.csv", sep = ",")
    y <- d2[, 2]
#   (2)
    dimd1 <- dim(hist1)
    nobs <- dimd1[1]
    npar <- dimd1[2]
    lamlim <- c(4.1, 6)
    ntbl <- 20
#   (3)
    y <- y - 4.086865 * (1:nobs) - 288.0848
#   (4)
    dist1 <- distnon(nobs, npar, hist1, y, lamlim, ntbl)
#   (5)
    print("info=")
    print(dist1$info)
#   (6)
    tbl <- dist1$tbl
    sparam <- tbl[, 1]
    gcv <- tbl[, 2]
    auxtbl <- dist1$auxtbl
#   (7)
    par(mfrow = c(1, 2), mai = c(2, 1, 1, 0.2),
      oma = c(4, 4, 4, 4))
    plot(10^sparam, gcv, log = "x", type = "n",
      xlim = c(10^4, 10^6), xlab = "smoothing parameter",
      ylab = "GCV", lwd = 2)
    points(10^sparam, gcv, pch = 17)
    lines(10^sparam, gcv, lwd = 2)
    sparagcv <- auxtbl[1, 1]
    gcvmin <- auxtbl[1, 2]
    points(10^sparagcv, gcvmin, pch = 15, cex = 1.2)
    text(10^sparagcv, gcvmin + 6, "optimal")
#   (8)
    xx <- seq(from = 15.25, to = 34.75, by = 0.5)
    plot(xx, dist1$coef, type = "n", xlab = "daily average
      temperature", ylab = "m(x)", lwd = 2)
    lines(xx, dist1$coef, lwd = 2)
    points(xx, dist1$coef, pch = 16, cex = 0.5)
}
```

(1) Data are retrieved. Since kahist1.txt is a vector whose column vector is a histogram of each year, a transpose is performed so that each row vector becomes a

histogram. The file `karice.csv` contains the data as follows: the data of the year and the rice yield positioned in the direction of a row are aligned in the direction of a column.

> 1951,208
> 1952,343
> 1953,335
> 1954,281
> 1955,314
> 1956,315
>
> $\vdots$ $\vdots$

Moreover, the data on rice yield are extracted from d2 and are saved as y.

(2) The number of data points is stored as `nobs`. The number of regression coefficients is saved as `npar`. The value of `nobs` is the number of row vectors of a design matrix. The value of `npar` is the number of its column vectors. The values of `lamlim` and `ntbl` are designated so that a total of 20 values are used as the values of the smoothing parameter to calculate the values of GCV; they are $10^{4.1}, 10^{4.2}, \ldots, 10^{6}$.

(3) The values given by linear regression are subtracted from the values of rice yields.

(4) The values obtained above by `distnon()` are used as arguments to carry out nonparametric regression with predictors represented as distributions.

(5) The value of `info` is displayed; `info` is a component of the list object (`dist1`) outputted from `distnon()`. The value of `info` means the following.

`info= 0`: Calculation ends normally.

`info= -3`: The value of `nnull` (`nnull` is the dimension of the null space of `sigma`) is too small. The dimension of null space is also called nullity. Even if this error occurs, the value of `nnull` is adjusted automatically to conduct an appropriate calculation.

`info= -2`: The second element of `lamlim` is too small; this element stands for the maximum value of $\log_{10}$(smoothing parameter). Hence, the result derived from the value of the second element of `lamlim` is employed as the optimal one. To obtain the optimal value for the smoothing parameter, the second element of `lamlim` is augmented for recalculation. Note that with the optimal regression equation ($m(t_j)$ in eq(5.3) (page 327), $DVR(\cdot)$ in eq(5.26) (page 335) or $a(x)$ in eq(5.32) (page 337)) is almost linear, and the optimal value of $\log_{10}$(smoothing parameter) is very large; hence, regression coefficients (`coef`) which supply -2 as the value of `info` are adopted as the optimal ones.

`info= -1`: The first element of `lamlim` is too large; this element stands for the minimum value of $\log_{10}$(smoothing parameter). Hence, the result derived from the value of the first element of `lamlim` is employed as the optimal one. To obtain the optimal value of the smoothing parameter, the first element of `lamlim` is reduced for recalculation.

`info= 1`: The number of elements of a vector or the size of a matrix is incorrect in arguments of `dsnsm()`. Since GCVPACK and LINPACK are written in FORTRAN77, erroneous specifications of the number of elements of a vector or of the size of a matrix cause malfunctions.

`info= 2`: The value of `lwa` (the number of elememts of `work` (a vector for workspace)) is too small. In this example, since the value of `lwa` is identified by other values of arguments in distnon(), correctly specified values of these arguments prevent errors.

info= 3: The value of liwa (the number of elements of iwork (a vector for workspace)) is too small. In this example, the value of liwa is supplied by other arguments in distnon(); correctly specified values of these arguments prevent errors.

info= 4: The value of ntbl or tau is wrong.

$100 <$ info < 200: ddcom, which dsnsm uses as one of its subroutines, encounters errors.

$200 <$ info < 300: dgcv, which dsnsm uses as one of its subroutines, encounters errors.

(6) The component dist1$tbl is transformed into a sole object tbl. Since the column vector located at the first column of tbl denotes $\log_{10}$(smoothing parameter) (the values specified by lamlim and ntbl in (2)), it is saved as sparam. The column vector positioned at the second column in tbl represents the values of GCV given by the values of sparam; it is stored as gcv. Note that the column vector seated at the third column in tbl contains the values of the residual sums of squares corresponding to the elements of sparam divided by the number of data points.

(7) The relationship between the 20 values of the smoothing parameter and the values of GCV is drawn on a graph. As in the case with dtpss() (Chapter 3), dsnsm() takes two steps to optimize the value of the smoothing parameter: (1) the values of GCV, designated by ntbl and lamlim, are calculated, and (2) the value of the smoothing parameter is optimized using the golden ratio search. Since the logarithm of the resultant value of the smoothing parameter to base 10 is contained in auxtbl[1, 1], and the corresponding value of GCV is contained in auxtbl[1, 2], these two values are saved as sparagcv and gcvmin, respectively. Then, the point given by sparagcv and gcvmin is plotted on a graph.

(8) The appearance given by the regression equation is drawn using the resulting optimal values of regression coefficient (coef).

(C) Object for conducting nonparametric regression to create an additive model with predictors represented as distributions using "dsnsm()"
Object: distadd()

```
function(nobs, npar1, npar2, hist2, y, lamlim, ntbl, rho, chi)
{
# (1)
    npar <- npar1 + npar2
    ldx <- max(nobs, npar)
    dimhist <- dim(hist2)
    if(dimhist[1] > ldx) {
        stop("The value of nobs must be the number of elements
            of the row vector which constitutes hist2. dsnsm as it
            is does not function. S-Plus possibly stops.")
    }
    if(dimhist[2] < npar) {
        stop("The value of (npar1+npar2) must be the numbers of
            elements of the row vector which constitutes hist2.
```

```
        dsnsm as it is does not function. S-Plus possibly stops.")
    }
    if(dimhist[1] < ldx) {
        hist2 <- rbind(hist2, matrix(rep(0, length =
        (ldx - dimhist[1]) * dimhist[2]), ncol = dimhist[2]))
    }
#   (2)
    dout <- rep(0, length = 5)
    iout <- rep(0, length = 3)
    info <- 0
    tbl <- matrix(rep(0, length = (ntbl * 3)), ncol = 3)
    coef <- rep(0, length = npar)
    auxtbl <- matrix(rep(0, length = 9), ncol = 3)
    ldsigm <- npar
    adiag <- rep(0, length = nobs)
    tau <- 10
    ldtbl <- ntbl
    nnull <- 0
    svals <- rep(0, length = npar - nnull)
    job <- 111
    lwa1 <- (npar - nnull) * (npar - 2 *
     nnull + 2 + nobs) + npar + nobs
    lwa2 <- (npar - nnull) * (npar - nnull) *
     (nobs - nnull) + npar - nnull + nobs
    lwa <- max(lwa1, lwa2)
    work <- rep(0, length = lwa)
    liwa <- 2 * npar - nnull
    iwork <- rep(0, length = liwa)
# (3)
    sigma <- matrix(rep(0, times = npar * npar), ncol = npar)
    sigma[1, 1:3] <- c(1, -2, 1)
    sigma[2, 1:4] <- c(-2, 5, -4, 1)
    for(ii in 3:(npar1 - 2)) {
        sigma[ii, (ii - 2):(ii + 2)] <- c(1, -4, 6, -4, 1)
    }
    sigma[npar1 - 1, (npar1 - 3):npar1] <- c(1, -4, 5, -2)
    sigma[npar1, (npar1 - 2):npar1] <- c(1, -2, 1)
# (4)
    sigma[(npar1 + 1), (npar1 + 1):(npar1 + 3)] <-
     c(1, -2, 1)
    sigma[(npar1 + 2), (npar1 + 1):(npar1 + 4)] <-
     c(-2, 5, -4, 1)
    for(ii in (npar1 + 3):(npar - 2)) {
        sigma[ii, (ii - 2):(ii + 2)] <- c(1, -4, 6, -4, 1)
    }
    sigma[npar - 1, (npar - 3):npar] <- c(1, -4, 5, -2)
    sigma[npar, (npar - 2):npar] <- c(1, -2, 1)
```

```
# (5)
    sigma <- sigma + diag(npar) * chi
# (6)
    result <- dsnsm(hist2, ldx, y, sigma, ldsigm, nobs,
    npar, nnull, adiag, tau, lamlim, ntbl, dout, iout,
    coef, svals, tbl, ldtbl, auxtbl, iwork, liwa, work,
    lwa, job, info)
# (7)
    tbl <- result$tbl
    auxtbl <- result$auxtbl
    coef <- result$coef
    info <- result$info
# (8)
    return(tbl, auxtbl, coef, info)
}
```

(1) This part examines whether the condition for the value of ldx (cf. (B)) is satisfied. If it is not satisfied, the execution of the object stops and the reason is revealed. Furthermore, if the value of ldx is larger than the number of row vectors which constitute hist2, a matrix that consists of 0 is combined with hist2.

(2) The values of arguments that require a minimal change and the sizes of workspaces are designated.

(3) The upper part of S_2 in eq(5.19) (page 332) (this part smooths $m_1(\cdot)$) is put into sigma.

(4) The lower part of S_2 in eq(5.19) (this part smooths $m_2(\cdot)$) is put into sigma.

(5) The fourth line of the right-hand side of eq(5.18) (page 332) is put into sigma.

(6) The object dsnsm() carries out the same calculations as dsnsm; the result is saved in result.

(7) The important objects (tbl, auxtbl, coef, and info) among the output arguments of dsnsm() are extracted and treated as sole objects.

(8) The objects tbl, auxtbl, coef, and info are outputted from distadd().

An example of an object to use distadd() (construction of figure 5.6 (page 334))

```
function()
{
# (1)
    d1 <- read.table("c:\\datasets\\kahist1.txt")
    d1 <- t(d1)
    d2 <- read.table("c:\\datasets\\karice.csv", sep = ",")
    y <- d2[, 2]
# (2)
    dimd1 <- dim(d1)
    nobs <- dimd1[1]
    npar1 <- dimd1[2]
    npar2 <- nobs
    rho <- 0.1
```

```
    chi <- 1e-010
    npar <- npar1 + npar2
    lamlim <- c(2.25, 10)
    ntbl <- 20
# (3)
    hist2 <- cbind(d1, diag(nobs) * rho)
# (4)
    dist <- distadd(nobs, npar1, npar2, hist2, y,
      lamlim, ntbl, rho, chi)
# (5)
    print("info=")
    print(dist$info)
    sparagcv <- dist$auxtbl[1, 1]
    print("sparagcv=")
    print(sparagcv)
    gcvmin <- dist$auxtbl[1, 2]
    print("gcvmin=")
    print(gcvmin)
    par(mfrow = c(1, 2), mai = c(2, 1, 1, 0.2),
      oma = c(4, 4, 4, 4))
    xx <- seq(from = 15, to = 34.5, by = 0.5)
    plot(xx, dist$coef[1:npar1] - 3.72, type = "n",
      xlab =  "daily average temperature", ylab =
      "m1'(x)", lwd = 2)
    points(xx, dist$coef[1:npar1] - 3.72, pch = 16, cex = 0.5)
    lines(xx, dist$coef[1:npar1] - 3.72)
    xx2 <- seq(from = 1951, to = 1992, by = 1)
    plot(xx2, dist$coef[(npar1 + 1):npar] * 0.1 + 342.2,
      type = "n", ylim = c(200, 500), xlab = "year",
      ylab = "m2'(x)", lwd = 2)
    points(xx2, dist$coef[(npar1 + 1):npar] * 0.1 + 342.2,
      pch = 16, cex = 0.5)
    lines(xx2, dist$coef[(npar1 + 1):npar] * 0.1 + 342.2)
}
```

(1) A histogram (kahist1.txt) is retrieved and saved as d1; then it is transposed. The data of the years and the rice yields (karice.csv) are retrieved and are saved as d2; the data of the rice yields are extracted and stored as y.

(2) A variety of values for regression is set up. The object rho denotes the values of ρ in eq(5.17) (page 331), and chi represents χ in eq(5.18) (page 332).

(3) The design matrix (eq(5.20) (page 333)) (hist2) is created using d1.

(4) The values obtained above are used as arguments of distadd(), and an additive model with predictors represented as distributions is constructed.

(5) The values of info, sparagcv (the logarithm of the optimal smoothing parameter to base 10), and gcvmin (i.e., the value of GCV corresponding to the regression equation supplied by sparagcv) are shown. Then, two regression curves are drawn.

REFERENCES

1. D.M. Bates, M.J. Lindstorm, G. Wahba, and B.S. Yandell (1987). GCVPACK - routines for generalized cross validation. *Communication in Statistics, Series B*, Vol. 16, No. 1, pp. 263-297.

2. R.F. Engle, C.W.J. Granger, J.A. Rice, and A. Weiss (1986). Semiparametric estimates of the relation between weather and electricty sales. *Journal of the American Statistical Association*, Vol. 81, No. 394, pp. 310-320.

3. T. Miwa, W. Takahashi, and S. Ninomiya (1998). Nonparametric regression analysis of remotely sensed spectroscopic data with high wavelength-resolustion. *Proceedings of the Environmentrics Conference–Biometry Toward 2000 Environment*, pp. 24-25.

4. T. Ogden, C.E. Miller, K. Takezawa, and S. Ninomiya (2002). Functional regression in crop lodging assessment with digital images. *Journal of Agricultural, Biological and Environmental Statistics*, Vol. 7, No. 3, pp. 389-402.

5. J.O. Ramsay and B.W. Silverman (1997). *Functional Data Analysis*, Springer-Verlag.

6. J.O. Ramsay and B.W. Silverman (2002). *Applied Functional Data Analysis*, Springer-Verlag.

7. K. Takezawa and Y. Tamuara (1991). Use of smoothing splines to estimate rates of development. *Agricultural and Forest Meteorology*, Vol. 57, pp. 129-145.

8. H. Tsukimoto, M. Kakimoto, C. Morita, Y. Kikuchi, E. Hatakeyama, and Y. Miyazaki (2000). Knowledge discovery from fMRI brain images by logical regression analysis. *Discovery Science: Third International Conference, Ds 2000, Kyoto, Japan, December 4-6, 2000, Proceedings (Lecture Notes in Computer Science, 1967)*, Setsuo Arikawa (Editor) and Shinichi Morishita (Editor), Springer-Verlag.

9. C.Y. de Wit, R. Bronwer, and F.W.T. Penning de Vries (1970). The simulation of photosynthetic systems. *Proceedings of the IBP/PP, Technical Meeting, Trebon (1969), PUDOC, Wageningen*, pp. 47-60.

Problems

5.1   Consider applying the concept of regression with predictors represented as distributions to simple regression.

(a) Derive an equation to be minimized for obtaining a simple regression equation on the assumption that the predictors are represented as histograms.

(b) Differentiate the equation obtained in (a) to be 0. This results in linear simultaneous equations.

(c) On the basis of the simultaneous equation derived in (b), create an S-Plus or R object for carrying out a simple regression with predictors represented as histograms.

(d) Produce an S-Plus or R object for calculating a hat matrix and the values of CV and GCV for this regression.

(e) Assume that nine sets of experiments are carried out. The value of the predictor is measured 10 times in each set of experiments (the value varies considerably during each experiment), and the value of a target variable is measured once in each experiment (the value indicates the amount of accumulation during each experiment). As a result, the frequencies of $\{X_i\}$ and the values of a target variable ($\{Y_i\}$) as below are obtained. Operate the object produced in (c) to these sets of data to derive a linear regression equation. In addition, compute a hat matrix and the values of CV and GCV using the object constructed in (d).

5.2   Consider applying the concept of regression with predictors represented as histograms to quadratic regression. Conduct (a), (b), (c), (d), and (e) of Problem **5.1** by adopting a quadratic regression equation in place of a simple regression equation.

5.3   Consider the concept of simple regression with predictors represented as a set of delta functions. The function to be minimized for obtaining regression coefficients is written in the form of eq(5.51) (page 343); the number of bases (M) is 2 (a constant term and a linear term).

(a) Specify an equation to be minimized for realizing simple regression on the assumption that the predictors are represented as a set of delta functions.

(b) Differentiate the equation obtained in (a) to be 0. This results in linear simultaneous equations.

(c) On the basis of the simultaneous equations derived in (b), create an S-Plus or R object for carrying out simple regression with predictors represented as a set of delta functions.

(d) Construct an S-Plus or R object for calculating a hat matrix and the values of CV and GCV for this regression.

(e) Assume that the original nine sets of data presented in **5.1 (e)** are as below. Operate the object produced in (c) to these sets of data to derive a linear regression equation. In addition, obtain a hat matrix and the values of CV and GCV using the object constructed in (d).

5.4   Consider the concept of regression with predictors represented as a set of delta functions to quadratic regression. Conduct (a), (b), (c), and (d) of Problem **5.3** by employing a quadratic regression equation in place of a simple regression equation.

5.5   Consider the following equation obtained by deleting the last term on the right-hand side of eq(5.19) (page 332). We have

	$5 \leq X_i < 10$	$10 \leq X_i < 15$	$15 \leq X_i < 20$	$20 \leq X_i < 25$
d1	5	2	3	0
d2	1	2	3	1
d3	0	2	3	5
d4	0	1	2	6
d5	0	2	0	4
d6	0	0	0	4
d7	0	0	1	1
d8	0	0	0	0
d9	0	0	0	0

	$25 \leq X_i < 30$	$30 \leq X_i < 35$	$35 \leq X_i < 40$	$40 \leq X_i < 45$	Y_i
d1	0	0	0	0	2.8
d2	2	1	0	0	5.4
d3	0	0	0	0	6.2
d4	1	0	0	0	6.0
d5	3	1	0	0	7.6
d6	4	1	0	1	8.2
d7	4	3	1	0	9.7
d8	3	4	2	1	11.2
d9	0	0	4	6	13.2

$$E_{two-dim}(\bar{\mathbf{m}}_2) = \| \mathbf{F}_2\bar{\mathbf{m}}_2 - \mathbf{y} \|^2 + \lambda\bar{\mathbf{m}}_2^t\mathbf{S}_2\bar{\mathbf{m}}_2. \tag{5.54}$$

(a) Show that differentiation of eq(5.54) to be 0 with respect to each element of $\bar{\mathbf{m}}_2$ supplies

$$(\mathbf{F}_2^t\mathbf{F}_2 + \lambda\mathbf{S}_2)\bar{\mathbf{m}}_2 = \mathbf{F}_2^t\mathbf{y}. \tag{5.55}$$

(b) Use $\{X_{i1}\} = \{2, 5, 7, 6, 8, 9, 1, 2, 2, 4, 2, 6, 7, 8, 9\}$ and $\{X_{i2}\} = \{3, 5, 8, 1, 2, 4, 4, 3, 2, 4, 6, 6, 1, 5, 7\}$ as values of two predictors, and $\{Y_i\} = \{1.6, 4.4, 10.3, 0.8, 1.3, 3.1, 2.5, 1.6, 0.9, 2.9, 5.7, 6.1, 0.9, 4.5, 8.1\}$ as values of a target variable. The values of these two predictors do not appear to be distributions at first sight. If the values of these two predictors are transformed into the following histograms, however, they can be treated as distributions.

Using the histograms displayed in the two tables below, create an S-Plus or R object for deriving regression coefficients ($\bar{\mathbf{m}}_2$) by solving eq(5.55) directly (using the Gauss-Jordan method, for example). Confirm that this method does not result in regression coefficients.

	Observed predictor values	Y_i
d1	6.23, 6.87, 7.11, 8.01, 9.42, 11.97, 12.89, 15.74, 17.28, 17.92	2.8
d2	7.09, 10.64, 12.79, 15.41, 16.37, 16.44, 21.62, 26.21, 29.94, 31.11	5.4
d3	12.97, 13.81, 17.31, 17.71, 19.73, 21.05, 21.96, 23.58, 23.72, 24.12	6.2
d4	14.30, 15.17, 19.15, 20.07, 20.33, 20.47, 23.17, 23.60, 23.79, 27.91	6.0
d5	11.29, 14.80, 20.56, 21.03, 24.65, 24.67, 27.58, 28.45, 29.21, 31.29	7.6
d6	20.36, 21.58, 22.60, 24.23, 25.62, 25.64, 26.36, 27.48, 33.45, 44.99	8.2
d7	17.36, 21.85, 25.54, 26.71, 28.44, 28.50, 31.54, 32.75, 32.81, 36.31	9.7
d8	25.30, 26.01, 29.91, 30.58, 31.69, 33.22, 34.27, 36.27, 38.59, 43.98	11.2
d9	35.56, 38.33, 38.37, 38.62, 40.16, 40.55, 41.06, 43.72, 43.97, 44.64	13.2

(**c**) Using the methods described in Problem **4.10** of Chapter 4, produce an S-Plus or R object for obtaining regression coefficients ($\bar{m}_2$) in eq(5.55), and run the object. Confirm that in some cases these methods do not perform well. In what situations does it occur?

$\{X_{i1}\}$	1	2	3	4	5	6	7	8	9
d1	0	1	0	0	0	0	0	0	0
d2	0	0	0	0	1	0	0	0	0
d3	0	0	0	0	0	0	1	0	0
d4	0	0	0	0	0	1	0	0	0
d5	0	0	0	0	0	0	0	1	0
d6	0	0	0	0	0	0	0	0	1
d7	1	0	0	0	0	0	0	0	0
d8	0	1	0	0	0	0	0	0	0
d9	0	1	0	0	0	0	0	0	0
d10	0	0	0	1	0	0	0	0	0
d11	0	1	0	0	0	0	0	0	0
d12	0	0	0	0	0	1	0	0	0
d13	0	0	0	0	0	0	1	0	0
d14	0	0	0	0	0	0	0	1	0
d15	0	0	0	0	0	0	0	0	1

$\{X_{i2}\}$	1	2	3	4	5	6	7	8
d1	0	0	1	0	0	0	0	0
d2	0	0	0	0	1	0	0	0
d3	0	0	0	0	0	0	0	1
d4	1	0	0	0	0	0	0	0
d5	0	1	0	0	0	0	0	0
d6	0	0	0	1	0	0	0	0
d7	0	0	0	1	0	0	0	0
d8	0	0	1	0	0	0	0	0
d9	0	1	0	0	0	0	0	0
d10	0	0	0	1	0	0	0	0
d11	0	0	0	0	0	1	0	0
d12	0	0	0	0	0	1	0	0
d13	1	0	0	0	0	0	0	0
d14	0	0	0	0	1	0	0	0
d15	0	0	0	0	0	0	1	0

CHAPTER 6

SMOOTHING OF HISTOGRAMS AND NONPARAMETRIC PROBABILITY DENSITY FUNCTIONS

6.1 INTRODUCTION

Most nonparametric regression techniques described in the previous chapters assume that the distance between predictor values of data and the estimates (estimated values) rendered by a regression equation (regression function) is measured by the residual sum of squares. These techniques are based on the assumption that the data are approximated by

$$Y_i = m(X_i, \boldsymbol{\theta}) + \epsilon_i \quad (1 \leq i \leq n), \tag{6.1}$$

where n is the number of data points, X_i is the predictor value of each data point, and Y_i is the target variable (object variable) value of each data point. $\{\epsilon_i\}$ are random errors; the mean is 0.0 and the variance is $\sigma^2(x)$. Furthermore, $m(X_i, \boldsymbol{\theta})$ is a regression function, and $\boldsymbol{\theta}$ denotes a vector: the regression coefficients. Since nonparametric regression means regression accompanying a large number of parameters, the number of elements of $\boldsymbol{\theta}$ is large. Least squares (least square method) yields the values of $\boldsymbol{\theta}$ by minimizing

$$E_{squares} = \sum_{i=1}^{N} (m(X_i, \boldsymbol{\theta}) - Y_i)^2. \tag{6.2}$$

Introduction to Nonparametric Regression, Kunio Takezawa.
Copyright © 2006 John Wiley & Sons, Inc.

This method coincides with the fitting of regression equations by a maximum likelihood procedure (maximum likelihood method) when $\{\epsilon_i\}$ obey a normal distribution and $\sigma^2(x)$ $(0 < \sigma(x))$ is a constant that does not depend on x (cf. Chapter 3). That is, the minimization of $E_{squares}$ is equivalent to the maximization of likelihood. The likelihood of $m(\cdot)$ in light of $\{X_i, Y_i\}$, which obey a normal distribution, is depicted as

$$L_1(\boldsymbol{\theta}) = \frac{1}{\sqrt{2\pi\sigma^2}} \exp\left(-\sum_{i=1}^{N} \frac{(m(X_i, \boldsymbol{\theta}) - Y_i)^2}{2\sigma^2} \right). \tag{6.3}$$

The term likelihood, $L_1(\boldsymbol{\theta})$, is ascribed to the fact that, when the values of $\{X_i, Y_i\}$ and σ are assumed to be fixed in $L_1(\boldsymbol{\theta})$, the values of $L_1(\boldsymbol{\theta})$ given by various values of $\boldsymbol{\theta}$ indicate the degrees of likelihood of $\boldsymbol{\theta}$ in light of $\{X_i, Y_i\}$. When certain values are given as $\boldsymbol{\theta}$, a small value of $\sum_{i=1}^{N}(m(X_i, \boldsymbol{\theta}) - Y_i)^2$ means that $\boldsymbol{\theta}$ is realistic. Hence, the value of $L_1(\boldsymbol{\theta})$ is large, that is, the likelihood is large. On the other hand, a large value of $\sum_{i=1}^{N}(m(X_i, \boldsymbol{\theta}) - Y_i)^2$ means that $\boldsymbol{\theta}$ is difficult to accept, namely, the likelihood is small. Since the value of σ is not involved in the maximization of the likelihood, the maximum likelihood procedure is viable without estimating σ. This is the essence of the least squares.

The form of the function to represent a likelihood relies on the assumed model. Regression by minimizing a value which does not take the form of eq(6.2) is required in some situations. This chapter treats problems in which $m(X_i, \boldsymbol{\theta})$ is a histogram or a probability density function (density function), because these instances are typical examples that need equations other than eq(6.2) and have high practical value. Note that a myriad of theories and methods on smoothing a histogram and a probability density function based on nonparametric regression have been suggested thus far. Hence, the range of algorithms for obtaining estimates and methods for optimizing smoothing parameters is wide. This chapter treats only a small part of these methods. This chapter aims chiefly at introducing examples of the methods in this field to exemplify nonparametric regression techniques which are not based on the least squares. It is not the intent of the author to recommend any of the methods described herein.

6.2   HISTOGRAM

When data are given as $\{X_i\}$ $(1 \leq i \leq n)$, a resulting estimate of a histogram $(\{\hat{F}_j\}$ $(1 \leq j \leq R))$ is defined as

$$\hat{F}_j = F_j, \tag{6.4}$$

where F_j stands for the number of $\{X_i\}$ that fall in the jth bin; this number is called a frequency. The data that fall in the jth bin indicate the data that satisfy

$$r_0 \leq X_i \leq r_0 + \Delta r \quad (j = 1),$$
$$r_0 + (j-1)\Delta r < X_i \leq r_0 + j \cdot \Delta r \quad (2 \leq j \leq R), \tag{6.5}$$

where r_0 denotes the position of the left anchor, and Δr indicates the binwidth of a histogram. Since all data should fall in one of the bins, all X_i must satisfy $r_0 \leq X_i \leq r_0 + R\Delta r$; the values of r_0, Δr, and R are determined to satisfy this condition. However, most histograms utilize the following equations instead of eq(6.5).

$$r_0 + (j - 1)\Delta r \leq X_i < r_0 + j \cdot \Delta r \quad (1 \leq j \leq R). \tag{6.6}$$

Eq(6.5) is adopted here because `hist()`, which is included in S-Plus by default, uses eq(6.5).

A histogram defined by eq(6.4) is an effective tool for grasping one-dimensional data intuitively. However, the histogram has a rough shape in many situations, while a data distribution usually has a smooth shape. Furthermore, as shown in Chapter 1, an alternation of the values of r_0 and Δr can change the impression of the shape of a histogram greatly; this could cause serious problems. In a similar manner in which characterisitics that are not exhibited by representing the data $\{X_i, Y_i\}$ as a scatter graph are often clarified by smoothing the data using nonparametric regression, the characteristics of the data depicted by eq(6.4) are expected to become clearer by smoothing and by replacing them with the resulting smooth curve. Such a treatment allows analysts to argue that the characteristics are based on sufficient grounds.

Then, if a histogram is smoothed in a fashion similar to eq(6.2), a possible method involves minimizing the value of

$$E'_{squares} = \sum_{j=1}^{R} (m(j) - F_j)^2, \tag{6.7}$$

where $m(j)$ is a parametric or nonparametric regression equation. Eq(6.7) functions effectively when an association between a regression equation ($\{m(j)\}$) and a resultant histogram ($\{F_j\}$) is assumed as

$$F_j = m(j) + \epsilon_j \quad (1 \leq j \leq R), \tag{6.8}$$

where $\{\epsilon_j\}$ are errors derived from a normal distribution; the mean is 0 and the variance does not depend on j. It is not appropriate to represent a relationship between the regression equation for depicting a histogram and a histogram given by data. There are at least two intuitive reasons for considering eq(6.8) an inappropriate hypothesis. First, the distribution of the errors added to $m(j)$ is not assumed to stretch out symmetrically. This is apparent when we simply think that, while the actual value of F_j does not assume a negative value, eq(6.8) implies that F_j can assume a negative value. Second, since the number of data points is supposed to be large in the region where the value of the real possibility density function is large, the discrepancy between the value of a histogram and that of a real distribution tends to be wide. Consequently, we cannot assume that the amount of error is independent of j. Hence, the use of eq(6.7) for smoothing a histogram can produce unreliable results.

An index different from the residual sum of squares is required to show the degree of likelihood of a histogram in light of data. To create such an index, we note that the number of data points corresponding to each bin obeys a Poisson distribution. A Poisson distribution appears when the numbers of events which randomly occur are observed in a certain time period. For example, if accidents in a factory are counted every week, the numbers obey a Poisson distribution. Note, however, that we suppose that each accident happens independently. If one accident induces another one, or one accident prevents another one from occurring because of cautious checkup after the accident, the resulting data do not obey a Poisson distribution. Hence, if the jth factory is compared to the jth bin, and F_j is compared to the number of accidents occurring in the jth factory, F_j is considered to be a realization of a Poisson distribution. Then, if the jth factory has characteristics similar to those of the $(j+1)$th factory, smoothing can be used to derive the estimate of the number of accidents per week in the jth factory. This process is expected to yield a better histogram than $\{F_j\}$.

The probability density function $(P(\tilde{F}))$ of a Poisson distribution is

$$P(\tilde{F}) = \exp(-\lambda) \cdot \frac{\lambda^{\tilde{F}}}{\tilde{F}!}, \tag{6.9}$$

where $\lambda \geq 0$, and $\tilde{F}$ is a positive integer or 0. $\tilde{F}!$ indicates the factorial of $\tilde{F}$; the factorial of 0 is 1. $P(\tilde{F})$ satisfies

$$
\begin{aligned}
\sum_{\tilde{F}=0}^{\infty} \tilde{F} P(\tilde{F}) &= \sum_{\tilde{F}=1}^{\infty} \tilde{F} \cdot \exp(-\lambda) \cdot \frac{\lambda^{\tilde{F}}}{\tilde{F}!} \\
&= \lambda \exp(-\lambda) \sum_{\tilde{F}=1}^{\infty} \frac{\lambda^{(\tilde{F}-1)}}{(\tilde{F}-1)!} \\
&= \lambda \exp(-\lambda) \exp(\lambda) \\
&= \lambda. \tag{6.10}
\end{aligned}
$$

That is, the expectation of $P(\tilde{F})$ is λ.

When the frequency of the jth bin obeys a Poisson distribution $(P_j(\tilde{F}_j))$, $P_j(\tilde{F}_j)$ is written as

$$P_j(\tilde{F}_j) = \exp(-\lambda_j) \cdot \frac{\lambda_j^{\tilde{F}_j}}{\tilde{F}_j!}. \tag{6.11}$$

Hence, when a histogram $(\{F_j\} \; (1 \leq j \leq R))$ is given, the degree of cogency of the assumption that the value of λ of a Poisson distribution in the jth bin is λ_j, in light of the data (F_j), is represented as the likelihood

$$L_2(\lambda_j) = \exp(-\lambda_j) \cdot \frac{\lambda_j^{F_j}}{F_j!}. \tag{6.12}$$

The likelihood for the entire histogram becomes

$$L_3(\boldsymbol{\lambda}) = \prod_{j=1}^{R} \exp(-\lambda_j) \cdot \frac{\lambda_j^{F_j}}{F_j!}, \tag{6.13}$$

where $\boldsymbol{\lambda} = (\lambda_1, \lambda_2, \ldots, \lambda_R)^t$ (superscript t denotes transpose) is defined.

When the logarithm of $L_3(\boldsymbol{\lambda})$ (log-likelihood) is defined as $L_4(\boldsymbol{\lambda})$, we have

$$L_4(\boldsymbol{\lambda}) = -\sum_{j=1}^{R} (\lambda_j - F_j \cdot \log(\lambda_j) + \log(F_j!)). \tag{6.14}$$

Since the values of data are fixed, the last term on the right-hand side does not influence the maximization of $L_4(\boldsymbol{\lambda})$. Hence, we maximize

$$L_4'(\boldsymbol{\lambda}) = -\sum_{j=1}^{R}(\lambda_j - F_j \cdot \log(\lambda_j)). \tag{6.15}$$

Then, the differentiation of $L_4'(\boldsymbol{\lambda})$ with respect to λ_j to be 0 leads to

$$\frac{\partial L_4'(\boldsymbol{\lambda})}{\partial \lambda_j} = -1 + \frac{F_j}{\lambda_j} = 0. \tag{6.16}$$

As a result, we have

$$\hat{\lambda}_j = F_j. \tag{6.17}$$

That is, when a Poisson distribution is fitted to the jth bin of a histogram, the value of log-likelihood is maximized by setting the value of λ_j at F_j. Eq(6.10) indicates that the expectation of the frequency of the jth bin is λ_j. Hence, the expectation given by a maximum likelihood procedure becomes F_j. This means that the histogram defined previously as eq(6.4) (page 360) is obtained by fitting a Poisson distribution using a maximum likelihood procedure, and the expectations of the resulting distributions are used as estimates.

When eq(6.1) (page 359) is assumed, obtaining a regression equation is considered to be the derivation of a function which is the result of subtracting ϵ_i from the right-hand side of eq(6.1) and is also considered to be an estimation of $E[y|x]$ (the expectation of y: the average of y when the value of x is fixed and experiments or observations are repeated). However, if a histogram is smoothed on the assumption of a Poisson distribution, the estimation of a regression equation is regarded as the estimation of $E[y|x]$ because of the impossibility of eq(6.1). That is, $E[y|x]$ is more general as the definition of a regression equation. From this point of view, a regression equation is also called an expectation function.

6.3 SMOOTHING A HISTOGRAM

The previous section implies that an ordinary histogram assumes that each bin obeys a Poisson distribution, and after Poisson distributions are fitted to data using a maximum likelihood procedure, the expectations of the resulting distributions are used as estimates. Since eq(6.10) indicates that the expectation of a Poisson distribution is λ, $\hat{\lambda}_j$ in eq(6.15) is replaced with $\hat{F}_j$ (an estimate of the jth bin). That is, we have

$$\hat{\mathbf{F}} = \hat{\boldsymbol{\lambda}}, \tag{6.18}$$

where $\hat{\mathbf{F}} = (\hat{F}_1, \hat{F}_2, \ldots, \hat{F}_R)^t$ is defined. To smooth a histogram, this $\hat{F}_j$ is required to vary smoothly with respect to j.

First, since smoothing a histogram is regarded as smoothig the data in which predictor values are equispaced, we think of using a tool similar to a binomial filter (cf. Chapter 2). When a binomial filter for the smallest degree of smoothing is employed, the corresponding log-likelihood is

$$
\begin{aligned}
L_5(\boldsymbol{\lambda}) \;=\; & -\frac{1}{2}(\lambda_1 - F_1 \cdot \log(\lambda_1)) - \frac{1}{4}(\lambda_1 - F_2 \cdot \log(\lambda_1)) \\
& + \sum_{j=2}^{R-1} \Big(-\frac{1}{4}(\lambda_j - F_{j-1} \cdot \log(\lambda_j)) - \frac{1}{2}(\lambda_j - F_j \cdot \log(\lambda_j)) \\
& \qquad -\frac{1}{4}(\lambda_j - F_{j+1} \cdot \log(\lambda_j)) \Big) \\
& -\frac{1}{4}(\lambda_R - F_{R-1} \cdot \log(\lambda_R)) - \frac{1}{2}(\lambda_R - F_R \cdot \log(\lambda_R)). \quad (6.19)
\end{aligned}
$$

This means that when the cogency (log-likelihood) of λ_j is defined, the cogencies in light of F_{j-1} and F_{j+1} are considered as well as those in light of F_j. To minimize this value, $L_5(\boldsymbol{\lambda})$ is differentiated with respect to λ_j $(1 \leq j \leq R)$ to be 0; we have

$$
\begin{aligned}
\frac{\partial L_5(\boldsymbol{\lambda})}{\partial \lambda_1} &= -\frac{1}{2}\left(1 - \frac{F_1}{\lambda_1}\right) - \frac{1}{4}\left(1 - \frac{F_2}{\lambda_1}\right) = 0, \\
\frac{\partial L_5(\boldsymbol{\lambda})}{\partial \lambda_j} &= -\frac{1}{4}\left(1 - \frac{F_{j-1}}{\lambda_j}\right) - \frac{1}{2}\left(1 - \frac{F_j}{\lambda_j}\right) - \frac{1}{4}\left(1 - \frac{F_{j+1}}{\lambda_j}\right) = 0 \\
& \qquad\qquad\qquad\qquad\qquad\qquad\qquad\qquad (2 \leq j \leq R-1), \\
\frac{\partial L_5(\boldsymbol{\lambda})}{\lambda_R} &= -\frac{1}{4}\left(1 - \frac{F_{R-1}}{\lambda_R}\right) - \frac{1}{2}\left(1 - \frac{F_R}{\lambda_R}\right) = 0. \quad (6.20)
\end{aligned}
$$

Hence, we obtain

$$
\begin{aligned}
\hat{F}_1 &= \hat{\lambda}_1 = \frac{1}{3}F_1 + \frac{2}{3}F_2, \\
\hat{F}_j &= \hat{\lambda}_j = \frac{1}{4}F_{j-1} + \frac{1}{2}F_j + \frac{1}{4}F_{j+1} \quad (2 \leq j \leq R-1), \\
\hat{F}_R &= \hat{\lambda}_R = \frac{2}{3}F_{R-1} + \frac{1}{3}F_R. \quad (6.21)
\end{aligned}
$$

The histogram obtained by maximizing $L_5(\boldsymbol{\lambda})$ is identical to the result of smoothing $\{F_j\}$ by a binomial filter, except at two ends of the data region. The summation of $\{\hat{F}_j\}$ in eq(6.21) results in

$$
\sum_{j=1}^{R} \hat{F}_j = \frac{1}{3}F_1 + \frac{11}{12}F_2 + \sum_{j=3}^{R-2} F_j + \frac{11}{12}F_{R-1} + \frac{1}{3}F_R. \quad (6.22)
$$

The sum of $\{\hat{F}_j\}$ is not necessarily the same as that of $\{F_j\}$. To improve this point, a possible strategy is that all $\{\hat{F}_j\}$ are multiplied by the same constant so that the sum of $\{\hat{F}_j\}$ is the same as that of $\{F_j\}$. The purpose of the histogram determines whether such a procedure is required.

Figure 6.1(left) shows a histogram of daily average temperatures observed on Ishigaki Island (cf. chapter 1). In eq(6.5) (page 360), $r_0 = 8$, $\Delta r = 0.25$, and $R = 110$ are used. The resulting estimates are defined as $\{F_j\}$ and the use of eq(6.21) yields $\{\hat{F}_j\}$; figure 6.1(right) illustrates these estimates. Apparently, some

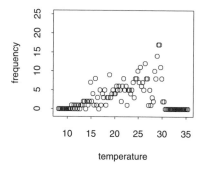
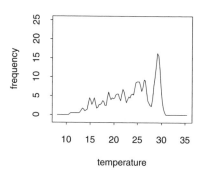

Figure 6.1 Histogram of daily average air temperatures observed on Ishigaki Island in 1986; $r_0 = 8$, $\Delta r = 0.25$, and $R = 110$ are used in eq(6.5) (page 360), (left). Histogram obtained by smoothing using eq(6.21), (right).

degree, but not a sufficient degree, of smoothing was applied to the histogram. Hence, the characteristics of the histogram are elucidated to a certain extent. Note that, while eq(6.9) (page 362) shows that even if the value of $\tilde{F}$ is very large, the value of $P(\tilde{F})$ does not become 0, the frequencies in figure 6.1(left) do not assume values larger than 365. When the total number of data points is a constant (365 in this example), the frequency of each bin does not obey a Poisson distribution in an exact sense. More correctly, the frequencies of the histogram as a whole obey a multinomial distribution. However, the assumption that the frequency of each bin obeys a Poisson distribution acts as a close approximation, except for a histogram in which most data are located in a small number of bins.

In a context similar to that in which a binomial filter is a beneficial tool for smoothing based on nonparametric regression but in which there are other desirable tools, various weights are possible in smoothing a histogram. Hence, $L_5(\hat{F}_j)$ is generalized to be

$$L_6(f(r)) = -\sum_{k=1}^{R} W(r, r(k))(f(r) - F_k \cdot \log(f(r))), \qquad (6.23)$$

where $L_6(f(r))$ is the weighted sum of the likelihoods of $f(\cdot)$ at r in light of $\{F_k\}$; the weights are $W(r, r(k))$. The function of $f(r)$ indicates a histogram defined to give an estimate for a continuous value ($r_0 + 0.5\Delta r \leq r \leq r_0 + (R - 0.5)\Delta r$). That is, the number of data points located in the region ($r - 0.5\Delta r \leq r \leq r + 0.5\Delta r$) is estimated to be $f(r)$. In addition, $r(k) = r_0 + (k - 0.5)\Delta r$ is assumed. $W(r, r(k))$ is a smooth function of the distance between the central points of the kth bin ($r(k)$) and r, and $W(r, r(k)) \geq 0$ is satisfied. Furthermore, the value of this function decreases as the value of $|r - r(k)|$ increases. This function is considered a kernel function. For example, the Gaussian function as follows is used:

$$W(r, r(k)) = \exp\left(-\frac{1}{2}\left(\frac{r - r(k)}{h}\right)^2\right), \qquad (6.24)$$

where h is a positive constant. Even if a positive constant is multiplied with $W(r, r(k))$, the result of the maximization of $L_6(f(r))$ is not affected; hence, a limitation such as $\int_{-\infty}^{\infty} W(r, r(k)) dr = 1$ is not required.

The generalization of $L_5(F_j)$ in eq(6.20) to $L_6(f(r))$ in eq(6.23) has two meanings. One is that $\{\hat{F}_j\}$ becomes $f(r)$. That is, the discrete histogram obtained is changed into continuous one. This procedure results in a continuous and smooth histogram. The other is that a general kernel function $W(r, r(k))$ has been utilized.

Calculus of variation with respect to $f(r)$ in (eq(6.23)) to be 0 yields

$$\frac{\delta L_6(f(r))}{\delta f(r)} = \sum_{k=1}^{R} W(r, r(k))\left(1 - \frac{F_k}{f(r)}\right) = 0. \qquad (6.25)$$

Since $f(r)$ in $L_6(f(r))$ is considered a variable, calculus of variation in eq(6.25) is regarded as differentiation with respect to the variable $f(r)$. Then, eq(6.25) yields the estimate

$$\hat{f}(r) = \frac{\sum_{k=1}^{R}(W(r, r(k)) \cdot F_k)}{\sum_{k=1}^{R} W(r, r(k))}. \qquad (6.26)$$

That is, $\hat{f}(r)$ is a weighted average of $\{F_k\}$. The weights are the values of the kernel function divided by a constant; the kernel function is used to calculate the likelihood of $f(r)$ in light of data.

A different function from $L_6(f(r))$ is a possible log-likelihood for deriving a smooth histogram. A function which approximates $f(r)$ in the neighborhood of a certain r^* is called $f'(r, r^*)$. To obtain $f'(r, r^*)$, first, the likelihood of the value of $f'(r, r^*)$ (i.e., $f'(r(k), r^*)$) at the central point of the kth bin in light of F_k is calculated; this procedure is conducted for each k. Then, the log-likelihood of $f'(r, r^*)$ is defined as the weighted sum of these log-likelihoods; the values of the weight decline as the position of the kth bin moves far from the position of r^*.

First, the log-likelihood of $f'(r, r^*)$ in light of F_k is

$$L_7(f'(r(k), r^*)) = -f'(r(k), r^*) + F_k \cdot \log(f'(r(k), r^*)). \qquad (6.27)$$

Next, when the log-likelihood of $f'(r, r^*)$ is defined, the log-likelihood in light of data located near F_k in addition to F_k is considered, and we have

$$L_8(f'(r, r^*)) = -\sum_{k=1}^{R} W(r^*, r(k))(f'(r(k), r^*) - F_k \cdot \log(f'(r(k), r^*))). \qquad (6.28)$$

$L_8(f'(r, r^*))$ is the result of deriving the cogency of $f'(r, r^*)$ by attaching larger weights to $\{F_j\}$ positioned close to r^*.

To obtain a smooth histogram by maximizing eq(6.28), the simplest function is assumed to be $f'(r, r^*)$: $f'(r, r^*) = \theta_0(r^*)$ ($\theta_0(r^*)$ is a function of r^*, but not of r). This assumption implies that the value of the estimate remains the same in the neighborhood of r^*. Hence, $L_8(f'(r, r^*))$ becomes

$$L_9(f'(r, r^*)) = -\sum_{k=1}^{R} W(r^*, r(k))(\theta_0(r^*) - F_k \cdot \log(\theta_0(r^*))). \qquad (6.29)$$

The differentiation of this equation with respect to $\theta_0(r^*)$ to be 0 yields

$$\hat{\theta}_0(r^*) = \hat{f}'(r, r^*) = \frac{\sum_{k=1}^{R}(W(r^*, r(k)) \cdot F_k)}{\sum_{k=1}^{R} W(r^*, r(k))}. \tag{6.30}$$

This is identical to eq(6.26), because eq(6.23) is obtained if $\theta_0(r^*)$ in eq(6.29) is defined as $f(r)$.

To smooth a histogram using eq(6.30) (i.e., eq(6.26)), the bandwidth (e.g., h in eq(6.24)) of $W(r, r(k))$ must be optimized. For this purpose, as in the case for regression by the least squares, cross-validation is used. Note, however, that the deviance defined by the following is used for a histogram:

$$D(\{\hat{F}_i\}; \{F_i\}) = 2\sum_{j=1}^{n}(L(\hat{F}_j^{max}; F_j) - L(\hat{F}_j; F_j)), \tag{6.31}$$

where $\hat{F}_j$ is an estimate for each j when $r^* = r_0 + (j - 0.5)\Delta r$ ($1 \leq j \leq R$) is assumed. $L(\hat{F}_j; F_j)$ indicates the log-likelihood; the likelihood of the estimates at the center of each bin ($\hat{F}_j$) is derived in light of the value of the bin (F_j, i.e., frequency). Then, we have

$$L(\hat{F}_j; F_j) = -\hat{F}_j + F_j \cdot \log(\hat{F}_j). \tag{6.32}$$

$\{\hat{F}_j^{max}\}$ denote estimates that maximize the log-likelihood; these values are mostly the same as $\{F_j\}$. The cogency of an estimate described as a log-likelihood such as $L(\hat{F}_j; F_j)$ is called a local log-likelihood. To emphasize that the log-likelihood does not have the form of the residual sum of squares, it is identified as a generalized local log-likelihood. Deviance is derived by the logarithmic transformation of the likelihood ratio, and the transformed value is multiplied by (-2); one set of estimates for the likelihood ratio is that which maximizes the log-likelihood. The model that maximizes the log-likelihood is called the saturated model. Therefore, deviance is an index for measuring the discrepancy between the saturated model and a model that is lower in fidelity to the data.

Hence, eq(6.31) becomes

$$D(\{\hat{F}_i\}; \{F_i\}) = 2\sum_{j=1}^{n}\left(-F_j + F_j \cdot \log(F_j) + \hat{F}_j - F_j \cdot \log(\hat{F}_j)\right). \tag{6.33}$$

When $\sum_{j=1}^{n} F_j = \sum_{j=1}^{n} \hat{F}_j$ is assumed, we obtain

$$\begin{aligned}
D(\{\hat{F}_i\}; \{F_i\}) &= 2\sum_{j=1}^{n}\left(F_j \cdot \log(F_j) - F_j \cdot \log(\hat{F}_j)\right) \\
&= 2\sum_{j=1}^{n}\left(F_j \cdot \log\left(\frac{F_j}{\hat{F}_j}\right)\right). \tag{6.34}
\end{aligned}$$

This value is twice as large as the Kullback-Leibler distance (a.k.a. Kullback-Leibler information). Note that $0 \log(0) = 0$ is assumed.

General deviance is defined as

$$D(\{\hat{F}_i\}; \{F_i\}) = 2\phi \sum_{j=1}^{n} (L(\hat{F}_j^{max}; F_j) - L(\hat{F}_j; F_j)). \tag{6.35}$$

The ϕ is called the dispersion parameter (scale parameter). The value of this parameter depends on the form of an assumed distribution. As for a Poisson distribution, $\phi = 1$ holds. Then, eq(6.31) is obtained. When the least squares is employed (i.e., a normal distribution is assumed), we have $\phi = \sigma^2$ (i.e., variance). Furthermore, the deletion of ϕ from the right-hand side of eq(6.35) yields a scaled deviance. Note, however, that this scaled deviance is identified as deviance in some cases (e.g., page 66 in [7]).

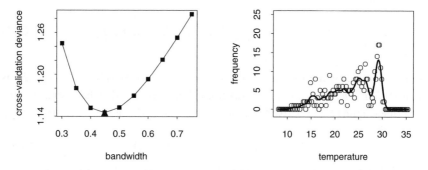

Figure 6.2   The histogram in figure 6.1(left) (page 365) is smoothed using eq(6.30). The locations of the bin break points are $\{8.0, 8.25, \ldots, 35.5\}$. Optimization of the bandwidth using a cross-validated deviance (left). The result of smoothing using the optimal bandwidth ($h = 0.45$) (right).

Cross-validation (CV) for the least squares is generalized. The cross-validated deviance is derived as

$$CV(\{\hat{F}_k\}; \{F_k\}) = \frac{1}{R} \sum_{j=1}^{R} D(\hat{F}_j^{-j}; F_j). \tag{6.36}$$

The value of bandwidth is determined to minimize the value of CV. Since $\hat{F}_j^{-j}$ denotes $\hat{F}_j$ which are obtained by omitting F_j, when we assume that an estimate of a histogram is a local constant (eq(6.30)), we obtain

$$\hat{F}_j^{-j} = \frac{\sum_{k=1(k \neq j)}^{R} (W(r(j), r(k)) \cdot F_k)}{\sum_{k=1(k \neq j)}^{R} W(r(j), r(k))}. \tag{6.37}$$

Figure 6.2 illustrates the result of smoothing the histogram (figure 6.1(left) (page 365)) on the basis of eq(6.30). The process of optimizing the bandwidth (h in eq(6.24) (page 365)) using a cross-validated deviance (eq(6.36)) is shown in figure 6.2(left). Since the figure indicates that $h = 0.45$ is optimal, smoothing with the bandwidth is conducted (figure 6.2(right)). The estimates are obtained at 200 equispaced points between $r_0 + 0.5\Delta r$ and $(r_0 + (R - 0.5) \cdot \Delta r)$; the two ends of the region are included.

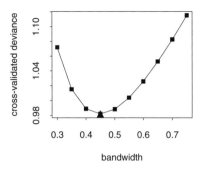
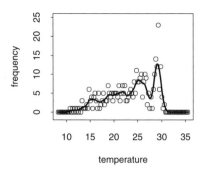

Figure 6.3 Result of smoothing using eq(6.30); the same data as those in figure 6.1 (left) are used and the positions of the bin break points are replaced with $\{8.15, 8.4, \ldots, 35.65\}$. Optimization of the bandwidth using a cross-validated deviance (left). Result of smoothing using the optimal bandwidth ($h = 0.45$) (right).

As discussed in Chapter 1, the appearance of a histogram is altered considerably by the position of anchors (r_0 in eq(6.5) (page 360)) in some situations. Figure 6.3 illustrates how smoothing ameliorates this problem. In this figure, the histogram created by shifting the positions of the bin break points to $(8.15, 8.4, \ldots, 35.65)$ is smoothed. The curve in figure 6.3(right), which is given by a procedure similar to that of figure 6.2, is very close to that in figure 6.2(right). This implies that smoothing has overcome the inessential differences between histograms and revealed the fundamental nature of the histograms.

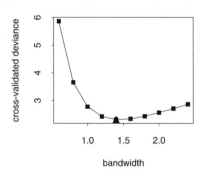
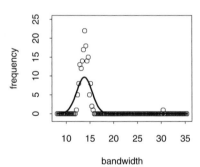

Figure 6.4 Example of a histogram with an outlier. Optimization of the bandwidth using a cross-validated deviance (left). Result of smoothing using the optimal bandwidth ($h = 1.4$) (right).

However, problems concerning the optimization using the cross-validated deviance have been pointed out [10]. One is that the optimal value of bandwidth can be altered markedly by slight changes in the values of data. The other is that the resulting smoothing tends to leave undesirable roughness in a histogram (e.g., [10]). Furthermore, when the data contain outliers, the optimization using the cross-validated deviance evidently yields a very large bandwidth. This problem is illustrated using a simple example (figure 6.4 and figure 6.5). The data in figure 6.5 are obtained by omitting the outlier in figure 6.4. The curve in figure 6.4(right) is apparently the result of oversmoothing. Eq(6.36) explains this problem. The deletion of an outlier in the cross-validation provides estimates that do not reflect the existence of the outlier. Hence, the deviance between the outliers and the corresponding estimates is very large. Since the only countermeasure against this large deviance is the enlargement of the bandwidth, the resulting bandwidth is unnecessarily large.

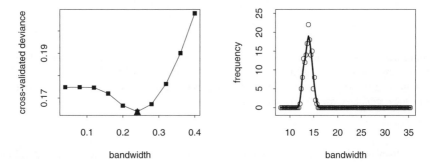

Figure 6.5 Smoothing of a histogram in which the outlier is omitted from the histogram used in figure 6.4. Optimization of the bandwidth using a cross-validated deviance (left). The result of smoothing using the optimal bandwidth ($h = 0.24$) (right).

Next, let us discuss the use of a local linear equation as $f'(r, r^*)$. If eq(6.29) (page 366) is employed to smooth a histogram, a downward bias is observed near local maxima, and an upward bias is observed near local minima. Since the fitting of a local linear equation or a local quadratic equation is an effective strategy to cope with a similar problem when smoothing based on the least squares is carried out, an analogous effect is expected for the smoothing of a histogram.

Local linear equations are fitted to a histogram on the basis of $L_8(f'(r, r^*))$ (eq(6.28) (page 366)). For this purpose, the value of the following is maximized.

$$L_{10}(f'(r, r^*)) = - \sum_{k=1}^{R} W(r^*, r(k)) (\theta_0 + \theta_1 r(k) - F_k \cdot \log(\theta_0 + \theta_1 r(k))).$$

(6.38)

The maximization of $L_{10}(f'(r, r^*))$ by adjusting the values of θ_0 and θ_1 is conducted using a nonlinear optimization method. The resulting values of θ_0 and θ_1 are defined as $\hat{\theta}_0(r^*)$ and $\hat{\theta}_1(r^*)$, respectively. $\hat{f}'(r, r^*)$ becomes

$$\hat{f}'(r, r^*) = \hat{\theta}_0(r^*) + \hat{\theta}_1(r^*)r.$$

(6.39)

Then, the value of the following is defined as an estimate at r^*.

$$\hat{f}'(r^*, r^*) = \hat{\theta}_0(r^*) + \hat{\theta}_1(r^*)r^*.$$

(6.40)

However, this method involves a problem. $\hat{f}'(r^*, r^*)$ may assume a negative value. A possible method of treating this problem is nonlinear optimization under the condition that $\hat{f}'(r^*, r^*)$ does not become negative. This method, however, causes another problem. If $\hat{f}'(r^*, r^*)$ takes a positive value but very close to 0, the absolute value of the second term of eq(6.38) becomes negative with a very large absolute value; this usually causes computational difficulties.

To avoid such a problem, $(\theta_0(r^*) + \theta_1(r^*)r)$ as an equation for local fitting is commonly replaced by $\exp(\theta_0(r^*) + \theta_1(r^*)r)$. This alteration displaces eq(6.38) with

$$
L_{11}(f(r, r^*)) = -\sum_{k=1}^{R} W(r^*, r(k)) \\
\cdot \left(\exp\left(\theta_0(r^*) + \theta_1(r^*)r(k) \right) - F_k \cdot \left(\theta_0(r^*) + \theta_1(r^*)r(k) \right) \right).
$$

(6.41)

Then, the resultant $\hat{f}'(r, r^*)$ becomes

$$
\log(\hat{f}'(r, r^*)) = \hat{\theta}_0(r^*) + \hat{\theta}_1(r^*)r.
$$

(6.42)

Consequently, the estimate at r^* is written as

$$
\hat{f}'(r^*, r^*) = \exp\left(\hat{\theta}_0(r^*) + \hat{\theta}_1(r^*)r^* \right).
$$

(6.43)

The derivation of $\hat{f}'(r, r^*)$ by maximizing the value of (6.41) is categorized as a form of generalized linear regression. In the context of generalized linear regressions, a function, such as the logarithmic function in eq(6.42), for transforming a target variable and for representing the transformed target variable as a linear combination of predictors is called a link function ($g(\cdot)$). Since the link function here is a logarithmic function, the inverse function ($g^{-1}(\cdot)$: a function for transforming the estimates given by a linear equation of predictors into primary estimates) is an exponential function. Let us define the value obtained by substituting the values of predictors in a linear equation as η (it is written as $\eta(\boldsymbol{\theta}, r^*)$ if the vector consisting of regression coefficients is called $\boldsymbol{\theta}$). Furthermore, let us define the estimates derived by substituting the predictor values of data in the same linear equation as $\boldsymbol{\eta} = (\eta_1, \ldots, \eta_n)^t$ (this is also written as $(\eta_1(\boldsymbol{\theta}, r^*), \ldots, \eta_n(\boldsymbol{\theta}, r^*))^t$). When eq(6.42) is employed, we have

$$
\eta = \hat{\theta}_0(r^*) + \hat{\theta}_1(r^*)r^*,
$$

(6.44)

$$
\eta_k = \hat{\theta}_0(r(k)) + \hat{\theta}_1(r(k))r(k).
$$

(6.45)

Next, the transformed value of η using the inverse function of the link function (the inverse function is an exponential function in this example) is denoted by μ. In addition, the transformed vector of $\boldsymbol{\eta}$ using the inverse function of the link function (for smoothing of a histogram, they are estimates obtained by smoothing) is called $\boldsymbol{\mu} = (\mu_1, \ldots, \mu_n)^t$. When eq(6.43) is used, we derive

$$
\mu = \exp(\eta) = \exp\left(\hat{\theta}_0(r^*) + \hat{\theta}_1(r^*)r^* \right),
$$

(6.46)

$$
\mu_k = \exp(\eta_k) = \exp\left(\hat{\theta}_0(r(k)) + \hat{\theta}_1(r(k))r(k) \right).
$$

(6.47)

Since this regression assumes that the logarithmic transformation of the target variable is represented by a linear combination of predictors, a regression equation of this kind is called a log-linear model. Additionally, since this regression is based on the assumption that the target variable obeys a Poisson distribution, it is also called a Poisson regression.

The procedure for deriving a regression equation in line with a generalized linear regression is as follows. This procedure is occasionally called adjusted independent variable regression.

(1) $g(y)$ (y is a data point) is expanded in the neighborhood of $g(\mu)$ (μ is an estimate corresponding to y) using a Taylor expansion to approximate it by taking terms up to the first degree. The result is

$$g(y) \approx g(\mu) + (y - \mu)\frac{\partial g(\mu)}{\partial \mu}. \tag{6.48}$$

When the right-hand side of this equation is represented as z, z is called a working value (working response, pseudo-response variable, or adjusted dependent variable). The substitution of each element of $\{\mu_i\}$ and $\{y_i\}$ (y_i is the ith data point, and μ_i is a corresponding estimate) into eq(6.48) yields the approximate equation

$$g(y_i) \approx g(\mu_i) + (y_i - \mu_i)\left(\frac{\partial g(\mu)}{\partial \mu}\right)_{\mu=\mu_i} \qquad (1 \le i \le n). \tag{6.49}$$

When the right-hand side of this equation is depicted as a vector of z ($= (z_1, z_2, \ldots, z_n)^t \approx (g(y_1), g(y_2), \ldots, g(y_n))^t$), this vector is called a working response vector. This value is an approximation of the transformed data point using a link function. When the link function is a logarithmic function, we have

$$
\begin{aligned}
z &= \eta + (y - \mu)\frac{\partial \eta}{\partial \mu} \\
&= \eta + \frac{y - \mu}{\mu} \\
&= \eta + (y - \mu)\exp(-\eta).
\end{aligned} \tag{6.50}
$$

(2) To calculate $\boldsymbol{\theta}$ ($\theta_0(r^*)$ and $\theta_1(r^*)$ if eq(6.45) is used) using these results, iterative calculations from (2.1) to (2.4) are carried out.

(2.1) The initial values of z are given. When eq(6.45) is utilized, $\theta_0(r^*)$ is set at $\log(\sum_{j=1}^{R} \frac{F_j}{R})$, and $\theta_1(r^*)$ is set at 0 to obtain the initial values of z; different initial values are possible.

(2.2) By carrying out weighted least squares (WLS) method, $\boldsymbol{\theta}$ (regression coefficients that yield η), which minimizes the value in the following, is obtained.

$$\sum_{j=1}^{R} w_j (z_j - \eta_j(\boldsymbol{\theta}, r^*))^2. \tag{6.51}$$

$\{w_j\}$ are written as

$$w_j = \frac{W(r^*, r(j))}{V_j}\left(\frac{\partial \mu_j}{\partial \eta_j}\right)^2, \tag{6.52}$$

where $W(r^*, r(j))$ denotes weight for each data point. Since $\{W(r^*, r(j))\}$ is a value giving the degree of significance of each data point prior to the iterative calculations, it is called a prior weight. V_j is the variance of μ_j at this stage. Note that the variance, which does not depend on the value of an estimate (μ_j) in a regression assuming a normal distribution, is a function of the estimate in the regression based on distributions such as a Poisson distribution. When eq(6.45) is used, eq(6.51) becomes

$$\sum_{j=1}^{R} w(r^*, r(j))(z_j - \theta_0(r^*) - \theta_1(r^*)r(j))^2. \tag{6.53}$$

$w(r^*, r(j))$ becomes

$$
\begin{aligned}
w(r^*, r(j)) &= \frac{W(r^*, r(j))}{V_j}(\exp(\eta_j))^2 \\
&= W(r^*, r(j))\exp(\eta_j) \\
&= W(r^*, r(j))\mu_j, \tag{6.54}
\end{aligned}
$$

where we utilize a characteristic of a Poisson distribution: $V_j = \mu_j = \exp(\eta_j)$.
(2.3) Using $\hat{\theta}$ obtained in (2.2) (when eq(6.45) is employed, $\hat{\theta}_0(r^*)$ and $\hat{\theta}_1(r^*)$ are used), η, μ, and z are updated.
(2.4) After the repetition of (2.2) and (2.3), $\hat{\theta}$ (when eq(6.45) is adopted, $\hat{\theta}_0(r^*)$ and $\hat{\theta}_1(r^*)$ are utilized) hardly moves; the iterative calculation terminates.

This method is considered to be an Iterative Reweighted Least Squares (IRLS) method. The iterative reweighted least squares method is used to optimize the weights of weighted least squares and also carry out generalized linear regression. Since this is a repetition of the weighted least squares, it is easy to carry out the method. A small number of iterative calculations converge these estimates in most cases. The derivation of regression coefficients of generalized linear regression using this method is equivalent to the Fisher scoring procedure (Fisher scoring algorithm). In this manner, a regression equation is obtained by using the second derivatives of the expectation of the log-likelihood. In addition, the link function, such as the logarithmic function in a Poisson regression, is called a natural link function (canonical link function). When a natural link function is employed, the iterative reweighted least squares method is identical to the derivation of a regression equation by maximizing the log-likelihood using the Newton-Raphson algorithm. As for the details of generalized linear regression, refer to previous reports such as [8], [3], [2], [5], and [9].

Figure 6.6 shows the results of smoothing using eq(6.41) (page 371); the original histogram is the same as that in figure 6.2. Figure 6.6(left) shows the results of optimizing the bandwidth using a cross-validated deviance. Furthermore, figure 6.6(right) shows the results of smoothing obtained using the optimal bandwidth ($h = 0.5$). In contrast, figure 6.7 reveals the results when the original histogram is the same as that in figure 6.3 (page 369). Both figures are almost equivalent to figure 6.2 (page 368) and figure 6.3.

Since using a cross-validated deviance greatly augments the calculation for optimizing the bandwidth, we need a method that demands computational costs lower than those for eq(6.36) (page 368) for optimizing the bandwidth. To this end, a hat

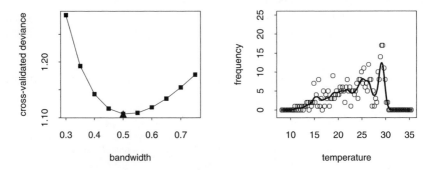

Figure 6.6   Smoothing of the same histogram shown in figure 6.2 (page 368) using generalized linear regression. The locations of the bin break points are $\{8.0, 8.25, \ldots, 35.5\}$. Optimization of the bandwidth using a cross-validated deviance (left). Result of smoothing using the optimal bandwidth ($h = 0.5$) (right).

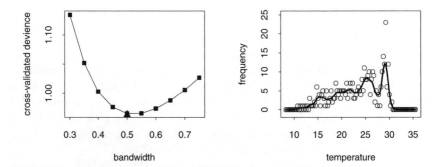

Figure 6.7   Smoothing using generalized linear regression; the original histogram is the same as that in figure 6.3 (page 369). The positions of the bin break points are $\{8.15, 8.4, \ldots, 35.65\}$. Optimization of the bandwidth using a cross-validated deviance (left). Result of smoothing using the optimal bandwidth ($h = 0.5$) (right).

matrix corresponding to generalized linear regression is defined. For this purpose, $r^* = r(k)$ is used in eq(6.53). Then, we have

$$\sum_{j=1}^{R} w(r(k), r(j)) \left(z_j - \theta_0(r(k)) - \theta_1(r(k))r(j) \right)^2. \tag{6.55}$$

The estimates supplied by the regression equation $(\hat{\theta}_0(r(k)) + \hat{\theta}_1(r(k))r(k))$ derived using eq(6.55) are defined as $\hat{\mathbf{z}}^{(k)}$ $(= (\hat{z}_1^{(k)}, \ldots, \hat{z}_n^{(k)})^t)$. The following equation holds.

$$\hat{\mathbf{z}}^{(k)} = \mathbf{H}^{IRLS(k)}\mathbf{z}. \tag{6.56}$$

This $\mathbf{H}^{IRLS(k)}$ is a hat matrix corresponding to the last regression in the iterative calculation. The hat matrix is obtained by

$$\mathbf{H}^{IRLS(k)} = \mathbf{X}(\mathbf{X}^t\mathbf{\Omega}\mathbf{X})^{-1}\mathbf{X}^t\mathbf{\Omega}, \tag{6.57}$$

where $\mathbf{\Omega}$ is a diagonal matrix (with a size of $R \times R$), and the value of the ith diagonal element is $w(r(k), r(j))$ (obtained by the iterative calculation). $\mathbf{X}$ is a design matrix for this regression. As for logarithmic linear regression using a local linear equation, the design matrix is written as

$$\mathbf{X} = \begin{pmatrix} 1 & r(1) \\ 1 & r(2) \\ \vdots & \vdots \\ 1 & r(R) \end{pmatrix}. \tag{6.58}$$

Then, a matrix with a size of $R \times R$ is defined as $\mathbf{H}^{hist}$; the kth row vector is extracted from $\{\mathbf{H}^{IRLS(k)}\}$ $(1 \leq k \leq R)$ and the resulting R row vectors are connected to construct $\mathbf{H}^{hist}$. The regression in which $\hat{\mathbf{z}}$ (not $\hat{\mathbf{z}}^{(k)}$ but $\hat{\mathbf{z}}$) is used as the value of the target variable yields a hat matrix $\mathbf{H}^{hist}$. This calculation is similar to that for local polynomial regression based on a normal distribution (cf. Chapter 3). Note that when a normal distribution is hypothesized, it is assumed that the variance of the target variable is constant and the identity transformation (identical transformation) (namely, $\{\eta_i\}$ is identical to $\{\mu_i\}$) is employed as a link function in most situations. In this case, the derivation of a hat matrix and estimates does not require iterative calculation.

Note that a hat matrix in generalized linear regression is usually defined as the following instead of eq(6.57).

$$\mathbf{H}^{general} = \mathbf{\Omega}^{1/2}\mathbf{X}(\mathbf{X}^t\mathbf{\Omega}\mathbf{X})^{-1}\mathbf{X}^t\mathbf{\Omega}^{1/2}. \tag{6.59}$$

$\mathbf{\Omega}^{1/2}$ is a matrix with a size of $R \times R$; each diagonal element is the positive square root of a corresponding diagonal element of $\mathbf{\Omega}$. Since the values of the diagonal elements of a hat matrix remain the same whichever definition of the hat matrix (eq(6.57) or eq(6.59)) may be used, the optimization of the bandwidth for smoothing a histogram is not influenced by the choice of definition. However, to examine the diverse characteristics of generalized linear regression, the use of the hat matrix defined in eq(6.59) is usually more advantageous.

Using $\mathbf{H}^{hist}$ and replacing the residual sum of squares by deviance in the definition of Generalized Cross-Validation (GCV), GCV as defined in the following is obtained.

$$GCV = \frac{2\sum_{k=1}^{R} F_k \log\left(\frac{F_k}{\hat{F}_k}\right)}{R \cdot \left(1 - \frac{\text{trace}(\mathbf{H}^{hist})}{R}\right)^2}. \tag{6.60}$$

The numerator on the right-hand side is the doubled Kullback-Leibler distance. If $\sum_{j=1}^{R} \hat{F}_j = \sum_{j=1}^{R} F_j$ is assumed, eq(6.34) (page 367) indicates that this is deviance. Hence, eq(6.60) is called the generalized cross-validated deviance. Eq(6.60) uses deviance in place of the residual sum of squares in GCV given by a normal distribution. This is not only because the deviance based on a normal distribution is identical to the residual sum of squares, but also because the deviance based on a Poisson distribution or the like obeys the distribution with the same form as the residual sum of squares; this residual sum of squares is obtained by creating a regression equation on the assumption that a normal distribution is used and the degrees of freedom are trace($\mathbf{H}^{last}$). Here, $\mathbf{H}^{last}$ is a hat matrix, such as $\mathbf{H}^{hist}$, which corresponds to the last least squares in the iterative calculation of the iterative reweighted least squares method. This background explains that deviance is regarded as the result of the generalization of the residual sum of squares.

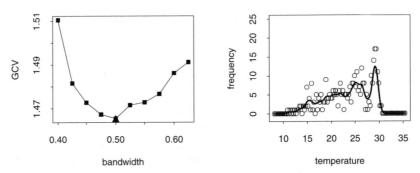

Figure 6.8 Smoothing the histogram in figure 6.2 (page 368) using generalized linear regression (eq(6.41)). Optimization of the bandwidth using GCV (eq(6.60)) (left). Result of smoothing using the optimal bandwidth ($h = 0.5$) (right).

Figure 6.8 illustrates the optimization of the bandwidth using GCV to smooth a histogram by generalized linear regression (eq(6.41) (page 371)). The data used here are the same as in the histogram used to create figure 6.2 (page 368).

On the other hand, the $AICc$ (Corrected AIC) [11] (refer to Chapter 3) for regression given by a linear estimator is defined as

$$AICc = \log\left(\frac{1}{n}\sum_{i=1}^{n}(\hat{m}(X_i) - Y_i)^2\right) + \frac{n + \text{trace}(\mathbf{H})}{n - \text{trace}(\mathbf{H}) - 2}. \tag{6.61}$$

In the first term on the right-hand side of this equation, the term in $\log()$ is the residual sum of squares divided by the number of data points. Hence, the residual

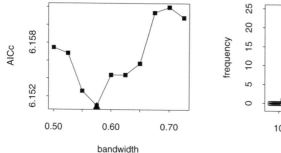
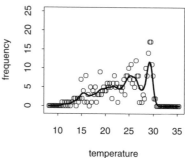

Figure 6.9   Smoothing the histogram in figure 6.2 (page 368) using generalized linear regression (eq(6.41)). Optimization of the bandwidth using $AICc$ (eq(6.62)) (left). Result of smoothing using the optimal bandwidth ($h = 0.575$) (right).

sum of squares in the $AICc$ is replaced with deviance, and the constant terms are omitted; then, we have

$$AICc = \log\left(2 \sum_{k=1}^{R} F_k \log\left(\frac{F_k}{\hat{F}_k}\right) \right) + \frac{R + \text{trace}(\mathbf{H}^{hist})}{R - \text{trace}(\mathbf{H}^{hist}) - 2}. \tag{6.62}$$

Figure 6.9 illustrates the optimization of the bandwidth using $AICc$ instead of GCV.

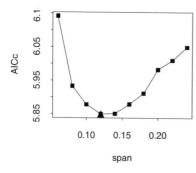
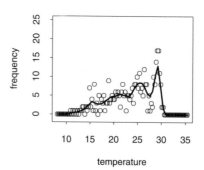

Figure 6.10   Smoothing the histogram in figure 6.2 using generalized nonparametric regression. Optimization of the span using $AICc$ (left). Result of smoothing using the optimal span ($span = 0.12$) (right).

Another option for a method for smoothing a histogram is the use of a generalized additive model [6]. In the same manner that the generalization of the assumption concerning the distribution of the target variable in simple regression or multiple regression (multiple linear regression) results in generalized linear regression, the generalization of an assumption concerning the distribution of the target variable of an additive model leads to a generalized additive model. Since the number of predictors is limited to one here, the technique should be called the generalized nonparametric regression (Chapter 6 of [4]) rather than the generalized additive model. However, the term generalized additive model is more deeply rooted, and this term is used as the name of an object in S-Plus (gam()) (Generalized Additive Model).

To smooth a histogram, a generalized additive model based on the assumption of a Poisson distribution is employed. The smoothing of a histogram is carried out under the assumption that the value of each bin of a histogram (the frequency of the data) is obtain from a Poisson distribution, and the expectation of a Poisson distribution corresponding to the nearby bins varies smoothly. LOESS or smoothing spline (smoothing splines) can be used as a smoothing method.

An example of smoothing a histogram using generalized linear regression is shown in figure 6.10. LOESS (cf. Chapter 3 and Chapter 4) is applied to construct a generalized additive model based on the assumption of a Poisson distribution. $AICc$ is used to optimize the bandwidth. Smoothing spline and polynomials are also used to implement smoothing of a histogram using generalized linear regression.

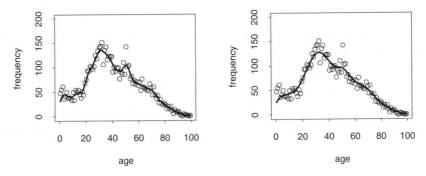

Figure 6.11   Smoothing a histogram which shows the composition of the population of Yamashita-cho in Naka ward of Yokohama city by age bracket. A kernal estimator is used for smoothing and a Gaussian function (eq(6.65)) is employed as a kernal function: $h = 2$ (left) and $h = 4$ (right).

A kernel estimator is thought of as a simple method for smoothing a histogram. The result of smoothing a histogram by this method is written as

$$\hat{\theta}_0(r) = \sum_{k=1}^{R} W(r, r(k)) F_k \Delta r, \tag{6.63}$$

where $\hat{\theta}_0(r)$ is an estimate at r. Although eq(6.30) (page 367) has a similar form, this $W(r, r(k))$ (the kernel function) must satisfy the condition

$$\int_{-\infty}^{\infty} W(r, r(k)) dr = 1. \tag{6.64}$$

Hence, the following Gaussian function, for instance, is employed as $W(r, r(k))$.

$$W(r, r(k)) = \frac{1}{\sqrt{2\pi h^2}} \exp\left(-\frac{1}{2}\left(\frac{r - r(k)}{h}\right)^2\right). \tag{6.65}$$

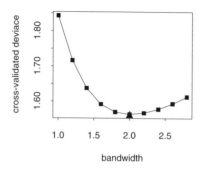 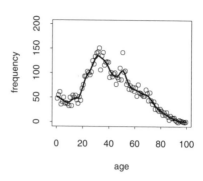

Figure 6.12   The histogram in figure 6.11 is smoothed in the same fashion as that in figure 6.2 (page 368).

This method is intuitively understandable because smoothing of this sort is readily understood by considering that the shape of bins of a histogram is shifted from a rectangle to the shape of a Gaussian function. However, the estimates near the two ends give rise to downward bias. This bias is considered to be a boundary bias. To see if it is real, data on the population by age bracket are utilized. The advantages of looking at the population by age bracket in a histogram are that the positions of bins are positive integers, and it is an appropriate example of data with a limited area because the value of age cannot be negative. Figure 6.11 shows the results of smoothing such a histogram using eq(6.63) and eq(6.65); the original histogram is created using the population of Yamashita-cho in Naka ward of Yokohama city in age brackets (September 30, 1999. These data were supplied courtesy of the municipal office of Yokohama city.). Figure 6.11(left) shows the result when $h = 2$, and figure 6.11(right) displays that when $h = 4$. The data used here are 100 integer values; each value corresponds to the frequency of an age from 0 to 99. This frequency is regarded as a histogram with $R = 100$ ($\{F_k\}$ ($1 \le k \le 100$)); the values of r in eq(6.63) are set at $\{0.5, 1, 1.5, \ldots, 99.5\}$, and $r(k) = (k - 0.5)\Delta r$ is configured to obtain estimates ($\hat{\theta}_0(r)$). In this graph, the value of $\hat{\theta}_0(0.5)$ is drawn at the position where the value of F_1 is plotted, and that of $\hat{\theta}_0(1.5)$ is drawn at the position where the value of F_2 is plotted, and other estimates are plotted in the same fashion. Both sets of estimates indicate that the values of the estimates close to the left end are somewhat smaller than what is expected. On the other hand, figure 6.12 shows the estimates obtained by the same method as that used in the case of figure 6.2 (page 368) (eq(6.30) (page 367)); the values of r^* are set at $\{0.5, 1, 1.5, \ldots, 99.5\}$ to derive the estimates ($\hat{\theta}_0(r^*)$). This graph is created in the same fashion as that in figure 6.11. Compared with the bias of the estimates drawn in figure 6.11, the bias of the estimates near the left end is small.

6.4 NONPARAMETRIC PROBABILITY DENSITY FUNCTION

A histogram to be smoothed can have a very large number of bins (R) and exhibit a very rough shape. Since the operation of eq(6.5) (page 360) loses information concerning the position of X_i in a certain bin, a histogram to be smoothed should have a large number of bins to minimize the loss of information. Even if the original histogram has a very rough shape, it should become sufficiently smooth by the operation of smoothing. Hence, the original histogram which contained the information in the original data as much as possible should be used as the material to be smoothed.

We then think of the use of a histogram with an extremely large number of bins. If the number of bins in a histogram is very large, every data is positioned in a different bin except for tied data. The frequency of each bin becomes 1. In this event, the following function is defined as $D_i(r)$.

$$D_i(r) = \begin{cases} \dfrac{1}{n\Delta r} & \text{if the bin corresponding to } r \text{ is identical to that corresponding } X_i \\ 0 & \text{otherwise.} \end{cases}$$

(6.66)

Then, $D_i(r)$ satisfies

$$\sum_{i=1}^{n} D_i(r(k)) = \begin{cases} \dfrac{1}{n\Delta r} & \text{if } F_k = 1 \\ 0 & \text{if } F_k = 0. \end{cases}$$

(6.67)

Furthermore, the following equation holds.

$$\int_{r_0}^{r_1} \left(\sum_{i=1}^{n} D_i(r) \right) dr = 1.$$

(6.68)

Hence, the smoothing with $\sum_{i=1}^{n} D_i(r)$ provides a nonparametric probability density function (nonparametric density function).

On the other hand, eq(6.28) (page 366) becomes

$$L_{12}(f'(r, r^*)) = \sum_{k=1}^{R} W(r^*, r(k)) \cdot L_{13}(f'(r(k), r^*)),$$

(6.69)

where $L_{13}(f'(r(k), r^*))$ indicates

$$L_{13}(f'(r(k), r^*)) = \begin{cases} -f'(r(k), r^*) + \dfrac{1}{n\Delta r} \log(f'(r(k), r^*)) \\[2mm] \qquad \text{if } \sum_{i=1}^{n} D_i(r(k)) = \dfrac{1}{n\Delta r} \;\; (\text{i.e., } F_k = 1) \\[4mm] -f'(r(k), r^*) \\[2mm] \qquad \text{if } \sum_{i=1}^{n} D_i(r(k)) = 0 \;\; (\text{i.e., } F_k = 0). \end{cases}$$

(6.70)

Then, $L_{12}(f'(r, r^*))$ is rewritten as

$$L_{12}(f'(r, r^*)) = -\sum_{k=1}^{R} W(r^* - r(k)) f'(r(k), r^*)$$

$$+ \frac{1}{n\Delta r} \sum_{\substack{k=1 (F_k=1)}}^{R} W(r^* - r(k)) \log(f'(r(k), r^*)).$$ (6.71)

NONPARAMETRIC PROBABILITY DENSITY FUNCTION

The second term on the right-hand side means that the summation is conducted only if k satisfies $F_k = 1$. Furthermore, $W(r^*, r(k))$ is assumed to be $W(r^* - r(k))$. Using eq(6.66), this equation becomes

$$
\begin{aligned}
L_{12}(f'(r, r^*)) &= -\sum_{k=1}^{R} W(r^* - r(k)) f'(r(k), r^*) \\
&+ \sum_{k=1}^{R} \sum_{i=1}^{n} D_i(r(k)) W(r^* - r(k)) \log(f'(r(k), r^*)).
\end{aligned}
$$

(6.72)

After this equation is multiplied by Δr, the limits of $\Delta r \to 0$ and $R \to \infty$ under the condition of $r_1 = r_0 + R \Delta r$ yield

$$
\begin{aligned}
L_{14}(f'(r, r^*)) &= -\int_{r_0}^{r_1} W(r^* - u) f'(u, r^*) du \\
&+ \frac{1}{n} \sum_{i=1}^{n} \int_{r_0}^{r_1} \delta(u - X_i) W(r^* - u) \log(f'(u, r^*)) du \\
&= -\int_{r_0}^{r_1} W(r^* - u) f'(u, r^*) du \\
&+ \frac{1}{n} \sum_{i=1}^{n} W(r^* - X_i) \log(f'(X_i, r^*)),
\end{aligned}
$$

(6.73)

where it is considered that when $\Delta r \to 0$, $D_i(r) = \frac{1}{n} \delta(r - X_i)$ holds. Be aware that $\delta(\cdot)$ is a delta function.

The simplest form of $f'(r, r^*)$ is

$$
f'(r, r^*) = \theta_0(r^*).
$$

(6.74)

In this case, $L_{14}(f'(r, r^*))$ becomes

$$
L_{15}(f'(r, r^*)) = -w_0(r^*)\theta_0(r^*) + \frac{1}{n} \sum_{i=1}^{n} W(r^* - X_i) \log(\theta_0(r^*)),
$$

(6.75)

where $w_0(r^*)$ is

$$
w_0(r^*) = \int_{r_0}^{r_1} W(r^* - u) du.
$$

(6.76)

The differentiation of $L_{15}(f'(r, r^*))$ with respect to $\theta_0(r^*)$ to be 0 provides an estimate at r^* as

$$
\hat{f}'(r^*, r^*) = \hat{\theta}_0(r^*) = \frac{\sum_{i=1}^{n} W(r^* - X_i)}{n \cdot w_0(r^*)}.
$$

(6.77)

A possible statistic for optimizing the bandwidth (h in eq(6.24) (page 365)) of a kernel function is likelihood cross-validation. The likelihood cross-validation for optimizing the bandwidth to estimate a probability density function is defined as

$$
LCV = \sum_{i=1}^{n} LCV_i.
$$

(6.78)

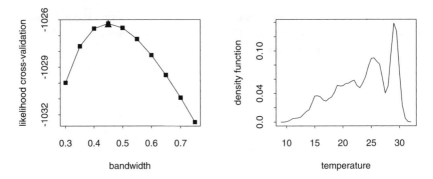

Figure 6.13   Nonparametric probability density function obtained using eq(6.77) and eq(6.81) ($r_0 = 9$ and $r_1 = 32$); the data are the same as that used in figure 6.1(left) (page 365). Optimization of the bandwidth by likelihood cross-validation (left). Nonparametric probability density function obtained using the optimal bandwidth ($h = 0.45$) (right).

When a nonparametric probability density function obtained by deleting X_i from the data is denoted by $\hat{f}^{-i}(\cdot)$, the essential part of the log-likelihood representing the cogency of $\hat{f}^{-i}(\cdot)$ in light of X_i is LCV_i. While the value of the cross-validated deviance is minimized to derive the optimal model, that of the likelihood cross-validation is defined so that it is maximized to obtain the optimal model. LCV_i is derived using $LCV_i'(r)$ as defined in

$$LCV_i'(r) = -\hat{f}^{-i}(r) + \frac{1}{n}\delta(r - X_i)\log(\hat{f}^{-i}(r)). \qquad (6.79)$$

By comparison with eq(6.32) (page 367), $LCV_i'(r)$ is shown to be the log-likelihood which indicates the cogency of $\hat{f}^{-i}(\cdot)$ at r in light of X_i. Then, the comprehensive cogency of $\hat{f}^{-i}(\cdot)$ in light of X_i is written as

$$\int_{r_0}^{r_1} LCV_i'(r)dr = -\int_{r_0}^{r_1} \hat{f}^{-i}(r)dr + \frac{1}{n}\log(\hat{f}^{-i}(X_i)). \qquad (6.80)$$

Since the first term on the right-hand side is approximately constant, LCV_i is defined by omitting this part and multiplying by n. Hence, eq(6.78) is written as

$$LCV = \sum_{i=1}^{n} \log(\hat{f}^{-i}(X_i)). \qquad (6.81)$$

This is the likelihood cross-validation for estimating a probability density function. The form of this equation implies that the likelihood cross-validation is strongly influenced by outliers contained in the data in the same way as the cross-validated deviance for smoothing a histogram.

Figure 6.13 illustrates a nonparametric probability density function given by the data shown in figure 6.1 (page 365). The optimization of the bandwidth using a cross-validated deviance (eq(6.78)) yields $h = 0.45$ as the optimal bandwidth (figure

6.13(left)). This bandwidth realizes the estimation of the nonparametric probability density function from eq(6.77); the result is shown in figure 6.13(right).

This method does not guarantee that the integration of the resulting probability density function in the region between r_0 and r_1 equals 1. This example, however, indicates that the sum of the resultant series $f(9), f(9.5), f(10), \ldots, f(31.5), f(32)$ is 2.000423. Hence, the integration of the probability density function in the region between r_0 and r_1 appears to be approximately 1.0. Nevertheless, if this limitation must strictly hold, a countermeasure is required, such as the multiplication of the resultant probability density function by a constant. Furthermore, the calculation here assumes $r_0 = 9$ and $r_1 = 32$. Since the minimum value of $\{X_i\}$ is 10.9 and the maximum value of $\{X_i\}$ is 30.5, the region is set to be slightly wider. Some phenomena, however, clearly indicate the region where data exist. In this situation, the values of r_0 and r_1 are determined clearly.

Next, let us use the following equation as $f'(r, r^*)$.

$$f'(r, r^*) = \exp(\theta_0(r^*) + \theta_1(r^*)r). \tag{6.82}$$

$L_{14}(f'(r, r^*))$ becomes

$$
\begin{aligned}
L_{16}(f'(r, r^*)) &= -\int_{r_0}^{r_1} W(r^* - u)\exp(\theta_0(r^*) + \theta_1(r^*)u)du \\
&\quad + \frac{1}{n}\sum_{i=1}^{n} W(r^* - X_i)(\theta_0(r^*) + \theta_1(r^*)X_i).
\end{aligned}
\tag{6.83}
$$

If r^* is fixed at a certain value and $L_{16}(f'(r, r^*))$ is maximized by a nonlinear optimization method, $\theta_0(r^*)$ and $\hat{\theta}_1(r^*)$ are obtained; they are called $\hat{\theta}_0(r^*)$ and $\hat{\theta}_1(r^*)$, respectively. Then, the estimate at r^* is written as

$$\hat{f}'(r^*, r^*) = \exp(\hat{\theta}_0(r^*) + \hat{\theta}_1(r^*)r^*). \tag{6.84}$$

A vector that consists of derivatives of a function with respect to each variable is called a gradient. The gradient of $L_{16}(f'(r, r^*))$ is

$$
\begin{pmatrix}
\dfrac{\partial L_{16}(f'(r, r^*))}{\partial \theta_0(r^*)} \\[2ex]
\dfrac{\partial L_{16}(f'(r, r^*))}{\partial \theta_1(r^*)}
\end{pmatrix}.
\tag{6.85}
$$

The elements of the vector are written as

$$
\begin{aligned}
\frac{\partial L_{16}(f'(r, r^*))}{\partial \theta_0(r^*)} &= -\int_{r_0}^{r_1} W(r^* - u)\exp(\theta_0(r^*) + \theta_1(r^*)u)du \\
&\quad + \frac{1}{n}\sum_{i=1}^{n} W(r^* - X_i),
\end{aligned}
\tag{6.86}
$$

$$
\begin{aligned}
\frac{\partial L_{16}(f'(r, r^*))}{\partial \theta_1(r^*)} &= -\int_{r_0}^{r_1} W(r^* - u)\exp(\theta_0(r^*) + \theta_1(r^*)u)udu \\
&\quad + \frac{1}{n}\sum_{i=1}^{n} W(r^* - X_i)X_i.
\end{aligned}
\tag{6.87}
$$

To obtain the values of $\theta_0(r^*)$ and $\theta_1(r^*)$ when $L_{16}(f'(r,r^*))$ is maximized, the values of eq(6.86) and eq(6.87) are set at 0. Then, we obtain

$$\frac{1}{n}\int_{r_0}^{r_1} W(r^*-u)\sum_{i=1}^{n}\delta(X_i-u)du = \int_{r_0}^{r_1} W(r^*-u)\exp(\theta_0(r^*)+\theta_1(r^*)u)du,$$

(6.88)

$$\frac{1}{n}\int_{r_0}^{r_1} W(r^*-u)\sum_{i=1}^{n}\delta(X_i-u)udu = \int_{r_0}^{r_1} W(r^*-u)\exp(\theta_0(r^*)+\theta_1(r^*)u)udu.$$

(6.89)

The left-hand side of eq(6.88) is the result of integration; the integrand is the probability density function ($\frac{1}{n}\sum_{i=1}^{n}\delta(X_i-u)$, smoothing is not conducted) multiplied by the kernel function ($W(r^*-u)$) which takes a maximum value at r^* (the point where the estimate is calculated). The right-hand side is also the result of integration: the integrand is the local probability density function ($\exp(\theta_0(r^*)+\theta_1(r^*)u)$ obtained by smoothing) multiplied by the kernel function ($W(r^*-u)$) which is the same as that on the left-hand side. Consequently, eq(6.88) means that smoothing does not change the value of the integration of the probability density function multiplied by $W(r^*-u)$. On the other hand, the integrand of the left-hand side of eq(6.89) is the probability density function ($\frac{1}{n}\sum_{i=1}^{n}\delta(X_i-u)$, smoothing is not conducted) multiplied by the value of the predictor (u) and the value of the kernel function ($W(r^*-u)$). The integrand of the right-hand side is the local probability density function ($\exp(\theta_0(r^*)+\theta_1(r^*)u)$ obtained by smoothing) multiplied by the value of the predictor (u) and the value of the kernel function ($W(r^*-u)$). That is, smoothing does not change the result of the other integration either; the integrand is a probability density function multiplied by the value of the predictor and the value of the kernel function. The same is true of the use of higher-degree polynomials (page 80 of [7]). It means that smoothing maintains the values of local moments of a probability density function.

A form of Hessian matrix is employed to represent the second derivatives of a function in some situations. The Hessian matrix of $L_{16}(f'(r,r^*))$ is

$$\begin{pmatrix} \dfrac{\partial^2 L_{16}(f'(r,r^*))}{\partial\theta_0(r^*)^2} & \dfrac{\partial^2 L_{16}(f'(r,r^*))}{\partial\theta_0(r^*)\partial\theta_1(r^*)} \\[2mm] \dfrac{\partial^2 L_{16}(f'(r,r^*))}{\partial\theta_1(r^*)\partial\theta_0(r^*)} & \dfrac{\partial^2 L_{16}(f'(r,r^*))}{\partial\theta_1(r^*)^2} \end{pmatrix}.$$

(6.90)

The elements of the matrix are written as

$$\frac{\partial^2 L_{16}(f'(r,r^*))}{\partial\theta_0(r^*)^2} = -\int_{r_0}^{r_1} W(r^*-u)\exp(\theta_0(r^*)+\theta_1(r^*)u)du, \quad (6.91)$$

$$\frac{\partial^2 L_{16}(f'(r,r^*))}{\partial\theta_0(r^*)\partial\theta_1(r^*)} = \frac{\partial^2 L_{16}(f'(r,r^*))}{\partial\theta_1(r^*)\partial\theta_0(r^*)}$$

$$= -\int_{r_0}^{r_1} W(r^*-u)\exp(\theta_0(r^*)+\theta_1(r^*)u)udu, \quad (6.92)$$

$$\frac{\partial^2 L_{16}(f'(r, r^*))}{\partial \theta_1 (r^*)^2} = -\int_{r_0}^{r_1} W(r^* - u)\exp(\theta_0(r^*) + \theta_1(r^*)u)u^2 du.(6.93)$$

To maximize the value of $L_{16}(f'(r, r^*))$ so as to obtain the values of $\theta_0(r^*)$ and $\theta_1(r^*)$, nonlinear optimization is required. There are various methods for conducting nonlinear optimization. Most utilize a gradient and a Hessian matrix. The gradient and the Hessian matrix corresponding to $L_{16}(f'(r, r^*))$ are calculated analytically as above; they are used for the optimization, because the use of an analytically calculated gradient and Hessian matrix decreases the amount of computation and leads to better results. Some functions to be optimized, however, possess characteristics that make the analytical calculation of the gradient and the Hessian matrix difficult. In such a case, numerical differentiation is utilized.

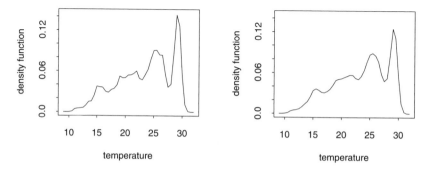

Figure 6.14 Nonparametric probability density function obtained using eq(6.83) and eq(6.84) ($r_0 = 9$, $r_1 = 32$, and $h = 0.4$); the data are the same as those shown in figure 6.1 (page 365) (left). Nonparametric probability density function obtained using $h = 0.575$ (the value is selected using $AICc$ in figure 6.9 (page 377)) (right).

Figure 6.14 shows the nonparametric probability density function obtained using eq(6.83) and eq(6.84) by the same data as those shown in figure 6.1 (page 365). Figure 6.14(left) shows the nonparametric probability density function ($r_0 = 9$, $r_1 = 32$, and $h = 0.4$). The adoption of $h = 0.575$ creates figure 6.14(right). These computations use an analytically calculated gradient and Hessian matrix. The setup of $h = 0.575$ is provided by the choice using $AICc$ when this histogram is smoothed using the local linear equation (figure 6.9). A great number of methods for optimizing the bandwidth of a nonparametric probability density function directly are known [1], [7], [10]. However, the use of the optimized bandwidth for smoothing a histogram obtained using the same data is a reasonable alternative.

6.5 EXAMPLES OF S-PLUS OBJECT

(A) Smoothing a histogram using a local constant (the bandwidth is optimized using a cross-validated deviance)
Object: `histcv1()`

```
function(xx, ex, bandw, ntrial, ydata, br)
{
#   (1)
    devf <- function(fr, est)
    {
        dev1 <- 2 * (fr * ifelse(is.finite(log(fr)),
         log(fr), -1e+100) - fr - fr * ifelse(is.
         finite(log(est)), log(est), -1e+100) + est)
        return(dev1)
    }
#   (2)
    hist1 <- hist(ydata, breaks = br, plot = F)
#   (3)
    nb <- length(xx)
    yy <- hist1$counts
#   (4)
    devt <- rep(0, length = ntrial)
#   (5)
    for(kk in 1:ntrial) {
        hh <- bandw[kk]
        py <- NULL
        for(ii in 1:nb) {
#   (6)
            xxdel <- xx[ - ii]
            yydel <- yy[ - ii]
#   (7)
            dwt <- exp(-0.5 * ((xxdel - xx[ii])/hh)^2)
            py <- c(py, sum(dwt * yydel)/sum(dwt))
        }
#   (8)
        devt[kk] <- sum(devf(yy, py))/nb
    }
#   (9)
    devmin <- min(devt)
    idevmin <- (1:ntrial)[devt == devmin]
    bandwdev <- bandw[idevmin]
#   (10)
    esthist1 <- function(ex1, xx1, yy1, band1)
    {
        dwt <- exp(-0.5 * ((xx1 - ex1)/band1)^2)
        est <- sum(dwt * yy1)/sum(dwt)
        return(est)
    }
    ey <- apply(matrix(ex, ncol = 1), 1, esthist1,
      xx1 = xx, yy1 = yy, band1 = bandwdev)
#   (11)
    return(devt, idevmin, yy, ey)
```

}

(1) A function for calculating deviance (eq(6.31) (page 367)) is defined as devf(). The object fr corresponds to $\{F_i\}$ ($= \{\hat{F}_i^{max}\}$) in eq(6.31) (page 367), and est corresponds to $\{\hat{F}_i\}$. If the argument of log() is 0, it yields $-\text{Inf}$; this can cause problems in subsequent computations. Hence, the output in this situation is set at -10^{100}.

(2) The histogram of ydata is created using hist(). The vector br designates the locations of the bin break points. plot = F indicates that a graph of the histogram is not drawn. If include.lowest = F is designated in hist(), the following equation is used to obtain frequencies in place of eq(6.5) (page 360):

$$r_0 + (j - 1)\Delta r < X_i \le r_0 + j \cdot \Delta r \quad (1 \le j \le R). \quad (6.94)$$

The object cut() also uses eq(6.94). Note that cut() yields the position of the bin in which each data point falls instead of the frequency.

An ordinary histogram, however, uses the definition

$$r_0 + (j - 1)\Delta r \le X_i < r_0 + j \cdot \Delta r \quad (1 \le j \le R). \quad (6.95)$$

When the definition above is adopted, (2) is replaced with (2)' as follows.

```
#  (2)'
dbr1 <- br1[2] - br1[1]
yyhist <- floor(ydata/dbr1) * dbr1 + dbr1 * 0.5
hist1 <- hist(yyhist, breaks = br1, plot = F)
```

That is, hist() is used after data are discretized.

(3) Since xx stands for the central points of bins, nb is the number of bins. The frequencies of bins are defined as yy.

(4) The object devt is prepared to save the values of cross-validated deviance for the bandwidths contained in bandw.

(5) The values of cross-validated deviance for the bandwidths contained in bandw are calculated. The number of bandwidths to be used is ntrial.

(6) A histogram obtained by omitting the ii-th bin is represented by xxdel and yydel.

(7) $\hat{F}_j^{-j}$ are obtained using eq(6.37) (page 368); the resulting values are connected to create py.

(8) $CV[\{\hat{F}_k\}; \{F_k\}]$ corresponding to each bandwidth is calculated using eq(6.36) (page 368); the resultant values are connected to construct devt.

(9) The minimum value among the elements of devt is called devmin. The bandwidth which gives devmin is saved as the idevmin-th element of bandw. The value of the optimal bandwidth is stored as bandwdev.

(10) The object esthist1(), which smooths a histogram on the basis of eq(6.30) (page 367), is used to smooth the histogram created using all data; the bandwidth is bandwdev. The result is saved as ey.

(11) The objects devt, idevmin, yy, and ey are the output arguments.

An example of an object to use histcv1() (construction of figure 6.2 (page 368))

```
function()
{
#   (1)
      nb <- 110
      r0 <- 8
      dr <- 0.25
      br <- seq(from = r0, by = dr, length = nb + 1)
      xx <- seq(from = r0 + dr * 0.5, by = dr, length = nb)
      ex <- seq(from = min(xx), to = max(xx), length = 200)
      ntrial <- 10
      bandw <- seq(from = 0.3, by = 0.05, length = ntrial)
#   (2)
      ydata <- scan("c:\\datasets\\ishi86.csv")
#   (3)
      fit.his <- histcv1(xx, ex, bandw, ntrial, ydata, br)
      devt <- fit.his$devt
      idevmin <- fit.his$idevmin
      yy <- fit.his$yy
      ey <- fit.his$ey
#   (4)
      devmin <- min(devt)
      bandwdev <- bandw[idevmin]
#   (5)
      par(mfrow = c(1, 2), mai = c(2, 1, 2, 0.5),
        oma = c(1, 1, 1, 1))
      plot(bandw, devt, type = "n", xlab = "bandwidth",
        ylab = "cross-validation deviance")
      points(bandw, devt, pch = 15, cex = 0.7)
      lines(bandw, devt)
      points(bandwdev, devmin, pch = 17, cex = 1.2)
#   (6)
      plot(xx, yy, type = "n", ylim = c(-1, 25),
        xlab = "temperature", ylab = "frequency")
      points(xx, yy, pch = 1, cex = 0.7)
      lines(ex, ey, lwd = 3)
}
```

(1) The number of bins (nb), the position of the left end of the histogram (r0), the width of a bin of the histogram (dr), the positions of the bin break points of the histogram (br), the positions of the central points of the bins (xx), the locations of the estimation points (ex), the number of bandwidths to be used (ntrial), and the values of bandwidths to be used (bandw) are provided.

(2) Data are retrieved and stored as ydata.

(3) Using histcv1(), the bandwidth is optimized using a cross-validated deviance, and the histogram is smoothed by the resultant optimal bandwidth. The results are saved as devt (the values of cross-validated deviance for various bandwidths),

idevmin (the bandwidth which gives the minimum cross-validated deviance is bandw [idevmin]), yy (the original histogram), and ey (a smoothed histogram).
(4) The minimum value of cross-validated deviance is called devmin, and its bandwidth is denoted by bandwdev.
(5) The relationship between the bandwidths and the values of cross-validated deviance is drawn in a graph.
(6) The result of smoothing a histogram using the optimal bandwidth is displayed in a graph.

(B) Smoothing of a histogram using local linear equations by generalized linear regression; the cross-validated deviance optimizes the bandwidth

```
function(xx, ex, bandw, ntrial, ydata, br)
{
#   (1)
    devf <- function(fr, est)
    {
        dev1 <- 2 * (fr * ifelse(is.finite(log(fr)),
        log(fr), -1e+100) - fr - fr * ifelse(is.
        finite(log(est)), log(est), -1e+100) + est)
        return(dev1)
    }
#   (2)
    hist1 <- hist(ydata, breaks = br, plot = F)
#   (3)
    nb <- length(xx)
    yy <- hist1$counts
#   (4)
    devt <- rep(0, length = ntrial)
#   (5)
    for(kk in 1:ntrial) {
        hh <- bandw[kk]
        py <- rep(0, length = nb)
        for(ii in 1:nb) {
#   (6)
            xxdel <- xx[ - ii]
            yydel <- yy[ - ii]
#   (7)
            dwt <- exp(-0.5 * ((xxdel - xx[ii])/hh)^2)
#   (8)
            assign("data1", data.frame(x = xxdel,
            y = yydel, wt = dwt), frame = 1)
            fit.glm <- glm(y ~ x, data = data1,
            family = poisson, weight = wt, control =
            glm.control(maxit = 50))
            py[ii] <- exp(fit.glm$coef[1.] + fit.glm$coef[2.]
            * xx[ii])
        }
```

```
#    (9)
         devt[kk] <- sum(devf(yy, py))/nb
     }
#    (10)
     devmin <- min(devt)
     idevmin <- (1:ntrial)[devt == devmin]
     bandwdev <- bandw[idevmin]
#    (11)
     esthist2 <- function(ex1, xx1, yy1, band1)
     {
         dwt <- exp(-0.5 * ((xx1 - ex1)/band1)^2)
         assign("data2", data.frame(x = xx1, y = yy1,
         wt = dwt), frame = 1)
         fit.glm <- glm(y ~ x, data = data2, family = poisson,
         weight = wt, control = glm.control(maxit = 50))
         ey <- exp(fit.glm$coef[1] + fit.glm$coef[2] * ex1)
         return(ey)
     }
     ey <- apply(matrix(ex, ncol = 1), 1, esthist2,
     xx1 = xx, yy1 = yy, band1 = bandwdev)
#    (12)
     return(devt, idevmin, yy, ey)
}
```

(1) The function devf() is defined to carry out the calculation of deviance (eq(6.31) (page 367)).

(2) A histogram for representing ydata is constructed using hist().

(3) The number of bins is set at nb. The frequencies of bins are denoted by yy.

(4) The object devt is prepared to store the values of cross-validated deviance corresponding to the elements of bandw.

(5) The values of cross-validated deviance obtained using the elements of bandw are calculated. The number of bandwidths to be used is ntrial.

(6) A histogram provided by omitting the ii-th bin is represented by xxdel and yydel.

(7) The values of dwt (eq(6.41) (page 371)) are set at the values of $W(r^*, r(k))$ ($r(k)$ is fixed at the central point of the ii-th bin).

(8) Data are assembled in a data frame called data1. The object of data1 is placed in the expression frame (top-level frame, frame 1); frame = 1 means the expression frame. The object glm() (an object for generalized linear regression), which is included in S-Plus by default, is used to maximize the value of eq(6.41) (page 371) accompanying data1; the resultant estimates are saved as ey. Since family = poisson is assigned, the regression based on a Poisson distribution (Poisson regression) is carried out. The assignment of family = allows, in addition to poisson, gaussian (normal distribution), binomial (binomial distribution), Gamma (gamma distribution), inverse.gaussian (inverse Gaussian distribution), and quasi (quasi-likelihood is employed). If the form of the link function is not specified, a typical link function for each distribution is adopted; for example, a log-

arithmic function is adopted for a Poisson distribution. When the link function is specified, an assignment such as binomial(link=probit) is used.

(9) The values of $CV(\{\hat{F}_k\}; \{F_k\})$ are calculated using eq(6.36) (page 368); the values are stored as devt (the values of cross-validated deviance for various bandwidths).

(10) The minimum value of devt is saved as devmin. The bandwidth given by devmin is the idevmin-th element of bandw. The optimal bandwidth is stored as bandwdev.

(11) The object esthist2(), which smooths a histogram on the basis of eq(6.41), smooths a histogram created using all data; the bandwidth is bandwdev. The result is saved as ey.

(12) The objects devt, idevmin, yy, and ey are outputted.

The object histcv2() is used in the same manner as histcv1() (figure 6.6 (page 374)).

(C) Smoothing a histogram using a local linear equation by generalized linear regression; the bandwidth is optimized using $AICc$.
Object: histaicc1()

```
function(xx, ex, bandw, ntrial, ydata, br)
{
#   (1)
    hist1 <- hist(ydata, breaks = br, plot = F)
#   (2)
    nb <- length(xx)
    yy <- hist1$counts
#   (3)
    aicct <- rep(0, length = ntrial)
#   (4)
    for(kk in 1:ntrial) {
        ey <- rep(0, length = nb)
        dhat <- rep(0, length = nb)
        hh <- bandw[kk]
#   (5)
        for(ii in 1:nb) {
            dwt <- exp(-0.5 * ((xx - xx[ii])/hh)^2)
            assign("data1", data.frame(x = xx, y = yy,
            wt = dwt), frame = 1)
            fit.glm <- glm(y ~ x, data = data1, family = poisson,
            weight = wt, control = glm.control(maxit = 200))
#   (6)
            ey[ii] <- fit.glm$fitted.values[ii]
            infl <- lm.influence(fit.glm)
            dhat[ii] <- infl$hat[ii]
        }
#   (7)
        aicct[kk] <- log(2 * sum(yy *
```

```
      ifelse(is.finite(log(yy/ey)), log(yy/ey),
      -1e+100))) + (nb + sum(dhat))/(nb - sum(dhat) - 2)
      }
#  (8)
      aiccmin <- min(aicct)
      iaiccmin <- (1:ntrial)[aicct == aiccmin]
      bandwaicc <- bandw[iaiccmin]
#  (9)
      esthist2 <- function(ex1, xx1, yy1, band1)
      {
          dwt <- exp(-0.5 * ((xx1 - ex1)/band1)^2)
          assign("data2", data.frame(x = xx1, y = yy1,
          wt = dwt), frame = 1)
          fit.glm <- glm(y ~ x, data = data2,
          family = poisson, weight = wt,
          control = glm.control(maxit = 200))
          ey <- exp(fit.glm$coef[1] + fit.glm$coef[2] * ex1)
          return(ey)
      }
      ey <- apply(matrix(ex, ncol = 1), 1, esthist2,
      xx1 = xx, yy1 = yy, band1 = bandwaicc)
#  (10)
      return(aicct, iaiccmin, yy, ey)
}
```

(1) A histogram of ydata is created using hist().

(2) The number of bins is set at nb. Furthermore, the frequencies for bins are called yy.

(3) The object aicct is prepared to store the values of $AICc$ corresponding to the elements of bandw.

(4) The values of $AICc$ corresponding to the elements of bandw are calculated. The number of bandwidths to be used is ntrial.

(5) Data are assembled in a data frame called data1. The object data1 is placed in the expression frame; frame = 1 means the expression frame. The object glm() calculates the estimates by generalized linear regression; the data frame is used as data.

(6) The estimates at bins (ey) and the diagonal elements of a hat matrix (dhat) are computed.

(7) The values of $AICc$ corresponding to the values of the bandwidth are calculated and stored as aicct. If the argument log() is 0, the output is set at -10^{100}.

(8) The minimum value among the elements of aicct is defined as aiccmin. The bandwidth which gives aiccmin is the iaiccmin-th element of bandw. The optimal bandwidth is saved as bandwaicc.

(9) The histogram created using all data is smoothed; the bandwidth is set at bandwaicc. The estimates at ex are calculated; the results are stored as ey.

(10) The objects aicct, iaiccmin, yy, and ey are outputted.

The object histaicc1() is used in the same fashion as histcv1() and histcv2() (figure 6.9 (page 377)).

(D) Smoothing a histogram using generalized nonparametric regression based on LOESS; $AICc$ optimizes the bandwidth
Object: histgam1()

```
function(xx, ex, span1, ntrial, ydata, br)
{
#   (1)
    hist1 <- hist(ydata, breaks = br, plot = F)
#   (2)
    nb <- length(br) - 1
    yy <- hist1$counts
    assign("data1", data.frame(x = xx, y = yy), frame = 1)
#   (3)
    aicct <- rep(0, length = ntrial)
#   (4)
    for(jj in 1:ntrial) {
        span2 <- span1[jj]
        formula.name <- substitute(y ~ lo(x, span = span3,
          degree = degree1), list(span3 = span2, degree1 = 1))
        fit.gam <- gam(formula.name, data = data1,
          family = "poisson", surface = "direct")
#   (5)
        df <- fit.gam$nl.df + 2
        aicct[jj] <- logb(fit.gam$deviance) +
        (1. + df/nb)/(1. - df/nb - 2./nb)
    }
#   (6)
    aiccmin <- min(aicct)
    iaiccmin <- (1:ntrial)[aicct == aiccmin]
#   (7)
    span2 <- span1[iaiccmin]
    formula.name <- substitute(y ~ lo(x, span = span1,
      degree = degree1), list(span1 = span2, degree1 = 1))
    fit.gam <- gam(formula.name, data = data1,
      family = "poisson", surface = "direct")
#   (8)
    data2 <- data.frame(x = ex)
    ey <- predict.gam(fit.gam, data2, type = "response")
#   (9)
    return(aicct, iaiccmin, yy, ey)
}
```

(1) A histogram is created by ydata; the result is stored as hist1.

(2) The number of bins is set at nb. The frequencies for bins are denoted by yy. Data are assembled in a data frame called data1. The object data1 is placed in the expression frame; frame = 1 means the expression frame (refer to page 514).

(3) The object aicct is prepared to save the values of $AICc$.

(4) The object gam() (aimed at creating a generalized additive model), which is included in S-Plus by default, smooths the histogram. In the previous example, the object (formula.nam e) specifying the form of the regression equation to be used in gam() is defined beforehand to be used as the first argument of gam(). This process allows the use of variables (in this example, span2) in assigning values to lo(). substitute(,) replaces the objects in the first argument with the objects in the second argument with the same name if the names of some objects in the first object are identical to the names of the components of the second argument (i.e., the object with the form of list). In this example, the first argument is y ~ lo(x, span = span3, degree = degree1), and the second is list(span3 = span2, degree1 = 1). The comparison between the two arguments shows that the names span3 and degree1 are shared by the two arguments. Therefore, the value of span2 is substituted into span3 (the first object), and the value of 1 is substituted into degree1. Since lo() is specified in gam(), a generalized nonparametric regression to which LOESS is applied is carried out. As an assignment for this part, various alternatives are allowed, such as s() (smoothing spline (smoothing splines)), bs() (B-spline (B-splines)), ns() (natural spline), and poly() (polynomial (polynomial function)).

(5) The values of $AICc$ for the ntrial values of span (span1) are computed; the results are saved as aicct. The values of fit.gam$nl.df denote the degrees of freedom of nonparametric regression equations in the generalized additive model created by gam(). In the calculation of LOESS based on a local linear equation, the values of the target variable are separated into the part which is represented by a linear equation and the remaining part, and after the latter is smoothed, the two parts are added; this method reduces computational costs. Hence, the degrees of freedom of the latter (fit.gam$nl.df) added by that of the former (= 2) is the degrees of freedom of this regression. This is the background of the definition of the degrees of freedom here (df): fit.gam$nl.df plus 2. When LOESS based on a local linear equation is replaced with a smoothing spline, the same measure is required.

(6) The minimum value among the elements of aicct is aiccmin. The value of span which gives aiccmin is saved as the iaiccmin-th object of span1.

(7) The value of span which minimizes the value of $AICc$ is defined as span2. A histogram created using all data is smoothed using the optimal span; the result is saved in fit.gam.

(8) A data frame (data2) in which the values of coordinates of estimation points are saved is created. The estimates are calculated using fit.gam; the results are saved as ey.

(9) The objects aicct, iaiccmin, yy, and ey are outputted.

The object histgam1() is utilized in the same manner as histaicc1(), histcv1(), and histcv2() (figure 6.10 (page 377)).

(E) Smoothing of a histogram using a kernel estimator (use of density(), construction of figure 6.11 (page 378))

```
function()
{
#   (1)
    nb <- 100
    from1 <- 0.5
    to1 <- 99.5
    ne <- 199
    dr <- 1
    wd <- 8
    xx <- seq(from = 0.5, to = nb - 0.5, length = nb)
#   (2)
    yy <- scan("c:\\datasets\\yamashi.csv")
#   (3)
    xdata <- NULL
    for(ii in 0:(nb - 1)) {
        xdata <- c(xdata, rep(ii + 0.5, length = yy[ii + 1]))
    }
#   (4)
    den1 <- density(xdata, n = ne, window = "g", width = wd,
     from = from1, to = to1)
    ex <- den1$x
    ey <- den1$y * length(xdata) * dr
#   (5)
    par(mfrow = c(1, 2), mai = c(2, 1, 2, 0.5),
     oma = c(1, 1, 1, 1))
    plot(xx, yy, type = "n", ylim = c(-1, 200), xlab = "age",
     ylab = "frequency")
    points(xx, yy, pch = 1, cex = 0.7)
    lines(ex, ey, lwd = 3)
#   (6)
    wd <- 16
    den1 <- density(xdata, n = ne, window = "g", width = wd,
     from = from1, to = to1)
    ex <- den1$x
    ey <- den1$y * length(xdata) * dr
#   (7)
    plot(xx, yy, type = "n", ylim = c(-1, 200),
     xlab = "age", ylab = "frequency")
    points(xx, yy, pch = 1, cex = 0.7)
    lines(ex, ey, lwd = 3)
}
```

(1) The number of bins (nb) in eq(6.63) (page 378), the region of estimation points (from from1 to to1), the number of estimates (ne), Δr (dr), the bandwidths (wd, the quadruple of h in eq(6.65)), and xx (positions of the centers of bins) are defined.
(2) The data are retrieved and saved as yy. The object yy contains the frequencies for the bins.

(3) The object yy is transformed into data with the original form and saved as xdata. The values of the elements of xdata are ones of $0.5, 1.5, \ldots, 99.5$; the number of elements of xdata is the same as the sum of the frequencies.

(4) The values of the probability density function at points from from1 to to1 $(0.5, 1, 1.5, \ldots, 99.5)$ are calculated using density(). The estimates are obtained as the resultant probability density function (den1\$y) multiplied by dr. The object den1\$x indicates the values of the predictor $(0.5, 1, 1.5, \ldots, 99.5)$ which show the estimation points. The form of the function $(W(r, r(k)))$ in eq(6.63) (page 378) (a kernel function) is specified as window =. The function is chosen from among the following:

window ="gaussian": Gaussian function (eq(6.65) (page 378))
window ="cosine": if $(r(k) - r)$ is defined as x, it takes the values of $\frac{1}{2h}(1 + \cos(\frac{\pi x}{h}))$ in the region between $(x - h)$ and $(x + h)$, and 0 in the remaining regions. Note that cosine kernel can mean something else.
window ="rectangular": it takes the value of $\frac{1}{2h}$ in the region between $(x - h)$ and $(x + h)$, and 0 in the remaining regions.
window ="triangular": it takes the value of $\frac{1}{h}(1 - \frac{|x|}{h})$ in the region between $(x - h)$ and $(x + h)$, and 0 in the remaining regions.

The initial letter is sufficient for specifying a function. Hence, window ="g" is assigned here (abbreviation of window ="g" gives the same result). The bandwidth of the kernel functions is defined as width =. A special notice is required to give the value of width. When window ="gaussian" is designated, the quadruple of h in eq (6.65) (page 378) is supposed to be entered. Furthermore, as for other kernels, the double of h is supposed to be entered; the value is identical to the width of the region where the kernel function take a nonzero value.

(5) The data and the estimates are displayed.
(6)(7) The same procedure as (4)(5) are carried out after wd is set at 16.

(F) Nonparametric probability density function using a local constant; the likelihood cross-validation optimizes the bandwidth
Object: const1()

```
function(nd, r0, r1, bandw, ntrial, xdata, br, ne, ex)
{
#   (1)
    estconst1 <- function(exs, xx, hh, r0, r1)
    {
#   (2)
        wt <- function(x2)
        {
            www <- exp(-0.5 * ((rr - x2)/hh)^2)
            return(www)
        }
#   (3)
        r0a <- r0
        if(r0a < exs - 10 * hh) r0a <- exs - 10 * hh
        r1a <- r1
```

```
        if(r1a > exs + 10 * hh) r1a <- exs + 10 * hh
        assign("hh", hh, frame = 1)
        assign("rr", exs, frame = 1)
        w0 <- integrate(wt, lower = r0a, upper = r1a)$integral
        th0 <- sum(wt(xx))/w0/length(xx)
        return(th0)
    }
#   (4)
    liket <- NULL
#   (5)
    for(kk in 1:ntrial) {
#   (6)
        like <- NULL
        hh <- bandw[kk]
#   (7)
        for(ii in 1:nd) {
            th0 <- estconst1(xdata[ii], xdata[ - ii],
            hh, r0, r1)
            like <- c(like, log(th0))
        }
#   (8)
        liket <- c(liket, sum(like))
    }
#   (9)
    likemax <- max(liket)
    ilikemax <- (1:ntrial)[liket == likemax]
#   (10)
    hh <- bandw[ilikemax]
    ey <- apply(matrix(ex, nrow = 1), 2, estconst1,
        xx = xdata, hh = hh, r0 = r0, r1 = r1)
#   (11)
    return(liket, ilikemax, ey)
}
```

(1) The function estconst1() for calculating estimates using eq(6.77) (page 381) is defined. The object exs denotes r^*, xx denotes $\{X_i\}$, hh denotes the bandwidth (h), r0 denotes r_0, and r1 denotes r_1.

(2) The function wt() is defined to calculate the value of $W(r^* - X_i)$ in eq(6.77). The object rr means r^*, and x2 means $\{X_i\}$.

(3) The value of th0 ($\hat{\theta}_0$) is obtained using eq(6.77). Since the unnecessarily wide region of integration in eq(6.76) (page 381) for obtaining $w_0(r^*)$ causes a decline in numerical accuracy, the integration is designed not to be carried out in the region where the value of the integrand (wt) is very small.

(4) The object liket is prepared to save the values of likelihood cross-validation corresponding to the elements of bandw.

(5) Since bandw contains ntrial bandwidths, the value of likelihood cross-validation is computed for each element.

(6) The object like is prepared to store the values of $\{\log(\hat{f}^{-i}(X_i))\}$ defined in eq(6.81) (page 382). The bandwidth (hh) is saved as bandw[kk].

(7) The values of $\theta_0(X_i)$ (i.e., $\hat{f}^{-i}(X_i)$) are obtained and saved as th0. The values of $\{\log(\hat{f}^{-i}(X_i))\}$ are calculated and stored as like.

(8) The sum of the values of the elements of like is computed for each bandwidth. These values are connected to create liket.

(9) The maximum value of likelihood cross-validation is saved as likemax. The position of the element of the optimal bandwidth in bandw is specified as ilikemax.

(10) The optimal bandwidth obtained in (9) is stored as hh. The estimates at ex (a vector) are calculated using estconst1(); the results are saved as ey.

(11) The objects liket, ilikemax, and ey are outputted.

An example of an object to use const1() (construction of figure 6.13 (page 382))

```
function()
{
#    (1)
     nd <- 365
     r0 <- 9
     r1 <- 32
     ntrial <- 10
     bandw <- seq(from = 0.3, by = 0.05, length = ntrial)
#    (2)
     xdata <- scan("c:\\datasets\\ishi86.csv")
#    (3)
     ne <- 47
     ex <- seq(from = r0, by = 0.5, length = ne)
#    (4)
     fit.den <- const1(nd, r0, r1, bandw, ntrial, xdata, br,
      ne, ex)
#    (5)
     liket <- fit.den$liket
     ilikemax <- fit.den$ilikemax
     ey <- fit.den$ey
#    (6)
     likemax <- max(liket)
     bandwlike <- bandw[ilikemax]
#    (7)
     par(mfrow = c(1, 2), mai = c(2, 1, 2, 0.5),
      oma = c(1, 1, 1, 1))
     plot(bandw, liket, type = "n", xlab = "bandwidth",
      ylab = "likelihood cross-validation")
     points(bandw, liket, pch = 15, cex = 0.7)
     lines(bandw, liket)
     points(bandwlike, likemax, pch = 17, cex = 1.2)
#    (8)
```

```
      plot(ex, ey, type = "n", xlab = "temperature",
        ylab = "density function")
      lines(ex, ey)
}
```

(1) The number of data (nd), the region of integration (from r0 to r1), the number of bandwidths to be used (ntrial), and the bandwidths to be used (bandw) are given.
(2) Data are retrieved and stored as xdata.
(3) The number of estimation points is defined as ne, and the values of the coordinates are specified as ex.
(4) The optimization of the bandwidth and the calculation of the nonparametric probability density function using the optimal bandwidth are carried out using const1(); the results are saved in fit.den.
(5) The components of fit.den are saved as liket, ilikemax, and ey.
(6) The maximum value of likelihood cross-validation (liket) is defined as likemax, and the bandwidth which gives likemax is specified as bandwlike.
(7) The relationship between the bandwidth and the values of likelihood cross-validation is drawn in a graph.
(8) The nonparametric probability density function obtained using the optimal bandwidth is shown in a graph.

(G) Nonparametric probability density function using a local linear equation
Object: lldens1()

```
function(nd, r0, r1, hh, xdata, ne, ex)
{
#   (1)
    estlldens1 <- function(exs, xx, hh, r0, r1, nd)
    {
#   (2)
          assign("rrf", exs, frame = 1)
          assign("hhf", hh, frame = 1)
          assign("r0f", r0, frame = 1)
          assign("r1f", r1, frame = 1)
          assign("ndf", nd, frame = 1)
          assign("xif", xx, frame = 1)
#   (3)
          formula1 <- function(th0, th1)
          {
              term1 <- function(th0, th1)
              {
                  integrand1 <- function(uu, th0, th1)
                  {
                    f <- exp(-0.5 * ((rrf - uu)/hhf)^2)
                      * exp(th0 + th1 * uu)
                    return(f)
                  }
                  r0a <- r0f
```

```
          if(r0a < rrf - 10 * hhf) r0a <- rrf - 10 * hhf
          r1a <- r1f
          if(r1a > rrf + 10 * hhf) r1a <- rrf + 10 * hhf
          f2 <- integrate(integrand1, lower = r0a,
           upper = r1a, th0 = th0, th1 = th1)$integral
          return(f2)
      }
```
(4)
```
      term2 <- function(th0, th1)
      {
          ss <- sum(exp(-0.5 * ((rrf - xif)/hhf)^2)
          * (th0 + th1 * xif))
          return(ss)
      }
```
(5)
```
      val1 <- term1(th0, th1) - term2(th0, th1)/ndf
```
(6)
```
      grad1 <- function(th0, th1)
      {
```
(7)
```
          integrand2 <- function(uu, th0, th1)
          {
            f <- exp(-0.5 * ((rrf - uu)/hhf)^2)
            * exp(th0 + th1 * uu)
            return(f)
          }
          r0a <- r0f
          if(r0a < rrf - 10 * hhf) r0a <- rrf - 10 * hhf
          r1a <- r1f
          if(r1a > rrf + 10 * hhf) r1a <- rrf + 10 * hhf
          t1 <- integrate(integrand2, lower = r0a,
           upper = r1a, th0 = th0, th1 = th1)$integral
          t2 <- sum(exp(-0.5 * ((rrf - xif)/hhf)^2))
          dx <- t1 - t2/ndf
```
(8)
```
          integrand3 <- function(uu, th0, th1)
          {
            f <- exp(-0.5 * ((rrf - uu)/hhf)^2)
            * uu * exp(th0 + th1 * uu)
            return(f)
          }
          r0a <- r0f
          if(r0a < rrf - 10 * hhf) r0a <- rrf - 10 * hhf
          r1a <- r1f
          if(r1a > rrf + 10 * hhf) r1a <- rrf + 10 * hhf
          t1 <- integrate(integrand3, lower = r0a,
           upper = r1a, th0 = th0, th1 = th1)$integral
```

```
                    t2 <- sum(exp(-0.5 * ((rrf - xif)/hhf)^2) * xif)
                    dy <- t1 - t2/ndf
#  (9)
                    md1 <- matrix(c(dx, dy), ncol = 2)
                    return(md1)
                }
#  (10)
                attr(val1, "gradient") <- grad1(th0, th1)
#  (11)
                hess1 <- function(th0, th1)
                {
#  (12)
                    integrand4 <- function(uu, th0, th1)
                    {
                      f <- exp(-0.5 * ((rrf - uu)/hhf)^2)
                       * exp(th0 + th1 * uu)
                      return(f)
                    }
                    r0a <- r0f
                    if(r0a < rrf - 10 * hhf) r0a <- rrf - 10 * hhf
                    r1a <- r1f
                    if(r1a > rrf + 10 * hhf) r1a <- rrf + 10 * hhf
                    t1 <- integrate(integrand4, lower = r0a,
                     upper = r1a, th0 = th0, th1 = th1)$integral
                    dxx <- t1
#  (13)
                    integrand5 <- function(uu, th0, th1)
                    {
                      f <- exp(-0.5 * ((rrf - uu)/hhf)^2)
                       * uu * exp(th0 + th1 * uu)
                      return(f)
                    }
                    r0a <- r0f
                    if(r0a < rrf - 10 * hhf) r0a <- rrf - 10 * hhf
                    r1a <- r1f
                    if(r1a > rrf + 10 * hhf) r1a <- rrf + 10 * hhf
                    t1 <- integrate(integrand5, lower = r0a,
                     upper = r1a, th0 = th0, th1 = th1)$integral
                    dxy <- t1
#  (14)
                    integrand6 <- function(uu, th0, th1)
                    {
                      f <- exp(-0.5 * ((rrf - uu)/hhf)^2)
                       * uu * uu * exp(th0 + th1 * uu)
                      return(f)
                    }
                    r0a <- r0f
```

```
                    if(r0a < rrf - 10 * hhf) r0a <- rrf - 10 * hhf
                    r1a <- r1f
                    if(r1a > rrf + 10 * hhf) r1a <- rrf + 10 * hhf
                    t1 <- integrate(integrand6, lower = r0a,
                      upper = r1a, th0 = th0, th1 = th1)$integral
                    dyy <- t1
#   (15)
                    md2 <- array(c(dxx, dxy, dxy, dyy),
                      c(1, 2, 2))
                    return(md2)
                }
#   (16)
                attr(val1, "hessian") <- hess1(th0, th1)
                return(val1)
            }
#   (17)
            start1 <- list(th0 = th0h, th1 = th1h)
            msout <- ms( ~ formula1(th0, th1), start = start1,
              trace = F)
#   (18)
            coef1 <- coef(msout)
            names(coef1) <- NULL
            assign("th0h", coef1[1], frame = 1)
            assign("th1h", coef1[2], frame = 1)
            eys <- exp(th0h + th1h * rrf)
            return(eys)
        }
#   (19)
        assign("th0h", 0.1, frame = 1)
        assign("th1h", 0.1, frame = 1)
        ey <- apply(matrix(ex, nrow = 1), 2, estlldens1,
          xx = xdata, hh = hh, r0 = r0, r1 = r1, nd = nd)
#   (20)
        return(ey)
    }
```

(1) The function estlldens1() is defined to calculate the value of a probability density function at exs. The object xx is a vector which contains data, hh is the bandwidth, r0 and r1 are the minimum and maximum values of the region where the probability density function takes a nonzero value, and nd is the number of data points.

(2) The objects exs, hh, r0, r1, nd, and xx are named rrf, hhf, r0f, r1f, ndf, and xif to be placed in the expression frame.

(3) The function to be minimized is defined as formula1(). The first term on the right-hand side of eq(6.83) (page 383) is denoted by term1().

(4) The second term on the right-hand side of eq(6.83) (page 383) is saved as term2().

(5) The value to be minimized is defined as val1. Since the value of eq(6.83) is to be maximized, the sign of val1 is opposite to that of eq(6.83).
(6) The function grad1() for calculating the gradient (eq(6.85) (page 383)) is defined.
(7) The first element of the gradient is computed and saved as dx.
(8) The second element of the gradient is computed and denoted by dy.
(9) The gradient vector is stored as md1 (a matrix with a size of 1×2).
(10) The fact that the gradient is md1 is defined as one of the attributes of val1.
(11) The function hess1() that yields a Hessian matrix (eq(6.90) (page 384)) is defined.
(12) The $1 \cdot 1$ element of the matrix defined in eq(6.90) is calculated and saved as dxx.
(13) The $1 \cdot 2$ element (= $2 \cdot 1$ element) of the matrix defined in eq(6.90) is computed and stored as dxy.
(14) The $2 \cdot 2$ element of the matrix defined in eq(6.90) is derived and saved as dyy.
(15) The Hessian matrix is stored as md2 (an array with a size of $1 \times 2 \times 2$).
(16) The fact that the Hessian matrix is md2 is given as one of the attributes of val1.
(17) The values of th0h and th1h which minimize formula1(th0, th1) are derived; the values of th0h and th1h are used as initial values. The initial values for obtaining the estimates of the first estimation are assigned to (19). The estimates of the second or later estimations are derived using the values th0h and th1h which are calculated in the process of obtaining the previous estimates.
(18) The estimates are derived and saved as eys and the outputs from estlldens1().
(19) The values of the probability density function at ex are derived using estlldens1(); the results are stored as ey (a vector).
(20) The object ey is outputted.

An example of an object of use lldens1() (construction of figure 6.14 (page 385))

```
function()
{
#    (1)
        nd <- 365
#    (2)
        xdata <- scan("c:\\datasets\\ishi86.csv")
#    (3)
        r0 <- 9
        r1 <- 32
        ne <- 47
        ex <- seq(from = 9, by = 0.5, length = ne)
#    (4)
        par(mfrow = c(1, 2), mai = c(2, 1, 2, 0.5),
          oma = c(1, 1, 1, 1))
#    (5)
        bandw <- c(0.4, 0.575)
        for(kk in 1:2) {
            hh <- bandw[kk]
#    (6)
```

```
            ey <- lldens1(nd, r0, r1, hh, xdata, ne, ex)
#    (7)
            plot(ex, ey, type = "n", ylim = c(0, 0.145),
             xlab = "temperature", ylab = "density function")
            lines(ex, ey)
        }
}
```

(1) The number of data points (nd) is provided.

(2) Data are retrieved.

(3) The region of integration (from r0 to r1), the number of estimation points (ne), and coordinates of estimation points (ex) are supplied.

(4) The area for drawing a graph is defined.

(5) The values to be used as bandwidths are saved as bandw.

(6) One of the elements of bandw is selected and stored as hh. The estimation of a nonparametric probability density function based on eq(6.83) (page 383) is conducted using lldens1(); the estimates are saved as ey.

(7) The estimates are drawn in a graph.

REFERENCES

1. K.P. Burnhan and D.R. Anderson (1998). *Model Selection and Inference - A Practical Information-Theoretic Approach*, Springer-Verlag.

2. J.M. Chambers and T.J. Hastie (1993). *Statistical Models in S*, Chapman & Hall/CRC.

3. A.J. Dobson (1990). *An Introduction to Generalized Linear Models*, Chapman & Hall/CRC.

4. J. Fox (2000). *Multiple and Generalized Nonparametric Regression*, Sage Publications.

5. P.J. Green and B.W. Silverman (1994). *Nonparametric Regression and Generalized Linear Models, A Roughness Penalty Approach*, Chapman & Hall/CRC.

6. T.J. Hastie and R.J. Tibshirani (1990). *Generalized Additive Models*, Chapman & Hall/CRC.

7. G. Loader (1999). *Local Regression and Likelihood*, Springer-Verlag.

8. P. McCullagh and J.A. Nelder (1983). *Generalized Linear Models*, Chapman & Hall/CRC.

9. R.H. Myers, D.C. Montgomery, and G.G. Vining (2001). *Generalized Linear Models: With Applications in Engineering and the Sciences*, John Wiley & Sons.

10. J. S. Simonoff (1996). *Smoothing Methods in Statistics*, Spinger-Verlag.

11. J.S. Simonoff (1998). Three sides of smoothing: categorical data smoothing, nonparametric regression, and density estimation. *International Statistical Review*, Vol. 66, No. 2, pp.137-156.

Problems

6.1   Confirm that the number of data within each bin of a histogram approximately obeys a Poisson distribution.

(a) Generate 100 data by obtaining realizations of $N(15, 5^2)$ (a normal distribution; the mean is 15 and the standard deviation is 5).

(b) Create a histogram of the data produced in **(a)**. The bin break points are $\{0, 1, 2, 3, 4, \ldots, 30\}$; the data outside this range are omitted.

(c) Repeat **(a)** and **(b)** 1000 times. This results in 1000 frequencies for each bin. Then, consider the distribution of the frequencies at a certain bin (e.g., the bin for the data ranging from 14 to 15 (15 is not included)). Calculate the mean and variance of the distribution. Does it pass for a Poisson distribution? Additionally, draw a histogram of the distribution. Does it appear like a Poisson distribution?

(d) In **(b)**, use the bin break points $\{0, 5, 10, \ldots, 30\}$; the data outside this range are omitted. Then, conduct **(c)**, and find the influence of the binwidth on the distributions; discuss its background.

6.2   Conduct the processes of Problem **6.1** using distributions other than a normal distribution, for example, a lognormal distribution, a Poisson distribution, and summation of plural distributions.

6.3   Consider the use of random sampling for obtaining the confidence intervals of the smoothed histogram derived.

(a) The data given below represent the population composition of Isezaki-cho in the Nakaward of Yokohama city by age bracket (observed on September 30, 2004, the data are supplied courtesy of the municipal office of Yokohama city). The first number (8) represents the 0-year-old population, the second number (11) represents the 1-year-old population, and the last number (1) represents the 94-year-old population.

> 8, 11, 15, 9, 9, 14, 13, 11, 9, 11, 9, 9, 11, 16, 9, 6, 5, 9, 16, 17, 31,
> 33, 25, 37, 48, 49, 59, 61, 43, 48, 41, 52, 46, 40, 35, 40, 53, 44, 33,
> 32, 37, 44, 44, 24, 28, 27, 40, 31, 20, 31, 27, 30, 30, 31, 29, 42, 36,
> 36, 14, 16, 21, 25, 18, 22, 17, 13, 20, 28, 21, 17, 16, 16, 19, 16, 11,
> 11, 7, 16, 10, 11, 16, 6, 5, 7, 8, 4, 4, 5, 1, 1, 4, 1, 0, 1, 1

Produce an S-Plus or R object for smoothing this histogram using the objects presented in (D) of this chapter, and run it.

(b) Create an S-Plus or R object for generating 100 sets of pseudo-data (values of bins); the estimates given by **(a)** are used as the approximations of the real values of the histogram and the value of each bin is given as a realization of a Poisson distribution (the mean and variance of the Poisson distribution are the estimates of each bin). Run the object.

(c) Construct an S-Plus or R object for smoothing the 100 sets of pseudo-data obtained in **(b)** using the value of the smoothing parameter optimized for the original data by the object created in **(a)**. Run the object: the sets of estimates imply the confidence interval of the original estimates.

(d) Another method is possible for estimating the confidence interval of the estimates calculated in **(a)**. It is random sampling of the original data: (1) The original data are restored by transforming the value of 8 at the bin "0" into "0.5, 0.5, 0.5, 0.5, 0.5, 0.5,

0.5, 0.5" for example. (2) A set of resampled data is obtained by sampling 2083 data at random with replacement from the restored data. (3) A histogram is derived from the resampled data. (4) The processes of (1), (2), and (3) are repeated using another pseudo-random number to outline the confidence interval of the estimates obtained in **(a)**. Construct an S-Plus or R object for this method, and run it.

6.4 Poisson distribution is a common tool for smoothing a histogram. The central limit theorem, however, indicates that a normal distribution effectively substitutes for a Poisson distribution under some conditions.

(a) Generate 200 realizations of $N(15, 3^2)$ (a normal distribution; the mean is 15 and the standard deviation is 3) to be called the vector **x**; the number of elements is 200. Also generate 100 realizations of $N(10, 1^2)$ (a normal distribution; the mean is 10 and the standard deviation is 1) to be called the vector **y**; the number of elements is 100. Then, create the vector **z** by combining the two vectors; the number of elements of **z** is 300, and the elements of **z** that are less than 0 or more than 30 are deleted. Generate **z** 1000 times using various seeds of pseudo-random numbers.

(b) Create a histogram for each of 1000 **z**'s. The bin break points are $\{0, 1, 2, 3, 4, \ldots, 30\}$. Examine the distributions of the values of bins of these histograms. Moreover, construct additional histograms of 1000 **z**'s; the bin break points are $\{0, 0.1, 0.2, 0.3, 0.4, \ldots, 30\}$. Investigate the distributions of the values of bins of these histograms. Which distributions of the values of bins of the two histograms seem to be approximately normal? Discuss the reason.

6.5 The likelihood cross-validation defined in eq(6.81) (page 382) is apparently affected by outliers more than expected. Countermeasures against the phenomenon are possible.

(a) Generate 50 realizations of $N(2, 2^2)$ (a normal distribution; the mean is 2 and the standard deviation is 2) to be called the vector **x**; the number of element is 50. Also generate 50 realizations of $N(6, 1^2)$ (a normal distribution; the mean is 6 and the standard deviation is 1) to be called the vector **y**; the number of elements is 50. Then, create **z** by combining the two vectors; the number of elements of **z** is 100, and omit the elements of **z** that are less than 0 or more than 10.

(b) Estimate a nonparametric probability density function by utilizing the object const1() presented in **(F)** of this chapter by utilizing **z** constructed in **(a)**.

(c) Add a data of 15 as an outlier to **z**, and carry out the same procedure as **(b)**. Check that the resulting nonparametric probability density function is too smooth, and determine its reason by displaying the values of $\log(\hat{f}^{-i}(X_i))$ in eq(6.81).

(d) Modify the object const1() presented in **(F)** in this chapter to solve the problem indicated in **(c)**. Possible countermeasures are as follows:

(1) The elements of **z** are sorted in ascending order, and about 5% of the elements positioned near the right end and about 5% of the elements positioned near the left end are deleted to use about 90% of the elements located in the middle region as data for optimizing the bandwidth.

(2) The values of $\log(\hat{f}^{-i}(X_i))$ in eq(6.81) are in ascending order, and about 5% of the elements positioned near the right end are deleted to obtain the likelihood cross-validation given in eq(6.81), which is robust to outliers.

(3) Other strategies are possible to make the likelihood cross-validation given in eq(6.81) robust to outliers.

6.6   By modifying the objects presented in (C) of this chapter, create an S-Plus or R object for smoothing a histogram using a local quadratic regression equation. Compare the estimates given by a local quadratic regression equation with those given by a local regression linear equation using the simulation data.

CHAPTER 7

PATTERN RECOGNITION

7.1 INTRODUCTION

Pattern recognition clears up the differences in the abilities of various methods of regression analysis. Utilization of diverse methods of regression analysis in the field of pattern recognition has helped develop and improve a great many techniques through friendly competition. Under such circumstances, the significant role of nonparametric regression in pattern recognition plainly demonstrates its high practical value. Hence, an immense amount of literature concerning pattern recognition treats nonparametric regression explicitly and implicitly; this fact convinces us of the great role of nonparametric regression in this discipline. It is no exaggeration to say the application of nonparametric regression to pattern recognition is synonymous with pattern recognition as a whole.

This chapter outlines the fundamental concepts of pattern recognition and gives an overview of pattern recognition that uses nonparametric regression techniques. These expositions emphasize the importance of understanding pattern recognition in view of nonparametric regression to elucidate and utilize a variety of techniques involving it. However, the descriptions here do not go far beyond an introduction of

Introduction to Nonparametric Regression, Kunio Takezawa.
Copyright © 2006 John Wiley & Sons, Inc.

basic concepts and simple methods. As for the wider spectrum of explanations and pioneering approaches, the reader is advised to refer to established books on pattern recongnition with a central focus on nonparametric regression (e.g., [1], [6], and [2]).

7.2 BAYES' DECISION RULE

Let us assume that the subjects of pattern recognition (objects, animals, plants, characters, and so forth) constitute a specific population. All elements of the population are assumed to belong to one of G classes ($\{C_1, C_2, \ldots, C_G\}$). Then, the result of sampling from the population is called a sample. By measuring or observing the sample, we obtain images, videos, sounds, measured values, observed values, results of a census, and so on; these are called patterns. By carrying out preprocessing (a.k.a. feature selection (feature extraction)), we obtain a vector $\mathbf{x}$ (the number of elements is m) which represents the characteristics of a pattern. The vector $\mathbf{x}$ is called a feature vector (feature pattern). Using this $\mathbf{x}$, a functional relationship is constructed to indicate the class to which $\mathbf{x}$ belongs. This process is called learning. When the learning process is conducted without considering at least tentatively the class to which each sample belongs, it is called unsupervised learning. In contrast, the learning process involving knowledge concerning the class to which each sample belongs is called supervised learning. Supervised learning is the only subject treated here, because supervised learning is closely associated with regression analysis. When an unidentified $\mathbf{x}$ is supplied, the class to which $\mathbf{x}$ belongs is estimated using the results of learning. Such an estimation is called classification. Note that hereafter the term discrimination is used as an equivalent term for classification.

First, the probability density function (density function) obeyed by the data (a member of C_g ($1 \leq g \leq G$)) is called $p_g(\mathbf{x})$ ($1 \leq g \leq G$). Hence, we have

$$\int p_g(\mathbf{x})d\mathbf{x} = 1 \quad (1 \leq g \leq G). \tag{7.1}$$

Then $\mathbf{x}$ is regarded as a member of C_k if the following relationship holds.

$$p_k(\mathbf{x}) > p_g(\mathbf{x}) \quad (g = 1, 2, \ldots, k-1, k+1, \ldots, G). \tag{7.2}$$

This is based on the philosophy that if the value of $p_k(\mathbf{x})$ is the maximum among $\{p_g(\mathbf{x})\}$ ($1 \leq g \leq G$), it is highly probable that $\mathbf{x}$ belongs to C_k.

Pattern recognition, however, aims to reduce the probability of classifying data into an incorrect class. For this purpose, the method just described is not appropriate. Here is a simple example: $G = 2$ and the element of the vector $\mathbf{x}$ is one (i.e., $\mathbf{x}$ is a scalar). Using the method described, figure 7.1 shows that, if x is located on the left-hand side of the boundary shown by x_p, x is classified as C_1; otherwise, it is classified as C_2.

To investigate the appropriateness of this method, the number of data points positioned in the region between x and $x + \Delta x$ which are members of C_1 is defined as $r_1(x)\Delta x$, and that of data points which are members of C_2 is designated as $r_2(x)\Delta x$. These assumptions lead to

$$\begin{aligned} r_1(x)\Delta x &= n_1\Delta x \cdot p_1(x), \\ r_2(x)\Delta x &= n_2\Delta x \cdot p_2(x). \end{aligned} \tag{7.3}$$

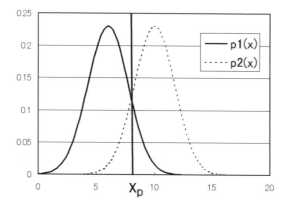

Figure 7.1   When x is considered to belong to a class which corresponds to the larger value between $p_1(\mathbf{x})$ and $p_2(\mathbf{x})$, x_p constructs a decision boundary.

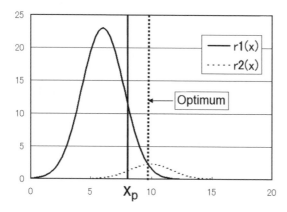

Figure 7.2   If the number of data points belonging to C_1 is large and that belonging to C_2 is small, x_p is not an optimal decision boundary.

Then we have

$$
\begin{aligned}
r_1(x) &= n_1 p_1(x), \\
r_2(x) &= n_2 p_2(x),
\end{aligned}
\tag{7.4}
$$

where n_1 is the number of data points that belong to C_1, and n_2 is the number of data points that belong to C_2. Hence, when a method classifies data, the number of data points that are classified into incorrect classes is

$$
\int_{\bar{\Omega}_1} n_1 p_1(x) dx + \int_{\bar{\Omega}_2} n_2 p_2(x) dx,
\tag{7.5}
$$

where $\bar{\Omega}_1$ denotes a region where this method does not classify a data point as C_1; that is, the data point is classified as C_2. The same is true of $\bar{\Omega}_2$. Such a boundary between regions is called a decision boundary.

If a large portion of elements of a population belong to C_1 and the number of elements that belong to C_2 is small, the distribution is illustrated in figure 7.2. If x_p is used, $\mathbf{x}$ that belongs to C_1 is mistakenly classified as C_2 on many occasions, while $\mathbf{x}$ that belongs to C_2 is incorrectly classified as C_1 rarely. Hence, the total number of data points that suffer from misclassification is significant. A value a little larger than x_p seems more appropriate as a decision boundary.

Let us formulate a method of pattern recognition to minimize the possibility of the occurrence of misclassification. First, the possibility of the classification of a pattern as a class that is different from C_g though the pattern is actually a member of C_g is defined as $p_g(error)$:

$$
p_g(error) = \int_{\bar{\Omega}_g} p_g(\mathbf{x}) d\mathbf{x} \quad (1 \le g \le G),
\tag{7.6}
$$

where $\bar{\Omega}_g$ indicates a region where a method classifies a data point as a class that is different from C_g. Then, the total number of data points which the method misclassifies is

$$
\sum_{g=1}^{G} n_g p_g(error) = \sum_{g=1}^{G} \int_{\bar{\Omega}_g} n_g p_g(\mathbf{x}) d\mathbf{x} \quad (1 \le g \le G),
\tag{7.7}
$$

where n_g is the number of data points which are members of C_g. Next, eq(7.1) generates the equation

$$
\int_{\bar{\Omega}_g} p_g(\mathbf{x}) d\mathbf{x} = 1 - \int_{\Omega_g} p_g(\mathbf{x}) d\mathbf{x} \quad (1 \le g \le G).
\tag{7.8}
$$

Ω_g shows the region where a method classifies a data point as C_g. Eq(7.6) and eq(7.8) result in

$$
p_g(error) = 1 - \int_{\Omega_g} p_g(\mathbf{x}) d\mathbf{x}.
\tag{7.9}
$$

Both sides of this equation are multiplied by n_g and summed with respect to g; we have

$$
\sum_{g=1}^{G} n_g p_g(error) = \sum_{g=1}^{G} n_g \left(1 - \int_{\Omega_g} p_g(\mathbf{x}) d\mathbf{x} \right).
\tag{7.10}
$$

Consequently, the number of misclassified data point is minimized by obtaining $\{\Omega_g\}$ $(1 \le g \le G)$, which maximizes the value of $\sum_{g=1}^{G} \int_{\Omega_g} n_g p_g(\mathbf{x}) \mathbf{dx}$. To achieve this, it is assumed that $\mathbf{x}$ is fixed at a certain point. Then, if $n_k p_k(\mathbf{x})$ is the maximum value of $\{n_g p_g(\mathbf{x})\}$, this $\mathbf{x}$ must be located in the region of Ω_k. To verify this, we carry out the following calculation.

$$\sum_{g=1}^{G} \int_{\Omega_g} n_g p_g(\mathbf{x}) \mathbf{dx} - \sum_{g=1}^{G} \int_{\Omega_g^*} n_g p_g(\mathbf{x}) \mathbf{dx}$$

$$= \sum_{g=1}^{G} \int_{\Omega_g} \max_j(\{n_j p_j(\mathbf{x})\}) \mathbf{dx} - \sum_{g=1}^{G} \int_{\Omega_g^*} n_g p_g(\mathbf{x}) \mathbf{dx}$$

$$= \int_{\text{all}} \max_j(\{n_j p_j(\mathbf{x})\}) \mathbf{dx} - \int_{\text{all}} \text{choice}_{\{\Omega_i^*\}}(\{n_j p_j(\mathbf{x})\}) \mathbf{dx}, \quad (7.11)$$

where $\{\Omega_g\}$ are obtained to satisfy the condition: the maximum value of $\{n_g p_g(\mathbf{x})\}$ is $n_k p_k(\mathbf{x})$ in the region of Ω_k. $\{\Omega_g^*\}$ are assumed to be regions that are different from $\{\Omega_g\}$. $\int_{\text{all}}$ means integration over the entire region where $\mathbf{x}$ exists. The expression $\max_j(\{n_j p_j(\mathbf{x})\})$ $(1 \le j \le G)$ indicates the maximum value among $\{n_j p_j(\mathbf{x})\}$ when the vector $\mathbf{x}$ is fixed. The expression $\text{choice}_{\{\Omega_i^*\}}(\{n_j p_j(\mathbf{x})\})$ takes the value $n_i p_i(\mathbf{x})$ if $\mathbf{x}$ is positioned in Ω_i^*. The definition of $\{\Omega_g\}$ assures that the first term on the first line of this equation is identical to the first term on the second line. The integrands in the terms on the third line show that the value of $\text{choice}_{\{\Omega_i^*\}}(\{n_j p_j(\mathbf{x})\})$ is equal to or larger than $\max(\{n_j p_j(\mathbf{x})\})$. Hence, the value of this equation is positive. Thus, the probability of misclassification is minimized by the classification: the maximum value of $\{n_j p_j(\mathbf{x})\}$ becomes $n_k p_k(\mathbf{x})$ in the region of Ω_k.

Using $\{\Omega_g\}$ obtained in this manner, the number of misclassified data points is

$$\sum_{g=1}^{G} n_j - \sum_{g=1}^{G} \int_{\Omega_j} n_j p_j(\mathbf{x}) \mathbf{dx}. \quad (7.12)$$

The first term is the total number of data points, and the second term is the number of data points that are classified correctly. The decision boundary shown in figure 7.2 carries out the classification by which data located in the region where the value of $n_1 p_1(x)$ is larger than that of $n_2 p_2(x)$ are classified as C_1, and data positioned in the region where the value of $n_1 p_1(x)$ is smaller than that of $n_2 p_2(x)$ are classified as C_2.

This method is expected to provide beneficial results in actual classifications if the ratio of data that belong to $\{C_g\}$ (these data were used for obtaining $\{\Omega_g\}$ (training data)) is not very much different from that of data for actual classification (test data). However, if the ratio of data belonging to $\{C_g\}$ (when the training data is used) is considerably different from that of the test data, the situations are different. Nevertheless, $\{p_g(\mathbf{x})\}$ of the test data are close to those of the training data. For example, in character recognition, while the number of patterns classified as "a" and the number of patterns classified as "b" depend heavily on the targeted series of characters, the probability density functions obeyed by random variables representing the characteristics of the characters remain almost the same if the same person writes the series of characters. Therefore, when test data are classified, only $\{n_g\}$ are altered according to the conditions of the test data.

In many situations, the values of $\{n_g\}$ are known beforehand from knowledge concerning the phenomenon that generates the data. For example, the approximate ratio of emergences of the character "a" is known if the language is specified. The values of $\{n_g\}$ obtained in this manner are divided by the value of $\sum_{g=1}^{G} n_g$; the resultant values are called prior probabilities in the context of Bayesian inference. They are hereafter denoted as $\{P_g\}$. Hence, for classification of test data, the probability that a pattern represented as $\mathbf{x}$ belongs to C_j is written as $\{q_g(\mathbf{x})\}$:

$$q_g(\mathbf{x}) = \frac{P_g p_g(\mathbf{x})}{\sum_g P_g p_g(\mathbf{x})}. \tag{7.13}$$

When a specific vector is given as $\mathbf{x}$, the g that maximizes the value of eq(7.13) is chosen; that is, the pattern is classified as C_g. The terms $\{q_g(\mathbf{x})\}$ are called posterior probabilities. Eq(7.13) is a form of Bayes' theorem. The strategy of choosing the class that maximizes posterior probability given by Bayes' theorem is equivalent to classification by the same method as that for the training data on the basis of estimation of $\{n_g\}$ in the test data.

Furthermore, in addition to the fact that classification conducted by eq(7.13) means that data are classified as the class which yields a maximum posterior probability value, the equation indicates that $\mathbf{x}$ is judged to be a member of class k if the following condition is satisfied:

$$P_k p_k(\mathbf{x}) \geq P_g p_g(\mathbf{x}) \qquad (g = 1, 2, \ldots, k - 1, k + 1, \ldots, G). \tag{7.14}$$

This method is called Bayes' decision rule. Functions for classification in the same manner as that of eq(7.14) are called discriminant functions, even if the functions do not have the form of $\{P_g p_g(\mathbf{x})\}$. When $\{f_g(\mathbf{x})\}$ are discriminant functions, $\mathbf{x}$ is classified as class k if the following condition is satisfied:

$$f_k(\mathbf{x}) \geq f_g(\mathbf{x}) \qquad (g = 1, 2, \ldots, k - 1, k + 1, \ldots, G). \tag{7.15}$$

When discriminant functions are derived directly without obtaining $\{p_g(\mathbf{x})\}$, the predictors for each $\{f_g(\mathbf{x})\}$ are not necessarily identical.

7.3   LINEAR DISCRIMINANT RULE AND QUADRATIC DISCRIMINANT RULE

Both $\{p_g(\mathbf{x})\}$ and $\{P_g\}$ are required for classification based on Bayes' decision rule. The terms $\{p_g(\mathbf{x})\}$ are generated from the very nature of the phenomenon that is the source of the data, while $\{P_g\}$ depends on the methods of experimentation, observations, or the goal of the classification. Therefore, the derivation of $\{p_g(\mathbf{x})\}$ is the central concern of classification.

The simplest method is the fitting of a normal distribution to each $\{p_g(\mathbf{x})\}$. That is, $p_g(\mathbf{x})$ is assumed to be

$$p_g(\mathbf{x}) = \left(\frac{1}{\sqrt{2\pi}}\right)^m |\mathbf{V}^{(g)}|^{-1/2} \exp\left(-\frac{1}{2}(\mathbf{x} - \boldsymbol{\mu}^{(g)})^t (\mathbf{V}^{(g)})^{-1}(\mathbf{x} - \boldsymbol{\mu}^{(g)})\right), \tag{7.16}$$

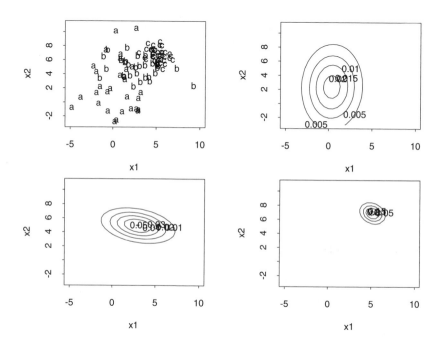

Figure 7.3 Data generated by pseudo-random numbers. The numbers of data points that belong to "a," "b," and "c" are 30, 30, and 40, respectively (top left). The results of fitting the data that belong to "a," "b," and "c" to normal distributions (eq(7.16)) are drawn (top right, bottom left, bottom right).

where the vector $\mu^{(g)}$ indicates the mean of a normal distribution; data that belong to C_g obey the normal distribution. $\mathbf{V}^{(g)}$ denotes the variance-covariance matrix of the normal distribution. $|\mathbf{V}^{(g)}|$ stands for the determinant of $\mathbf{V}^{(g)}$ and is called the generalized variance. The likelihood for an actual feature vector is

$$L(p_g(\mathbf{x}), \{\mathbf{x}_i^{(g)}\})$$
$$= \left(\frac{1}{\sqrt{2\pi}}\right)^{mn_j} |\hat{\mathbf{V}}^{(g)}|^{-n_j/2} \prod_{i=1}^{n_j} \exp\left(-\frac{1}{2}(\mathbf{x}_i^{(g)} - \mu^{(g)})^t (\mathbf{V}^{(g)})^{-1}(\mathbf{x}_i^{(g)} - \mu^{(g)})\right),$$
$$(7.17)$$

where $\mathbf{x}_i^{(g)} (1 \leq i \leq n_j)$ is a feature vector of a data point which is a member of C_j. By maximizing this value, $\mu^{(g)}$ and $\mathbf{V}^{(g)}$ are written in the following; as for proof, the reader is advised to refer to textbooks on multivariate probability density functions.

$$\hat{\mu}^{(g)} = \frac{1}{n_g} \sum_{i=1}^{n_g} \mathbf{x}_i^{(g)}, \qquad (7.18)$$

$$\hat{\mathbf{V}}^{(g)} = \frac{1}{n_g} \sum_{i=1}^{n_g} \sum_{l=1}^{n_g} (\mathbf{x}_i^{(g)} - \hat{\mu}^{(g)})(\mathbf{x}_l^{(g)} - \hat{\mu}^{(g)})^t. \qquad (7.19)$$

The mean and maximum likelihood estimate of the variance-covariance matrix of $p_g(\mathbf{x})$ coincide with the mean and variance-covariance matrix of the data. The mean given by eq(7.18) is the sample mean, and the covariance provided by eq(7.19) is the sample variance. Note that $\mathbf{V}^{(g)}$ given by eq(7.19) is not an unbiased estimator. The unbiased estimator of $\mathbf{V}^{(g)}$ is

$$\hat{\mathbf{V}}^{(g)} = \frac{1}{n_g - 1} \sum_{i=1}^{n_g} \sum_{l=1}^{n_g} (\mathbf{x}_i^{(g)} - \hat{\mu}^{(g)})(\mathbf{x}_l^{(g)} - \mu^{(g)})^t. \qquad (7.20)$$

Furthermore, if $\hat{\mathbf{V}}^{(g)}$ (the unbiased estimators) for all g are identical, the matrix is represented as $\hat{\mathbf{V}}$:

$$\hat{\mathbf{V}} = \frac{1}{n - G} \sum_{g=1}^{G} \sum_{i=1}^{n_g} \sum_{l=1}^{n_g} (\mathbf{x}_i^{(g)} - \hat{\mu}^{(g)})(\mathbf{x}_l^{(g)} - \mu^{(g)})^t. \qquad (7.21)$$

The resulting $\log\left(P_g p_g(\mathbf{x})\right)$ is written as

$$\log\left(P_g p_g(\mathbf{x})\right)$$
$$= \log\left(P_g \left(\frac{1}{\sqrt{2\pi}}\right)^{n_g} |\hat{\mathbf{V}}|^{-1/2} \exp\left(-\frac{1}{2}(\mathbf{x} - \hat{\mu}^{(g)})^t \hat{\mathbf{V}}^{-1}(\mathbf{x} - \hat{\mu}^{(g)})\right)\right)$$
$$= \log(P_g) + n_g \log\left(\frac{1}{\sqrt{2\pi}}\right) - \frac{1}{2}\log(|\hat{\mathbf{V}}|)$$
$$- \frac{1}{2}(\mathbf{x} - \hat{\mu}^{(g)})^t \hat{\mathbf{V}}^{-1}(\mathbf{x} - \hat{\mu}^{(g)})$$

$$= \log(P_g) + n_g \log\left(\frac{1}{\sqrt{2\pi}}\right) - \frac{1}{2}\log(|\hat{\mathbf{V}}|)$$
$$- \frac{1}{2}(\mathbf{x}^t\hat{\mathbf{V}}^{-1}\mathbf{x} - 2\mathbf{x}^t\hat{\mathbf{V}}^{-1}\hat{\boldsymbol{\mu}}^{(g)} + \hat{\boldsymbol{\mu}}^{(g)t}\hat{\mathbf{V}}^{-1}\hat{\boldsymbol{\mu}}^{(g)}). \tag{7.22}$$

The equality between the fourth line and the sixth line is derived on the basis of the fact that $\hat{\mathbf{V}}^{-1}$ is a symmetric matrix, because the inverse matrix of a symmetric matrix is also a symmetric matrix.

When classification is carried out for a certain vector $\mathbf{x}$, the maximum value is selected among $\{P_g p_g(\mathbf{x})\}$. For this purpose, essential parts are extracted from eq(7.22). We then have

$$2\log(P_g) - \hat{\boldsymbol{\mu}}^{(g)t}\hat{\mathbf{V}}^{-1}\hat{\boldsymbol{\mu}}^{(g)} + 2\mathbf{x}^t\hat{\mathbf{V}}^{-1}\hat{\boldsymbol{\mu}}^{(g)}. \tag{7.23}$$

The value of g which maximizes the value of eq(7.23) is defined as k. By this process, a pattern that has a feature vector $\mathbf{x}$ is classified as C_k. This method is called the linear discriminant rule. The term originates from the fact that eq(7.23) is a linear equation with respect to $\mathbf{x}$.

Figure 7.3(top left) shows the data generated by pseudo-random numbers. Data points that are members of "a," "b," and "c" are distributed. The number of data points are 30, 30, and 40, respectively. A normal distribution (eq(7.16)) is fitted to the data of each class; the results are displayed in figure 7.3(top right), figure 7.3(bottom left) and figure 7.3(bottom right). Classification achieved using the linear discriminant rule is carried out for these distributions; figure 7.4 shows the results. In figure 7.4(left), prior probabilities corresponding to "a," "b," and "c" are set at 0.3, 0.3, and 0.4, respectively; they are proportional to the number of data points. On the other hand, the values $17/20$, $2/20$, and $1/20$ are adopted in figure 7.4(right). The area of "a" is wide in figure 7.4(right); this is the result of the values of prior probabilities.

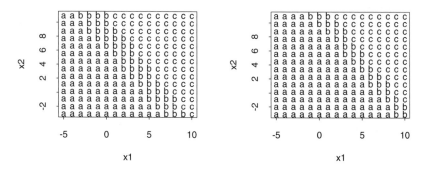

Figure 7.4   Results of classification supplied by the linear discriminant rule; the data set is shown in figure 7.3(top left). The prior probabilities are 0.3, 0.3, and 0.4 for "a," "b," and "c", respectively (left). They are $17/20$, $2/20$, and $1/20$, respectively (right).

In contrast, when the variance-covariance matrix depends on class, eq(7.16) (page 414) and eq(7.14) (page 414) indicate that the value of g which maximizes the value in the following equation is chosen:

$$2\log(P_g) - \log(|\hat{\mathbf{V}}^{(g)}|) - (\mathbf{x} - \hat{\boldsymbol{\mu}}^{(g)})^t\hat{\mathbf{V}}^{(g)-1}(\mathbf{x} - \hat{\boldsymbol{\mu}}^{(g)}), \tag{7.24}$$

where $|\hat{\mathbf{V}}^{(g)}|$ is the determinant of $\hat{\mathbf{V}}^{(g)}$. The term $\hat{\mathbf{V}}^{(g)-1}$ is the inverse matrix of $\hat{\mathbf{V}}^{(g)}$. This method is known as the quadratic discriminant rule. This term comes from the fact that eq(7.24) is a quadratic equation with respect to $\mathbf{x}$.

Figure 7.5 shows the results of the quadratic discriminant rule; the data used here are shown in figure 7.3(top left). In the same fashion as that of figure 7.4, the prior probabilities for "a," "b," and "c" are 0.3, 0.3, and 0.4 in the left graph; they are $17/20$, $2/20$, and $1/20$ in the right graph.

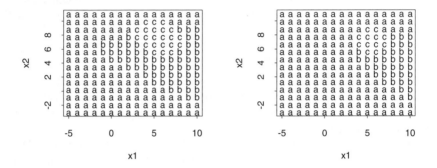

Figure 7.5 Results of classification supplied by the quadratic discriminant rule; the data are shown in figure 7.3(top left). The prior probabilities are $0.3, 0.3$, and 0.4 for "a," "b," and "c", respectively (left). They are $17/20$, $2/20$, and $1/20$, respectively (right).

7.4 CLASSIFICATION USING NONPARAMETRIC PROBABILITY DENSITY FUNCTION

Figure 7.6 exemplifies a nonparametric probability density function (nonparametric density function); a Gaussian function is used as the kernel function. The data are shown in figure 7.3(top left) (page 415). The bandwidth is fixed at 8 for all classes and both axes (x_1 and x_2). This bandwidth is determined intuitively by the appearance of nonparametric probability density functions. These nonparametric probability density functions are utilized for the classification by Bayes' decision rule; the prior probabilities for "a," "b," and "c" are 0.3, 0.3, and 0.4, respectively. Figure 7.7(left) exhibits the results. When the prior probabilities are configured at $17/20$, $2/20$, and $1/20$, figure 7.7(right) is obtained.

Since the bandwidth plays an important role in deriving a nonparametric probability density function, the misclassification rate is estimated using 10-fold cross-validation. It is assumed that the bandwidth of x_1 is the same as that of x_2, and this bandwidth is applied to the three nonparametric probability density functions. Figure 7.8 shows the result. The vertical axis indicates the misclassification rate. The optimal bandwidth is 3.5. The characteristics of nonparametric probability density functions based on the bandwidth ($= 3.5$) (figure 7.9) are substantially different from those based on the bandwidth ($= 8$). Nevertheless, the performance of the two sets of nonparametric probability density functions makes a negligible difference.

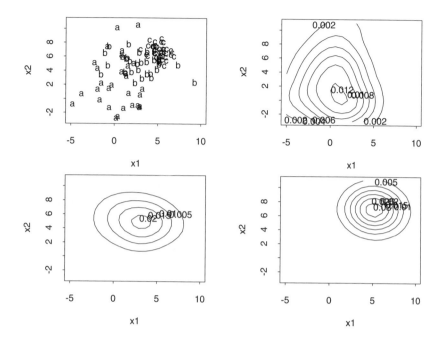

Figure 7.6   Data that are the same as those in figure 7.3(top left) (page 415), (top left). The numbers of data points that belong to "a," "b," and "c" are 30, 30, and 40, respectively. Nonparametric probability density functions supplied by data that are members of "a," "b," and "c" are illustrated in graphs (top right, bottom left, and bottom right, respectively). The bandwidth is fixed at 8 for all classes and axes (x_1 and x_2).

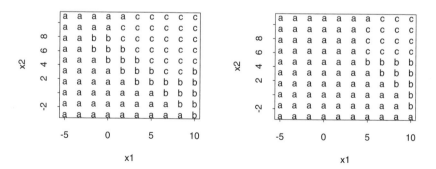

Figure 7.7   Results of classification based on the nonparametric probability density functions displayed in figure 7.6(top right), figure 7.6(bottom left), and figure 7.6(bottom right). The prior probabilities for the right graph are 0.3, 0.3, and 0.4 for "a," "b," and "c", respectively. They are $17/20$, $2/20$, and $1/20$ for the left graph.

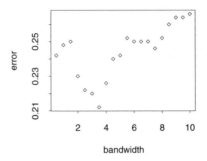

Figure 7.8 Relationship between the bandwidth and the misclassification rate given by 10-fold cross-validation. The classification is carried out using the nonparametric probability density functions; the data exhibited in figure 7.3(top left) (page 415) are used here.

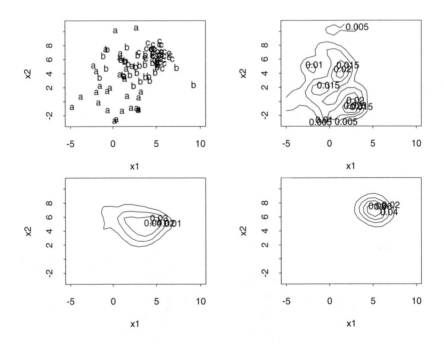

Figure 7.9 Data generated by pseudo-random numbers; the numbers of data points that belong to "a," "b," and "c" are 30, 30, and 40, respectively (top left). The nonparametric probability density functions based on data that are members of "a," "b," and "c" are illustrated in graphs (top right, bottom left, and bottom right, respectively). The bandwidth is fixed at 3.5 for all classes and axes (x_1 and x_2).

7.5 LOGISTIC REGRESSION

A random variable r is supposed to be either 1 or 0; the probability of it being 1 is t, and that of it being 0 is $(1-t)$. The probability density function $(d_{Ber}(s))$ is written as

$$d_{Ber}(s) = t^s(1-t)^{(1-s)} \qquad (s = 0, 1). \qquad (7.25)$$

This distribution is called a Bernoulli distribution.

When a feature vector ($\mathbf{x}$) is given, it is assumed that the corresponding data point belongs to one of the classes: "a," "b," or "c". Furthermore, if a data point is a member of "a," the situation is represented as $s = 1$; the probability of the occurrence of the event is defined as t. In addition, if the data point is a member of "b" or "c," it is denoted by $s = 0$; the probability is defined as $(1-t)$. Hence, when a random number r is defined to indicate that $\mathbf{x}$ is a member of "a," r obeys a probability density function given by eq(7.25). Since t is a function of $\mathbf{x}$, r is also a function of $\mathbf{x}$. That is, r obeys a Bernoulli distribution which is characterized by t. Furthermore, $E[r]$ (the expectation of r) is t.

The generalization of a Bernoulli distribution results in a binomial distribution, the probability density of which $(d_{binom}(s))$ is

$$d_{binom}(s) = {}_lC_s t^s(1-t)^{(l-s)} \qquad (s = 0, 1, \ldots, l), \qquad (7.26)$$

where ${}_lC_s$ is a binomial factor. For example, when the number of feature vectors having a certain value is l, and the number of data points that belong to "a" is represented as a random variable r', r' obeys a probability density function $(d_{binom}(s))$. Only one data point has a feature vector which is a certain vector; the probability density function obeyed by the number of data belonging to "a" is represented as a Bernoulli distribution. If one or more than one data point has such a feature vector, the probability density function is represented as a binomial distribution. Furthermore, $E[r]$ is t, and $E[r']$ is lt. Therefore, a binomial distribution contains a Bernoulli distribution as a special case.

In the field of pattern recognition, l indicates the number of data points which are located in a narrow region where $\mathbf{x}$ is positioned. Hence, l is represented as $l(\mathbf{x})$. Since r' is a function of $\mathbf{x}$, it is written as $r'(\mathbf{x})$. Then the value of the target variable of a data point that belongs to the class g is defined as 1, and the value of the target variable of a data point that does not belong to the class g is defined as 0; the resultant $\frac{E[r'(\mathbf{x})]}{l(\mathbf{x})}$ is denoted by $\mu_g(\mathbf{x})$. The term $\mu_g(\mathbf{x})$ denotes the ratio of the number of data points that belong to g in the narrow region (where $\mathbf{x}$ is located) to the total number of data points positioned in the same area. Logistic regression represents the logit transformation of $\mu_g(\mathbf{x})$ as a linear combination of each element of $\mathbf{x}$ plus a constant (as a linear model) under the condition that $r'(\mathbf{x})$ obeys a binomial distribution. It is a regression that uses a regression equation:

$$\log\left(\frac{\mu_g(\mathbf{x})}{1 - \mu_g(\mathbf{x})}\right) = \beta_0^{(g)} + \sum_{j=1}^m \beta_j^{(g)} x_j, \qquad (7.27)$$

where x_j is the jth element of $\mathbf{x}$. This regression equation is called a logistic model (a.k.a. logit model). In the context of a generalized linear regression, a function for a transformation, such as a logit transformation, which provides the target variable of a linear model is called a link function.

Eq(7.27) is rewritten as

$$
\begin{aligned}
\mu_g(\mathbf{x}) &= \frac{\exp(\beta_0^{(g)} + \sum_{j=1}^{m} \beta_j^{(g)} x_j)}{1 + \exp(\beta_0^{(g)} + \sum_{j=1}^{m} \beta_j^{(g)} x_j)} \\
&= \frac{1}{1 + \exp(-\beta_0^{(g)} - \sum_{j=1}^{m} \beta_j^{(g)} x_j)}.
\end{aligned}
\tag{7.28}
$$

To construct a regression equation with the form of eq(7.28), the value of a target variable of a data point that belongs to "a" is defined as 1, for example; the value of a target variable of a data point that does not belong to "a" is defined as 0. Then, when $\mathbf{x}$ is fixed, the value of $\mu_1(\mathbf{x})$ indicates that, if the number of data points existing in the narrow region near $\mathbf{x}$ is $l(\mathbf{x})$, the number of data points that belong to "a" is $l(\mathbf{x})\mu_1(\mathbf{x})$. The same is true of other classes. Therefore, a classification based on the principle that a data point is classified into a class which supplies the maximum value among $\{\mu_g(\mathbf{x})\}$ means that prior probabilities are derived from $\{n_g\}$, which are given by the data for creating regression equations. However, the summation of $\{\mu_g(\mathbf{x})\}$ for a fixed $\mathbf{x}$ is not guaranteed to be 1. In this sense, $\mu_g(\mathbf{x})$ is not considered to be equivalent to $q_g(\mathbf{x})$ (eq(7.13) (page 414)).

Note that the value of $\mu_g(\mathbf{x})$ is between 0 and 1, regardless of the values of $\{\beta_j^{(g)}\}$ and $\{x_j\}$. In this respect, the value fulfills one of the conditions that $q_g(\mathbf{x})$ is supposed to satisfy. This is one benefit of logit transformation. Probit transformation shares this property and yields similar results to those obtained by logit transformation (page 122 of [5]).

However, even if a linear combination of $\{x_j\}$ plus a constant is employed as $\mu_g(\mathbf{x})$, some conditions for the values of $\{\beta_j^{(g)}\}$ and $\{x_j\}$ restrict the value of $\mu_g(\mathbf{x})$ from 0 through 1. Furthermore, when $\mu_g(\mathbf{x})$ is utilized as a discriminant function (eq(7.15) (page 414)), it would appear that realization of beneficial classification is more important than the condition that the values are between 0 and 1. In conclusion, the choice among logit transformation, another transformation, and no transformation depends on the goodness of fitting, the fulfillment of purposes, and other issues (page 165 of [5]).

Creation of a logistic model or linear models based on transformations other than a logistic transformation (an identity transformation is also possible) on the assumption of a binomial distribution is a generalized linear regression. The Iterative Reweighted Least Squares (IRLS) method is equivalent to the Fisher scoring procedure (Fisher scoring algorithm); this characteristic is shared by the Poisson regression for smoothing a histogram.

Figure 7.10 shows the results of a classification given by logistic regression; the data are the same as those shown in figure 7.3(top left) (page 415). The estimates of logistic models for "a," "b," and "c" are showed in figure 7.10(top left), figure 7.10(top right), and figure 7.10(bottom left), respectively. These three functions are used as discriminant functions. Classification is conducted based on the philosophy that a class is selected if the corresponding function gives the maximum value among competitors. Figure 7.10(bottom right) shows the results. As expected, the decision boundary is a straight line.

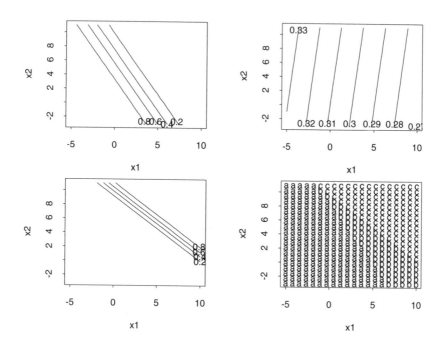

Figure 7.10 Results of classification rendered by logistic regression; the data are the same as that shown in figure 7.3(left top) (page 415). Estimates by logistic models for "a," "b," and "c" ($\{\mu_g(\mathbf{x})\}$) are drawn in graphs (left top, right top, and left bottom, respectively). The results of the classification given by the three discriminant functions are illustrated in a graph (right bottom).

Since logistic regression imposes strict limitations, more flexible regression equations deserve examination. One possibility is that the result of logit transformation is depicted as an additive model instead of a linear model. The model as follows is employed.

$$\log\left(\frac{\mu_g(\mathbf{x})}{1 - \mu_g(\mathbf{x})}\right) = \sum_{j=1}^{m} \psi_j^{(g)}(x_j). \tag{7.29}$$

Each $\{\psi_j^{(g)}(x_j)\}$ is a smooth nonparametric function. This model is a generalization of eq(7.27) (page 421). Eq(7.29) is a form of generalized additive model. This equation is transformed into

$$\mu_g(\mathbf{x}) = \frac{\exp(\sum_{j=1}^{m} \psi_j^{(g)}(x_j))}{1 + \exp(\sum_{j=1}^{m} \psi_j^{(g)}(x_j))}. \tag{7.30}$$

To derive a generalized additive model, the local likelihood is maximized. However, the maximization of the expectation of local likelihood instead of local likelihood itself is recommended (Chapter 6.13 of [1]). This is called a local-scoring procedure.

Figure 7.11 illustrates the results of a classification using the generalized additive model; the data are the same as those displayed in figure 7.3(top left) (page 415). Figure 7.11(top left), figure 7.11(top right), and figure 7.11(bottom left) show the estimates provided by the generalized additive models corresponding to "a," "b," and "c", respectively. The results of classification into the class that renders the maximum value of $\mu_g(\mathbf{x})$ are displayed in figure 7.11(right bottom).

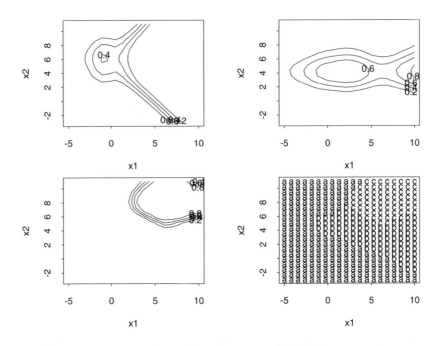

Figure 7.11   Results of classification provided by generalized additive models; the data are the same as those shown in figure 7.3(top left) (page 415). The estimates provided by generalized additive models for "a," "b," and "c" are displayed in graphs (top left, top right, and left bottom, respectively). The result of the three discriminant functions is shown in a graph (bottom right).

As previously stated, the values of $\{\mu_g(\mathbf{x})\}$ given by eq(7.27) or eq(7.28) are not guaranteed to sum up to the exact value of 1. Hence, each value is not regarded as $q_g(\mathbf{x})$ in a strict sense. This could be advantageous because this characteristic enables the treatment of data points that can belong to more than one class. However, it is not desirable when the number of classes to which each data point belongs is definitely 1. The method of logistic discrimination is free of this problem, because this method represents $q_g(\mathbf{x})$ as

$$q_g(\mathbf{x}) = \frac{s_g(\mathbf{x})}{\sum_{g'=1}^{G} s_{g'}(\mathbf{x})}. \tag{7.31}$$

In the context of neural networks, this form is called a softmax function. To calculate $\{q_g(\mathbf{x})\}$, the ratios among $\{s_g(\mathbf{x})\}$ are required. Even if each $\{s_g(\mathbf{x})\}$ is multiplied by a nonzero constant to be used as $\{s_g(\mathbf{x})\}$, $\{q_g(\mathbf{x})\}$ are not affected. Hence, $s_1(\mathbf{x}) \equiv 1$ is imposed without loss of generality. Hence, we have

$$q_1(\mathbf{x}) = \frac{1}{1 + \sum_{g'=2}^{G} s_{g'}(\mathbf{x})}, \tag{7.32}$$

$$q_g(\mathbf{x}) = \frac{s_g(\mathbf{x})}{1 + \sum_{g'=2}^{G} s_{g'}(\mathbf{x})} \qquad (2 \le g \le G). \tag{7.33}$$

A simple method employs the following functions as $\{s_g(\mathbf{x})\}$ $(2 \le g \le G)$.

$$s_g(\mathbf{x}) = \exp(\beta_0^{(g)} + \sum_{j=1}^{m} \beta_j^{(g)} x_j). \tag{7.34}$$

By substituting eq(7.34) into eq(7.32) and eq(7.33), we obtain

$$q_1(\mathbf{x}) = \frac{1}{1 + \sum_{g'=2}^{G} \exp(\beta_0^{(g')} + \sum_{j=1}^{m} \beta_j^{(g')} x_j)}, \tag{7.35}$$

$$q_g(\mathbf{x}) = \frac{\exp(\beta_0^{(g)} + \sum_{j=1}^{m} \beta_j^{(g)} x_j)}{1 + \sum_{g'=2}^{G} \exp(\beta_0^{(g')} + \sum_{j=1}^{m} \beta_j^{(g')} x_j)}.$$
$$(2 \le g \le G). \tag{7.36}$$

Under the conditions of a Bernoulli distribution, the maximum likelihood procedure (maximum likelihood method) provides the regression coefficients of eq(7.35) and eq(7.36). Furthermore, when $\mathbf{x}$ is fixed at a certain vector, the value of g which corresponds to the maximum value of $\{q_g(\mathbf{x})\}$ is selected as the class that contains $\mathbf{x}$ as a member. This method is called multinominal logistic regression. This is different from the classification rendered by logistic regression in spite of the similarity of terms.

Figure 7.12 illustrates the result of logistic discrimination. The data are the same as those shown in figure 7.3(bottom left) (page 415). Estimates given by $q_g(\mathbf{x})$ which correspond to "a," "b," and "c" are displayed in figure 7.12(top left), figure 7.12(top right), and figure 7.12(bottom left), respectively. Neural networks are used to obtain regression coefficients in $\{q_g(\mathbf{x})\}$. As a result of classification given by the three functions, figure 7.12(bottom right) is obtained.

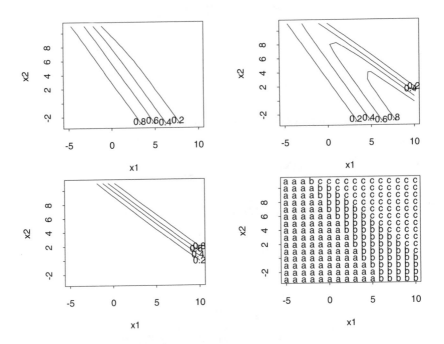

Figure 7.12 Results of classification provided by logistic discrimination; the data are the same as those shown in figure 7.3(top left) (page 415). The estimates provided by discriminant functions for "a," "b," and "c" are displayed in graphs (top left, top right, and bottom left, respectively). The result of the three discriminant functions is shown in a graph (bottom right).

7.6   NEURAL NETWORKS

Neural networks realize more flexible discriminant functions ($\{q_g(\mathbf{x})\}$). For example, in order that $\{q_g(\mathbf{x})\}$ for a fixed vector of $\mathbf{x}$ sum to 1, the output of neural networks is formulated as

$$q_g(\mathbf{x}) = \frac{\exp(r_g(\mathbf{x}))}{\sum_{g'} \exp(r_{g'}(\mathbf{x}))} \qquad (1 \le g \le G). \tag{7.37}$$

Each $\{r_g(\mathbf{x})\}$ is a smooth function. The sum of G outputs becomes 1. Eq(7.37) is a more flexible softmax function than eq(7.35) and eq(7.36).

The probability of $\mathbf{x}$ belonging to C_k is defined as $q_k(\mathbf{x})$. Hence, the likelihood of $\{q_g(\mathbf{x})\}$ in light of a data point indicating that $\mathbf{x}_i$ is a member of C_k is $q_k(\mathbf{x}_i)$. Then the likelihood of $\{q_g(\mathbf{x})\}$ in light of all data becomes

$$\prod_{g=1}^{G} \prod_{i=1}^{n,\mathbf{x}_i \in C_g} q_g(\mathbf{x}_i) = \prod_{g=1}^{G} \prod_{i=1}^{n} q_g(\mathbf{x}_i)^{t_{ig}}, \tag{7.38}$$

where n is the total number of data points. $\prod_{i=1}^{n,\mathbf{x}_i \in C_g}$ indicates that the multiplication is carried out only when $\mathbf{x}_i$ belongs to C_g. The term t_{ig} takes the value of 1 if $\mathbf{x}_i$ belongs to C_g, and it takes the value of 0 if $\mathbf{x}_i$ does not belong to C_g. Hence, the log-likelihood is

$$\sum_{g=1}^{G} \sum_{i=1}^{n} t_{ig} \log(q_g(\mathbf{x}_i)). \tag{7.39}$$

To maximize this value, the weights of the neural networks are optimized.

Using the data shown in figure 7.3(top left) (page 415), a neural network that outputs softmax functions in the form of eq(7.37) conducts classification; figure 7.13 shows the results. Estimates of $q_g(\mathbf{x})$ for "a," "b," and "c" are shown in figure 7.13(top left), figure7.13(top right), and figure 7.13(bottom left), respectively. The number of units located in the hidden layer of the neural networks is 5. A technique called weight decay is employed to adjust the smoothness of the estimates. The value of the regularization parameter in this situation is configured at 0.1. The value is due to the result of estimating the misclassification rate yielded by 10-fold cross-validation (figure 7.14).

Another possible use of neural networks is that the number of neural networks is increased to G and the number of outputs of each network is reduced to 1. This corresponds to an alteration by which the right-hand side of a logistic regression (eq(7.28)) (page 422) is replaced by the estimate (output) of a neural network. Furthermore, this setting assumes that the estimate of each neural network obeys a binomial distribution. Then, when the output of each neural network is defined as $f_g(\mathbf{x})$ $(1 \le g \le G)$, the likelihood of each $\{f_g(\mathbf{x})\}$ in light of the data, such as "$\mathbf{x}_i$ belongs to C_k, and it does not belong to other classes," is depicted as follows.

$$\begin{cases} f_g(\mathbf{x}_i) & \text{if } g = k \\ (1 - f_g(\mathbf{x}_i)) & \text{if } g \neq k. \end{cases} \tag{7.40}$$

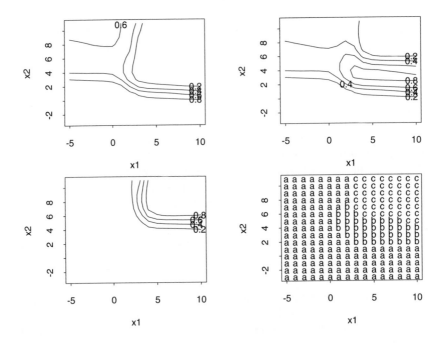

Figure 7.13   Results of classification given by neural networks; the data are the same as those shown in figure 7.3(top left) (page 415). Estimates by softmax functions for "a," "b," and "c" ($\{\mu_g(\mathbf{x})\}$) are drawn in graphs (top left, top right, and bottom left, respectively). The results of the classification generated by the three discriminant functions are illustrated in a graph (bottom right).

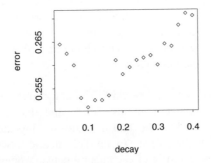

Figure 7.14   Relationship between the value of the regularization parameter and the misclassification rate estimated by 10-fold cross-validation; a neural network that output softmax functions is used. The data are the same as those displayed in figure 7.3(left top) (page 415).

Hence, the likelihood of each $\{f_g(\mathbf{x})\}$ in light of all data is

$$
\prod_{i=1}^{n,\mathbf{x}_i \in C_g} f_g(\mathbf{x}_i) \prod_{i=1}^{n,\mathbf{x}_i \notin C_g} (1 - f_g(\mathbf{x}_i)) = \prod_{i=1}^{n} f_g(\mathbf{x}_i)^{t_{ig}} (1 - f_g(\mathbf{x}_i))^{(1-t_{ig})}
$$

$$
(1 \le g \le G), \qquad (7.41)
$$

where $\prod_{i=1}^{n,\mathbf{x}_i \notin C_g}$ indicates that the multiplication is carried out only if $\mathbf{x}_i$ does not belong to C_g. Therefore, the log-likelihood of $\{f_g(\mathbf{x})\}$ is

$$
\sum_{i=1}^{n} \left(t_{ig}\log(f_g(\mathbf{x}_i)) + (1 - t_{ig})\log(1 - f_g(\mathbf{x}_i)) \right) \qquad (1 \le g \le G). \quad (7.42)
$$

To maximize these G values, the weights of the neural networks are optimized.

7.7 TREE-BASED MODEL

A tree-based model is another helpful tool for classification. While a tree-based model for the purpose of regression is constructed on the basis of reducing the residual sum of squares, another statistics is required for classification.

It is assumed that the number of data points that reach the sth terminal node as a result of inputting data into a tree-base model is N_s, and the number of data points that belong to the g-class among them is N_{sg}. Hence, we have

$$
N_s = \sum_{g=1}^{G} N_{sg} \qquad (1 \le s \le |T|), \qquad (7.43)
$$

where $|T|$ is the number of terminal nodes. As a characteristic of a tree-based model, a certain value among $\{N_{sg}\}$ $(1 \le g \le G)$ should be large and the others should be close to 0. Therefore, the desirability of $\{N_{sg}\}$ is measured by the distance from the uniform distribution of $\{\frac{N_s}{G}\}$ $(1 \le g \le G)$.

Furthermore, each $\{N_{sg}\}$ approximately obeys a Poisson distribution. Each $\{N_{sg}\}$ does not obey a Poisson distribution in the exact sense because each $\{N_{sg}\}$ has a higher limit (i.e., they cannot surpass N_s), while the region where the value of a Poisson distribution is not 0 does not have a higher limit. This situation is close to that of frequencies of a histogram; they obey approximately a Poisson distribution, but they do not obey it exactly.

To measure the distance between $\{N_{sg}\}$ and $\{\frac{N_s}{G}\}$, the statistic below is used; it measures the distance between realizations of a Poisson distribution and corresponding estimates:

$$
Kull(\{\hat{F}_i\}; \{F_i\}) = \sum_{j=1}^{n} \left(F_j \cdot \log(F_j) - F_j \cdot \log(\hat{F}_j) \right)
$$

$$
= \sum_{j=1}^{n} \left(F_j \cdot \log\left(\frac{F_j}{\hat{F}_j}\right) \right), \qquad (7.44)
$$

where $\{F_i\}$ are realizations of a Poisson distribution (i.e., data). $\{\hat{F}_i\}$ are estimates that correspond to the data. It is assumed that $0 \log(0) = 0$. $Kull(\{\hat{F}_i\}; \{F_i\})$ is called the Kullback-Leibler distance (it is also called the Kullback-Leibler information). The distribution of data ($\{N_{sg}\}$) that arrive at the sth split is substituted into $\{F_i\}$ in eq(7.44), and a uniform distribution ($\{\frac{N_s}{G}\}$) at s is substituted into $\{\hat{F}_i\}$. As a result, we have

$$Kull_s\left(\left\{\frac{N_s}{G}\right\}; \{N_{sg}\}\right) = \sum_{g=1}^{G}\left(N_{sg} \cdot \log\left(\frac{N_{sg}G}{N_s}\right)\right). \tag{7.45}$$

A splitting rule created to augment the value of this equation provides a beneficial split.

The value which is $Kull_s$ divided by N_s is defined as the deviance (a.k.a. the cross-entropy) at the sth terminal node. That is,

$$D_s = \sum_{g=1}^{G}\left(\frac{N_{sg}}{N_s} \cdot \log\left(\frac{N_{sg}}{N_s}\right)\right). \tag{7.46}$$

The essential meaning of this equation is identical to that of the generalized linear regression. D_s multiplied by (-1) is defined as I_s. Then we have

$$I_s = -\sum_{g=1}^{G}\left(\frac{N_{sg}}{N_s} \cdot \log\left(\frac{N_{sg}}{N_s}\right)\right). \tag{7.47}$$

I_s is called the impurity at the sth node. A large value of deviance, or a small value of impurity, means a desirable node. Construction of a splitting rule at a node creates two nodes. The impurity (I_{t+u}) of the two nodes is defined as

$$
\begin{aligned}
I_{t+u} &= \frac{N_t}{N_s}I_t + \frac{N_u}{N_s}I_u \\
&= -\frac{N_t}{N_s}\sum_{g=1}^{G}\left(\frac{N_{tg}}{N_t} \cdot \log\left(\frac{N_{tg}}{N_t}\right)\right) - \frac{N_u}{N_s}\sum_{g=1}^{G}\left(\frac{N_{ug}}{N_u} \cdot \log\left(\frac{N_{ug}}{N_u}\right)\right) \\
&= -\sum_{g=1}^{G}\left(\frac{N_{tg}}{N_s} \cdot \log\left(\frac{N_{tg}}{N_t}\right) + \frac{N_{ug}}{N_s} \cdot \log\left(\frac{N_{ug}}{N_u}\right)\right). \tag{7.48}
\end{aligned}
$$

The amount of reduction in impurity by setting out nodes is called improvement in some instances. The improvement by building the sth node is

$$
\begin{aligned}
\Delta I_s &= I_s - I_{t+u} \\
&= I_s - \frac{N_t}{N_s}I_t - \frac{N_u}{N_s}I_u. \tag{7.49}
\end{aligned}
$$

A splitting rule at the sth node is constructed to maximize this value. The repetition of this procedure creates a tree-based model for the purpose of classification.

Use of eq(7.49) as a statistic to set out a split aims to increase the differences among the numbers of members of classes at a newly developed terminal node. If a new terminal node becomes a terminal node of the final tree-based model, this statistic may not be desirable. Since each terminal node classifies data points at the node as members of the class having the maximum number of members, the splitting rule should be created to minimize the number of data points that belong to the other classes. The process of constructing a tree-based mode, however, is that after producing a tree-based model with an unnecessarily large number of terminal nodes, the number of terminal nodes is reduced by pruning several times to complete an appropriate model. Therefore, each node must function as both a splitting node and a terminal node. Furthermore, if all splitting nodes are constructed to maximize the number of data points that belong to one class and minimize the number of data points that belong to other classes, the splits located ahead possibly may not function well. Therefore, when splits are created, we employ a statistic that makes the numbers of members of classes as inhomogeneous as possible.

Other statistics for this purpose are possible. The Gini diversity index is commonly used. Its original definition is

$$1 - \sum_{g=1}^{G} p_g^2, \tag{7.50}$$

where each $\{p_g\}$ indicates the ratio of data points that are classified as C_g (the gth class). Hence, $\sum_{g=1}^{G} p_g = 1$ is assumed. Therefore, eq(7.50) is rewritten as

$$1 - \sum_{g=1}^{G} p_g^2 = \sum_{g=1}^{G} p_g - \sum_{g=1}^{G} p_g^2 = \sum_{g=1}^{G} (1 - p_g) p_g. \tag{7.51}$$

Furthermore, it is also represented as

$$
\begin{aligned}
1 - \sum_{g=1}^{G} p_g^2 &= 1 - \sum_{i=1}^{G} \sum_{j=1}^{G} p_i p_j + \sum_{i,j=1(i \neq j)}^{G,G} p_i p_j \\
&= 1 - \sum_{i=1}^{G} p_i \sum_{j=1}^{G} p_j + \sum_{i,j=1(i \neq j)}^{G,G} p_i p_j \\
&= 1 - 1 \cdot 1 + \sum_{i,j=1(i \neq j)}^{G,G} p_i p_j \\
&= \sum_{i,j=1(i \neq j)}^{G,G} p_i p_j,
\end{aligned} \tag{7.52}
$$

where $\sum_{i,j=1(i \neq j)}^{G,G} p_i p_j$ indicates that the summation is conducted with respect to i and j (both range from 1 to G), but the case with $i = j$ is omitted.

An impurity based on the Gini diversity index is defined as follows in place of eq(7.47).

$$I_s' = 1 - \sum_{g=1}^{G} \left(\frac{N_{sg}}{N_s} \right)^2. \tag{7.53}$$

The improvement is derived by replacing I_s, I_t, and I_u of eq(7.49) (page 430) with I_s', I_t', and I_u'.

The meaning of measuring the desirability of the sth terminal node using the Gini diversity index is understood from the viewpoint of the distance from the distribution in which the numbers of members in G classes are equal, namely, $N_{sg} = \frac{N_s}{G}$ ($1 \leq g \leq G$), because the squared sum of the distances between a uniform distribution and $\{N_{sg}\}$ is

$$\sum_{g=1}^{G}\left(\frac{1}{G} - \frac{N_{sg}}{N_s}\right)^2 = \sum_{g=1}^{G}\left(\frac{1}{G^2} - \frac{2N_{sg}}{GN_s} + \left(\frac{N_{sg}}{N_s}\right)^2\right)$$

$$= \frac{1}{G} - \frac{2}{G} + \sum_{g=1}^{G}\left(\frac{N_{sg}}{N_s}\right)^2. \tag{7.54}$$

The essence of this value multiplied by (-1) is the same as that of eq(7.53). Consequently, if the Kullback-Leibler distance is used to measure the distance between distributions with the same values of all classes, $\left\{\dfrac{N_{sg}}{N_s}\right\}$ ($1 \leq g \leq G$) is an impurity defined by eq(7.47). On the other hand, the residual sum of squares employed for this evaluation is also an impurity and is presented in eq(7.53). Several other definitions have been developed for deviance in tree-based models. However, the form of the function used as deviance slightly affects the resulting tree-based models (page 38 in [2]).

The cost-complexity measure is used to construct the optimal tree-based model corresponding to each number of the terminal nodes when the purpose of the tree-based model is classification; it is carried out in the same fashion as when the purpose is regression. The cost-complexity measure for classification is defined as

$$D_\alpha(T) = D(T) + \alpha|T|, \tag{7.55}$$

where α is the complexity parameter. The best number of terminal nodes is chosen by the optimization of the value of the complexity parameter. To optimize the value of the complexity parameter, the misclassification rate estimated by 10-fold cross-validation is used. Figure 7.15 shows the relationship between the value of the complexity parameter and the misclassification rate given by 10-fold cross-validation; the data are the same as those shown in figure 7.3(top left) (page 415). When the value of the complexity parameter is 15, the misclassification rate is small; the tree-based model with this setting and the results of classification supplied by the tree-based model are illustrated in figure 7.16.

In this example, weights are not attached to data. In other words, all weights are 1. This is equivalent to the assumption that the numbers of members of classes in the data are proportional to the prior probabilities corresponding to these classes. If the prior probabilities are different from the ones in this scenario, the weights of data depend on the class.

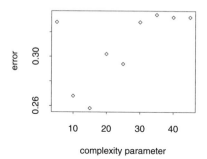

Figure 7.15 Relationship between the value of the complexity parameter and the misclassification rate given by 10-fold cross-validation; this treatment aims to optimize the number of terminal nodes in the tree-based model for the data shown in figure 7.3(top left) (page 415).

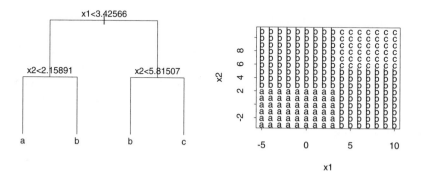

Figure 7.16 Tree-based model (the value of the complexity parameter is 15); the data are the same as those shown in figure 7.3(top left) (page 415), and the results of figure 7.15 are used.

7.8 k-NEAREST-NEIGHBOR CLASSIFIER

A supposed intuitive method for classification is that a feature vector, which is similar to an unidentified feature vector, is chosen among the vectors whose classes are known. Then the class of the selected feature vector is estimated to be the class in which the unidentified feature vector is a member. This method is described as follows.

When a feature vector $\mathbf{x}$ is fixed, the volume of the neighboring region is assumed to be $V(\mathbf{x})$. The neighboring region may be defined, for example, by the condition that the euclidean distance from $\mathbf{x}$ is less than or equal to a specific value. The number of C_g data in the region is defined as $K_g(\mathbf{x})$ $(1 \leq g \leq G)$. Then we define

$$p_g(\mathbf{x}) = \frac{K_g(\mathbf{x})}{n_g V(\mathbf{x})}. \tag{7.56}$$

In addition, prior probabilities are approximated to be

$$P_g = \frac{n_g}{n}, \tag{7.57}$$

where $\sum_{g=1}^{G} n_g = n$ holds.

Hence, Bayes' decision rule classifies $\mathbf{x}$ as the class k if the condition in the following is satisfied:

$$\frac{K_k(\mathbf{x})}{V(\mathbf{x})n} \geq \frac{K_g(\mathbf{x})}{V(\mathbf{x})n} \qquad (k \neq g). \tag{7.58}$$

This condition is equivalent to

$$K_k(\mathbf{x}) \geq K_g(\mathbf{x}) \qquad (k \neq g). \tag{7.59}$$

That is, when an unidentified feature vector $\mathbf{x}$ is supplied, $\mathbf{x}$ is determined to be a member of a certain class if the number of data points, located in the neighboring region, which belong to the class, is maximum among numbers of members of the classes in the region. If more than one class corresponds to the maximum number of data points in the neighboring region, the class for the feature vector can be chosen by a random number. Or if a data point is closest to one of the data points which belongs to the nominated classes, $\mathbf{x}$ can be classified as the class in which the closest data point is a member.

When the neighboring region is defined as the region where only one data point exists and the euclidean distance between the data point and $\mathbf{x}$ is the smallest, the class to which the data point belongs is identified as the class of the data. If more than one data point corresponds to the smallest distance, and these data points do not belong to the same class, a possible countermeasure is selection using a random number or reference to data points located a little further from the unidentified data point. When the data points that are positioned in a wider region are referred to rather than only the one data point with the smallest euclidean distance, the data in the wider region are laid out in ascending order according to the euclidean distance, and the first several data points in the row are used for the classification given by eq(7.59). Furthermore, if each element of the feature vector has different characteristics, the feature vectors must be transformed to be used for classification. Thus, the method for classification using eq(7.59), which uses a specific number of data points in the neighborhood of **x**, is called the k-nearest-neighbor classifier.

Figure 7.17 illustrates the results of classification using the k-nearest-neighbor classifier; the data are the same as those shown in figure 7.3(top left) (page 415). Figure 7.17(left) displays the results when the neighboring region is defined as the region that contains only one data point whose euclidean distance from **x** is the smallest. Figure 7.17(right) shows the results when 8 data points whose euclidean distances from **x** are small are used.

As shown in eq(7.57), this method employs the numbers of members of classes as prior probabilities for classification. When different prior probabilities are adopted, bagging is available. The term bagging comes from "Bootstrap AGGregatING." In its primary form, it derives a regression equation using the equation

$$m_{bag}(x) = \frac{1}{B} \sum_{i=1}^{B} m_i^*(x), \qquad (7.60)$$

where each $\{m_i^*(x)\}$ is a regression equation obtained from randomly sampled data (bootstrap sample or bootstrap data) from the entire data set; the number of data points in the bootstrap sample is the same as in the original data set. The B sets of bootstrap samples are created using different initial values of random numbers, and a regression equation for each set is derived. Bagging uses the means of estimates derived from $\{m_i^*(x)\}$ as final estimates. When bagging is applied to the k-nearest-neighbor classifier, a maximum vote determines the final decision of classification if the bootstrap sample does not conclude with unanimous agreement. If the maximum vote is tied, a random number chooses one of the classes at random. If the prior probabilities are not proportional to the numbers of members of classes, the numbers of bootstrap samples are set out to be proportional to the prior probabilities. This procedure results in classifications accompanied by diverse prior probabilities.

Figure 7.18 exemplifies the results of this method. The data are the same as those shown in figure 7.3(top left) (page 415). Figure 7.18(left) assumes 0.3, 0.3, and 0.4 as the prior probabilities of "a," "b," and "c", respectively; the prior probabilities are proportional to the numbers of members of classes. Figure 7.18 (right) is based on the prior probabilities $17/20$, $2/20$, and $1/20$.

7.9   NONPARAMETRIC REGRESSION BASED ON THE LEAST SQUARES

To derive $\{q_g(\mathbf{x})\}$ $(1 \leq g \leq G)$ defined in eq(7.13) (page 414), the basic procedure consists of two steps: (1) $\{p_g(\mathbf{x})\}$ $(1 \leq g \leq G)$ are derived, and (2) $\{P_g\}$ $(1 \leq g \leq G)$ are given. However, another procedure is worthy of consideration: derivation of $\{f_g(\mathbf{x})\}$ $(1 \leq g \leq G)$ (eq(7.15) (page 414)) without estimation of $\{q_g(\mathbf{x})\}$ rendered by Bayes' theorem. To obtain $\{f_g(\mathbf{x})\}$ for this purpose, a possible method is minimization of each of $\{E_g\}$ $(1 \leq g \leq G)$ defined as below.

$$E_g = \sum_{i=1}^{n_g} (f_g(\mathbf{x}_i) - y_{ig})^2, \qquad (7.61)$$

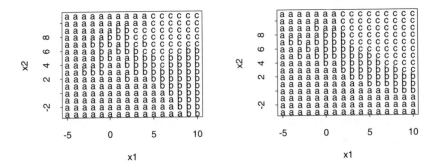

Figure 7.17 Results of classification provided by the k-nearest-neighbor classifier; the data are the same as those shown in figure 7.3(top left) (page 415): 1-nearest-neighbor classifier (left); 8-nearest-neighbor classifier (right).

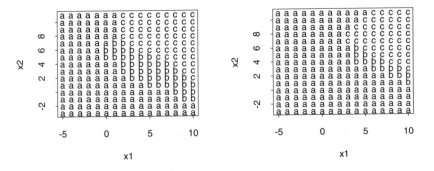

Figure 7.18 Results of classification generated by the k-nearest-neighbor classifier accompanied by bagging; the data are the same as those shown in figure 7.3(top left) (page 415). The prior probabilities of "a," "b," and "c" are 0.3, 0.3, and 0.4 (left). The prior probabilities of "a," "b," and "c" are 17/20, 2/20, and 1/20 (right).

where $\{y_{ig}\}$ are depicted as

$$
y_{ig} = \begin{cases} 1 & \text{if} \quad \text{the data accompanied by the feature vector } \mathbf{x}_i \text{ belong to } C_g, \\ 0 & \text{if} \quad \text{the data accompanied by the feature vector } \mathbf{x}_i \text{ do not belong to } C_g. \end{cases}
$$
(7.62)

Various functions are tested as $\{f_g(\mathbf{x})\}$. Then each $\{f_g(\mathbf{x})\}$ may take a negative value. It is not guaranteed that $\sum_{g=1}^{G} f_g(\mathbf{x}) = 1$ holds. That is, $\{f_g(\mathbf{x})\}$ may not satisfy the conditions which $\{q_g(\mathbf{x})\}$, given by Bayes' theorem, must satisfy. Therefore, $\{f_g(\mathbf{x})\}$ derived by this method usually do not bear the feature of classification yielded by Bayes' theorem based on $\{q_g(\mathbf{x})\}$: if the prior probabilities are correct, misclassification is minimized. Estimation of $\{f_g(\mathbf{x})\}$ using eq(7.61) is, however, an easy task in most cases, because nonparametric regression by the least squares (least square method) requires simpler algorithms and smaller computational costs compared with the estimation of nonparametric probability density functions. When the number of elements of a feature vector is large, this difference is prominent. Therefore, sophisticated $\{f_g(\mathbf{x})\}$ given by nonparametric regression by the least squares may realize beneficial classifications, compared to $\{q_g(\mathbf{x})\}$ which are based on Bayes' theorem and are constructed by nonparametric probability density functions generated by an unsophisticated method.

Note that when the functions to be used as $\{f_g(\mathbf{x})\}$ are written as follows, a characteristic that suggests the justification of the use of the least squares is found.

$$
f_g(\mathbf{x}) = \alpha_{g0} + \sum_{j=1}^{P} \alpha_{gj} \phi_j(\mathbf{x}),
$$
(7.63)

where $\{\phi_j(\mathbf{x})\}$ $(1 \leq j \leq P)$ are bases, and $\{\alpha_{gj}\}$ $(0 \leq j \leq P)$ are regression coefficients. Then, all of the elements of $\sum_{g=1}^{G} \hat{\mathbf{y}}^{(g)}$ are 1. The term $\hat{\mathbf{y}}^{(g)}$ is a column vector created by connecting the estimates given by calculating the values of the target variable (object variable) using feature vectors of data when $f_g(\mathbf{x})$ in the form of eq(7.63) is derived using $\mathbf{y}^{(g)}$. The proof is as follows.

First, the vector $\mathbf{y}^{(g)}$ is defined as

$$
\mathbf{y}^{(g)} = (y_{1g}, y_{2g}, \dots, y_{ng})^t \quad (1 \leq g \leq G).
$$
(7.64)

Then, we obtain

$$
\hat{\mathbf{y}}^{(g)} = \mathbf{H} \mathbf{y}^{(g)} \quad (1 \leq g \leq G).
$$
(7.65)

The hat matrix $(\mathbf{H})$ does not depend on g. Hence, we have the equation

$$
\sum_{g=1}^{G} \mathbf{H}^{-1} \hat{\mathbf{y}}^{(g)} = \sum_{g=1}^{G} \mathbf{y}^{(g)} = \mathbf{1},
$$
(7.66)

where $\mathbf{1}$ is a column vector in which all elements are 1; the number of elements is n. Furthermore, eq(7.66) is transformed into

$$
\sum_{g=1}^{G} \hat{\mathbf{y}}^{(g)} = \mathbf{H}\mathbf{1}.
$$
(7.67)

If **1** is used as $\mathbf{y}^{(g)}$, and $f_g(\mathbf{x})$ in the form of eq(7.63) is derived on the supposition that all elements of $\mathbf{y}^{(g)}$ are 1, it is found that the estimates of the target variable given by feature vectors of data are written as $\sum_{g=1}^{G} \hat{\mathbf{y}}^{(g)}$. If all elements of $\mathbf{y}^{(g)}$ are 1, fitting to eq(7.63) by the least squares results in the following solution: $\alpha_{g0} = 1$, $\alpha_{gj} = 0$ $(1 \leq j \leq P)$, because the residual sum of the squares is 0. Therefore, all elements of $\sum_{g=1}^{G} \hat{\mathbf{y}}^{(g)}$ are 1 (Q.E.D.).

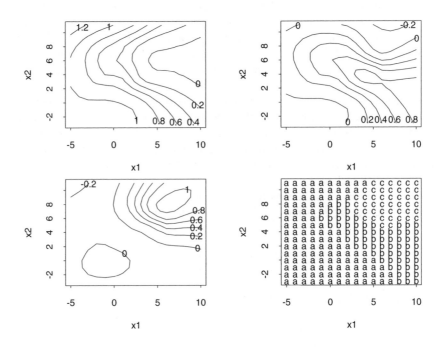

Figure 7.19 Discriminant functions represented by thin plate smoothing splines with two predictors given by $\{y_{ig}\}$ using eq(7.62); the data are the same as those shown in figure 7.3(top left) (page 415). Discriminant functions for "a," "b," and "c" are illustrated in graphs (top left, top right, bottom left, respectively). The results of classification by the three discriminant functions are charted in a graph (bottom right).

Furthermore, to adjust the values of prior probabilities according to the characteristics of the data for estimation yielded by Bayes' theorem, eq(7.61) is replaced with

$$E_g = \sum_{i=1}^{n_g} w_{ig'} \left(f_g(\mathbf{x}_i) - y_{ig} \right)^2, \tag{7.68}$$

where $w_{ig'}$ is the weight defined as

$$w_{ig'} = \frac{P_{g'}}{n_{g'}} \quad \text{if } \mathbf{x}_i \text{ belongs to } C_{g'}. \tag{7.69}$$

Figure 7.19 illustrates the results of classification using discriminant functions produced by thin plate smoothing splines. The values of the target variable are supplied by eq(7.62) (page 437). In place of eq(7.61) (page 435), equations of thin plate smoothing splines with two predictors are used. Weights are not attached to the data; this means that the values which are proportional to the numbers of members of classes are employed as prior probabilities. Discriminant functions for "a," "b," and "c" are shown in figure 7.19(top left), figure 7.19(top right) and figure 7.19(bottom left). The resulting classifications given by the three discriminant functions are shown in figure 7.19(bottom right).

7.10   TRANSFORMATION OF FEATURE VECTORS

Transformation of feature vectors leads to a beneficial classification in some cases. Variable selection, namely, pattern recognition based on selection of elements of feature vectors, is considered to be a form of transformation of feature vectors. The most well known technique of transformation for pattern recognition is the transformation given by canonical variates.

To obtain canonical variates, the class means ($\{\bar{\mathbf{x}}^{(g)}\}$) of feature vectors ($\{\mathbf{x}_i^{(g)}\}$), corresponding to data of various classes, and the mean of all feature vectors ($\bar{\mathbf{x}}$) are defined as

$$\bar{\mathbf{x}}^{(g)} = \frac{1}{n_g} \sum_{i=1}^{n_g} \mathbf{x}_i^{(g)}, \tag{7.70}$$

$$\bar{\mathbf{x}} = \frac{1}{n} \left(\sum_{g=1}^{G} \sum_{i=1}^{n_g} \mathbf{x}_i^{(g)} \right), \tag{7.71}$$

where n is the total number of data points, namely,

$$n = \sum_{g=1}^{G} n_g. \tag{7.72}$$

Furthermore, the elements of the within-group covariance matrix (a matrix of $\mathbf{W}$ with size of $m \times m$) and those of the between-group covariance matrix (a matrix of $\mathbf{B}$ with size of $m \times m$) are defined as

$$[\mathbf{W}]_{kl} = \frac{1}{n} \sum_{g=1}^{G} \sum_{i=1}^{n_g} (x_{ik}^{(g)} - \bar{x}_k^{(g)})(x_{il}^{(g)} - \bar{x}_l^{(g)}), \tag{7.73}$$

$$[\mathbf{B}]_{kl} = \frac{1}{n} \sum_{g=1}^{G} n_g (\bar{x}_k^{(g)} - \bar{x}_k)(\bar{x}_l^{(g)} - \bar{x}_l), \tag{7.74}$$

where $x_{ik}^{(g)}$ is the kth element of $\mathbf{x}_i^{(g)}$. The term $\bar{x}_k^{(g)}$ is the kth element of $\bar{\mathbf{x}}^{(g)}$. The term $\bar{x}_k$ is the kth element of $\bar{\mathbf{x}}$. The relationship among the total covariance (a matrix of $\mathbf{A}$ with size of $m \times m$), the within-group covariance matrix, and the between-group covariance matrix is

$$[\mathbf{W}]_{kl} + [\mathbf{B}]_{kl} = \frac{1}{n} \sum_{g=1}^{G} \left(\sum_{i=1}^{n_g} \left(x_{ik}^{(g)} x_{il}^{(g)} - x_{ik}^{(g)} \bar{x}_l^{(g)} - \bar{x}_k^{(g)} x_{il}^{(g)} + \bar{x}_k^{(g)} \bar{x}_l^{(g)} \right) \right.$$

$$+ n_g \bar{x}_k^{(g)} \bar{x}_l^{(g)} - n_g \bar{x}_k^{(g)} \bar{x}_l - n_g \bar{x}_k \bar{x}_l^{(g)} + n_g \bar{x}_k \bar{x}_l \Bigg)$$

$$= \frac{1}{n} \sum_{g=1}^{G} \Bigg(\sum_{i=1}^{n_g} \left(x_{ik}^{(g)} x_{il}^{(g)} \right) - n_g \bar{x}_k^{(g)} \bar{x}_l^{(g)} - n_g \bar{x}_k^{(g)} \bar{x}_l^{(g)}$$

$$+ n_g \bar{x}_k^{(g)} \bar{x}_l^{(g)} + n_g \bar{x}_k^{(g)} \bar{x}_l^{(g)} - n_g \bar{x}_k^{(g)} \bar{x}_l - n_g \bar{x}_k \bar{x}_l^{(g)}$$

$$+ n_g \bar{x}_k \bar{x}_l \Bigg)$$

$$= \frac{1}{n} \sum_{g=1}^{G} \sum_{i=1}^{n_g} \left(x_{ik}^{(g)} - \bar{x}_k \right) \left(x_{il}^{(g)} - \bar{x}_l \right)$$

$$= [\mathbf{A}]_{kl}. \tag{7.75}$$

Using these results, the transformation rendered by the equations below summarize feature vectors with m elements to obtain a scalar ($z_i^{(g)}$).

$$z_i^{(g)} = \sum_{k=1}^{m} a_k x_{ik}^{(g)} \qquad (1 \le g \le G, \ 1 \le i \le n). \tag{7.76}$$

The term $\mathbf{a} = (a_1, \dots, a_m)^t$ is called the discriminant coordinate. $\{z_i^{(g)}\}$ is a canonical variate because it is closely associated with canonical correlation analysis.

Therefore, when $\{\mathbf{z}_i^{(g)}\}$ are used as feature vectors, the within-group variance and the between-group variance are

$$\frac{1}{n} \sum_{g=1}^{G} \sum_{i=1}^{n_g} (z_i^{(g)} - \bar{z}^{(g)})(z_i^{(g)} - \bar{z}^{(g)})$$

$$= \frac{1}{n} \sum_{g=1}^{G} \Bigg(\sum_{i=1}^{n_g} \Big(\sum_{k=1}^{m} a_k x_{ik}^{(g)} - \frac{1}{n_g} \sum_{k=1}^{m} \sum_{l=1}^{n_g} a_k x_{lk}^{(g)} \Big)$$

$$\cdot \Big(\sum_{k=1}^{m} a_k x_{ik}^{(g)} - \frac{1}{n_g} \sum_{k=1}^{m} \sum_{l=1}^{n_g} a_k x_{lk}^{(g)} \Big) \Bigg)$$

$$= \frac{1}{n} \sum_{k=1}^{m} \sum_{k'=1}^{m} \Bigg(a_k \sum_{g=1}^{G} \sum_{i=1}^{n_g} \Big(x_{ik}^{(g)} - \frac{1}{n_g} \sum_{l=1}^{n_g} x_{lk}^{(g)} \Big) \Big(x_{ik'}^{(g)} - \frac{1}{n_g} \sum_{l=1}^{n_g} x_{lk'}^{(g)} \Big) a_{k'} \Bigg)$$

$$= \mathbf{a}^t \mathbf{W} \mathbf{a}, \tag{7.77}$$

$$\frac{1}{n} \sum_{g=1}^{G} \Bigg(n_g \Big(\sum_{i=1}^{n_g} z_i^{(g)} - \frac{1}{G} \sum_{g'=1}^{G} \sum_{i=1}^{n_{g'}} z_i^{(g')} \Big) \Big(\sum_{i=1}^{n_g} z_i^{(g)} - \frac{1}{G} \sum_{g'=1}^{G} \sum_{i=1}^{n_{g'}} z_i^{(g')} \Big) \Bigg)$$

$$= \frac{1}{n} \sum_{g=1}^{G} \Bigg(n_g \Big(\sum_{i=1}^{n_g} \sum_{k=1}^{m} a_k x_{ik}^{(g)} - \frac{1}{G} \sum_{g'=1}^{G} \sum_{i=1}^{n_{g'}} \sum_{k=1}^{m} a_k x_{ik}^{(g')} \Big)$$

$$\cdot \Big(\sum_{i=1}^{n_g} \sum_{k=1}^{m} a_k x_{ik}^{(g)} - \frac{1}{G} \sum_{g'=1}^{G} \sum_{i=1}^{n_{g'}} \sum_{k=1}^{m} a_k x_{ik}^{(g')} \Big) \Bigg)$$

$$= \frac{1}{n} \sum_{k=1}^{m} \sum_{k'=1}^{m} \left(a_k \sum_{g=1}^{G} n_g \left(\sum_{i=1}^{n_g} x_{ik}^{(g)} - \frac{1}{G} \sum_{g'=1}^{G} \sum_{i=1}^{n_{g'}} x_{ik}^{(g')} \right) \right.$$

$$\left. \cdot \left(\sum_{i=1}^{n_g} x_{ik'}^{(g)} - \frac{1}{G} \sum_{g'=1}^{G} \sum_{i=1}^{n_{g'}} x_{ik'}^{(g')} \right) a_{k'} \right)$$

$$= \mathbf{a}^t \mathbf{B} \mathbf{a}. \tag{7.78}$$

The within-group variance is the statistic representing the broadness of the distribution of data, and the between-group variance is that representing the degree of separation between classes. Hence, if the transformation of feature vectors augments the value of the between-group variance compared with the value of the within-group variance, an appropriate transformation for classification may be expected to be realized. Then, transformation to maximize the value of the within-group variance is conducted under the condition that the value of the within-group variance is 1. To obtain a transformation of this kind, the method of Lagrange multipliers (Lagrange multiplier method) is used. The equation below is differentiated with respect to $\mathbf{a}$ and the Lagrange multiplier (λ) to be 0.

$$\mathbf{a}^t \mathbf{B} \mathbf{a} - \lambda (\mathbf{a}^t \mathbf{W} \mathbf{a} - 1). \tag{7.79}$$

Considering the fact that both $\mathbf{B}$ and $\mathbf{W}$ are symmetric matrices, this value is differentiated with respect to each element of $\mathbf{a}$ and λ to be 0. Then we have

$$\mathbf{B} \mathbf{a} - \lambda \mathbf{W} \mathbf{a} = \mathbf{0}, \tag{7.80}$$

$$\mathbf{a}^t \mathbf{W} \mathbf{a} - 1 = 0, \tag{7.81}$$

where $\mathbf{0}$ is a vector whose m elements are 0. Eq(7.80) is transformed into

$$\mathbf{W}^{-1} \mathbf{B} \mathbf{a} = \lambda \mathbf{a}. \tag{7.82}$$

The vector $\mathbf{a}$ that satisfies this equation is the eigenvector of $\mathbf{W}^{-1}\mathbf{B}$. The corresponding eigenvalues (characteristic values) are defined as $\{\lambda_i\}$ ($\lambda_1 > \lambda_2 > \cdots > \lambda_m$). In addition, the eigenvectors (corresponding to $\{\lambda_i\}$) multiplied by a constant to satisfy eq(7.81) are defined as $\{\mathbf{a}_i\}$ ($1 \leq i \leq m$). These $\{\mathbf{a}_i\}$ satisfy both eq(7.80) and eq(7.81).

Both sides of eq(7.82) are multiplied by $\mathbf{a}^t\mathbf{W}$ from the left; we arrive at

$$\mathbf{a}^t \mathbf{B} \mathbf{a} = \lambda \mathbf{a}^t \mathbf{W} \mathbf{a}. \tag{7.83}$$

This λ is the ratio of the between-group variance to the within-group variance. Therefore, among $\{\mathbf{a}_i\}$, $\mathbf{a}_1$ maximizes the ratio of the between-group variance to the within-group variance; the maximum value is λ_1. Furthermore, characteristics of eigenvectors indicate that $\{\mathbf{a}_i\}$ are linearly independent.

If $\mathbf{a}_1$ does not result in a sufficient classification, another discriminant coordinate is set out. The discriminant coordinate is assumed to be orthogonal to $\mathbf{W}\mathbf{a}_1$. That is, the following equation is assumed.

$$\mathbf{a}^t \mathbf{W} \mathbf{a}_1 = 0. \tag{7.84}$$

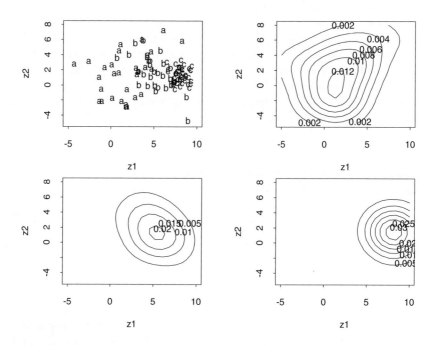

Figure 7.20   Results of transformation of data (figure 7.3(top left) (page 415)) rendered by a canonical variate. The nonparametric probability density functions generated by data that belong to "a," "b," and "c" are displayed in graphs (top right, right bottom, and left bottom, respectively). The bandwidth is fixed at 8 for both of z_1 and z_2.

In addition, the length of **a** is 1. Among values of **a** which satisfy these conditions, the second discriminant coordinate is set out to maximize $\mathbf{a}^t\mathbf{Ba}$.

$\{\mathbf{a}_k\}$ $(2 \leq k \leq m)$ is proved to satisfy eq(7.84) as follows.

The definition of $\mathbf{a}_1$ leads to

$$\mathbf{Ba}_1 = \lambda_1\mathbf{Wa}_1. \tag{7.85}$$

Both sides of this equation are multiplied by $\mathbf{a}_k^t$ $(k \neq 1)$ from the left; we have

$$\mathbf{a}_k^t\mathbf{Ba}_1 = \lambda_1\mathbf{a}_k^t\mathbf{Wa}_1. \tag{7.86}$$

On the other hand, the equation below is obtained.

$$\mathbf{Ba}_k = \lambda_k\mathbf{Wa}_k. \tag{7.87}$$

Both sides of this equation are multiplied by $\mathbf{a}_1^t$ from the left; we obtain

$$\mathbf{a}_1^t\mathbf{Ba}_k = \lambda_k\mathbf{a}_1^t\mathbf{Wa}_k. \tag{7.88}$$

Since **B** and **W** are symmetric matrices, transpose of both sides of eq(7.88) yields

$$\mathbf{a}_k^t\mathbf{Ba}_1 = \lambda_k\mathbf{a}_k^t\mathbf{Wa}_1. \tag{7.89}$$

From both sides of eq(7.86) are subtracted both sides of eq(7.89) to arrive at

$$(\lambda_k - \lambda_1)\mathbf{a}_k^t\mathbf{Wa}_1 = 0. \tag{7.90}$$

Thus, since $\lambda_k \neq \lambda_1$ is assumed, $\mathbf{a}_k$ is orthogonal to $\mathbf{Wa}_1$ (Q.E.D.).

Therefore, any linear combination of $\{\mathbf{a}_i\}$ $(2 \leq i \leq m)$ is orthogonal to $\mathbf{Wa}_1$. Furthermore, since $\{\mathbf{a}_i\}$ $(1 \leq i \leq m)$ are linearly independent, $\{\mathbf{a}_i\}$ $(2 \leq i \leq m)$ are also linearly independent. Then, the vector space spanned by $\mathbf{Wa}_1$ is defined as W_1, and the vector space spanned by $\{\mathbf{a}_i\}$ $(2 \leq i \leq m)$ is A_{m-1}. As a result, we obtain

$$W_1 \oplus A_{m-1} = R^m, \tag{7.91}$$

where R^m indicates the entire vector space spanned by m dimensional vectors. $\oplus$ stands for the direct sum (a.k.a. orthogonal sum) of two functional spaces. Hence, any vector that belongs to the space orthogonal to W_1 is represented as a linear combination of $\{\mathbf{a}_i\}$ $(2 \leq i \leq m)$. Then, the vectors that are depicted by linear combinations of $\{\mathbf{a}_i\}$ $(2 \leq i \leq m)$ and have a length of 1 are defined as **a**. The following equation is obtained.

$$\mathbf{a} = \sum_{i=2}^{m} \alpha_i\mathbf{a}_i. \tag{7.92}$$

The second discriminant coordinate is the vector that is represented as this equation.

When **a** given by eq(7.92) maximizes $\mathbf{a}^t\mathbf{Ba}$ under the condition that $\mathbf{a}^t\mathbf{Wa} = 1$ is satisfied, the necessary conditions which **a** must satisfy are eq(7.80) (page 441) and eq(7.81) (page 441). Then, when **a** given by eq(7.92) is substituted into eq(7.80), we derive

$$\sum_{i=2}^{m} \mathbf{B}\alpha_i\mathbf{a}_i - \lambda\sum_{i=2}^{m} \mathbf{W}\alpha_i\mathbf{a}_i = 0. \tag{7.93}$$

The space spanned by $\{\mathbf{a}_i\}$ $(2 \leq i \leq m)$ is a subspace of m dimensional vector space. Hence, the necessary conditions for eq(7.93) are that $\sum_{i=2}^{m} \alpha_i \mathbf{a}_i$ is identical to one of $\{\mathbf{a}_i\}$ $(1 \leq i \leq m)$, and $\lambda = \lambda_i$ holds when $\sum_{i=2}^{m} \alpha_i \mathbf{a}_i$ is $\mathbf{a}_i$. Furthermore, $\{\mathbf{a}_i\}$ $(1 \leq i \leq m)$ satisfy eq(7.81). This implies that, if the space spanned by $\{\mathbf{a}_i\}$ $(2 \leq i \leq m)$ is the entire m dimensional vector space, the necessary and sufficient conditions for eq(7.93) are that $\sum_{i=2}^{m} \alpha_i \mathbf{a}_i$ is identical to one of $\{\mathbf{a}_i\}$ $(1 \leq i \leq m)$, and $\lambda = \lambda_i$ holds when $\sum_{i=2}^{m} \alpha_i \mathbf{a}_i$ is $\mathbf{a}_i$.

Since the only prospect of $\sum_{i=2}^{m} \alpha_i \mathbf{a}_i$ is $\{\mathbf{a}_i\}$ $(1 \leq i \leq m)$, each $\{\mathbf{a}_i\}$ $(1 \leq i \leq m)$ is examined. The vector $\mathbf{a}_1$ cannot be represented as $\sum_{i=2}^{m} \alpha_i \mathbf{a}_i$, because $\{\mathbf{a}_i\}$ $(1 \leq i \leq m)$ are linearly independent. Hence, $\mathbf{a}_1$ does not fill the requirements for $\mathbf{a}$. Any $\{\mathbf{a}_i\}$ $(2 \leq i \leq m)$ is depicted as $\sum_{i=2}^{m} \alpha_i \mathbf{a}_i$. Then, considering eq(7.83) (page 441), among these vectors, the vector that maximizes $\mathbf{a}^t \mathbf{B} \mathbf{a}$ is proved to be $\mathbf{a}_2$. The corresponding value of $\mathbf{a}^t \mathbf{B} \mathbf{a}$ is λ_2.

Discussion along the same lines augments the number of discriminant coordinates in the order of $\mathbf{a}_3, \mathbf{a}_4, \ldots$. A large number of discriminant coordinates do not guarantee a beneficial classification. An approximate number of discriminant coordinates are required. Note that slightly different equations are used for $\mathbf{B}$ and $\mathbf{W}$ in some cases, and the results are hardly altered.

The procedure for deriving the discriminant coordinates is similar to that for principal component analysis. In principal component analysis, however, eigenvectors are orthogonal to each other, while the eigenvectors for creating the discriminant coordinates are not necessarily orthogonal to each other.

Figure 7.20 shows the result of the transformation of data (figure 7.3(top left) (page 415)) given by a canonical variate, and the resulting nonparametric probability density function. The three classes are separated clearly to a considerable degree in the direction of z_1.

7.11   EXAMPLES OF S-PLUS OBJECT

(A) Fitting of a normal distribution with two predictors
Object: ndens1()

```
function(xx, ex1, ex2)
{
#   (1)
    exx <- expand.grid(ex1, ex2)
    exx <- as.matrix(exx)
    npoint1 <- length(ex1) * length(ex2)
#   (2)
    nd <- dim(xx)[1]
    ndim <- dim(xx)[2]
#   (3)
    cov1 <- (nd - 1)/nd * var(xx, na.method = "omit")
    mean1 <- rep(0, length = ndim)
    for(jj in 1:ndim) {
        mean1[jj] <- mean(xx[, jj], na.rm = T)
    }
```

```
#   (4)
    dif1 <- t(t(exx) - mean1)
#   (5)
    dens1fun <- function(jj, ndim, cov1, dif1)
    {
        dens1val <- (2 * pi)^( - ndim * 0.5)/sqrt(prod(
        eigen(cov1)$values)) * exp(-0.5 * t(dif1[jj,  ])
        %*% solve(cov1) %*% (dif1[jj,  ]))
    }
    dens1 <- apply(matrix(1:npoint1, ncol = 1), 1, dens1fun,
        ndim = ndim, cov1 = cov1, dif1 = dif1)
#   (6)
    return(dens1)
}
```

(1) The object for containing the estimates at grid points (where the values of probability density functions are calculated) is defined as exx and the form of a matrix is given to it. The number of grid points is npoint1.

(2) The number of data points (feature vector) is denoted by nd, and that of the elements of feature vectors is defined as ndim.

(3) The variance-covariance matrix given by the maximum likelihood procedure is cov1. The means are defined as mean1.

(4) The distances between the means of data and the grid points are calculated and saved as dif1.

(5) The function dens1fun() is defined to compute estimates. The estimates at those grid points are derived and stored as dens1.

(6) The object dens1 is outputted.

An example of an object to use ndens1() (construction of figure 7.3 (page 415))

```
function()
{
#   (1)
    ncl <- 3
    ndim <- 2
#   (2)
    set.seed(39)
    nda <- 30
    cla <- rep("a", length = nda)
    xxa <- matrix(rep(0., length = nda * 2.), ncol = 2.)
    xxa[, 1.] <- rnorm(nda, mean = 1., sd = 2.)
    xxa[, 2.] <- rnorm(nda, mean = 2., sd = 4.)
    ndb <- 30
    clb <- rep("b", length = ndb)
    xxb <- matrix(rep(0., length = ndb * 2.), ncol = 2.)
    xxb[, 1.] <- rnorm(ndb, mean = 3., sd = 3.)
    xxb[, 2.] <- rnorm(ndb, mean = 5., sd = 2.)
```

```
    ndc <- 40
    clc <- rep("c", length = ndc)
    xxc <- matrix(rep(0., length = ndc * 2.), ncol = 2.)
    xxc[, 1.] <- rnorm(ndc, mean = 5., sd = 1.)
    xxc[, 2.] <- rnorm(ndc, mean = 7., sd = 1.)
#   (3)
    xxall <- rbind(xxa, xxb, xxc)
    clall <- c(cla, clb, clc)
#   (4)
    x1min <- -5
    x1max <- 10
    x2min <- -3
    x2max <- 11
    par(mfrow = c(2., 2.), mai = c(1., 0.7, 0.5, 0.5),
      oma = c(1., 1., 1., 1.))
#   (5)
    plot(xxall, type = "n", xlab = "x1", ylab = "x2", xlim =
      c(x1min, x1max), ylim = c(x2min, x2max))
    text(xxall, labels = clall)
#   (6)
    ex1 <- seq(from = x1min, to = x1max, by = 0.5)
    ex2 <- seq(from = x2min, to = x2max, by = 0.5)
    d1 <- ndens1(xxa, ex1, ex2)
    contour(ex1, ex2, matrix(d1, ncol = length(ex2)))
#   (7)
    d1 <- ndens1(xxb, ex1, ex2)
    contour(ex1, ex2, matrix(d1, ncol = length(ex2)))
#   (8)
    d1 <- ndens1(xxc, ex1, ex2)
    contour(ex1, ex2, matrix(d1, ncol = length(ex2)))
}
```

(1) The number of classes (ncl) is set at 3, and the number of elements of the feature vector (ndim) is configured at 2.
(2) Data are generated. The feature vectors of the three classes are xxa, xxb, and xxc. The three classes are named a, b, and c.
(3) The data obtained in (2) are connected to create xxall and clall.
(4) The regions where estimates are calculated are defined. The area for drawing a graph is prepared.
(5) The data are plotted in the graph.
(6) The estimates of the probability density function of a normal distribution, which is obeyed by the data belonging to the class a, are calculated at grid points and a contour is drawn.
(7) The estimates of the probability density function of a normal distribution obeyed by the data belonging to the class b are calculated at grid points and a contour is drawn.

(8) The estimates of probability density function of a normal distribution, which the data belonging to the class c obey, are calculated at grid points and a contour is drawn.

(B) Classification by linear discriminant rule
Object: mylda2()

```
function(xxall, clall, x1min, x1max, x2min, x2max, ne1, ne2,
 prior1)
{
#   (1)
    clu <- unique(clall)
    ncl <- length(clu)
    nd <- dim(xxall)[1]
    ndim <- dim(xxall)[2]
#   (2)
    f1array <- array(ne1 * ne2 * ncl, c(ne1, ne2, ncl))
    ex1 <- seq(from = x1min, to = x1max, length = ne1)
    ex2 <- seq(from = x2min, to = x2max, length = ne2)
    exx <- expand.grid(ex1, ex2)
    exx <- as.matrix(exx)
#   (3)
    v1 <- matrix(rep(0, length = ndim * ndim), ncol = ndim)
    sumpr1 <- sum(prior1)
    for(kk in 1:ncl) {
        xx <- xxall[clall == clu[kk],  ]
        nmem <- dim(xx)[1]
        v1 <- v1 + (nmem/nd * var(xx) * (nmem - 1))/nmem
    }
    v1 <- (v1 * nd)/(nd - ncl)
#   (4)
    v1inv <- solve(v1)
    for(kk in 1:ncl) {
        xx <- xxall[clall == clu[kk],  ]
        avxx <- apply(xx, 2, mean)
        print(avxx)
        for(ii in 1:(ne1 * ne2)) {
            ii2 <- ii %% ne1
            if(ii2 == 0)
                ii2 <- ne1
            jj2 <- floor(ii/ne1 - 1e-009) + 1
            f1array[ii2, jj2, kk] <- 2 * log(prior1[kk]/sumpr1)
                - matrix(avxx, nrow = 1) %*% v1inv %*%
                matrix(avxx, ncol = 1) + 2 * t(exx[ii,  ]) %*%
                v1inv %*% matrix(avxx, ncol = 1)
        }
    }
#   (5)
    ecl <- matrix(rep(0, length = ne1 * ne2), ncol = ne2)
```

```
#   (6)
    for(ii in 1:ne1) {
        for(jj in 1:ne2) {
            cl1 <- which(max(f1array[ii, jj,  ]) ==
            f1array[ii, jj,])
            if(length(cl1) > 1)
                ecl[ii, jj] <- "N"
            else ecl[ii, jj] <- clu[cl1]
        }
    }
#   (7)
    return(ex1, ex2, ecl)
}
```

(1) The names of classes are defined as clu, the number of classes is ncl, the number of data points is denoted by nd, and the number of elements of the feature vector is defined as ndim.

(2) The object f1array is prepared to save the values of the discriminant function (eq(7.23) (page 417)) at grid points. The values of the coordinates of the grid points are stored as exx.

(3) $\hat{V}$ is calculated using eq(7.21) (page 416) and defined as v1.

(4) The values of the three discriminant functions at grid points are computed and saved as f1array.

(5) The object ecl (a matrix) is prepared to store the results of the classification at the grid points.

(6) The estimates given by the three regression equations are compared at a grid point and the class which supplies the maximum value is designated as the class at the point. When the number of classes which corresponds to the maximum value is greater than one, N is assigned. This result is stored as ecl.

(7) The objects ex1, ex2, and ecl are outputted.

S-Plus includes the object discrim() to carry out classifications generated by the linear discriminant rule or quadratic discriminant rule. When this object is employed, the use of mylda3() below in place of mylda2() yields the same result.
Object: mylda3()

```
function(xxall, clall, x1min, x1max, x2min, x2max, ne1, ne2,
  prior1)
{
#   (1)
    clu <- unique(clall)
    ncl <- length(clu)
    ndim <- dim(xxall)[2]
    nd <- dim(xxall)[1]
    prior1 <- prior1/sum(prior1)
#   (2)
    clalln <- rep(0, length = nd)
    clnum <- seq(from = 1, to = ncl, by = 1)
```

```
    dnum <- seq(from = 1, to = nd, by = 1)
    for(ii in 1:ncl) {
        clalln <- replace(clalln, dnum[clall == clu[ii]],
        clnum[ii])
    }
    data1 <- data.frame(cl = factor(clalln, labels = clu),
    x1 = xxall)
#   (3)
    dis.out <- discrim(cl ~ x1.1 + x1.2, data = data1,
    family = Classical("homoscedastic"), prior = prior1)
#   (4)
    ex1 <- seq(from = x1min, to = x1max, length = ne1)
    ex2 <- seq(from = x2min, to = x2max, length = ne2)
    exx <- expand.grid(ex1, ex2)
    exx <- as.matrix(exx)
#   (5)(6)
    data2 <- data.frame(x1.1 = exx[, 1], x1.2 = exx[, 2])
    ey <- predict(dis.out, newdata = data2)
    ecl <- matrix(ey[, 1], ncol = ne2)
#   (7)
    return(ex1, ex2, ecl)
}
```

In (3) of the object above, family = Classical("homoscedastic") indicates that the variances of members of classes are identical. Hence, the linear discriminant rule is applied.

In the process of (2), the names of classes are transformed from characters to numbers $(1, 2, \ldots)$. Furthermore, the data are put together as data1 to use discrim().

An example of an object to use mylda2() (construction of figure 7.4 (page 417))

```
function()
{
#   (1)
    set.seed(39)
    nda <- 30
    cla <- rep("a", length = nda)
    xxa <- matrix(rep(0., length = nda * 2.), ncol = 2.)
    xxa[, 1.] <- rnorm(nda, mean = 1., sd = 2.)
    xxa[, 2.] <- rnorm(nda, mean = 2., sd = 4.)
    ndb <- 30
    clb <- rep("b", length = ndb)
    xxb <- matrix(rep(0., length = ndb * 2.), ncol = 2.)
    xxb[, 1.] <- rnorm(ndb, mean = 3., sd = 3.)
    xxb[, 2.] <- rnorm(ndb, mean = 5., sd = 2.)
    ndc <- 40
    clc <- rep("c", length = ndc)
```

```
    xxc <- matrix(rep(0., length = ndc * 2.), ncol = 2.)
    xxc[, 1.] <- rnorm(ndc, mean = 5., sd = 1.)
    xxc[, 2.] <- rnorm(ndc, mean = 7., sd = 1.)
#   (2)
    xxall <- rbind(xxa, xxb, xxc)
    clall <- c(cla, clb, clc)
#   (3)
    x1min <- -5
    x1max <- 10
    x2min <- -3
    x2max <- 11
#   (4)
    ne1 <- 16
    ne2 <- 14
    prior1 <- c(3/10, 3/10, 4/10)
    r1 <- mylda2(xxall, clall, x1min, x1max, x2min, x2max,
    ne1, ne2, prior1)
#   (5)
    ex1 <- r1$ex1
    ex2 <- r1$ex2
    exx <- expand.grid(ex1, ex2)
    exx <- as.matrix(exx)
    ecl <- as.vector(r1$ecl)
#   (6)
    par(mfrow = c(1., 2.), mai = c(2., 0.7, 1.5, 0.5),
    oma = c(1., 1., 1., 1.))
    plot(exx, type = "n", xlab = "x1", ylab = "x2", xlim =
    c(x1min, x1max), ylim = c(x2min, x2max))
    text(exx, labels = ecl)
#   (7)
    prior1 <- c(17/20, 2/20, 1/20)
    r1 <- mylda2(xxall, clall, x1min, x1max, x2min, x2max,
    ne1, ne2, prior1)
    ex1 <- r1$ex1
    ex2 <- r1$ex2
    ecl <- r1$ecl
    plot(exx, type = "n", xlab = "x1", ylab = "x2",
    xlim = c(x1min, x1max), ylim = c(x2min, x2max))
    text(exx, labels = as.character(ecl))
}
```

(1) Data are generated. The feature vectors of the three classes of data are xxa, xxb, and xxc. The three classes are named a, b, and c.

(2) The data created in (1) are connected to construct xxall and clall.

(3) The region where estimates are calculated is supplied. The area for drawing a graph is prepared.

(4) The number of estimation points (ne1·ne2) and the values of prior probabilities (prior1) are given, and classification is carried out using mylda2().
(5) The positions of grid points (exx) and the estimated classes (ecl) are extracted.
(6) The results of classification are drawn in a graph.
(7) The processes (4), (5), and (6) are conducted using different values of prior probabilities.

(C) Classification by the use of the quadratic discriminant rule
Object: myqda2()

```
function(xxall, clall, x1min, x1max, x2min, x2max, ne1, ne2,
 band1, prior1)
{
#   (1)
    clu <- unique(clall)
    ncl <- length(clu)
    ndim <- dim(xxall)[2]
#   (2)
    f1array <- array(ne1 * ne2 * ncl, c(ne1, ne2, ncl))
    ex1 <- seq(from = x1min, to = x1max, length = ne1)
    ex2 <- seq(from = x2min, to = x2max, length = ne2)
    exx <- expand.grid(ex1, ex2)
    exx <- as.matrix(exx)
    sumpr1 <- sum(prior1)
#   (3)
    v1array <- array(ndim * ndim * ncl, c(ndim, ndim, ncl))
    v1arrayinv <- array(ndim * ndim * ncl, c(ndim, ndim, ncl))
    for(kk in 1:ncl) {
        xx <- xxall[clall == clu[kk],  ]
        v1array[,  , kk] <- var(xx)
        print(v1array[,  , kk])
        v1arrayinv[,  , kk] <- solve(v1array[,  , kk])
    }
#   (4)
    for(kk in 1:ncl) {
        xx <- xxall[clall == clu[kk],  ]
        avxx <- apply(xx, 2, mean)
        for(ii in 1:(ne1 * ne2)) {
            ii2 <- ii %% ne1
            if(ii2 == 0) ii2 <- ne1
            jj2 <- floor(ii/ne1 - 1e-009) + 1
            eigenv1 <- prod(eigen(v1array[,  , kk])$values)
            f1array[ii2, jj2, kk] <- 2 * log(prior1[kk]/sumpr1)
             - log(eigenv1) - t(exx[ii,  ] - matrix(avxx,
            ncol = 1)) %*% v1arrayinv[,  , kk] %*%
            (exx[ii,  ] - matrix(avxx, ncol = 1))
        }
    }
```

```
#   (5)
    ecl <- matrix(rep(0, length = ne1 * ne2), ncol = ne2)
#   (6)
    for(ii in 1:ne1) {
        for(jj in 1:ne2) {
            cl1 <- which(max(flarray[ii, jj,  ]) ==
            flarray[ii, jj,])
            if(length(cl1) > 1)
                ecl[ii, jj] <- "N"
            else ecl[ii, jj] <- clu[cl1]
        }
    }
#   (7)
    return(ex1, ex2, ecl)
}
```

(1) The names of classes are defined as clu, the number of classes is ncl, and the number of elements of the feature vector is denoted by ndim.

(2) The object flarray is prepared to store the values of discriminant functions (eq(7.24) (page 417)) at grid points. The values of the coordinates of the grid points are saved as exx. Furthermore, the sum of prior probabilities is saved as sumpr1.

(3) $\hat{V}^{(g)}$ is calculated using eq(7.20) (page 416) and stored as v1array. The inverse matrices are saved as v1arrayinv. The object cov.mve() or cov.mcd() (they are included in S-Plus by default) enables calculation of a variance-covariance matrix using a method that is robust to outliers. Furthermore, mahalanobis(), which S-Plus includes by default, is suitable for obtaining the value of the third term of eq(7.24).

(4) The values of discriminant functions at the grid points are calculated and stored as flarray.

(5) The object ecl (a matrix) is prepared to store the results of classifications at the grid points.

(6) The values supplied by the three regression equations at the grid points are compared and the points are classified as the class that corresponds to the maximum value. When the number of classes that corresponds to the maximum value is more than one, N is assigned. The result is saved as ecl.

(7) The objects ex1, ex2, and ecl are outputted.

The object myqda2() is used in the same manner as mylda2() (figure 7.5 (page 418)).

Furthermore, if (3) in mylda3() is replaced by the following, it performs as an object that carries out classification using discrim() based on the quadratic discriminant rule.

```
#   (3)'
    dis.out <- discrim(cl ~ x1.1 + x1.2, data = data1, family =
    Classical("heteroscedastic"), prior = prior1)
```

The assignment of family = Classical("heteroscedastic") means that the variances of classes are not equal. Hence, the classification rendered by the quadratic discriminant rule is carried out.

(D) Calculation of nonparametric probability density functions
Object: nond1()

```
function(xx, x1min, x1max, x2min, x2max, ne, band1)
{
#   (1)
    library(MASS)
#   (2)
    x1 <- xx[, 1]
    x2 <- xx[, 2]
#   (3)
    result1 <- kde2d(x1, x2, h = band1, n = ne, lims =
      c(x1min, x1max, x2min, x2max))
#   (4)
    ex1 <- result1$x
    ex2 <- result1$y
    ey <- result1$z
#   (5)
    return(ex1, ex2, ey)
}
```

(1) The library "MASS" is set out for use.
(2) The elements of feature vectors are extracted and defined as x1 and x2.
(3) The nonparametric probability density function given by the Gaussian function is derived using kde2d().
(4) The positions of grid points along each axis are defined as ex1 and ex2. The estimates at the grid points are denoted by ey.
(5) The objects ex1, ex2, and ey are outputted.

An example of an object to use nond1() (construction of figure7.6 (419page))

```
function()
{
#   (1)
    ncl <- 3
    ndim <- 2
#   (2)
    set.seed(39)
    nda <- 30
    cla <- rep("a", length = nda)
    xxa <- matrix(rep(0., length = nda * 2.), ncol = 2.)
    xxa[, 1.] <- rnorm(nda, mean = 1., sd = 2.)
    xxa[, 2.] <- rnorm(nda, mean = 2., sd = 4.)
    ndb <- 30
    clb <- rep("b", length = ndb)
```

```
    xxb <- matrix(rep(0., length = ndb * 2.), ncol = 2.)
    xxb[, 1.] <- rnorm(ndb, mean = 3., sd = 3.)
    xxb[, 2.] <- rnorm(ndb, mean = 5., sd = 2.)
    ndc <- 40
    clc <- rep("c", length = ndc)
    xxc <- matrix(rep(0., length = ndc * 2.), ncol = 2.)
    xxc[, 1.] <- rnorm(ndc, mean = 5., sd = 1.)
    xxc[, 2.] <- rnorm(ndc, mean = 7., sd = 1.)
#   (3)
    xxall <- rbind(xxa, xxb, xxc)
    clall <- c(cla, clb, clc)
#   (4)
    x1min <- -5
    x1max <- 10
    x2min <- -3
    x2max <- 11
    par(mfrow = c(2., 2.), mai = c(1., 0.7, 0.5, 0.5),
      oma = c(1., 1., 1., 1.))
#   (5)
    plot(xxall, type = "n", xlab = "x1", ylab = "x2", xlim =
      c(x1min, x1max), ylim = c(x2min, x2max))
    text(xxall, labels = clall)
#   (6)
    ne <- 20
    band1 <- c(8, 8)
    r1 <- nond1(xxa, x1min, x1max, x2min, x2max, ne, band1)
    ex1 <- r1$ex1
    ex2 <- r1$ex2
    d1 <- r1$ey
    contour(ex1, ex2, d1, xlab = "x1", ylab = "x2")
#   (7)
    r1 <- nond1(xxb, x1min, x1max, x2min, x2max, ne, band1)
    ex1 <- r1$ex1
    ex2 <- r1$ex2
    d1 <- matrix(r1$ey, ncol = length(ex2))
    contour(ex1, ex2, d1, xlab = "x1", ylab = "x2")
#   (8)
    r1 <- nond1(xxc, x1min, x1max, x2min, x2max, ne, band1)
    ex1 <- r1$ex1
    ex2 <- r1$ex2
    d1 <- r1$ey
    contour(ex1, ex2, d1, xlab = "x1", ylab = "x2")
}
```

(1) The number of classes (ncl) is set at 3, and the number of elements of the feature vector (ndim) is configured at 2.

(2) Data are generated. The feature vectors of the three classes are xxa, xxb, and xxc. The three classes are named as a, b, and c.

(3) The data generated in (2) are connected to create xxall and clall.

(4) The region where estimates are calculated is given. The area for drawing a graph is prepared.

(5) The data are plotted in a graph.

(6) The values of the nonparametric probability density functions generated by the data that are members of the class a are calculated at grid points. The bandwidths are fixed at 8 for both axes. A contour is charted.

(7) The values of the nonparametric probability density functions given by the data that are members of the class b are calculated at grid points. The bandwidths are fixed at 8 for both axes. A contour is drawn.

(8) The values of the nonparametric probability density functions based on the data that are members of the class c are calculated at grid points. The bandwidths are fixed at 8 for both axes. A contour is charted.

(E) Classification by nonparametric probability density function
Object: nondis1()

```
function(xxall, clall, x1min, x1max, x2min, x2max, ne,
 band1, prior1)
{
#    (1)
    library(MASS)
#    (2)
    clu <- unique(clall)
    ncl <- length(clu)
#    (3)
    eyarray <- array(ne * ne * ncl, c(ne, ne, ncl))
    for(kk in 1:ncl) {
        xx <- xxall[clall == clu[kk],  ]
        x1 <- xx[, 1]
        x2 <- xx[, 2]
        result1 <- kde2d(x1, x2, h = band1, n = ne,
         lims = c(x1min, x1max, x2min, x2max))
        ex1 <- result1$x
        ex2 <- result1$y
        ey <- result1$z
        eyarray[,  , kk] <- ey * prior1[kk]
    }
#    (4)
    ecl <- matrix(rep(0, length = ne * ne), ncol = ne)
#    (5)
    for(ii in 1:ne) {
        for(jj in 1:ne) {
            cl1 <- which(max(eyarray[ii, jj,  ]) ==
             eyarray[ii, jj,  ])
            if(length(cl1) > 1)
```

```
                ecl[ii, jj] <- "N"
            else ecl[ii, jj] <- clu[cl1]
        }
    }
# (6)
    return(ex1, ex2, ecl)
}
```

(1) The library "MASS" is set out for use.
(2) The names of classes are defined as clu. The number of classes is ncl.
(3) The nonparametric probability density functions are calculated using kde2d()
and saved as eyarray.
(4) The object ecl (a matrix) is prepared to store the results of classifications at the
grid points.
(5) Classes at estimation points are estimated at the grid points using nonparametric
probability density functions. When the number of classes which corresponds to the
maximun value is more than one, N is assigned.
(6) The objects ex1, ex2, and ecl are outputted.

An example of an object to use nondis1() (construction of figure 7.7 (page
419))

```
function()
{
# (1)
    ncl <- 3
    ndim <- 2
# (2)
    set.seed(39)
    nda <- 30
    cla <- rep("a", length = nda)
    xxa <- matrix(rep(0., length = nda * 2.), ncol = 2.)
    xxa[, 1.] <- rnorm(nda, mean = 1., sd = 2.)
    xxa[, 2.] <- rnorm(nda, mean = 2., sd = 4.)
    ndb <- 30
    clb <- rep("b", length = ndb)
    xxb <- matrix(rep(0., length = ndb * 2.), ncol = 2.)
    xxb[, 1.] <- rnorm(ndb, mean = 3., sd = 3.)
    xxb[, 2.] <- rnorm(ndb, mean = 5., sd = 2.)
    ndc <- 40
    clc <- rep("c", length = ndc)
    xxc <- matrix(rep(0., length = ndc * 2.), ncol = 2.)
    xxc[, 1.] <- rnorm(ndc, mean = 5., sd = 1.)
    xxc[, 2.] <- rnorm(ndc, mean = 7., sd = 1.)
# (3)
    xxall <- rbind(xxa, xxb, xxc)
    clall <- c(cla, clb, clc)
# (4)
```

```
        x1min <- -5
        x1max <- 10
        x2min <- -3
        x2max <- 11
        par(mfrow = c(1., 2.), mai = c(2., 0.7, 1.5, 0.5),
          oma = c(1., 1., 1., 1.))
#   (5)
        ne <- 10
        band1 <- c(8, 8)
        prior1 <- c(3/10, 3/10, 4/10)
        r1 <- nondis1(xxall, clall, x1min, x1max, x2min, x2max,
          ne, band1, prior1)
#   (6)
        ex1 <- r1$ex1
        ex2 <- r1$ex2
        exx <- expand.grid(ex1, ex2)
        exx <- as.matrix(exx)
        ecl <- as.vector(r1$ecl)
#   (7)
        plot(exx, type = "n", xlab = "x1", ylab = "x2",
          xlim = c(x1min, x1max), ylim = c(x2min, x2max))
        text(exx, labels = ecl)
#   (8)
        prior1 <- c(17/20, 2/20, 1/20)
        r1 <- nondis1(xxall, clall, x1min, x1max, x2min, x2max,
          ne, band1, prior1)
        ex1 <- r1$ex1
        ex2 <- r1$ex2
        ecl <- r1$ecl
        plot(exx, type = "n", xlab = "x1", ylab = "x2",
          xlim = c(x1min, x1max), ylim = c(x2min, x2max))
        text(exx, labels = as.character(ecl))
}
```

(1) The number of classes (ncl) is set at 3, and the number of elements of the feature vector (ndim) is configured at 2.

(2) Data are generated. The feature vector of the three feature vectors are xxa, xxb, and xxc. The three classes are named a, b, and c.

(3) The data generated in (2) are connected to construct xxall and clall.

(4) The region where estimates are calculated is defined. The area for drawing a graph is prepared.

(5) The number of grid points for each direction to configure estimation points is designated as ne. The object band1 denotes the bandwidth, and prior1 stands for a prior probability.

(6) The positions of grid points where classification is conducted are defined as exx. The form of exx is designated as a matrix. Estimated classes at the grid points where the row vectors constitute exx are defined as ecl.

(7) The classes estimated at the grid points are charted on a graph.
(8) The procedures (6) and (7) are carried out using different prior probabilities.

(F) Classification by the use of logistic regression
Object: glmdis1()

```
function(xxall, clall, ex1, ex2)
{
#   (1)
    clu <- unique(clall)
    ncl <- length(clu)
    nd <- dim(xxall)[1]
    ndim <- dim(xxall)[2]
#   (2)
    nex1 <- length(ex1)
    nex2 <- length(ex2)
    grid <- expand.grid(ex1, ex2)
#   (3)
    eyarray <- array(rep(0, length = nex1 *
     nex2 * ncl), c(nex1, nex2, ncl))
#   (4)
    for(kk in 1:ncl) {
        y <- rep(0, length = nd)
        y[clall == clu[kk]] <- rep(1, length =
         sum(clall == clu[kk]))
        data1 <- data.frame(x1 = xxall[, 1],
         x2 = xxall[, 2], y1 = y)
        glm.out <- glm(y1 ~ x1 + x2, data =
         data1, family = "binomial")
        data2 <- data.frame(x1 = ex1, x2 = ex2)
        margin.fit <- predict.gam(glm.out,
         newdata = data2, type = "terms")
        eylink <- outer(margin.fit[, 1],
         margin.fit[, 2], "+")
        eylink <- eylink + attr(margin.fit, "constant")
        ey <- binomial()$inverse(eylink)
        eymat <- matrix(ey, nrow = nex1)
        eyarray[, , kk] <- eymat
    }
#   (5)
    ecl <- matrix(rep(NA, length = nex1 * nex2),
     nrow = nex1)
    for(ii in 1:nex1) {
        for(jj in 1:nex2) {
            cl1 <- which(max(eyarray[ii,jj, ])
             == eyarray[ii, jj, ])
            if(length(cl1) > 1)
                ecl[ii, jj] <- "N"
```

```
            else ecl[ii, jj] <- clu[cl1]
        }
    }
#   (6)
    return(ecl, eyarray)
}
```

(1) The names of classes are denoted by clu. The number of classes is defined as ncl. The number of data points is nd. The number of elements of the feature vector is identified as ndim.

(2) The number of grid points for estimation is set at nex1· nex2. Corresponding coordinates are defined as grid.

(3) The object eyarray is prepared to save the values obtained using the logistic model.

(4) The logistic regression is carried out using glm(); the outputs result in estimates at the grid points. Assignment of type = "terms" in predict.gam() calculates the estimates of each term at the grid points. The sum of the estimates at each grid point plus a constant is transformed by an inverse function of a link function. This strategy saves computational costs compared to direct calculation of estimates at the grid points using predict.gam().

(5) Classification is conducted using estimates obtained in (4).

(6) The objects ecl and eyarray are outputted.

The object glmdis1() is used in almost the same fashion as nondis1() (figure 7.10 (page 423)). Furthermore, when eq(7.30) (page 423) is used, gam() is employed in place of glm() (figure7.11 (page 424)).

(G) Classification generated by neural networks
Object: netd1()

```
function(xxall, clall, ex1, ex2)
{
#   (1)
    library(nnet)
#   (2)
    clu <- unique(clall)
    ncl <- length(clu)
    nex1 <- length(ex1)
    nex2 <- length(ex2)
    nd <- dim(xxall)[1]
    ndim <- dim(xxall)[2]
#   (3)
    exx <- expand.grid(ex1, ex2)
#   (4)
    data1 <- data.frame(x1 = xxall[, 1], x2 = xxall[, 2],
      y1 = clall)
    nn.out <- nnet(y1 ~ x1 + x2, data = data1, softmax = T,
      size = 5, entropy = T, decay = 0.1, maxit = 300)
    data2 <- data.frame(x1 = exx[, 1], x2 = exx[, 2])
```

```
    ey <- predict(nn.out, newdata = data2)
    eyarray <- array(rep(0, length = nex1 * nex2 * ncl),
     c(nex1, nex2, ncl))
    for(kk in 1:ncl) {
        eyarray[,  , kk] <- matrix(ey[, kk], nrow = nex2)
    }
#   (5)
    ecl <- matrix(rep(NA, length = nex1 * nex2), nrow
     = nex1)
    for(ii in 1:nex1) {
        for(jj in 1:nex2) {
            print(sum(eyarray[ii, jj,  ]))
            cl1 <- which(max(eyarray[ii, jj, ]) ==
             eyarray[ii, jj,  ])
            if(length(cl1) > 1)
                ecl[ii, jj] <- "N"
            else ecl[ii, jj] <- clu[cl1]
        }
    }
#   (6)
    return(ecl, eyarray)
}
```

(1) The library "net" is set out for use.

(2) The names of classes are defined as clu. The number of classes is ncl. The number of elements of the feature vector is set at ndim. The number of grid points is designated as nex1·nex2 . The number of data points is configured at nd.

(3) The coordinates of grid points to be classified are fixed at exx.

(4) A neural network is created using nnet(). The resultant neural network leads to estimates (the values of softmax function) at the grid points using predict().

(5) Classification is conducted using the estimates given by (4).

(6) The objects ecl and eyarray are outputted.

The object netd1() is utilized in a manner similar to that of nondis1() and glmdis1() (figure 7.13).

(H) Optimization of the complexity parameter by the use of the 10-fold cross-validation to create a tree-based model for classification

Object: trdcv3()

```
function(xxall, clall, lam1, ncv)
{
#   (1)
    clu <- unique(clall)
    ncl <- length(clu)
    nd <- dim(xxall)[1]
    ndim <- dim(xxall)[2]
#   (2)
    minsize1 <- 2
```

```
#   (3)
    press <- 0
    for(k in 1:ncv) {
        set.seed(45 * k)
        rand <- sample(10, nd, replace = T)
        for(i in unique(rand)) {
            newdatx <- xxall[rand != i,  , drop = F]
            newdaty <- clall[rand != i]
            testdatx <- xxall[rand == i,  , drop = F]
            testdaty <- clall[rand == i]
            ntest <- length(testdaty)
            assign("data1", data.frame(x1 =  newdatx[, 1],
            x2 = newdatx[ , 2], y1 = newdaty), frame = 1)
            yy.tr <- tree(y1 ~ x1 + x2, data1, minsize =
            minsize1)
            yy.pr <- prune.tree(yy.tr, k = lam1)
            data2 <- data.frame(x1 = testdatx[, 1], x2 =
            testdatx[, 2])
            ecl <- predict(yy.pr, newdata = data2, type =
            "class")
            press <- press + sum(ecl != testdaty)
        }
    }
#   (4)
    press <- press/ncv
#   (5)
    return(press)
}
```

(1) The names of classes are defined as `clu`. The number of classes is defined as `ncl`. The number of data points is set at `nd`. The number of elements of the feature vector is configured at `ndim`.

(2) The minimum number of data points that reach a terminal node is identified as 2; this value is applied to the tree-based model with the maximum number of terminal nodes in the first step of this procedure.

(3) The procedure of 10-fold cross-validation is carried out `ncv` times to estimate the number of misclassifications when `lam1` is used as α in eq(7.55) (page 432). The number of misclassifications is defined as `press`.

(4) The sum of `press` is divided by `ncv`; the result is identified as the number of misclassifications by 10-fold cross-validation.

(5) The object `press` is outputted.

An example of an object to use `trdcv3()` (construction of figure 7.15 (page 433))

```
function()
{
#   (1)
```

```
      ncl <- 3
      ndim <- 2
#  (2)
      set.seed(39)
      nda <- 30
      cla <- rep("a", length = nda)
      xxa <- matrix(rep(0., length = nda * 2.), ncol = 2.)
      xxa[, 1.] <- rnorm(nda, mean = 1., sd = 2.)
      xxa[, 2.] <- rnorm(nda, mean = 2., sd = 4.)
      ndb <- 30
      clb <- rep("b", length = ndb)
      xxb <- matrix(rep(0., length = ndb * 2.), ncol = 2.)
      xxb[, 1.] <- rnorm(ndb, mean = 3., sd = 3.)
      xxb[, 2.] <- rnorm(ndb, mean = 5., sd = 2.)
      ndc <- 40
      clc <- rep("c", length = ndc)
      xxc <- matrix(rep(0., length = ndc * 2.), ncol = 2.)
      xxc[, 1.] <- rnorm(ndc, mean = 5., sd = 1.)
      xxc[, 2.] <- rnorm(ndc, mean = 7., sd = 1.)
#  (3)
      xxall <- rbind(xxa, xxb, xxc)
      clall <- c(cla, clb, clc)
#  (4)
      x1min <- -5
      x1max <- 10
      x2min <- -3
      x2max <- 11
#  (5)
      ntrial1 <- 9
      ncv <- 5
      lam1t <- rep(0, length = ntrial1)
      error1t <- rep(0, length = ntrial1)
      for(kk in 1:ntrial1) {
          lam1t[kk] <- kk * 5
          lam1 <- lam1t[kk]
          error1 <- trdcv3(xxall, clall, lam1, ncv)
          error1t[kk] <- error1 * 0.01
      }
#  (6)
      par(mfrow = c(1, 1), mai = c(2, 3, 2, 3),
       oma = c(2, 2, 2, 2))
      plot(lam1t, error1t, type = "n", xlab =
       "complexity parameter", ylab = "error")
      points(lam1t, error1t, pch = 5, cex = 0.8)
}
```

(1) The number of classes (ncl) is set at 3. The number of elements of the feature vector (ndim) is configured at 2.

(2) Data are generated. The feature vectors of the members of the three classes are xxa, xxb, and xxc. The three classes of the feature vectors are named a, b, and c.

(3) The data constructed in (2) are connected to create xxall and clall.

(4) The area for obtaining estimates is defined.

(5) The total of ntrial1 values are attempted as α in eq(7.48) (page 430). The values are saved as lam1t. The misclassification rate for each of lam1t is calculated; the result is stored as error1t.

(6) The results of 10-fold cross-validation are drawn in a graph.

(I) k-nearest-neighbor classifier accompanied by bagging
Obejct: knnbag1 ()

```
function(xxall, clall, x1min, x1max, x2min, x2max, ne1,
    ne2, k1, prior1, nsample)
{
#   (1)
    library(class)
#   (2)
    clu <- unique(clall)
    ncl <- length(clu)
    ndim <- dim(xxall)[2]
    nmen <- seq(0, length = ncl)
    for(ii in 1:ncl) {
        nmen[ii] <- sum(clall == clu[ii])
    }
#   (3)
    ex1 <- seq(from = x1min, to = x1max, length = ne1)
    ex2 <- seq(from = x2min, to = x2max, length = ne2)
    exx <- expand.grid(ex1, ex2)
    exx <- as.matrix(exx)
#   (4)
    rarray <- array(ne1 * ne2 * nsample, c(ne1, ne2,
    nsample))
#   (5)
    for(kk in 1:nsample) {
        newxx <- NULL
        newcl <- NULL
        set.seed(45 * kk)
        for(ii in 1:ncl) {
            xx <- xxall[clall == clu[ii],  ]
            cl <- clall[clall == clu[ii]]
            size1 <- round(nmen[ii] * prior1 * ncl)
            rand1 <- sample(c(1:nmen[ii]), size =
             round(nmen[ii] * prior1[ii] * ncl), replace = T)
            newxx <- rbind(newxx, xx[rand1, , drop = F])
            newcl <- c(newcl, cl[rand1])
```

```
            }
        rarray[,   , kk] <- as.character(knn(newxx,
            exx, newcl, k = k1))
    }
#   (6)
    clmat <- matrix(rep(0, length = ne1 * ne2), ncol = ne2)
    for(ii in 1:ne1) {
        for(jj in 1:ne2) {
            vote1 <- rep(0, length = ncl)
            for(kk in 1:ncl) {
                vote1[kk] <- sum(rarray[ii, jj,  ] == clu[kk])
            }
            cl1 <- clu[vote1 == max(vote1)]
            if(length(cl1) > 1)
                cl1 <- sample(cl1, 1)
            clmat[ii, jj] <- cl1
        }
    }
#   (7)
    return(clmat)
}
```

(1) The library "class" is set out for use.

(2) The names of classes are defined as clu. The number of classes is ncl. The number of elements of the feature vector is set at ndim. The numbers of members of classes are denoted by nmen.

(3) The coordinates of grid points where classification is carried out are denoted by exx.

(4) The object rarray is prepared to store the results of classification by the bootstrap sample.

(5) A bootstrap sample is generated. Classification using the k-nearest-neighbor classifier is conducted on the basis of the sample. With different values of random numbers, this procedure is performed nsample times.

(6) The majority decision using the results of the classification in (5) is made; the result is saved as clmat. When the number of classes which corresponds to the maximum vote is more than one, the class is selected at random by a random number.

(7) The object ecl is outputted.

An example of an object to use knnbag1() (construction of figure 7.18 (page 436))

```
function()
{
#   (1)
    ncl <- 3
    ndim <- 2
#   (2)
    set.seed(39)
```

```
    nda <- 30
    cla <- rep("a", length = nda)
    xxa <- matrix(rep(0., length = nda * 2.), ncol = 2.)
    xxa[, 1.] <- rnorm(nda, mean = 1., sd = 2.)
    xxa[, 2.] <- rnorm(nda, mean = 2., sd = 4.)
    ndb <- 30
    clb <- rep("b", length = ndb)
    xxb <- matrix(rep(0., length = ndb * 2.), ncol = 2.)
    xxb[, 1.] <- rnorm(ndb, mean = 3., sd = 3.)
    xxb[, 2.] <- rnorm(ndb, mean = 5., sd = 2.)
    ndc <- 40
    clc <- rep("c", length = ndc)
    xxc <- matrix(rep(0., length = ndc * 2.), ncol = 2.)
    xxc[, 1.] <- rnorm(ndc, mean = 5., sd = 1.)
    xxc[, 2.] <- rnorm(ndc, mean = 7., sd = 1.)
#   (3)
    xxall <- rbind(xxa, xxb, xxc)
    clall <- c(cla, clb, clc)
#   (4)
    x1min <- -5
    x1max <- 10
    x2min <- -3
    x2max <- 11
#   (5)
    ne1 <- 16
    ne2 <- 15
    ex1 <- seq(from = x1min, to = x1max, length = ne1)
    ex2 <- seq(from = x2min, to = x2max, length = ne2)
    exx <- as.matrix(expand.grid(ex1, ex2))
#   (6)
    prior1 <- c(1/3, 1/3, 1/3)
    k1 <- 5
    nsample <- 100
    yy.knn <- knnbag1(xxall, clall, x1min, x1max, x2min,
     x2max, ne1, ne2, k1, prior1, nsample)
#   (7)
    par(mfrow = c(1., 2.), mai = c(2., 0.7, 1.5, 0.5),
     oma = c(1., 1., 1., 1.))
#   (8)
    plot(exx, type = "n", xlab = "x1", ylab = "x2", xlim =
     c(x1min, x1max), ylim = c(x2min, x2max))
    text(exx, labels = as.character(yy.knn))
#   (9)
    prior1 <- c(17/20, 2/20, 1/20)
    yy.knn <- knnbag1(xxall, clall, x1min, x1max, x2min,
     x2max, ne1, ne2, k1, prior1, nsample)
    plot(exx, type = "n", xlab = "x1", ylab = "x2",
```

```
    xlim = c(x1min, x1max), ylim = c(x2min, x2max))
    text(exx, labels = as.character(yy.knn))
}
```

(1) The number of classes (ncl) is configured at 3. The number of elements of the feature vector (ndim) is set at 2.

(2) Data are generated. The feature vectors of the three classes are xxa, xxb, and xxc. The three classes are named a, b, and c.

(3) The data obtained in (2) are connected to create xxall and clall.

(4) The region where estimation is conducted is defined.

(5) The values of the coordinates of grid points for classifiction are defined as exx.

(6) The prior probabilities for a, b, and c are set at 0.3, 0.3, and 0.4, respectively, and classification using the k-nearest-neighbor classifier accompanied by bagging is carried out.

(7) The area for drawing a graph is prepared.

(8) The results of (6) are drawn in a graph.

(9) The prior probabilities for a, b, and c are configured at $17/20$, $2/20$, and $1/20$, respectively, and the procedures, which are the same as the procedures (6) and (7), are carried out.

(J) Transformation of feature vectors by the use of canonical variates
Object: cano1()

```
function(xxall, clall)
{
#   (1)
    clu <- unique(clall)
    ncl <- length(clu)
#   (2)
    nd <- dim(xxall)[1]
    ndim <- dim(xxall)[2]
#   (3)
    w1 <- matrix(rep(0, length = ndim * ndim), ncol = ndim)
#   (4)
    avxmat <- matrix(rep(0, length = ndim * ncl), ncol = ncl)
#   (5)
    for(kk in 1:ncl) {
        xx <- xxall[clall == clu[kk],  ]
        ndx <- dim(xx)[1]
        avx <- apply(xx, 2, mean)
        avxmat[, kk] <- avx
        dif1 <- t(as.matrix(xx)) - matrix(rep(avx,
          times = dim(xx)[1]), nrow = 2)
        w1 <- w1 + (dif1 %*% t(dif1))
    }
    w1 <- w1/nd
#   (6)
    avxall <- apply(xxall, 2, mean)
```

```
#   (7)
    b1 <- matrix(rep(0, length = ndim * ndim), ncol = ndim)
    for(kk in 1:ncl) {
        xx <- xxall[clall == clu[kk],  ]
        dif2 <- t(avxmat - matrix(rep(avxall, times = ncl),
          nrow = ndim))
        b1 <- b1 + (t(dif2) %*% dif2 * dim(xx)[1])
    }
    b1 <- b1/nd
#   (8)
    eigen1 <- eigen(solve(w1) %*% b1)
    amat1 <- eigen1$vector
#   (9)
    return(amat1)
}
```

(1) The names of classes are defined as clu. The number of classes is ncl.
(2) The total number of data points is nd. The number of elements of the feature vector is denoted by ndim.
(3) The object w1 (a matrix) is prepared to save $\mathbf{W}$ (eq(7.73) (page 439)).
(4) The object avxmat (a vector) is prepared to save a vector whose element is $\bar{x}_k^{(g)}$ in eq(7.74) (page 439).
(5) $\mathbf{W}$ is computed using eq(7.73).
(6) A vector whose element is $\bar{x}_k$ in eq(7.74) is computed, and the result is saved as avxall.
(7) $\mathbf{B}$ given by eq(7.74) is calculated.
(8) The discriminant coordinates are obtained using eq(7.82) (page 441); a matrix, which consists of the column vectors representing the discriminant coordinates, is defined as amat1.
(9) The object amat1 is outputted.

An example of an object to use cano1() (construction of figure 7.20 (page 442))

```
function()
{
#   (1)
    ncl <- 3
    ndim <- 2
#   (2)
    set.seed(39)
    nda <- 30
    cla <- rep("a", length = nda)
    xxa <- matrix(rep(0., length = nda * 2.), ncol = 2.)
    xxa[, 1.] <- rnorm(nda, mean = 1., sd = 2.)
    xxa[, 2.] <- rnorm(nda, mean = 2., sd = 4.)
    ndb <- 30
    clb <- rep("b", length = ndb)
```

```
    xxb <- matrix(rep(0., length = ndb * 2.), ncol = 2.)
    xxb[, 1.] <- rnorm(ndb, mean = 3., sd = 3.)
    xxb[, 2.] <- rnorm(ndb, mean = 5., sd = 2.)
    ndc <- 40
    clc <- rep("c", length = ndc)
    xxc <- matrix(rep(0., length = ndc * 2.), ncol = 2.)
    xxc[, 1.] <- rnorm(ndc, mean = 5., sd = 1.)
    xxc[, 2.] <- rnorm(ndc, mean = 7., sd = 1.)
#   (3)
    xxall <- rbind(xxa, xxb, xxc)
    clall <- c(cla, clb, clc)
#   (4)
    amat1 <- cano1(xxall, clall)
#   (5)
    xxall2 <- xxall %*% amat1
    xxa2 <- xxa %*% amat1
    xxb2 <- xxb %*% amat1
    xxc2 <- xxc %*% amat1
#   (6)
    x1min <- -5
    x1max <- 10
    x2min <- -5
    x2max <- 8
    par(mfrow = c(2., 2.), mai = c(1., 0.7, 0.5, 0.5),
      oma = c(1., 1., 1., 1.))
#   (7)
    plot(xxall2, type = "n", xlab = "z1", ylab = "z2",
      xlim = c(x1min, x1max), ylim = c(x2min, x2max))
    text(xxall2, labels = clall)
#   (8)
    ne <- 20
    band1 <- c(8, 8)
    r1 <- nond1(xxa2, x1min, x1max, x2min, x2max, ne, band1)
    ex1 <- r1$ex1
    ex2 <- r1$ex2
    d1 <- matrix(r1$ey, ncol = length(ex2))
    contour(ex1, ex2, d1, xlab = "z1", ylab = "z2")
#   (9)
    r1 <- nond1(xxb2, x1min, x1max, x2min, x2max, ne, band1)
    ex1 <- r1$ex1
    ex2 <- r1$ex2
    d1 <- matrix(r1$ey, ncol = length(ex2))
    contour(ex1, ex2, d1, xlab = "z1", ylab = "z2")
#   (10)
    r1 <- nond1(xxc2, x1min, x1max, x2min, x2max, ne, band1)
    ex1 <- r1$ex1
    ex2 <- r1$ex2
```

```
    d1 <- matrix(r1$ey, ncol = length(ex2))
    contour(ex1, ex2, d1, xlab = "z1", ylab = "z2")
}
```

(1) The number of classes (ncl) is set at 3. The number of elements of the feature vector (ndim) is fixed at 2.

(2) Data are generated. The feature vectors of the three classes are xxa, xxb, and xxc. The three classes are named a, b, and c.

(3) The data obtained in (2) are connected to create xxall and clall.

(4) The discriminant coordinates are given by cano1(), and the matrix, which consists of vectors representing the discriminant coordinates, is named amat1.

(5) The components of xxall, xxa, xxb, and xxc in the directions of column vectors constituting amat1 are obtained; they are defined as xxall2, xxa2, xxb2, and xxc2.

(6) The region where estimates are calculated is defined. The area for drawing a graph is prepared.

(7) The data are charted in a graph; the coordinates of the graph are discriminant coordinates.

(8) The values of the probability density function obeyed by data belonging to a are derived at the grid points, and the contour is displayed in a graph in which the coordinates are discriminant coordinates.

(9) The values of the probability density function obeyed by data belonging to b are derived at the grid points, and the contour is displayed in a graph in which the coordinates are discriminant coordinates.

(10) The values of the probability density function obeyed by data belonging to c are derived at the grid points, and the contour is displayed in a graph in which the coordinates are discriminant coordinates.

REFERENCES

1. C.M. Bishop (1995). *Neural Networks for Pattern Recognition*, Oxford University Press.

2. L.Breiman, J.H. Friedman, R.A. Olsen, and J.S. Stone (1984). *Classification and Regression Trees*, CRC Press.

3. T.J. Hastie and R.J. Tibshirani (1990). *Generalized Additive Models*, Chapman & Hall/CRC.

4. T. Hastie, R. Tibshirani, and J.H. Friedman (2001). *The Elements of Statistical Learning: Data Mining, Inference, and Prediction*, Springer-Verlag.

5. R.H. Myers, D.C. Montgomery, and G.G. Vining (2001). *Generalized Linear Models: With Applications in Engineering and the Sciences*, John Wiley & Sons.

6. B.D. Ripley (1996). *Pattern Recognition and Neural Network*, Cambridge University Press.

Problems

7.1   Assume that $\mathbf{V}$ is a symmetric matrix. Prove that the inverse matrix of $\mathbf{V}$ is also symmetric.

7.2   Consider the linear discriminant rule and the quadratic discriminant rule.

(a) Create an S-Plus or R object for generating artificial feature vectors, and run it. The feature vectors belong to one of "a," "b," and "c"; the numbers of data are 50, 60, and 40, respectively. These data are created as follows.

"a": The first elements (along the x_1-axis) are the realizations of $N(1, 2^2)$ (a normal distribution; the mean is 1 and the standard deviation is 2), and the second elements (along the x_2-axis) are the realizations of $N(2, 4^2)$ (a normal distribution; the mean is 2 and the standard deviation is 4).

"b": The first elements (along the x_1-axis) are the realizations of $N(3, 3^2)$ (a normal distribution; the mean is 3 and the standard deviation is 3), and the second elements (along the x_2-axis) are the realizations of $N(5, 2^2)$ (a normal distribution; the mean is 5 and the standard deviation is 2).

"c": The first elements (along the x_1-axis) are the realizations of $N(5, 1^2)$ (a normal distribution; the mean is 5 and the standard deviation is 1), and the second elements (along the x_2-axis) are the realizations of $N(7, 1^2)$ (a normal distribution; the mean is 7 and the standard deviation is 1).

(b) Produce an S-Plus or R object for estimating the misclassification rates of the two methods using the 10-fold cross-validation; the simulation data created in **(a)** are employed. Run the object to compare the results of the two methods. The object `mylda2()` (or `mylda3()`) presented in **(B)** of this chapter and the object `myqda2()` presented in **(C)** of this chapter are useful.

(c) Construct S-Plus or R objects for estimating the misclassification rates of the two methods using bootstrap samples from the data. Run the objects to compare the results of the two methods, and the results obtained in **(b)**.

7.3   The regression based on a binomial distribution can use a function other than the logit transformation as a link function. Then, modify the object `glmdis1()` shown in **(F)** of this chapter to be able to use another link function; the S-Plus object `glm()` allows probit transformation (specified as `binomial(link=probit)` in `glm()`) and complementary log-log transformation (specified as `binomial(link=cloglog)` in `glm()`) as a link function for the regression based on a binomial distribution. In addition, compare the misclassification rates using another set of data given by the same equations with other pseudo-random numbers to determine the influence of the type of link function; the data presented in Problem **7.2(a)** is used for this purpose.

7.4   The purpose of bagging in eq(7.60) (page 435) is originally to enhance the predictability of a regression equation. The predictability of a tree-based model can be augmented by bagging. Then, create an S-Plus or R object to obtain estimates based on eq(7.60) using a set of tree-based models; α in eq(7.55) (page 432) is optimized using the original data and is used for all resampled data. The data generated in Problem **7.2(a)** are employed for this purpose. Furthermore, determine the effect of bagging by another set of data given by the same equations with other pseudo-random numbers.

7.5 The performance of the k-nearest-neighbor classifier is influenced by the multiplication of elements of feature vectors using a nonzero constant.

(a) Produce an S-Plus or R object to optimize the nonzero constant for multiplying the second element of the feature vector using 10-fold cross-validation, and run it using the artificial data shown in Problem **7.2(a)**.

(b) Verify the effect of the transformation of feature vectors by the object created in **(a)** using another set of data given by the same equations with other pseudo-random numbers.

APPENDIX A

CREATION AND APPLICATIONS OF

B-SPLINE BASES

A.1 INTRODUCTION

The equation using the Green function (Green's function) is chiefly adopted to represent a spline function in a finite region in Chapter 3. However, the equation using B-spline bases can be understood more viscerally and easily. Furthermore, estimates (estimated values) of high precision are obtained even if no particular ingenuity is exercised to carry out numerical calculation for deriving estimates. Then, we discuss the methods of constructing B-spline bases and extend the discussion to the use of B-spline in order to conduct calculations of spline function, natural spline, and smoothing spline (smoothing splines). In addition, objects for executing these calculations using S-Plus are introduced. Since special emphasis is placed on the description of the outline and the use of the B-spline basis, sufficient consideration of computational cost is not taken.

A.2 METHOD TO CREATE B-SPLINE BASIS

Using the B-spline basis, a cubic spline in a finite region ($s_{finite}(x)$) is represented as

$$s_{finite}(x) = \sum_{j=1}^{p+2} \beta_j B_j(x) \qquad (\xi_4 \le x \le \xi_{p+3}). \qquad (A.1)$$

$\{B_j(x)\}$ are B-spline bases constructed using $(p + 6)$ knots (breakpoints). The positions of $\{\xi_j\}$ ($\xi_1 \le \xi_2 \le \cdots \le \xi_{i-1} \le \xi_i \le \xi_{i+1} \le \cdots \le \xi_{p+5} \le \xi_{p+6}$) denote knots when B-spline is employed. $B_j(x)$ takes a nonzero value only in the region of $\xi_j \le x \le \xi_{j+4}$. As a result, $s_{finite}(x)$ turns out to be a cubic spline in a finite region (between ξ_4 and ξ_{p+3}). More generally than B-spline bases, if an arbitrary degree of spline is represented as a linear combination of bases in the form of eq(A.1), the bases are called spline bases. Note that the simple expression "spline" means a cubic spline hereafter.

When a spline in a finite region is represented using the Green functions (eq(3.164) (page 155)), its knots are identical to $\{\xi_4, \xi_5, \ldots, \xi_{p+3}\}$. B-spline bases require six new knots: $\xi_1, \xi_2, \xi_3, \xi_{p+4}, \xi_{p+5}$, and ξ_{p+6}. The positions of the six knots affect the shapes of B-splines. However, the locations of the six knots only require the conditions $\xi_1 \le \xi_2 \le \xi_3 \le \xi_4$ and $\xi_{n+3} \le \xi_{n+4} \le \xi_{n+5} \le \xi_{n+6}$. Only if these conditions are met, does the adjustment of the values of $\{\beta_j\}$ realize identical $s_{finite}(x)$ in the region of $\xi_4 \le x \le \xi_{p+3}$.

Other bases are also available for representing a spline in a finite region with the form of eq(A.1). However, the B-spline basis has the special feature of local support; support indicates the region where a function takes a value other than 0. Therefore, the fact that $B_j(x)$ has a local support means that the region where $B_j(x)$ takes a value other than 0 is limited to a narrow region. Furthermore, B-spline bases do not take negative values. Use of bases of this sort enables intuitive understanding of the shapes of functions, because when a value of β_j is large, the value of $s_{finite}(x)$ is large in the region where the value of $B_j(x)$ is large. Furthermore, when numerical calculation to obtain the values of $\{\beta_j\}$ is carried out, estimates of high precision are derived even if calculation is executed on the basis of mathematical equations as they are; this is another adavantage of bases of this kind.

Now, let us introduce a typical method of constructing B-spline bases, called the de Boor-Cox algorithm. There are other methods available for creating B-spline bases. However, since numerical calculation of high accuracy is realized regardless of the positions of knots, this algorithm is the most popular for deriving B-spline bases. $\{M_{kj}(x)\}$ denotes kth degree B-spline bases hereafter. Note that although the description here is limited to derivation of cubic spline bases, higher degree B-spline bases are realized by simple extension of this procedure.

To prove that the bases which the algorithm below yields satisfy the conditions of bases used to construct the spline, two points must be proved: (1) each basis is a spline, and (2) any spline located in the region of $\xi_4 \le x \le \xi_{p+3}$ is represented by a linear combination of the bases (this is true if these bases are linearly independent). For the proof itself, refer to, for example, [3].

When we conduct interpolation by B-spline bases in which data points coincide with knots, the relationship between data points and knots is set out as below to obtain

$s_{finite}(x)$.

$$\xi_{j+3} = X_j \quad (1 \le j \le n). \tag{A.2}$$

Furthermore, the equation below must be satisfied to realize interpolation.

$$s_{finite}(X_j) = Y_j \quad (1 \le j \le n). \tag{A.3}$$

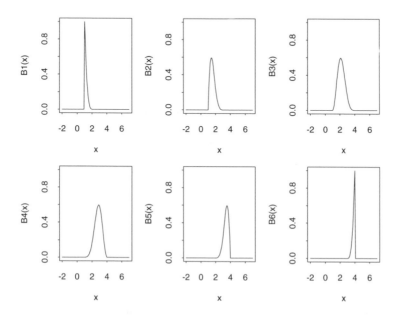

Figure A.1   All cubic B-spline bases for knots located at $\{1, 1, 1, 1, 2, 3, 4, 4, 4, 4\}$.

When splines defined by eq(A.1) are created, the position of the six knots (ξ_1, ξ_2, ξ_3, ξ_{n+4}, ξ_{n+5}, and ξ_{n+6}) must be determined. However, as discussed above, if the positions of the six points satisfy $\xi_1 \le \xi_2 \le \xi_3 \le \xi_4$ and $\xi_{n+3} \le \xi_{n+4} \le \xi_{n+5} \le \xi_{n+6}$, the resultant $s_{finite}(x)$ (eq(A.1)) can be represented as $s_{finite}(x)$, which uses other sets of six values that satisfy the conditions above. Hence, these six values can be set out uniquely. Other sets of values are not required to be examined. Moreover, while eq(A.2) supplies n conditions, the number of $\{\beta_j\}$ in eq(A.1) is $(n + 2)$ and hence there are an infinite number of combinations of the values of $\{\beta_j\}$ that satisfy eq(A.2) and eq(A.3) even if values of ξ_1, ξ_2, ξ_3 and values of $\xi_{n+4}, \xi_{n+5}, \xi_{n+6}$ are fixed.

If the value of ξ_4 is not identical to that of X_1 and $\xi_3 \le \xi_4 < X_1$ is satisfied, an estimate at x ($\xi_3 < x < \xi_4$) is obtained. That is, extrapolation is realized. However, the resulting estimates are identical to those obtained by interpolation based on $\xi_4 = X_1$ and by extension of the defined region ($\xi_4 \le x \le \xi_5$) of the cubic equation to the region of $x < \xi_4$. If ξ_{n+3} does not equal X_n and $X_n < \xi_{n+3} \le \xi_{n+4}$ is satisfied, the same discussion is possible.

While figure A.1 is obtained assuming that $\xi_1 = \xi_2 = \xi_3 = \xi_4 = 1$ and $\xi_{n+3} = \xi_{n+4} = \xi_{n+5} = \xi_{n+6} = 4$ ($n = 4$ in this example), figure A.2 is obtained when

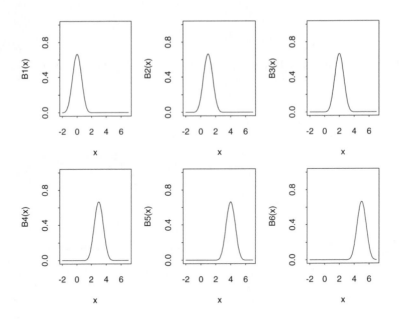

Figure A.2   All cubic B-spline bases when knots are positioned at $\{-2, -1, 0, 1, 2, 3, 4, 5, 6, 7\}$.

$\xi_1 = -2$, $\xi_2 = -1$, $\xi_3 = 0$, $\xi_4 = 1$, $\xi_{n+3} = 4$, $\xi_{n+4} = 5$, $\xi_{n+5} = 6$, and $\xi_{n+6} = 7$ are set. Whichever $\{\xi_j\}$ may be chosen to construct $s_{finite}(x)$ using $\{B_j(x)\}$, there are an infinite number of splines that satisfy eq(A.2) (page 475) and eq(A.3) (page 475), and a $s_{finite}(x)$ created using one set of $\{\xi_j\}$ can be represented as $s_{finite}(x)$, which is composed using another set of $\{\xi_j\}$. This is another way of saying that the functional space obtained by linear combinations of each set of bases remains the same, though alternations between the values of ξ_1, ξ_2, ξ_3 and $\xi_{n+4}, \xi_{n+5}, \xi_{n+6}$ result in sets of bases with different shapes. This functional space consists of splines, whose knots are positioned at $\{X_i\}$, in a finite region ($X_1 \leq x \leq X_n$).

To obtain a set of cubic B-spline bases, $\{M_{0j}(x)\}$ ($2 \leq j \leq p$) is defined as below. Each $\{M_{0j}(x)\}$ is a piecewise constant.

$$M_{0j}(x) = \begin{cases} \dfrac{1}{\xi_j - \xi_{j-1}} & \text{if} \quad \xi_{j-1} \leq x < \xi_j, \\ & \text{or the value of } x \text{ is identical to the maximum value of} \\ & \{\xi_j\}. \\ 0 & \text{if} \quad x \text{ takes another value} \\ & \quad (2 \leq j \leq p). \end{cases} \tag{A.4}$$

This definition is valid when some values in $\{\xi_j\}$ are tied.

Next, $\{M_{1j}(x)\}$ ($3 \leq j \leq p+2$) is derived as follows. Each $\{M_{1j}(x)\}$ turns out to be a linear B-spline.

$$M_{1j}(x) = \frac{(x - \xi_{j-2}) M_{0,j-1}(x) + (\xi_j - x) M_{0j}(x)}{\xi_j - \xi_{j-2}} \quad (3 \leq j \leq p+2). \tag{A.5}$$

Furthermore, $\{M_{2j}(x)\}$ $(4 \leq j \leq p+4)$ is calculated as below. Each $\{M_{2j}(x)\}$ turns out to be a quadratic B-spline.

$$M_{2j}(x) = \frac{(x - \xi_{j-3})M_{1,j-1}(x) + (\xi_j - x)M_{1j}(x)}{\xi_j - \xi_{j-3}} \quad (4 \leq j \leq p+4). \quad \text{(A.6)}$$

Furthermore, $\{M_{3j}(x)\}$ $(5 \leq j \leq p+6)$ is computed as follows. Each $\{M_{3j}(x)\}$ turns out to be a cubic spline.

$$M_{3j}(x) = \frac{(x - \xi_{j-4})M_{2,j-1}(x) + (\xi_j - x)M_{2j}(x)}{\xi_j - \xi_{j-4}} \quad (5 \leq j \leq p+6). \quad \text{(A.7)}$$

When a specific value is given as x $(\xi_j \leq x \leq \xi_{j+1})$, the procedure to obtain $\{M_{3j}(x)\}$ $(5 \leq j \leq p+6)$ is illustrated below. To obtain $M_{kj}(x)$ listed in each column, the two values at the upper and lower positions of the left neighboring column are utilized.

0-degree	linear	quadratic	cubic
		0	
	0		$M_{3j}(x)$
0		$M_{2j}(x)$	
	$M_{1j}(x)$		$M_{3,j+1}(x)$
$M_{0j}(x)$		$M_{2,j+1}(x)$	
	$M_{1,j+1}(x)$		$M_{3,j+2}(x)$
0		$M_{2,j+2}(x)$	
	0		$M_{3,j+3}(x)$
		0	

That is, the procedure from (a) to (j) described below is carried out by turns. (a) When a specific value is given as x, only one function among $\{M_{0j}(x)\}$ defined by eq(A.4) takes a value other than 0; a nonzero value $(M_{0j}(x))$ is derived. (b) The value of $M_{1j}(x)$ is calculated using eq(A.5). Note that the value of $M_{0,j-1}(x)$ is 0. (c) The value of $M_{1,j+1}(x)$ is computed using eq(A.5). Note that the value of $M_{0,j+1}(x)$ is 0. (d) $M_{2j}(x)$ is calculated using eq(A.6). Note that the value of $M_{1,j-1}(x)$ is 0. (e) The value of $M_{2,j+1}(x)$ is computed using eq(A.6). (f) The value of $M_{2,j+2}(x)$ is computed using eq(A.6). Note that the value of $M_{1,j+2}(x)$ is 0. (g) The value of $M_{3j}(x)$ is calculated using eq(A.7). Note that the value of $M_{2,j-1}(x)$ is 0. (h) The value of $M_{3,j+1}(x)$ is obtained using eq(A.7). (i) The value of $M_{3,j+2}(x)$ is obtained using eq(A.7). (j) The value of $M_{3,j+3}(x)$ is obtained using eq(A.7). Note that the value of $M_{2,j+3}(x)$ is 0.

The table above indicates that when a specific value is given for x, four values among $\{M_{3j}(x)\}$ $(5 \leq j \leq p+6)$ are nonzero. This means that when x is positioned in the region of $\xi_j \leq x \leq \xi_{j+1}$, the value of spline in this region is determined using four values among $\{\beta_j\}$ in eq(A.1). This corresponds to the fact that the cubic equation which represents a spline in this region has four coefficients.

If $\{B_j(x)\}$ used in eq(A.1) satisfies $\sum_{j=1}^{p+2} B_j(x) = 1$ $(\xi_4 \leq x \leq \xi_{p+3})$, the value of $s_{finite}(x)$ when the value of x is fixed is considered to be a weighted average of four values extracted from $\{\beta_j\}$. Hence, $\{B_j(x)\}$ below is usually used as cubic

B-spline bases:

$$B_j(x) = (\xi_{j+4} - \xi_j)M_{3,j+4}(x) \quad (1 \le j \le p+2), \tag{A.8}$$

where the suffix of $B_j(x)$ is renumbered in order that j satisfies $1 \le j \le p+2$.

Next, let us discuss a method of fitting a spline to the data $(\{(X_i, Y_i)\} (1 \le i \le n))$ using the least squares (least square method). When B-spline is employed, the value below is minimized.

$$E_{B-spline} = \sum_{i=1}^{n} \left(Y_i - \sum_{j=1}^{p+2} \beta_j B_j(X_i) \right)^2. \tag{A.9}$$

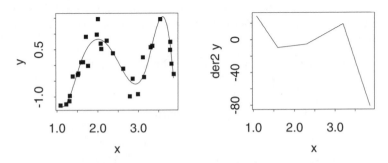

Figure A.3   Fitting of B-spline by the least squares. The knots are set at $\{1, 1, 1, 1, 1.6, 2.3, 3.2, 4, 4, 4, 4\}$. Data and result of interpolation (left). Second derivative of the interpolation function (right).

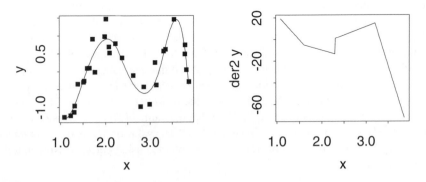

Figure A.4   Fitting of B-spline by the least squares. The knots are set at $\{1, 1, 1, 1, 1.6, 2.3, 2.3, 3.2, 4, 4, 4, 4\}$. Data and result of interpolation (left). Second derivative of the interpolation function (right).

Figure A.3 shows the results of fitting B-spline with knots positioned at $\{1, 1, 1, 1, 1.6, 2.3, 3.2, 4, 4, 4, 4\}$. As mentioned before, making knots quadruple at both ends has no particular purpose. Knots placed at $\{-9, -5, -1, 1, 1.6, 2.3, 3.2, 4, 7, 8, 10\}$,

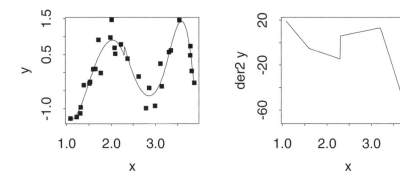

Figure A.5   Fitting of B-spline by the least squares.   The knots are set at $\{1, 1, 1, 1,$ $1.6, 2.3, 2.3, 2.3, 2.3, 3.2, 4, 4, 4, 4\}$. Data and result of interpolation (left). Second derivative of the interpolation function (right).

for instance, lead to the same results. On the other hand, when the same data are used with the knots placed at $\{1, 1, 1, 1, 1.6, 2.3, 2.3, 3.2, 4, 4, 4, 4\}$, figure A.4 is obtained. The influence of doubling the knot at 2.3 is not perceived in the graph of the estimates, although the existence of a double knot is ascertained by the fact that the second derivative function is discontinuous at 2.3. Furthermore, when the knots are positioned at $\{1, 1, 1, 1, 1.6, 2.3, 2.3, 2.3, 2.3, 3.2, 4, 4, 4, 4\}$, figure A.5 is derived. Since the knot positioned at 2.3 is quadruple, the estimate itself changes discontinuously at this point. As just described, multiple knots located at positions other than the two ends alter the appearance of estimates.

A.3 NATURAL SPLINE CREATED BY B-SPLINE

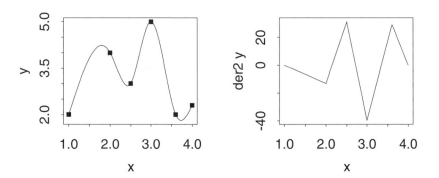

Figure A.6   Interpolation by natural spline. Data and result of interpolation (left). Second derivative of the function obtained by interpolation.

B-spline can be used when interpolation is carried out using a natural spline. If eq(A.2) (page 475) and eq(A.3) (page 475) are used only as conditions to interpolate $\{(X_i, Y_i)\}$ $(1 \leq i \leq n)$, an infinite number of splines are realized because $s_{finite}(x)$

(eq(A.1) (page 474)) contains $(n+2)$ regression coefficients ($\{\beta_i\}$ $(1 \leq i \leq n+2)$), while the number of data is n. Hence, the conditions below are added.

$$\left[\frac{d^2 ns(x)}{dx^2}\right]_{x=X_1} = 0, \tag{A.10}$$

$$\left[\frac{d^2 ns(x)}{dx^2}\right]_{x=X_N} = 0. \tag{A.11}$$

When $s_{finite}(x)$ satisfies the two conditions above (natural boundary conditions), the spline obtained is called a natural spline; it is represented as $ns(x)$ here. Addition of these two conditions to eq(A.2) yields a unique $s_{finite}(x)$ (i.e., $ns(x)$). Figure A.6 exemplifies interpolation by this method. Although knots in this example are positioned as described by eq(A.2), other locations of knots for interpolation are possible.

The estimates obtained with natural spline are written as

$$ns(x) = \sum_{j=1}^{n} \nu_j N_j(x). \tag{A.12}$$

$\{N_j(x)\}$ are bases, and $\{\nu_j\}$ are coefficients for the bases. To make $ns(x)$ satisfy eq(A.10) and eq(A.11) regardless of $\{\nu_j\}$, all $\{N_j(x)\}$ must be splines that fulfill natural boundary conditions. Furthermore, to ensure that $\{\nu_j\}$ is uniquely determined when $ns(x)$ is given, $\{N_j(x)\}$ is required to be linearly independent. Such $\{N_j(x)\}$ are called natural-spline bases (also called N-splines). Even when the locations of knots are fixed, various sets of natural-spline bases are possible. This is because when a set of natural-spline bases are at hand, a linear combination of the bases provides another set of natural-spline bases.

A set of natural-spline bases which satisfy the conditions below are created here.

$$N_j(X_i) = \begin{cases} 1 & \text{if } i = j \\ 0 & \text{if } i \neq j. \end{cases} \tag{A.13}$$

Not unique to natural spline, spline bases that satisfy eq(A.13) are addressed as cardinal-spline bases; however, the term cardinal splines has another meaning (page 24 in [1]). Cardinal-spline bases that are also natural-spline bases satisfy

$$ns(x) = \sum_{j=1}^{n} Y_j N_j(x). \tag{A.14}$$

This equation directly expresses the influence of each $\{Y_j\}$ on $ns(x)$. That is, when interpolation by natural spline is carried out under with the conditions of eq(A.10) (page 480) and eq(A.11) (page 480) as well as the conditions of eq(A.2) (page 475) and eq(A.3) (page 475), an equivalent kernel corresponding to Y_j is $N_j(x)$. Hence, $\{N_j(x)\}$ are obtained in the same manner as that for equivalent kernels. More specifically, when Y_k is 1 and other values of a target variable are 0 in $\{Y_j\}$, $ns(x)$ becomes $N_k(x)$, and hence the procedures with $k = 1, 2, \ldots, n$ yield $\{N_j(x)\}$. Figure A.7 displays $\{N_1(x), N_2(x), N_3(x), N_4(x), N_5(x), N_6(x)\}$ when knots are positioned at $\{1, 1, 1, 1, 2, 3, 4, 5, 6, 6, 6, 6\}$.

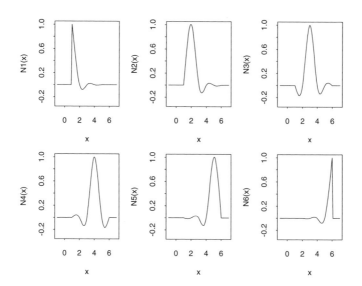

Figure A.7   Example of natural-spline bases (also regarded as cardinal-spline bases). The knots are located at $\{1, 1, 1, 1, 2, 3, 4, 5, 6, 6, 6, 6\}$.

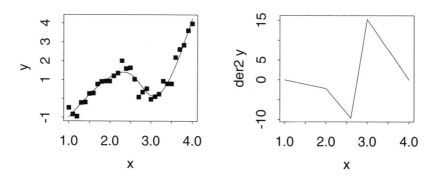

Figure A.8   Fitting of natural spline by the least squares. The locations of knots are $\{1, 1, 1, 1, 2, 2.6, 3, 4, 4, 4, 4\}$.

Bases obtained in this way can also be used to realize smoothing by fitting a natural spline to data determined with the least squares. When data are $\{(X_i, Y_i)\}$ $(1 \leq i \leq n)$ and knots are $\{\xi_i\}$ $(1 \leq i \leq p+6)$, knots must be positioned to satisfy

$$\xi_1 \leq \xi_2 \leq \xi_3 \leq \xi_4 \leq X_1 < X_2 < \cdots < X_n \leq \xi_{p+3} \leq \xi_{p+4} \leq \xi_{p+5} \leq \xi_{p+6}.$$
(A.15)

Furthermore, $n \geq p$ must hold. Additionally, the locations of knots must satisfy the Schoenberg-Whitney conditions [3].

Then, the regression equation to be created is

$$m(x) = \sum_{j=1}^{p} \nu_j N_j(x).$$
(A.16)

$\{N_j(x)\}$ are natural-spline bases. In order for $\{N_j(x)\}$ to be regarded as cardinal-spline bases, the conditions below are attached.

$$N_j(\xi_{i+3}) = \begin{cases} 1 & \text{if } i = j \\ 0 & \text{if } i \neq j \end{cases}$$
$$(1 \leq i \leq p, 1 \leq j \leq p).$$
(A.17)

By referring to eq(A.16), we have

$$\nu_j = m(\xi_{j+3}) \qquad (1 \leq j \leq p).$$
(A.18)

That is, $\{\nu_j\}$ is an estimate at a knot other than the six knots located at the two ends.

Furthermore, when estimates at X_i are defined as $\hat{Y}_i$, the equation below is obtained.

$$\hat{Y}_i = \sum_{j=1}^{p} \nu_j N_j(X_i) \qquad (1 \leq i \leq p).$$
(A.19)

Hence, the residual sum of squares ($E_{natural}$) is

$$E_{natural} = \sum_{i=1}^{n} \left(Y_i - \sum_{j=1}^{p} \nu_j N_j(X_i) \right)^2.$$
(A.20)

In the same manner as that of fitting a multiple regression equation (multiple linear regression equation) and a polynomial (polynomial function) by the least squares (least square method), estimation of the values of $\{\nu_j\}$ leads to eq(A.16). Figure A.8 exemplifies the fitting of a natural spline using this method. If $\{N_j(x)\}$ are natural-spline bases, fitting of a natural spline using eq(A.20) is accomplished even if eq(A.17) does not hold. The natural-spline bases created by ns() in S-Plus, for example, are not cardinal-spline bases. Moreover, there are natural-spline bases obtained directly from truncated power functions (page 121 of [2]). Alternatively, a set of natural-spline bases are calculated by integrating B-splines.

A.4 APPLICATION TO SMOOTHING SPLINE

There are several methods applicable for smoothing by smoothing spline. First, let us introduce a method that utilizes natural-spline bases. Although its practical value is not high, it is useful to understand the meaning of smoothing spline.

Smoothing spline is obtained by minimizing

$$E_{ss} = \sum_{i=1}^{n}(Y_i - m(X_i))^2 + \lambda \int_{-\infty}^{\infty} \left(\frac{d^2 m(x)}{dx^2}\right)^2 dx. \tag{A.21}$$

Using $\{N_j(x)\}$ (considered to be both natural-spline bases and cardinal-spline bases), this equation is rewritten as

$$E_{ss} = (\mathbf{y} - \mathbf{N}\boldsymbol{\nu})^2 + \lambda \boldsymbol{\nu}^t \boldsymbol{\Phi} \boldsymbol{\nu}, \tag{A.22}$$

where $\mathbf{y}$ and $\boldsymbol{\nu}$ are

$$\mathbf{y} = (Y_1, Y_2, \ldots, Y_n)^t, \tag{A.23}$$

$$\boldsymbol{\nu} = (\nu_1, \nu_2, \ldots, \nu_n)^t. \tag{A.24}$$

Furthermore, the ij-element of $\mathbf{N}$ is $N_j(X_i)$ $(1 \le i \le n, 1 \le j \le n)$. The ij-element of $\boldsymbol{\Phi}$ is $\int_{-\infty}^{\infty} \left(\frac{d^2 N_i(x)}{dx^2}\right)\left(\frac{d^2 N_j(x)}{dx^2}\right) dx$ $(1 \le i \le n, 1 \le j \le n)$.

Differentiation of (A.22) to be 0 with respect to each $\boldsymbol{\nu}$ gives

$$\hat{\boldsymbol{\nu}} = (\mathbf{I}_n + \lambda \boldsymbol{\Phi})^{-1} \mathbf{y}. \tag{A.25}$$

Therefore, this $\boldsymbol{\Phi}$ is identical to $\mathbf{L}$ in eq(3.231)(page 176). Since $\hat{\boldsymbol{\nu}}$ is an estimate corresponding to $\mathbf{y}$, $\boldsymbol{\Phi}$ is considered to be a matrix for creating the roughness penalty; the roughness penalty is represented as a quadratic form using estimates directly. Furthermore, the corresponding hat matrix is

$$\mathbf{H} = (\mathbf{I}_n + \lambda \boldsymbol{\Phi})^{-1}. \tag{A.26}$$

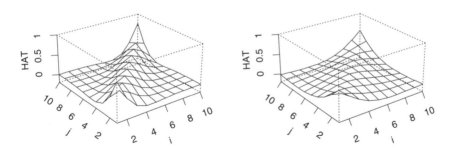

Figure A.9   The values of elements of the hat matrix given by eq(A.26). $\{X_i\} = \{1, 2, \cdots, 10\}$. $\lambda = 1$ (left). $\lambda = 10$ (right).

Figure A.9 displays the values of the hat matrix when the values of the predictor are $\{1, 2, \ldots, 10\}$ and the value of λ is 1 or 10. Eq(3.226) (page 176) yields the same hat matrix.

Since k_{ij} $\left(= \int_{-\infty}^{\infty} \left(\frac{d^2 B_i(x)}{dx^2}\right)\left(\frac{d^2 B_j(x)}{dx^2}\right) dx\right)$ defined in eq(3.239) (page 180) is a band matrix, computational cost is saved. On the other hand, $\boldsymbol{\Phi}$ does not have such an advantage. Additionally, calculation to obtain $\{N_j(x)\}$ is required after the

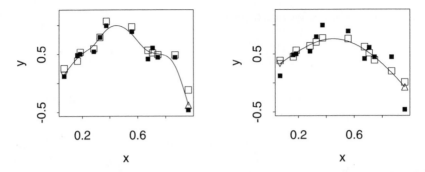

Figure A.10   Result of smoothing; smoothing spline based on B-spline is adopted ($n = 10$). ■ are data.   The solid line indicates estimates obtained by smoothing spline. □ denotes $\{(X_i, \beta_{i+1})\}$ $(1 \leq i \leq n)$. ∇ represents (X_1, β_1). $\triangle$ shows (X_n, β_{n+2}). $\lambda = 0.0001$ (left). $\lambda = 0.01$ (right).

construction of B-spline bases. Therefore, the practical value of eq(A.25) is not high.

With the use of B-spline, a small amount of calculation realizes smoothing by smoothing spline. That is, $g_j(x)$ in eq(3.235) (page 179) is employed as $B_j(x)$. This strategy enjoys another advantage: it deals more easily with data that contain tied data than does the equation with the Green functions (Green's function). Figure A.10 exemplifies the result of smoothing by smoothing spline. The estimates at x are the weighted averages of $(n + 2)$ values $(\beta_1, \{\beta_j\}$ $(2 \leq j \leq n + 1), \beta_{n+2})$ with the weights of $\{B_j(x)\}$ $(1 \leq j \leq n + 2)$.

A.5 EXAMPLES OF S-PLUS OBJECT

(A) Creation of bases of B-spline
Object: mybb1()

```
function(xn, ex)
{
#   (1)
    bb0 <- matrix(rep(0, length = length(xn) * length(ex)),
      ncol = length(xn))
    for(ii in 1:length(ex)) {
        ex1 <- ex[ii]
        for(jj in 2:(length(xn))) {
            if(xn[jj] != max(xn)) {
                if(xn[jj - 1] > ex1 || ex1 >= xn[jj])
                  bb0[ii, jj] <- 0
                else if(xn[jj] != xn[jj - 1])
                  bb0[ii, jj] <- 1/(xn[jj] - xn[jj - 1])
            }
            else {
```

```
                if(xn[jj - 1] > ex1 || ex1 > xn[jj])
                  bb0[ii, jj] <- 0
                else if(xn[jj] != xn[jj - 1])
                  bb0[ii, jj] <- 1/(xn[jj] - xn[jj - 1])
            }
        }
    }
#   (2)
    bb1 <- matrix(rep(0, length = length(xn) * length(ex)),
     ncol = length(xn))
    for(ii in 1:length(ex)) {
        ex1 <- ex[ii]
        for(jj in 3:(length(xn))) {
            if(xn[jj] != xn[jj - 2])
                bb1[ii, jj] <- ((ex1 - xn[jj - 2])
                * bb0[ii, jj - 1] + (xn[jj] - ex1)
                * bb0[ii, jj])/(xn[jj] - xn[jj - 2])
        }
    }
#   (3)
    bb2 <- matrix(rep(0, length = length(xn) * length(ex)),
     ncol = length(xn))
    for(ii in 1:length(ex)) {
        ex1 <- ex[ii]
        for(jj in 4:(length(xn))) {
            if(xn[jj] != xn[jj - 3])
                bb2[ii, jj] <- ((ex1 - xn[jj - 3])
                * bb1[ii, jj - 1] + (xn[jj] - ex1) * bb1[ii, jj])
                /(xn[jj] - xn[jj - 3])
        }
    }
#   (4)
    bb3 <- matrix(rep(0, length = length(xn) * length(ex)),
     ncol = length(xn))
    for(ii in 1:length(ex)) {
        ex1 <- ex[ii]
        for(jj in 5:(length(xn))) {
            bb3[ii, jj] <- ((ex1 - xn[jj - 4]) * bb2[ii, jj - 1] +
             (xn[jj] - ex1) * bb2[ii, jj])/(xn[jj] - xn[jj - 4])
            bb3[ii, jj] <- bb3[ii, jj] * (xn[jj] - xn[jj - 4])
        }
    }
#   (5)
    bs1 <- bb3[, 5:length(xn)]
#   (6)
    return(bs1)
}
```

(1) A zero-degree B-spline ($\{M_{0j}(x)\}$) (bb0) is derived.
(2) A linear B-spline ($\{M_{1j}(x)\}$) (bb1) is calculated.
(3) A quadratic B-spline ($\{M_{2j}(x)\}$) (bb2) is obtained.
(4) After a cubic B-spline ($\{M_{3j}(x)\}$) is computed, bb3 is derived by normalization.
(5) Shift of the suffix of bb3 leads to bs1; it is $\{B_j(x)\}$.
(6) The object bs1 is outputted.

An example of an object to use mybb1() (construction of figure A.1 (page 475))

```
function()
{
#    (1)
     xn <- c(1, 1, 1, 1, 2, 3, 4, 4, 4, 4)
     ex <- seq(from = -2, to = 7, by = 0.1)
#    (2)
     bs1 <- mybb1(xn, ex)
#    (3)
     par(mfrow = c(2, 3), mai = c(1, 1, 0.5, 0.1),
       oma = c(2, 2, 2, 2))
     plot(ex, bs1[, 1], type = "n", ylim = c(0, 1),
       xlab = "x", ylab = "B1(x)")
     lines(ex, bs1[, 1])
     plot(ex, bs1[, 2], type = "n", ylim = c(0, 1),
       xlab = "x", ylab = "B2(x)")
     lines(ex, bs1[, 2])
     plot(ex, bs1[, 3], type = "n", ylim = c(0, 1),
       xlab = "x", ylab = "B3(x)")
     lines(ex, bs1[, 3])
     plot(ex, bs1[, 4], type = "n", ylim = c(0, 1),
       xlab = "x", ylab = "B4(x)")
     lines(ex, bs1[, 4])
     plot(ex, bs1[, 5], type = "n", ylim = c(0, 1),
       xlab = "x", ylab = "B5(x)")
     lines(ex, bs1[, 5])
     plot(ex, bs1[, 6], type = "n", ylim = c(0, 1),
       xlab = "x", ylab = "B6(x)")
     lines(ex, bs1[, 6])
}
```

(1) The positions of knots (xn) and the locations where values of B-spline are derived (ex) are specified.
(2) Values of B-spline bases are calculated using mybb1(xn, ex); the result is saved as bs1. Each column of bs1 corresponds to a different basis.
(3) The shapes of the six resultant bases are illustrated.

(B) Fitting of B-splines by the least squares
Object: mybsp1()

```
function(xx, yy, xn, ex)
{
#   (1)
    gg <- mybb1(xn, xx)
#   (2)
    beta1 <- solve(t(gg) %*% gg, t(gg) %*%
     matrix(yy, ncol = 1))
#   (3)
    gg2 <- mybb1(xn, ex)
    ey <- gg2 %*% matrix(beta1, ncol = 1)
#   (4)
    return(ey)
}
```

(1) A design matrix (gg) is derived using mybb1().
(2) A normal equation is solved, and the resulting regression coefficients are stored as beta1.
(3) The values of B-spline are calculated at the estimation points (ex); the result is saved as gg2. Furthermore, multiplication by beta1 (obtained in (2)) yields estimates (ey).
(4) The object ey is outputted.

An example of an object to use mybsp1() (construction of figure A.3 (page 478))

```
function()
{
#   (1)
    set.seed(92)
    xx <- runif(30, min = 1, max = 4)
    xx <- sort(xx)
    yy <- sin(xx * 4) + rnorm(30, mean = 0, sd = 0.4)
#   (2)
    ex <- seq(from = min(xx), to = max(xx), by = 0.01)
#   (3)
    xn <- c(rep(xx[1], length = 4), 1.6, 2.3, 3.2,
     rep(xx[length(xx)], length = 4))
#   (4)
    ey <- mybsp1(xx, yy, xn, ex)
#   (5)
    d2 <- NULL
    xn2 <- unique(xn[4:(length(xn) - 3)])
    for(ii in 1:(length(xn2) - 1)) {
        if(ii == (length(xn2) - 1)) {
            x1 <- ex[ex >= xn2[ii] & ex <= xn2[ii + 1]]
            y1 <- ey[ex >= xn2[ii] & ex <= xn2[ii + 1]]
        }
        else {
```

```
                    x1 <- ex[ex >= xn2[ii] & ex < xn2[ii + 1]]
                    y1 <- ey[ex >= xn2[ii] & ex < xn2[ii + 1]]
            }
            des1 <- cbind(x1, x1^2, x1^3)
            ls1 <- lsfit(des1, y1)
            coef1 <- ls1$coef
            d2 <- c(d2, 2 * coef1[3] + 6 * coef1[4] * x1)
    }
#   (6)
    par(mfrow = c(1, 2), mai = c(2.5, 1.5, 2, 0.5),
      oma = c(1, 1, 1, 1))
    plot(ex, ey, ylim = c(min(ey), max(ey)), type = "n",
      xlab = "x", ylab = "y", cex = 1.1)
    points(xx, yy, pch = 15)
    lines(ex, ey)
#   (7)
    plot(ex, d2, ylim = c(min(d2), max(d2)), type = "n",
      xlab = "x", ylab = "der2 y", cex = 1.1)
    lines(ex, d2)
}
```

(1) Simulation data are generated.
(2) The estimation points are defined as ex. When the number of elements of ex located between knots at different positions is small, the calculation of the second derivative in (5) causes some trouble.
(3) The positions of knots are given as xn. Even if the knots at the two ends are not quadruple, the results remain the same; this is ascertained easily. Furthermore, some knots ($\{\xi_j\}$ $(5 \le j \le p + 2)$) can be multiple knots.
(4) Using mybsp1 (), estimates are derived by fitting a spline by the least squares.
(5) The values at ex, which the second derivative function given by the resulting spline yields, are calculated.
(6) The estimates are illustrated.
(7) The values supplied by the second derivative function of the resultant spline are plotted.

(C) Interpolation by natural spline
Object: myni1()

```
function(xx, yy, ex)
{
#   (1)
    nd <- length(xx)
    xn <- c(xx[1], xx[1], xx[1], xx, xx[nd], xx[nd],
      xx[nd])
#   (2)
    xxl <- seq(from = xn[4], to = xn[5], length = 4)
    xxr <- seq(from = xn[length(xn) - 4],
      to = xn[length(xn) - 3], length = 4)
```

```
    ggl <- mybb1(xn, xxl)
    ggr <- mybb1(xn, xxr)
    designl <- cbind(xxl, xxl^2, xxl^3)
    designr <- cbind(xxr, xxr^2, xxr^3)
#   (3)
    lsl <- lsfit(designl, ggl)
    lsr <- lsfit(designr, ggr)
    coefl <- t(lsl$coef[3:4,  ])
    coefr <- t(lsr$coef[3:4,  ])
#   (4)
    derl <- coefl %*% matrix(c(2, 6 * xn[4]), ncol = 1)
    derr <- coefr %*% matrix(c(2, 6 * xn[length(xn) - 3]),
      ncol = 1)
#   (5)
    gg <- mybb1(xn, xx)
#   (6)
    gga <- rbind(gg, t(derl), t(derr))
#   (7)
    beta1 <- solve(gga, rbind(matrix(yy, ncol = 1), 0, 0))
#   (8)
    gg2 <- mybb1(xn, ex)
#   (9)
    ey <- gg2 %*% matrix(beta1, ncol = 1)
#   (10)
    return(ey)
}
```

(1) The number of data is defined as nd. The positions of knots are set at xn.

(2) The interval between the fourth and the fifth knots from the left is divided equally into three regions, and the four coordinates that constitute the edges of the three regions are saved as xxl. The same procedure is carried out between the fourth and the fifth knots from the right. The resultant four coordinates are stored as xxr. The values of B-spline bases in these two regions are computed; they are named ggl and ggr. Furthermore, two design matrices corresponding to the four points in each region are defined as designl and designr.

(3) The matrices that consist of the regression coefficients of the cubic equations in the two regions near the two ends (defined in (2)) are saved as coefl and coefr; the cubic equations represent B-splines in the two regions.

(4) Using coefl and coefr, the derivatives of B-spline bases at the two ends of the region where data exist are listed in derl and derr, respectively.

(5) The values of B-spline bases at xx are calculated and are combined in gg.

(6) By taking into account the conditions on the second derivatives of the regression function at the two ends ($x = X_1, x = X_n$) (natural boundary conditions), gga is created using gg.

(7) The values of $\{\beta_j\}$ are obtained on the basis of the condition that the second derivatives at the two ends are 0; the results are saved as beta1.

(8) To calculate the estimates at ex, the value of B-spline is derived at each point of ex, and the results are stored as gg2.

(9) Estimates at ex are obtained using gg2, and they are named ey.

(10) The object ey is outputted.

An example of an object to use `myni1()` (construction of figure A.6 (page 479))

```
function()
{
#   (1)
    xx <- c(1, 2, 2.5, 3, 3.6, 4)
    yy <- c(2, 4, 3, 5, 2, 2.3)
#   (2)
    ex <- seq(from <- min(xx), to = max(xx), by = 0.02)
#   (3)
    ey <- myni1(xx, yy, ex)
#   (4)
    d2 <- NULL
    for(ii in 1:(length(xx) - 1)) {
        if(ii == (length(xx) - 1)) {
            x1 <- ex[ex >= xx[ii] & ex <= xx[ii + 1]]
            y1 <- ey[ex >= xx[ii] & ex <= xx[ii + 1]]
        }
        else {
            x1 <- ex[ex >= xx[ii] & ex < xx[ii + 1]]
            y1 <- ey[ex >= xx[ii] & ex < xx[ii + 1]]
        }
        des1 <- cbind(x1, x1^2, x1^3)
        ls1 <- lsfit(des1, y1)
        coef1 <- ls1$coef
        d2 <- c(d2, 2 * coef1[3] + 6 * coef1[4] * x1)
    }
#   (5)
    par(mfrow = c(1, 2), mai = c(2.5, 1.5, 2, 0.5),
      oma = c(1, 1, 1, 1))
    plot(ex, ey, ylim = c(min(ey), max(ey)), type = "n",
      xlab = "x", ylab = "y", cex = 1.1)
    points(xx, yy, pch = 15)
    lines(ex, ey)
#   (6)
    plot(ex, d2, ylim = c(min(d2), max(d2)), type = "n",
      xlab = "x", ylab = "der2 y", cex = 1.1)
    lines(ex, d2)
}
```

(1) The predictor values of data are defined as xx, and those of a target variable (object variable) are named yy.

(2) The positions of estimation points are stored as ex.

(3) Interpolation by natural spline using myni1() is performed; the obtained estimates are stored as ey.

(4) The second derivatives of the resulting natural spline are calculated, and they are saved as d2.

(5) The resultant natural spline is illustrated.

(6) The second derivative function of the derived natural spline is charted.

(D) Creation of natural-spline bases
Object: mynb1()

```
function(xn, ex)
{
#    (1)
    xn2 <- xn[4:(length(xn) - 3)]
    diag1 <- diag(length(xn2))
#    (2)
    nspline <- matrix(seq(0, length = length(xn2)
     * length(ex)), ncol = length(xn2))
#    (3)
    xxl <- seq(from = xn[4], to = xn[5], length = 4)
    xxr <- seq(from = xn[length(xn) - 4],
     to = xn[length(xn) - 3], length = 4)
    ggl <- mybb1(xn, xxl)
    ggr <- mybb1(xn, xxr)
    designl <- cbind(xxl, xxl^2, xxl^3)
    designr <- cbind(xxr, xxr^2, xxr^3)
    lsl <- lsfit(designl, ggl)
    lsr <- lsfit(designr, ggr)
    coefl <- t(lsl$coef[3:4,  ])
    coefr <- t(lsr$coef[3:4,  ])
    derl <- coefl %*% matrix(c(2, 6 * xn[4]), ncol = 1)
    derr <- coefr %*% matrix(c(2, 6 * xn[length(xn) - 3]),
     ncol = 1)
    gg <- mybb1(xn, xn2)
    gga <- rbind(gg, t(derl), t(derr))
    for(ii in 1:length(xn2)) {
        yunit <- diag1[ii,  ]
        beta1 <- solve(gga, rbind(matrix(yunit, ncol = 1), 0, 0))
        gg2 <- mybb1(xn, ex)
        eyunit <- gg2 %*% matrix(beta1, ncol = 1)
        nspline[, ii] <- eyunit
    }
#    (4)
    return(nspline)
}
```

(1) The positions of knots, which are obtained by omitting three knots at each end from the set of knots (xn), are stored as xn2. The identity matrix (the size is (the number of elements of xn2) × (the number of elements of xn2)) is defined as diag1.
(2) The object nspline is prepared for storing the calculated values of natural-spline bases at ex. Each column of nspline corresponds to each natural-spline basis.
(3) By the same process as that described from (2) to (9) of myni1() in (C), a natural spline that uses each row of diag1 as a target variable is derived; the result is saved as nspline.
(4) The object nspline is outputted.

The object mynb1() is used in a similar manner as that of mybb1() (figure A.7 (page 481)).

(E) Fitting of a natural spline by the least squares
Object: mynfit1()

```
function(xx, yy, xn, ex)
{
#    (1)
     nspline <- mynb1(xn, xx)
#    (2)
     beta2 <- solve(t(nspline) %*% nspline, t(nspline) %*% yy)
#    (3)
     nspline2 <- mynb1(xn, ex)
#    (4)
     ey <- nspline2 %*% beta2
#    (5)
     return(ey)
}
```

(1) The values of natural-spline bases at xx are calculated; the result is stored as nspline. The positions of knots for this natural spline are xn.
(2) Regression by the least squares is executed; the resultant regression coefficients ($\{\beta_j\}$) are named beta2.
(3) The values of natural-spline bases at ex are calculated when the knots remain at xn; the result is saved as nspline2.
(4) Estimates are calculated using beta2 and nspline2; the result is named ey.
(5) The object ey is outputted.

An example of an object to use mynfit1() (construction of figure A.8 (page 481))

```
function()
{
#    (1)
     set.seed(133)
     xx <- seq(from = 1, to = 4, by = 0.1)
     yy <- 0.5 * (xx - 1.5)^2 + cos(xx * 3)
        + rnorm(length(xx), mean = 0, sd = 0.4)
```

```
#   (2)
    xn <- c(1, 1, 1, 1, 2, 2.6, 3, 4, 4, 4, 4)
    ex <- seq(from = 1, to = 4, by = 0.05)
#   (3)
    ey <- mynfit1(xx, yy, xn, ex)
#   (4)
    d2 <- NULL
    for(ii in 4:(length(xn) - 4)) {
        if(ii == (length(xn) - 4)) {
            x1 <- ex[ex >= xn[ii] & ex <= xn[ii + 1]]
            y1 <- ey[ex >= xn[ii] & ex <= xn[ii + 1]]
        }
        else {
            x1 <- ex[ex >= xn[ii] & ex < xn[ii + 1]]
            y1 <- ey[ex >= xn[ii] & ex < xn[ii + 1]]
        }
        des1 <- cbind(x1, x1^2, x1^3)
        ls1 <- lsfit(des1, y1)
        coef1 <- ls1$coef
        d2 <- c(d2, 2 * coef1[3] + 6 * coef1[4] * x1)
    }
#   (5)
    par(mfrow = c(1, 2), mai = c(2.5, 1.5, 2, 0.5),
      oma = c(1, 1, 1, 1))
    plot(ex, ey, ylim = c(min(ey), max(ey)), type = "n",
      xlab = "x", ylab = "y", cex = 1.1)
    points(xx, yy, pch = 15)
    lines(ex, ey)
#   (6)
    plot(ex, d2, ylim = c(min(d2), max(d2)), type = "n",
      xlab = "x", ylab = "der2 y", cex = 1.1)
    lines(ex, d2)
}
```

(1) Simulation data are generated.

(2) The locations of knots are xn, and the positions of estimation points are ex.

(3) Using mynfit1(), a natural spline is fitted by the least squares; the result is stored as ey.

(4) The values of the second derivative function of the resultant natural spline are saved as d2.

(5) The natural spline obtained above is plotted.

(6) The second derivatives of the resultant natural spline are displayed.

(F) Creation of a matrix to construct a roughness penalty used in smoothing spline

Object: myphi1()

```
function(xx)
```

```
{
#   (1)
    xn <- c(rep(xx[1], length = 3), xx,
     rep(xx[length(xx)], length = 3))
#   (2)
    nd <- length(xn) - 6
#   (3)
    ex1 <- rep(0, length = nd * 3 - 2)
    for(ii in 4:(nd + 2)) {
        ex1[(ii * 3 - 11):(ii * 3 - 9)] <- c(xn[ii],
         (2 * xn[ii] + xn[ii + 1])/3, (xn[ii] + 2
         * xn[ii + 1])/3)
    }
    ex1[nd * 3 - 2] <- xn[length(xn)]
#   (4)
    bs1 <- mynb1(xn, ex1)
#   (5)
    coef2 <- matrix(rep(0, length = (nd - 1) * nd),
     ncol = nd)
    coef3 <- matrix(rep(0, length = (nd - 1) * nd),
     ncol = nd)
#   (6)
    for(jj in 1:nd) {
        bsa <- bs1[, jj]
        for(ii in 1:(nd - 1)) {
            xa <- ex1[(ii * 3 - 2):(ii * 3 + 1)]
            ya <- bsa[(ii * 3 - 2):(ii * 3 + 1)]
            data1 <- data.frame(x1 = xa, y1 = ya)
            lm1 <- lm(y1 ~ poly(x1, degree = 3),
             data = data1)
            lm1.poly <- poly(xa, degree = 3)
            coefa <- poly.transform(lm1.poly, coef(lm1))
            coef2[ii, jj] <- coefa[3]
            coef3[ii, jj] <- coefa[4]
        }
    }
#   (7)
    phi1 <- matrix(rep(0, length = nd * nd), ncol = nd)
    for(ii in 1:nd) {
        for(jj in 1:nd) {
            integ1 <- 0
            for(kk in 1:(nd - 1)) {
                integ1 <- integ1
                + 12 * coef3[kk, ii] * coef3[kk, jj] *
                (xn[kk + 4]^3 - xn[kk + 3]^3)
                + 6 * coef3[kk, ii] * coef2[kk, jj] *
                (xn[kk + 4]^2 - xn[kk + 3]^2)
```

```
            + 6 * coef2[kk, ii] * coef3[kk, jj] *
            (xn[kk + 4]^2 - xn[kk + 3]^2)
            + 4 * coef2[kk, ii] * coef2[kk, jj] *
            (xn[kk + 4] - xn[kk + 3])
        }
        phi1[ii, jj] <- integ1
      }
    }
#   (8)
    return(phi1)
}
```

(1) The positions of knots are named xn.
(2) Subtraction of 6 from the number of knots renders nd.
(3) The points obtained by bringing together knots and the points given by dividing the region between knots into three parts are saved as ex1.
(4) The values of B-spline bases at ex1 are computed by mynb1(), and the result is saved as bs1.
(5) The object coef2 is prepared for storing the coefficients of quadratic terms in the cubic equations corresponding to intervals between knots. The object coef3 is prepared to accommodate the coefficients of cubic terms in the cubic equations corresponding to intervals between knots.
(6) The objects coef2 and coef3 are calculated. In these objects, use of lm() and poly() reduces errors caused by numerical ill-conditions in the derivation of cubic equations.
(7) Φ (eq(A.22)(page 483)) is obtained and defined as phi1.
(8) The object phi1 is outputted.

An example of an object to obtain a hat matrix by the use of myphi1() (construction of figure A.9 (page 483))

```
function()
{
#   (1)
    nd <- 10
    xx <- seq(from = 1, to = nd, by = 1)
#   (2)
    phi1 <- myphi1(xx)
#   (3)
    lam <- 1
    hatmat1 <- solve((diag(nd) + lam * phi1))
#   (4)
    lam <- 10
    hatmat2 <- solve((diag(nd) + lam * phi1))
#   (5)
    par(mfrow = c(1, 2), mai = c(1.5, 1.5, 0.5, 0.5),
      oma = c(1, 1, 1, 1))
    persp(hatmat1, zlim = c(-0.1, 1), xlab = "i",
```

```
    ylab = "j", zlab = "HAT", lab = c(3, 3, 3))
  persp(hatmat2, zlim = c(-0.1, 1), xlab = "i",
    ylab = "j", zlab = "HAT", lab = c(3, 3, 3))
}
```

(1) The values of a predictor (xx) are set at $\{1, 2, \ldots, 10\}$.
(2) The object of Φ (eq(A.22) (page 483)) is calculated by myphi1(); the result is saved as phi1.
(3) The value of a smoothing parameter (λ) is set at 1. A hat matrix (eq(A.26)(page 483)) is obtained and stored as hatmat1.
(4) The value of a smoothing parameter (λ) is fixed at 10. A hat matrix (eq(A.26)) is computed and saved as hatmat2.
(5) The values of hatmat1 and hatmat2 are plotted.

(G) Derivation of smoothing spline using B-spline
Object: mysmbb1()

```
function(xx, yy, lam, ex)
{
#   (1)
    kk1fun <- function(xn)
    {
#   (2)
        nd <- length(xn) - 6
        ex1 <- NA
        for(ii in 4:(nd + 2)) {
            ex1[(ii * 3 - 11):(ii * 3 - 9)] <- c(xn[ii],
            (2 * xn[ii] + xn[ii + 1])/3,
            (xn[ii] + 2 * xn[ii + 1])/3)
        }
        ex1[nd * 3 - 2] <- xn[nd + 3]
#   (3)
        bs1 <- mybb1(xn, ex1)
#   (4)
        coef2 <- matrix(rep(0, length = (nd - 1) *
        (nd + 2)), ncol = nd + 2)
        coef3 <- matrix(rep(0, length = (nd - 1) *
        (nd + 2)), ncol = nd + 2)
        for(ii in 1:(nd - 1)) {
            for(jj in ii:(ii + 3)) {
                bsa <- bs1[, jj]
                xa <- ex1[(ii * 3 - 2):(ii * 3 + 1)]
                ya <- bsa[(ii * 3 - 2):(ii * 3 + 1)]
                data1 <- data.frame(x1 = xa, y1 = ya)
                lm1 <- lm(y1 ~ poly(x1, degree = 3),
                 data = data1)
                lm1.poly <- poly(xa, degree = 3)
                coefa <- poly.transform(lm1.poly, coef(lm1))
```

```
                coef2[ii, jj] <- coefa[3]
                coef3[ii, jj] <- coefa[4]
            }
        }
#   (5)
        kk1 <- matrix(rep(0, length = (nd + 2) *
          (nd + 2)), ncol = nd + 2)
        for(ii in 1:(nd + 2)) {
            for(jj in 1:(nd + 2)) {
                if(abs(ii - jj) <= 7) {
                    integ1 <- 0
                    for(kk in 1:(nd - 1)) {
                      if(abs(ii - jj) <= 3) {
                        integ1 <- integ1
                        + 12 * coef3[kk, ii] * coef3[kk, jj] *
                        (xn[kk + 4]^3 - xn[kk + 3]^3)
                        + 6 * coef3[kk, ii] * coef2[kk, jj] *
                        (xn[kk + 4]^2 - xn[kk + 3]^2)
                        + 6 * coef2[kk, ii] * coef3[kk, jj] *
                        (xn[kk + 4]^2 - xn[kk +3]^2)
                        + 4 * coef2[kk, ii] * coef2[kk, jj] *
                        (xn[kk + 4] - xn[kk + 3])
                      }
                    }
                    kk1[ii, jj] <- integ1
                }
            }
        }
#   (6)
        return(kk1)
    }
#   (7)
    xxu <- unique(xx)
    xn <- c(rep(xxu[1], length = 3), xxu,
      rep(xxu[length(xxu)], length = 3))
#   (8)
    kk1 <- kk1fun(xn)
#   (9)
    gg1 <- mybb1(xn, xx)
#   (10)
    beta1 <- solve((t(gg1) %*% gg1 + lam * kk1),
      t(gg1) %*% yy)
#   (11)
    gg2 <- mybb1(xn, ex)
    ey <- gg2 %*% beta1
#   (12)
    return(ey, beta1)
```

```
}
```

(1) A function (kk1fun()) is defined to obtain **K**; the elements of this matrix are given by eq(3.239) (page 180).

(2) The knots and the points obtained by dividing the neighboring knots into three regions are brought together, and they are saved as ex1.

(3) The values of B-spline bases at ex1 are derived by mybb1() and stored as bs1.

(4) The coefficients of the quadratic term of cubic equations between knots are obtained and saved as coef2. The coefficients of the cubic term of cubic equations between knots are computed and saved as coef3.

(5) The object **K** is derived; the result is stored as kk1.

(6) The variable kk1 is an output argument of kk1fun.

(7) Positions of knots (xn) are provided; multiple knots are avoided except the six knots at the two ends. By this strategy, this object can treat data including tied data.

(8) The object **K** computed by kk1fun() is stored as kk1.

(9) The values of B-spline bases are calculated by mybb1(), and the result is saved as gg1.

(10) The values of coefficients of a regression equation constructed by B-splines are obtained; the result is saved as beta1.

(11) The values of B-spline bases at ex are computed by mybb1(), and the result is stored as gg2. Estimates (ey) are computed by gg2.

(12) The objects ey and beta1 are outputted.

An example of an object to use mysmbb1() **(construction of figure A.10) (page 484))**

```
function()
{
#   (1)
    set.seed(84)
    nd <- 12
    xx <- runif(nd, min = 0, max = 1)
    xx <- sort(xx)
    yy <- sin(pi * xx) + rnorm(length(xx),
      mean = 0, sd = 0.3)
    ex <- seq(from = min(xx), to = max(xx),
      length = 100)
#   (2)
    lam <- 0.0001
    mysmbb1.out <- mysmbb1(xx, yy, lam, ex)
    ey <- mysmbb1.out$ey
    beta1 <- mysmbb1.out$beta1
#   (3)
    par(mfrow = c(1, 2), mai = c(1.5, 1.5, 0.5, 0.5),
      oma = c(6, 1, 6, 1))
    plot(xx, beta1[2:(nd + 1)], ylim = c(-0.5, 1.2),
      type = "n", xlab = "x", ylab = "y", cex = 1.05)
    points(xx, yy, pch = 15)
```

```
    points(xx[1], beta1[1], pch = 6, cex = 1.1)
    points(xx, beta1[2:(nd + 1)], pch = 0, cex = 1.1)
    points(xx[nd], beta1[nd + 2], pch = 2, cex = 1.1)
    lines(ex, ey)
#   (4)
    lam <- 0.01
    mysmbb1.out <- mysmbb1(xx, yy, lam, ex)
    ey <- mysmbb1.out$ey
    beta1 <- mysmbb1.out$beta1
#   (5)
    plot(xx, beta1[2:(nd + 1)], ylim = c(-0.5, 1.2),
      type = "n", xlab = "x", ylab = "y", cex = 1.05)
    points(xx, yy, pch = 15)
    points(xx[1], beta1[1], pch = 6, cex = 1.1)
    points(xx, beta1[2:(nd + 1)], pch = 0, cex = 1.1)
    points(xx[nd], beta1[nd + 2], pch = 2, cex = 1.1)
    lines(ex, ey)
}
```

(1) Simulation data (xx, yy) are generated. The values of coordinates of estimation points are defined as ex.

(2) The value of a smoothing parameter (λ) is fixed at 0.0001. The calculation of smoothing spline by mysmbb1() is carried out; the result is saved in mysmbb1.out. The values of $\{\beta_i\}$ are extracted from mysmbb1.out and they are saved as beta1. The estimates are extracted from the same object, and they are stored as ey.

(3) The values of xx, yy, beta1, and ey are plotted.

(4) The value of a smoothing parameter (λ) is set at 0.01, and the same procedure as described in (2) is performed.

(5) The values of xx, yy, beta1, and ey are illustrated.

REFERENCES

1. T.J. Hastie and R.J. Tibshirani (1990). *Generalized Additive Models*, Chapman & Hall/CRC.

2. T. Hastie, R. Tibshirani, and J.H. Friedman (2001). *The Elements of Statistical Learning: Data Mining, Inference, and Prediction*, Spinger-Verlag.

3. M.J. Powell (1981). *Approximation Theory and Methods*, Cambridge University Press.

APPENDIX B

R OBJECTS

B.1 INTRODUCTION

This chapter presents objects of "R"; their functions are almost equivalent to those of objects of S-Plus, which are shown in each chapter. The first objects listed in each item ((A), (B), . . .) of "Examples of S-Plus Object" are chiefly targeted. All R objects are verified to run on R1.7.1. The results of the two computer languages, however, can be somewhat different.

B.2 TRANSFORMATION OF S-PLUS OBJECTS IN CHAPTER 2

(A) Smoothing by the moving average
    An S-Plus object "move1()" is used as an R object as it is.

(B) Weights of the moving average
    An S-Plus object "moveh1()" is used as an R object as it is.

(C) Smoothing by the binomial filter

Introduction to Nonparametric Regression, Kunio Takezawa.
Copyright © 2006 John Wiley & Sons, Inc.

An S-Plus object "binom1()" is used as an R object as it is.

(D) Fitting of polynomials (polynomial functions) (a method for polynomial equation using lm())
The S-Plus object is used as an R object as it is.

(E) Fitting of a polynomial, and calculation of estimates (estimated values) which are placed at the positions where no data exist
The S-Plus object, in which the part of (2) is modified as below, is used as an R object.

```
#   (2)
    data1 <- data.frame(x = xx, y = yy)
```

(F) Fitting the spline function (a method of deriving the spline function represented by B-spline (B-splines) by applying lm())
The S-Plus object, in which library(splines) is added at the top, is used as an R object.

(G) Local linear regression for equispaced predictor
An S-Plus object "lline()" is used as an R object as it is.

(H) Smoothing spline (smoothing splines) for an equispaced predictor
The S-Plus object "smspe()" is used as an R object as it is.

(I) Calculation of eigenvalues (characteristic value) of a hat matrix corresponding to moving average
The S-Plus object is used as an R object as it is.

B.3   TRANSFORMATION OF S-PLUS OBJECTS IN CHAPTER 3

(A) Smoothing by using the Nadaraya-Watson estimator.
The S-Plus object, in which library(modreg) is added at the top, is used as an R object as it is.

(B) Calculation of CV and GCV of smoothing by Nadaraya-Watson estimator
The S-Plus object "kscv1()," in which library(modreg) is added at the top, is used as an R object.

(C) Calculation of CV and GCV in the case of smoothing by local linear regression
The S-Plus object "llcv1()" is used as an R object as it is.

(D) Smoothing by local linear regression
The S-Plus object "lll()" is used as an R object as it is.

(E) Interpolation using B-spline
The S-Plus object "intsp1()" is modified as follows to be used as an R object.

```
function(x1, y1, intkn, ex)
{
    library(splines)
#   (1)
    data1 <- data.frame(x = x1, y = y1)
    fit.lm <- lm(y ~ bs(x, knots = intkn, degree = 3),
    data = data1)
#   (2)
    data2 <- data.frame(x = ex)
    ey <- predict(fit.lm, newdata = data2)
#   (3)
    return(ey)
}
```

(F) Interpolation using natural spline

The S-Plus object "intns1()" is used as an R object after changing part (1) as below.

```
#   (1)
    exy <- spline(x1, y1, n = ne, method = "natural")
```

The S-Plus object "intns2()," in which library(modreg) is added at the top, is used as an R object.

The S-Plus object "intns3()" is modified as follows to be used as an R object.

```
function(x1, y1, ex)
{
    library(splines)
#   (1)
    data1 <- data.frame(x = x1, y = y1)
#   (2)
    kn1 <- x1[2:(length(x1) - 1)]
#   (3)
    fit.lm <- lm( y ~ ns(x, knots = kn1), data = data1)
#   (4)
    data2 <- data.frame(x = ex)
    ey <- predict(fit.lm, newdata = data2)
#   (5)
    return(ey)
}
```

(G) An object to realize the same calculation as dtpss (a subroutine of GCV-PACK) in S-Plus

The S-Plus object "dtpss()," in which part (1) is altered as follows, is used as an R object.

```
#   (1)
```

```
dyn.load("c:\\datasets\\gcvpack.dll")
```

(H) Smoothing using smoothing spline

R also includes an object for carrying out the calculation of smoothing spline by default: `smooth.spline()`. However, `sreg()`, which is included in the library "fields," is more useful because `sreg()` allows the direct designation of the value of the smoothing parameter. Note that `sreg()` calculates the value of GCV but does not calculate that of CV. The following object uses GCV to optimize the value of the smoothing parameter. Furthermore, the values of the smoothing parameter to be assigned in `sreg()` as `lam` = are identical to the value of the smoothing parameter in GCVPACK divided by the number of data points. Since the object below utilizes `sreg()`, the package "fields" must be installed beforehand.

```
function()
{
    library(fields)
#   (1)
    ll <- list(r1 = 0, r2 = 0)
    mm <- scan("c:\\datasets\\test31.csv", ll, sep = ",")
    xx <- mm$r1
    yy <- mm$r2
    nd <- length(xx)
#   (2)
    ntrial <- 46
    sparam <- 10^seq(from = -4, by = 0.1, length = ntrial)
    gcv <- rep(0, length = ntrial)
#   (3)
    for(k in c(1:ntrial)) {
#   (4)
        rlam <- sparam[k]/nd
        fit.sp <- sreg(xx, yy, lam = rlam)
#   (5)
        gcv[k] <- fit.sp$gcv.grid[1,"GCV"]
    }
#   (6)
    par(mfrow = c(1, 2), mai = c(2, 0.5, 2, 0.5),
      oma = c(1, 1, 1, 1))
    plot(log10(sparam), gcv, ylim = c(0.04, 0.2),
      xlab = "log10(lambda)", ylab = "GCV", type = "n")
    points(log10(sparam), gcv, pch = 3)
    lines(log10(sparam), gcv, lwd = 2)
    gcvmin <- min(gcv)
    igcvmin <- (1:ntrial)[gcv == gcvmin]
    sparacv <- sparam[igcvmin]
    points(log10(sparam[igcvmin]), gcvmin, pch = 15)
#   (7)
    rlam <- sparacv/nd
```

```
    fit.sp <- sreg(xx, yy, lam = rlam)
    ex <- seq(from = 0.1, to = 3, by = 0.02)
    ey <- predict(fit.sp, ex)
    plot(xx, yy, xlab = "x", ylab = "y", type = "n")
    points(xx, yy, pch = 5)
    lines(ex, ey, lwd = 2)
}
```

(l) Calculation of CV and GCV for LOESS

The S-Plus object "locv1()" is modified as follows to be used as an R object.

```
function(x1, y1, nd, span1, ntrial)
{
    library(modreg)
#   (1)
    locvgcv <- function(sp, x1, y1)
    {
        nd <- length(x1)
#   (2)
        data1 <- data.frame(xx1 = x1, yy1 = y1)
        fit.lo <- loess(yy1 ~ xx1, data = data1,
          span = sp, family = "gaussian", degree = 1)
        data1 = data.frame(xx1 = x1, yy1 = y1)
        fit.lo <- loess(yy1 ~ xx1, data = data1, span = sp,
          family = "gaussian", degree = 1,
          control=loess.control(surface = "direct"))
        res <- residuals(fit.lo)
#   (3)
        dhat2 <- function(x1, sp)
        {
            nd2 <- length(x1)
            diag1 <- diag(nd2)
            dhat <- rep(0, length = nd2)
#   (4)
            for(jj in 1:nd2) {
                y2 <- diag1[, jj]
                data1 <- data.frame(xx1 = x1, yy1 = y2)
                fit.lo <- loess(yy1 ~ xx1, data = data1,
                  span = sp, family = "gaussian", degree = 1,
                  control=loess.control(surface = "direct"))
                ey <- fitted.values(fit.lo)
                dhat[jj] <-  ey[jj]
            }
            return(dhat)
        }
#   (5)
        dhat <- dhat2(x1, sp)
```

```
            trhat <- sum(dhat)
            sse <- sum(res^2)
#   (6)
            cv <- sum((res/(1 - dhat))^2)/nd
            gcv <- sse/(nd * (1 - (trhat/nd))^2)
#   (7)
            return(cv, gcv)
        }
#   (8)
        cvgcv <- lapply(as.list(span1), locvgcv, x1 = x1,
          y1 = y1)
        cvgcv <- unlist(cvgcv)
        cv <- cvgcv[attr(cvgcv, "names") == "cv"]
        gcv <- cvgcv[attr(cvgcv, "names") == "gcv"]
#   (9)
        return(cv, gcv)
}
```

(J) Smoothing by supersmoother

The S-Plus object "zsu1()," in which `library(modreg)` is added at the top, is used as an R object.

(K) Smoothing by LOWESS (robust version of LOESS) and by LOESS

The following R object is used.

```
function()
{
    library(modreg)
#   (1)
    nd <- 100
    set.seed(123)
    xx <- seq(from = 1, by = 1, length = nd)
    yy <- sin(0.04 * pi * xx) + rnorm(nd, mean = 0, sd = 0.5)
    yy[30] <- 5
    yy[50] <- 10
    yy[70] <- 5
#   (2)
    ex <- seq(from = 1, to = nd, by = 0.1)
#   (3)
    data1 <- data.frame(x = xx, y = yy)
    data2 <- data.frame(x = ex)
#   (4)
    par(mfrow = c(2, 2), mai = c(1, 1, 0.5, 0.5),
      oma = c(2, 2, 1, 1))
#   (5)
    fit.low <- loess(y ~ x, data = data1, span = 0.1,
      family = "symmetric", degree=1, surface = "direct")
```

```
    ey <- predict.loess(fit.low, newdata = data2)
    plot(xx, yy, type = "n", xlab = "x",
     ylab = "y (LOWESS, span=0.1)")
    points(xx, yy, pch = 15, cex = 0.3)
    lines(ex, ey, lwd = 2)
#   (6)
    fit.low <- loess(y ~ x, data = data1, span = 0.2,
     family = "symmetric", degree=1, surface = "direct")
    ey <- predict.loess(fit.low, newdata = data2)
    plot(xx, yy, type = "n", xlab = "x",
     ylab = "y(LOWESS, span=0.2)")
    points(xx, yy, pch = 15, cex = 0.3)
    lines(ex, ey, lwd = 2)
#   (7)
    fit.lo <- loess(y ~ x, data = data1, span = 0.1,
     family = "gaussian", degree=1, surface = "direct")
    ey <- predict.loess(fit.lo, newdata = data2)
    plot(xx, yy, type = "n", xlab = "x",
     ylab = "y (LOESS, span=0.1)")
    points(xx, yy, pch = 15, cex = 0.3)
    lines(ex, ey, lwd = 2)
#   (8)
    fit.lo <- loess(y ~ x, data = data1, span = 0.2,
     family = "gaussian", degree=1, surface = "direct")
    ey <- predict.loess(fit.lo, newdata = data2)
    plot(xx, yy, type = "n", xlab = "x",
     ylab = "y, (LOESS, span=0.2)")
    points(xx, yy, pch = 15, cex = 0.3)
    lines(ex, ey, lwd = 2)
}
```

B.4   TRANSFORMATION OF S-PLUS OBJECTS IN CHAPTER 4

(A) Calculation of CV and GCV for local linear regression with two predictors
The S-Plus object "llin2dcv()" is used as an R object as it is.

(B) Local linear regression with two predictors
The S-Plus object "llin2d()" is used as an R object as it is. Note that the following object, for example, is employed to use llin2d(). In this example, thin plate smoothing splines accompanied by a very small value of the smoothing parameter (10^{-10}) are used, in place of Akima's method, as an example of interpolation. The package "fields" must be installed beforehand to use thin plate smoothing splines. As for Tps(), refer to (C).

```
function()
{
    library(fields)
```

```
#  (1)
   nd <- 100
   set.seed(2525)
   xx1 <- runif(nd, min = 0, max = 10)
   xx2 <- runif(nd, min = 0, max = 10)
   yy <- (xx1 - 4)^2 + (xx2 - 6)^2 + rnorm(nd,
    mean = 0, sd = 3)
#  (2)
   hx1 <- 0.7
   hx2 <- 0.8
#  (3)
   nex1 <- 11
   gridx1 <- seq(from = 0, to = 10, length = nex1)
   nex2 <- 21
   gridx2 <- seq(from = 0, to = 10, length = nex2)
   grid <- expand.grid(gridx1, gridx2)
#  (4)
   ey <- apply(grid, 1, llin2d, nd = nd, xx1 = xx1,
    xx2 = xx2, yy = yy, hx1 = hx1, hx2 = hx2)
#  (5)
   par(mfrow = c(1, 2), mai = c(2, 0.5, 2, 0.5),
    oma = c(1, 1, 1, 1))
   contour(gridx1, gridx2, matrix(ey, nrow = nex1),
    xlab = "x1", ylab = "x2")
#  (6)
   Tps.out <- Tps(cbind(xx1, xx2), yy, lambda=1e-10)
   Tps.pred <- predict(Tps.out, grid)
   contour(gridx1, gridx2, matrix(Tps.pred, nrow = nex1),
    xlab = "wx1", ylab = "x2")
}
```

(C) Smoothing using the thin plate smoothing splines for data with two predictors

The package "fields" for R includes Tps() to conduct calculations with thin plate smoothing splines. Therefore, installation of the package "fields" and the use of Tps() allow calculation that is almost equivalent to that by GCVPACK. Since Tps() optimizes the value of the smoothing parameter automatically, the following object extracts the values of GCV given by the values of the smoothing parameter ranging from 0.01 to 1 to display them. Furthermore, in a manner similar to sreg(), the smoothing parameter of Tps() is identical to that of GCVPACK divided by the number of data points.

In addition, the object dptss() can also be used in the same manner as that of Section B.3 (G).

```
function()
{
   library(fields)
```

```
#   (1)
    nobs <- 100
    set.seed(2525)
    xx1 <- runif(nobs, min = 0, max = 10)
    xx2 <- runif(nobs, min = 0, max = 10)
    yy <- (xx1 - 4)^2 + (xx2 - 6)^2 + rnorm(nobs, mean = 0,
      sd = 3)
#   (2)
    des <- cbind(xx1, xx2)
#   (3)
    nex1 <- 11
    gridx1 <- seq(from = 0, to = 10, length = nex1)
    nex2 <- 21
    gridx2 <- seq(from = 0, to = 10, length = nex2)
    grid <- expand.grid(gridx1, gridx2)
#   (4)(5)(6)
    Tps.out <- Tps(des, yy)
#   (7)(8)(9)(10)(11)
    par(mfrow = c(1, 2), mai = c(2, 0.5, 2, 0.5),
      oma = c(1, 1, 1, 1))
    spara <- Tps.out$gcv.grid[,"lambda"] * nobs
    gcv <- Tps.out$gcv.grid[,"GCV"]
    choice1 <- spara > 0.01 & spara < 1
    spara <- spara[choice1]
    gcv <- gcv[choice1]
    plot(log10(spara), gcv, type = "n", xlab =
      "log10(smoothing parameter)", ylab = "GCV")
    points(log10(spara), gcv, pch = 3)
    lines(log10(spara), gcv, lwd = 2)
    gcvmin <- min(gcv)
    sparagcv <- spara[which(gcvmin==gcv)]
    points(log10(sparagcv), gcvmin, pch = 15, cex = 1.2)
#   (12)
    Tps.pred <- predict(Tps.out, grid)
    contour(gridx1, gridx2, matrix(Tps.pred, nrow = nex1),
      xlab = "x1", ylab = "x2")
}
```

(D) Smoothing by LOESS with two predictors

The following R object is used.

```
function()
{
    library(modreg)
#   (1)
    nd <- 100
    set.seed(2525)
```

```
      xx1 <- runif(nd, min = 0, max = 10)
      xx2 <- runif(nd, min = 0, max = 10)
      yy <- (xx1 - 4)^2 + (xx2 - 6)^2 + rnorm(nd,
        mean = 0, sd = 3)
#   (2)
      nex1 <- 11
      nex2 <- 11
      gridx1 <- seq(from = 0, by = 1, length = nex1)
      gridx2 <- seq(from = 0, by = 1, length = nex2)
      grid <- expand.grid(gridx1, gridx2)
      data1 <- data.frame(x1 = xx1, x2 = xx2, y1 = yy)
#   (3)
      par(mfrow = c(1, 2),  mai = c(2, 0.8, 2, 0.1),
        oma = c(1, 1, 1, 1))
#   (4)
      sp1 <- 0.1
      fit.lo <- loess(y1 ~ x1 + x2, data = data1,
        span = sp1, family = "gaussian", degree = 1,
        control=loess.control(surface = "direct"))
      data2 <- data.frame(x1 = grid[, 1], x2 = grid[, 2])
      ey <- predict.loess(fit.lo, newdata = data2)
#   (5)
      grid.es <- NULL
      grid.es$x <- gridx1
      grid.es$y <- gridx2
      grid.es$z <- matrix(ey, nrow = nex1)
      contour(grid.es, xlab = "x1", ylab = "x2",
        zlim=c(0,90))
#   (6)
      sp1 <- 0.2
      fit.lo <- loess(y1 ~ x1 + x2, data = data1,
        span = sp1, family = "gaussian", degree = 1,
        control=loess.control(surface = "direct"))
      data2 <- data.frame(x1 = grid[, 1], x2 = grid[, 2])
      ey <- predict.loess(fit.lo, newdata = data2)
      grid.es <- NULL
      grid.es$x <- gridx1
      grid.es$y <- gridx2
      grid.es$z <- matrix(ey, nrow = nex1)
      contour(grid.es, xlab = "x1", ylab = "x2",
        zlim=c(0,90))
      invisible()
}
```

(E) Ordinary kriging with one predictor

Part (1) of the S-Plus object "mykrig1()" is altered as follows for use as an R object.

```
#    (1)
     dist1 <- rdist(xx)
     cc <- exp( - dist1^2/bw^2)
```

(F) Smoothing by simple kriging with one predictor

Part (1) of the S-Plus object "mykrigs()" is altered as follows for use as an R object.

```
#    (1)
     dist1 <- rdist(xx)
     cc <- exp( - dist1^2/bw^2)
```

(G) Universal kriging with one predictor (a rough trend is represented by a polynomial equation)

Part (1) of the S-Plus object "mykrig2()" is altered as follows for use as an R object.

```
#    (1)
     dist1 <- rdist(xx)
     cc <- exp( - dist1^2/bw^2)
```

(H) Universal kriging with two predictors (a rough trend is represented by a polynomial equation)

The following R object is used. Since this object utilizes Krig(), the package "fields" must be installed beforehand. Krig() performs various calculations such as kriging accompanied by smoothing and thin plate smoothing splines.

```
function()
{
#    (1)
     library(fields)
#    (2)
     xx1 <- c(1, 2, 3, 4, 5, 6, 7, 8, 9, 10)
     xx2 <- c(7, 5, 4, 2, 9, 5, 3, 4, 2, 9)
     yy <- c(2, 7, 4, 3, 5, 2, 9, 4, 1, 8)
#    (3)
     par(mfrow = c(2, 2), mai = c(0.5, 0.5, 0.25, 0.25),
       oma = c(1, 1, 1, 1))
#    (4)
#    (5)
     g.cov <- function(x1, x2, theta){exp(-(rdist(x1,
       x2)/theta)**2)}
     krig1 <- Krig(cbind(xx1, xx2), yy, cov.function=g.cov,
       theta=0.2, lambda=1e-10, m=3)
#    (6)
     ex1 <- seq(1, 10, by = 0.5)
     ex2 <- seq(1, 10, by = 0.5)
```

```
    grid1 <- expand.grid(ex1, ex2)
    ey <- predict(krig1, grid1)
    eymat <- matrix(ey, ncol=19)
    persp(x = ex1, y = ex2, z=eymat, zlim = c(0, 12),
    xlab = "x1", ylab = "x2", zlab = "y", r=10,
    theta=-30, phi=30, lab = c(3, 3, 3))
#   (7)
    krig1 <- Krig(cbind(xx1, xx2), yy, cov.function=g.cov,
      theta=0.5, lambda=1e-10, m=3)
    ey <- predict(krig1, grid1)
    eymat <- matrix(ey, ncol=19)
    persp(x = ex1, y = ex2, z=eymat, zlim = c(0, 12),
     xlab = "x1", ylab = "x2", zlab = "y",r=10,
     theta=-30, phi=30, lab = c(3, 3, 3))
    krig1 <- Krig(cbind(xx1, xx2), yy, cov.function=g.cov,
      theta=1, lambda=1e-10, m=3)
    ey <- predict(krig1, grid1)
    eymat <- matrix(ey, ncol=19)
    persp(x = ex1, y = ex2, z=eymat, zlim = c(0, 12),
     xlab = "x1", ylab = "x2", zlab = "y", r=10,
     theta=-30, phi=30, lab = c(3, 3, 3))
    krig1 <- Krig(cbind(xx1, xx2), yy, cov.function=g.cov,
      theta=2, lambda=1e-10, m=3)
    ey <- predict(krig1, grid1)
    eymat <- matrix(ey, ncol=19)
    persp(x = ex1, y = ex2, z=eymat, zlim = c(0, 12),
     xlab = "x1", ylab = "x2", zlab = "y", r=10, theta=-30,
     phi=30, lab = c(3, 3, 3))
    rm(g.cov, pos =1)
}
```

(l) Universal kriging with two predictors (a rough trend is represented by a polynomial equation. The broadness of the theoretical correlogram (correlogram model) is determined by reference to data)

The following R object is used. The values of an empirical correlogram and those of an empirical variogram supplied by this R object are slightly different from those given by the corresponding S-Plus objects.

```
function()
{
#   (1)
    library(fields)
#   (2)
    dd <- 0.8
#   (3)
    nd <- 100
    set.seed(199)
    xx1 <- runif(nd, min = 0, max = 10)
```

```
    xx2 <- runif(nd, min = 0, max = 10)
    yy <- 8 + 5 * sin(xx1^2/10) + 3 * cos(xx2 * 2) +
     3 * sin(xx2 * 2.5) + xx2 * 0.6
#  (4)
    par(mfrow = c(2, 2), mai = c(0.8, 0.8, 0.5, 0.5),
     oma = c(1, 1, 1, 1))
#  (5)
    g.cov  <- function(x1, x2, theta){exp(-(rdist(x1,
     x2)/theta)**2)}
    krig1 <- Krig(cbind(xx1, xx2), yy, cov.function=g.cov,
     theta=dd, lambda=1e-10, m=2)
#  (6)
    corvar1 <- function(x1, x2, y1, krig1, t1, dt, nt){
        ey <- krig1$fitted.values.null
        eps1 <- y1 - ey
        dis1 <- rdist(cbind(x1, x2))
        nd <- length(x1)
        correl1t <- rep(0, length = nt)
        vario1t <- rep(0, length = nt)
        for(kk in 1:nt){
            choice1 <- (t1 + dt * (kk - 1)<= dis1) &
             (dis1 <  t1 + dt * kk)
            ipos1 <- which(choice1==T) %% nd
            ipos1[ipos1 == 0] <- nd
            ipos2 <- ceiling(which(choice1==T)/nd)
            if(length(ipos1) == 0){
                correl1t[kk] <- 0
                vario1t[kk] <- 0
            }
            else {
                correl1t[kk] <- cor(eps1[ipos1], eps1[ipos2])
                vario1t[kk] <- var(eps1[ipos1]) * (1-
                 correl1t[kk])
            }
        }
        return(correl1t, vario1t)
    }
    t1 <- 0.0
    dt <- 0.3380719
    nt <- 35
    corvar1 <- corvar1(xx1, xx2, yy, krig1, t1, dt, nt)
    cor1 <- corvar1$correl1t
    var1 <- corvar1$vario1t
    xax1 <- seq(t1, t1+ dt * (nt-1), length = nt)
    plot(xax1, cor1, type = "n", xlab = "distance",
     ylab = "cor")
    points(xax1, cor1, pch = 15, cex = 0.8)
```

```
#   (7)
    rr <- seq(from = 0, to = 12, by = 0.2)
    lines(rr, exp(- (rr/dd)^2), lwd = 3)
#   (8)
    plot(xax1, var1, type = "n", xlab = "distance",
     ylab = "var")
    points(xax1, var1, pch = 15, cex = 0.8)
#   (9)
    ex1 <- seq(from = 1, to = 10, by = 0.5)
    ex2 <- seq(from = 1, to = 10, by = 0.5)
    gridex1 <- expand.grid(ex1, ex2)
    ey <- predict(krig1, gridex1)
    persp(ex1, ex2, matrix(ey, ncol=19), zlim = c(0, 18),
     xlab = "x1", ylab = "x2", zlab = "y",   r=10,
     theta=-30, phi=30,lab = c(3, 3, 3))
    rm(g.cov, pos =1)
}
```

(J) Derivation of an additive model by solving a normal equation using the Gauss-Seidel method (polynomial equations are used as equations ($m_1(x_1)$, $m_2(x_2)$) to transform predictors)

The S-Plus object is used as an R object as it is.

(K) Additive model given by solving a normal equation using the Gauss-Seidel method (a smoothing spline is employed for transforming each predictor)

The following R object is used. Since this object uses bruto(), the package "mda" must be installed beforehand. The object bruto(), if not specified otherwise, optimizes the value of the smoothing parameter using GCV. The values used for the smoothing parameter are outputted as fit.bruto$lambda in the following example.

Object: sadd1()

```
function(nd, ne, xx1, xx2, yy, ex1, ex2, ey1)
{
    library(mda)
#   (1)
#   (2)
    fit.bruto <- bruto(x = cbind(xx1, xx2), y = yy)
#   (3)
    fit.tmd <- predict.bruto(fit.bruto, x = cbind(xx1, xx2),
     type = "terms")
#   (4)
    eyd1 <- fit.tmd$xx1$y
    eyd2 <- fit.tmd$xx2$y
#   (5)
    data3 <- data.frame(x1 = ex1, x2 = ex2)
    fit.tme <- predict.bruto(fit.bruto, x = cbind(ex1,ex2),
     type = "terms")
```

```
#   (6)
    ey1 <- fit.tme$ex1$y
    ey2 <- fit.tme$ex2$y
#   (7)
    return(eyd1, eyd2, ey1, ey2)
}
```

(L) Derivation of regression equation in the form of ACE (each regression equation is a polynomial equation) using canonical correlation analysis

The S-Plus object "acecan1()," in which library(mva) is added at the top, is used as an R object.

(M) Derivation of a regression equation in the form of ACE by solving an eigenvalue problem using iterative calculation (each regression equation is a polynomial equation)

The S-Plus object "aceit1()" is used as an R object as it is.

(N) Derivation of a regression equation in the form of ACE using ACE algorithm (each regression equation is a polynomial equation)

The S-Plus object "aceit2()" is altered as follows for use as an R object.

```
function(nd, it, npx1, npx2, npy, xx1, xx2, yy){
#   (1)
    yyst <- (yy - mean(yy))/sqrt(sum((yy - mean(yy))^2)/nd)
#   (2)
    for(ii in 1:it) {
#   (3)
        data1 <- data.frame(x1 = xx1, x2 = xx2, y = yyst)
        fit.lm <- lm(yyst ~ poly(xx1, degree = npx1) +
         poly(xx2, degree = npx2), data = data1)
        yhat <- fitted.values(fit.lm)
#   (4)
        data2 <- data.frame(x = yy, y = yhat)
        fit.lm <- lm(yhat ~ poly(yy, degree = npy), data = data2)
        eta <- fitted.values(fit.lm)
#   (5)
        yyst <- eta/sqrt(sum((eta - mean(eta))^2)/nd)
        print(sqrt(sum((yyst - mean(yyst))^2)/nd))
    }
#   (6)
    return(yyst)
}
```

(O) Derivation of a regression equation by the use of ACE based on a super-smoother

After the package "acepack" is installed and library(acepack) is added at the top of the S-Plus object, the S-Plus object is used as an R object.

(P) Derivation of a regression equation by projection pursuit regression

The following R object is used.

```
function()
{
    library(modreg)
#   (1)
    nd <- 100
    set.seed(100)
    xx1 <- runif(nd, min = 0, max = 10)
    xx2 <- runif(nd, min = 0, max = 10)
    yy <- 2 + cos(xx1 * 0.6 + xx2 * 0.4) +
      rnorm(nd, mean = 0, sd = 0.1)
#   (2)
    xmat <- cbind(xx1, xx2)
    fit.pro <- ppr(xmat, yy, nterm = 1, min.term = 1,
      max.term = 7)
    print(fit.pro$gofn)
#   (3)
    nex1 <- 21
    nex2 <- 21
    ex1 <- seq(from = 0, by = 0.5, length = nex1)
    ex2 <- seq(from = 0, by = 0.5, length = nex2)
    grid <- expand.grid(ex1, ex2)
    grid12 <- cbind(grid[, 1], grid[, 2])
#   (4)
    fit.pro <- ppr(xmat, yy, nterm=2, min.term = 1,
      max.term = 7, xpred = grid12)
    ey1 <- matrix(predict(fit.pro, grid12), ncol = nex2)
#   (5)
    ff1x <- xx1 * fit.pro$alpha[1, 1] + xx2
      * fit.pro$alpha[2, 1]
    ff1y <- fit.pro$fitted.values
#   (6)
    exy <- NULL
    exy$x <- ex1
    exy$y <- ex2
    exy$z <- ey1
#   (7)
    par(mfrow = c(1, 2), mai = c(0.5, 0.5, 0.5, 0.5),
      oma = c(8, 1, 8, 1))
    plot(ff1x, ff1y, type = "n", xlab = "z", ylab = "m1(z)")
    points(ff1x, ff1y, pch = 15)
#   (8)
    persp(x = exy$x, y = exy$y, z = exy$z, xlab = "x1",
      ylab = "x2", zlab = "y", r=10, theta=-30,
      phi=30, lab = c(3, 3, 1))
    invisible()
```

}

B.5   TRANSFORMATION OF S-PLUS OBJECTS IN CHAPTER 5

(A) An object for realizing the same calculation as dsnsm (a subroutine of GCVPACK) in S-Plus

The S-Plus object "dsnsm()," in which part (1) is altered as follows, is used as an R object.

```
#   (1)
    dyn.load("c:\\datasets\\gcvpack.dll")
```

(B) Object for carrying out nonparametric regression with predictors represented as distributions using "dsnsm()"

The S-Plus object "distnon()" can be used as an R object as it is.

(C) Object for conducting nonparametric regression to create an additive model with predictors represented as distributions using "dsnsm()"

The S-Plus object "distadd()" can be used as an R object as it is.

B.6   TRANSFORMATION OF S-PLUS OBJECTS IN CHAPTER 6

(A) Smoothing of a histogram using a local constant (the bandwidth is optimized by cross-validated deviance)

The S-Plus object "histcv1()" is used as an R object as it is.

(B) Smoothing of a histogram using local linear equations given by generalized linear regression; the cross-validated deviance optimizes the bandwidth

Parts (8) and (11) of the S-Plus object "histcv2()" are altered for use as an R object.

```
#   (8)
                data1 <- data.frame(x = xxdel, y = yydel, wt = dwt)
                fit.glm <- glm(y ~ x, data = data1,
                 family = poisson, weight = wt, control =
                 glm.control(maxit = 50))
                py <- c(py, exp(fit.glm$coef[1] +
                 fit.glm$coef[2] * xx[ii]))
            }

#   (11)
    esthist2 <- function(ex1, xx1, yy1, band1)
    {
        dwt <- exp(-0.5 * ((xx1 - ex1)/band1)^2)
        data2 <- data.frame(x = xx1, y = yy1, wt = dwt)
        fit.glm <- glm(y ~ x, data = data2, family = poisson,
```

```
        weight = wt, control = glm.control(maxit = 50))
    ey <- exp(fit.glm$coef[1] + fit.glm$coef[2] * ex1)
    return(ey)
}
ey <- apply(matrix(ex, ncol = 1), 1, esthist2,
 xx1 = xx, yy1 = yy, band1 = bandwdev)
```

(C) Smoothing a histogram using a local linear equation given by generalized linear regression; the bandwidth is optimized using $AICc$

The S-Plus object "histaicc1()" is altered as follows for use as an R object.

```
function(xx, ex, bandw, ntrial, ydata, br)
{
#   (1)
    hist1 <- hist(ydata, breaks = br, plot = F)
#   (2)
    nb <- length(xx)
    yy <- hist1$counts
#   (3)
    aicct <- rep(0, length = ntrial)
#   (4)
    for(kk in 1:ntrial) {
        ey <- rep(0, length = nb)
        dhat <- rep(0, length = nb)
        hh <- bandw[kk]
        diag1 <- diag(nb)
#   (5)
        for(ii in 1:nb) {
            dwt <- exp(-0.5 * ((xx - xx[ii])/hh)^2)
            data1 <- data.frame(x = xx, y = yy, wt = dwt)
            fit.glm <- glm(y ~ x, data = data1, family =
             poisson, weight = wt, control = glm.control
             (maxit = 200))
#   (6)
            ey[ii] <- fit.glm$fitted.values[ii]
            yyd <- diag1[, ii]
            data1 <- data.frame(x = xx, y = yyd, wt = dwt)
            fit.glm <- glm(y ~ x, data = data1, family =
             poisson, weight = wt, control = glm.control
             (maxit = 200))
            hatii <- fit.glm$fitted.values[ii]
            dhat[ii] <- hatii
        }
#   (7)
    aicct[kk] <-  log(2 * sum(yy *
     ifelse(is.finite(log(yy/ey)), log(yy/ey),
     -1e+100))) + (nb + sum(dhat))/(nb - sum(dhat) - 2)
```

```
    }
#  (8)
    aiccmin <- min(aicct)
    iaiccmin <- (1:ntrial)[aicct == aiccmin]
    bandwaicc <- bandw[iaiccmin]
#  (9)
    esthist2 <- function(ex1, xx1, yy1, band1)
    {
        dwt <- exp(-0.5 * ((xx1 - ex1)/band1)^2)
        data2 <- data.frame(x = xx1, y = yy1, wt = dwt)
        fit.glm <- glm(y ~ x, data = data2,
          family = poisson, weight = wt,
          control = glm.control(maxit = 200))
        ey <- exp(fit.glm$coef[1] + fit.glm$coef[2] * ex1)
        return(ey)
    }
    ey <- apply(matrix(ex, ncol = 1), 1, esthist2,
      xx1 = xx, yy1 = yy, band1 = bandwaicc)
#  (10)
    return(aicct, iaiccmin, yy, ey)
}
```

(D) Smoothing a histogram using generalized nonparametric regression based on LOESS; $AICc$ optimizes the bandwidth

The S-Plus object "histgam1()" is altered as follows for use as an R object. Note that since gam(), included in the library "mgcv," is the object carrying out generalized nonparametric regression based on a smoothing spline, the functions of the object are considerably different from those of gam().

```
function(xx, ex, sp1, ntrial,ydata, br)
{
#  (1)
    library(mgcv)
#  (2)
    hist1 <- hist(ydata, breaks = br, plot = F)
#  (3)
    nb <- length(br) - 1
    yy <- hist1$counts
    data1 <- data.frame(x = xx, y = yy)
#  (4)
    kn1 <- seq(from = br[1], to = br[length(br)], length = 20)
#  (5)
    aicct <- rep(0, length = ntrial)
#  (6)
    for(jj in 1:ntrial) {
        sp2 <- sp1[jj]
        fit.gam <- gam(y ~ s(x, k = length(kn1)),
```

```
                 data = data1, family = "poisson",
                 sp = sp2, knots = list(x = kn1) )
                 dev1 <- fit.gam$deviance
                 df1 <- fit.gam$edf
                 aicct[jj] <- log(dev1) + (1 + df1/nb)/(1 - df1/nb
                 - 2/nb)
        }
#   (7)
        iaiccmin <- which(aicct == min(aicct))
        data2 <- data.frame(x = ex)
        ey <- predict.gam(fit.gam, data2, type = "response",
        sp = aicct(iaiccmin) )
#   (8)
        return(aicct, iaiccmin, yy, ey)
}
```

(1) The library "mgcv" is set out for use.

(2) A histogram is created using `ydata`; the result is saved in `hist1`.

(3) The number of bins is defined as `nb`. The frequencies are stored as `yy`. The data used for constructing a regression equation are saved in a data frame `data1`.

(4) The knots (breakpoints) (including two endpoints) of a spline function are identified as `kn1`. In this example, the number of knots is set to divide the histogram equally into 20 regions.

(5) The object `aicct` is prepared to save the values of $AICc$.

(6) Generalized additive models based on a smoothing spline are created. The elements of `sp1` (a vector) are used as the values of the smoothing parameter. The results lead to the values of $AICc$.

(7) The smooothing of a histogram is carried out using a smoothing parameter that minimizes the value of $AICc$; the result is stored in `fit.gam`.

(8) The objects `aicct`, `iaiccmin`, `yy`, and `ey` are outputted.

This object results in figure B.1. This is the result of generalized nonparametric regression based on smoothing spline instead of LOESS; the number of knots including the two ends is fixed at 20. Nevertheless, the resulting estimates do not differ significantly from those shown in figure 6.10(right) (page 377).

Furthermore, in this example, `gam()` of R is used to calculate the values of estimates, deviance (`fit.gam$deviance`), and the degrees of freedom (`fit.gam$edf`). However, the use of `gam()` without assigning the value of the smoothing parameter optimizes the smoothing parameter using GCV or other statistics.

(E) Smoothing of a histogram using a kernel estimator
The S-Plus object is used as an R object as it is.

(F) Nonparametric probability density function using a local constant; the likelihood cross-validation optimizes the bandwidth
Part (3) of S-Plus object "`const1()`" is altered as follows to be used as an R object.

```
#   (3)
        r0a <- r0
```

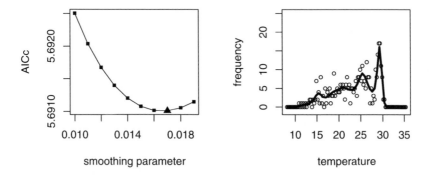

Figure B.1   Smoothing of a histogram, which is the same as that in figure 6.2. Optimizing the smoothing parameter using $AICc$ (left); the result of smoothing given by the optimal span (sp= 0.017) (right).

```
        if(r0a < exs - 10 * hh) r0a <- exs - 10 * hh
        r1a <- r1
        if(r1a > exs + 10 * hh) r1a <- exs + 10 * hh
        rr <- exs
        w0 <- integrate(wt, lower = r0a, upper = r1a,
          subdivisions=200)$value
        th0 <- sum(wt(xx))/w0/length(xx)
        return(th0)
    }
```

(G) Nonparametric probability density function using a local linear equation
The S-Plus object "lldens1()" is altered as follows for use as an R object.

```
function(nd, r0, r1, hh, xdata, ne, ex)
{
#  (1)
    estlldens1 <- function(exs, xx, hh, r0, r1, nd, th0h, th1h)
    {
#  (2)
        rrf <- exs
        hhf <- hh
        r0f <- r0
        r1f <- r1
        ndf <- nd
        xif <- xx
#  (3)
        formula1 <- function(th01)
        {
```

```
             th0 <- th01[1]
             th1 <- th01[2]
             term1 <- function(th0, th1)
             {
                 integrand1 <- function(uu, th0, th1)
                 {
                   f <- exp(-0.5 * ((rrf - uu)/hhf)^2) *
                    exp(th0 + th1 * uu)
                   return(f)
                 }
                 r0a <- r0f
                 if(r0a < rrf - 10 * hhf) r0a <- rrf - 10 * hhf
                 r1a <- r1f
                 if(r1a > rrf + 10 * hhf) r1a <- rrf + 10 * hhf
                 f2 <- integrate(integrand1, lower = r0a,
                  upper = r1a, th0 = th0, th1 = th1)$value
                 return(f2)
             }
# (4)
             term2 <- function(th0, th1)
             {
                 ss <- sum(exp(-0.5 * ((rrf - xif)/hhf)^2) *
                 (th0 + th1 * xif))
                 return(ss)
             }
# (5)
             val1 <- term1(th0, th1) - term2(th0, th1)/ndf
             return(val1)
         }
# (6)
             grad1 <- function(th01)
             {
                 th0 <- th01[1]
                 th1 <- th01[2]
# (7)
                 integrand2 <- function(uu, th0, th1)
                 {
                   f <- exp(-0.5 * ((rrf - uu)/hhf)^2) *
                    exp(th0 + th1 * uu)
                   return(f)
                 }
                 r0a <- r0f
                 if(r0a < rrf - 10 * hhf) r0a <- rrf - 10 * hhf
                 r1a <- r1f
                 if(r1a > rrf + 10 * hhf) r1a <- rrf + 10 * hhf
                 t1 <- integrate(integrand2, lower = r0a,
                  upper = r1a, th0 = th0, th1 = th1)$value
```

```
                  t2 <- sum(exp(-0.5 * ((rrf - xif)/hhf)^2))
                  dx <- t1 - t2/ndf
#   (8)
                  integrand3 <- function(uu, th0, th1)
                  {
                    f <- exp(-0.5 * ((rrf - uu)/hhf)^2) *
                    uu * exp(th0 + th1 * uu)
                    return(f)
                  }
                  r0a <- r0f
                  if(r0a < rrf - 10 * hhf) r0a <- rrf - 10 * hhf
                  r1a <- r1f
                  if(r1a > rrf + 10 * hhf) r1a <- rrf + 10 * hhf
                  t1 <- integrate(integrand3, lower = r0a,
                   upper = r1a, th0 = th0, th1 = th1)$value
                  t2 <- sum(exp(-0.5 * ((rrf - xif)/hhf)^2) * xif)
                  dy <- t1 - t2/ndf
#   (9)
                  md1 <- matrix(c(dx, dy), ncol = 2)
                  return(md1)
              }
#   (10)(11)(12)(13)(14)(15)(16)(17)
        start1 <- list(th0 = th0h, th1 = th1h)
        th01h <- c(th0h, th1h)
        optout <- optim(th01h, formula1, gr = grad1)
#   (18)
        coef1 <- optout$par
        th0h <- coef1[1]
        th1h <- coef1[2]
        eys <- exp(th0h + th1h * rrf)
        return(eys)
    }
#   (19)
    th0h <- 0.1
    th1h <- 0.1
    ey <- apply(matrix(ex, nrow = 1), 2, estlldens1,
     xx = xdata, hh = hh, r0 = r0, r1 = r1, nd = nd,
     th0h = th0h, th1h = th1h)
#   (20)
    return(ey)
}
```

B.7   TRANSFORMATION OF S-PLUS OBJECTS IN CHAPTER 7

(A) Fitting of a normal distribution with two predictors

In the S-Plus object "ndens1()," `cov1 <- (nd - 1)/nd * var(xx, na.method = "omit")` is replaced by `cov1 <- (nd - 1)/nd * var(xx, na.rm = T)`; then, a corresponding R object is obtained.

(B) Classification by linear discriminant rule

The S-Plus object "mylda2()" is used as an R object as it is.

(C) Classification by the use of the quadratic discriminant rule

The S-Plus object "myqda2()" is used as an R object as it is.

(D) Calculation of nonparametric probability density function

The S-Plus object "nond1()" is altered as follows for use as an R object. Note that the bandwidth must be one-fourth that used in the S-Plus object "nond1()." For example, when the bandwidth in the S-Plus object "nond1()" is 8, the same result is obtained by setting it as 2 in this object.

```
function(xx, x1min, x1max, x2min, x2max, ne, band1)
{
#   (1)
    library(sm)
#   (2)
    x1 <- xx[, 1]
    x2 <- xx[, 2]
#   (3)
    eval1 <- seq(from = x1min, to = x1max, length = ne)
    eval2 <- seq(from = x2min, to = x2max, length = ne)
    eval12 <- expand.grid(eval1, eval2)
    result1 <- sm.density(cbind(x1, x2), h = band1,
     n = ne, display = "none", eval.points = eval12)
#   (4)
    ex1 <- eval1
    ex2 <- eval2
    ey <- result1$estimate
#   (5)
    return(ex1, ex2, ey)
}
```

(E) Classification by nonparametric probability density function

The S-Plus object "nondis1()" is altered as follows for use as an R object. As in the case of "nond1()" in (D), the bandwidth must be one-fourth that for the S-Plus object "nondis1()" to derive the same result as that given by the S-Plus object "nondis1()."

```
function(xxall, clall, x1min, x1max, x2min, x2max, ne,
 band1, prior1)
{
#   (1)
    library(sm)
```

```
#   (2)
    clu <- unique(clall)
    ncl <- length(clu)
#   (3)
    eyarray <- array(ne * ne * ncl, c(ne, ne, ncl))
    for(kk in 1:ncl) {
        xx <- xxall[clall == clu[kk],  ]
        x1 <- xx[, 1]
        x2 <- xx[, 2]
        eval1 <- seq(from = x1min, to = x1max, length = ne)
        eval2 <- seq(from = x2min, to = x2max, length = ne)
        eval12 <- expand.grid(eval1, eval2)
        result1 <- sm.density(cbind(x1, x2), h = band1, n = ne,
         display = "none", eval.points = eval12)
        ex1 <- eval1
        ex2 <- eval2
        ey <- matrix(result1$estimate, ncol = ne)
        eyarray[,  , kk] <- ey * prior1[kk]
    }
#   (4)
    ecl <- matrix(rep(0, length = ne * ne), ncol = ne)
#   (5)
    for(ii in 1:ne) {
        for(jj in 1:ne) {
            cl1 <- which(max(eyarray[ii, jj,  ]) ==
             eyarray[ii, jj,  ])
            if(length(cl1) > 1)
                ecl[ii, jj] <- "N"
            else ecl[ii, jj] <- clu[cl1]
        }
    }
#   (6)
    return(ex1, ex2, ecl)
}
```

(F) Classification by the use of logistic regression

Part (4) of the S-Plus object "glmdis1()" is replaced with the following for use as an R object.

```
#   (4)
    for(kk in 1:ncl) {
        y <- rep(0, length = nd)
        y[clall == clu[kk]] <- rep(1, length = sum(clall ==
         clu[kk]))
        data1 <- data.frame(x1 = xxall[, 1],
         x2 = xxall[, 2], y1 = y)
        glm.out <- glm(y1 ~ x1 + x2, data = data1,
         family = "binomial")
```

```
            data2 <- data.frame(x1 = ex1, x2 = ex2)
            margin.fit <- predict(glm.out, newdata = data2,
             type = "terms")
            eylink <- outer(margin.fit[, 1], margin.fit[, 2], "+")
            eylink <- eylink + attr(margin.fit, "constant")
            ey <- exp(eylink)/(1 + exp(eylink))
            eymat <- matrix(ey, nrow = nex1)
            eyarray[,  , kk] <- eymat
      }
```

(G) Classification generated by neural networks
    In the S-Plus object "netd1()," library(MASS) is replaced by library(nnet),
and softmax = T is omitted from the arguments of nnet() for use as an R object.

(H) Optimization of the complexity parameter by the use of the 10-fold cross-validation to create a tree-based model for classification
    In the S-Plus object "trdcv3()," library(tree) is added at the top, and
assign("data1", data.frame(x1 = newdatx[, 1], x2 = newdatx
[, 2], y1 = newdaty), frame = 1) is replaced by data1 <- data.frame
(x1 = newdatx[, 1], x2 = newdatx[, 2], y1 = newdaty) for use as an R
object.

(I) k-nearest-neighbor classifier accompanied by bagging
    The S-Plus object "knnbag1()" is used as an R object as it is.

(J) Transformation of feature vectors by the use of canonical variates
    The S-Plus object "cano1()" is used as an R object as it is.

B.8   TRANSFORMATION OF S-PLUS OBJECTS IN APPENDIX A

(A) Creation of bases of B-spline
    The S-Plus object "mybb1()" is used as an R object as it is.

(B) Fitting of B-splines by the least squares
    The S-Plus object "mybsp1()" is used as an R object as it is.

(C) Interpolation by natural spline
    The S-Plus object "myni1()" is used as an R object as it is.

(D) Creation of natural-spline bases
    The S-Plus object "mynb1()" is used as an R object as it is.

(E) Fitting of a natural spline by the least squares
    The S-Plus object "mynfit1()" is used as an R object as it is.

(F) Creation of a matrix to construct a roughness penalty used in a smoothing spline

Part (6) of the S-Plus object "myphi1()" is altered as follows for use as an R object.

```
#  (6)
   for(jj in 1:nd) {
       bsa <- bs1[, jj]
       for(ii in 1:(nd - 1)) {
           xa <- ex1[(ii * 3 - 2):(ii * 3 + 1)]
           ya <- bsa[(ii * 3 - 2):(ii * 3 + 1)]
           data1 <- data.frame(x1 = xa, y1 = ya)
           lm1 <- lm(y1 ~ poly(x1, degree = 3), data = data1)
           xa2 <- seq(from  = ex1[1], to = ex1[length(ex1)],
            length = 4)
           eya <- predict(lm1, data.frame(x1 = xa2))
           exmat <- cbind(c(1, 1, 1, 1), xa2, xa2^2, xa2^3)
           coefa <- solve(exmat, eya)
           coef2[ii, jj] <- coefa[3]
           coef3[ii, jj] <- coefa[4]
       }
   }
```

(G) Derivation of a smoothing spline using B-spline

Part (4) of the S-Plus "mysmbb1()" is altered as follows for use as an R object.

```
#  (4)
          coef2 <- matrix(rep(0, length = (nd - 1) *
          (nd + 2)), ncol = nd + 2)
          coef3 <- matrix(rep(0, length = (nd - 1) *
          (nd + 2)), ncol = nd + 2)
          for(ii in 1:(nd - 1)) {
              for(jj in ii:(ii + 3)) {
                  bsa <- bs1[, jj]
                  xa <- ex1[(ii * 3 - 2):(ii * 3 + 1)]
                  ya <- bsa[(ii * 3 - 2):(ii * 3 + 1)]
                  data1 <- data.frame(x1 = xa, y1 = ya)
                  lm1 <- lm(y1 ~ poly(x1, degree = 3),
                   data = data1)
                  xa2 <- seq(from  = ex[1], to = ex[length(ex)],
                   length = 4)
                  eya <- predict(lm1, data.frame(x1 = xa2))
                  exmat <- cbind(c(1, 1, 1, 1), xa2, xa2^2,
                   xa2^3)
                  coefa <- solve(exmat, eya)
                  coef2[ii, jj] <- coefa[3]
                  coef3[ii, jj] <- coefa[4]
              }
          }
```

APPENDIX C

FURTHER READINGS

There are currectly a great number of valuable textbooks on nonparametric regession. The author expects that readers will move on to more advanced learning on nonparametric regression, S-Plus, and related areas by referring to the books below, for example.

M.J.D. Powell (1981). *Approximation Theory and Methods*, Cambridge University Press.

L. Breiman, J.H. Friedman, J.H. Olshen, and C.J. Stone (1984). *Classification and Regression Trees*, Chapman & Hall/CRC.

B.W. Silverman (1986). *Density Estimation for Statistics and Data Analysis*, Chapman & Hall/CRC.

Hans-Georg Müller (1988). *Nonparametric Regression Analysis of Longitudinal Data,* Springer-Verlag.

A. Ullah(Editor) (1989). *Semiparametric and Nonparametric Econometrics*, Springer-Verlag.

K. Fukunaga (1990). *Introduction to Statistical Pattern Recognition*, 2nd edition, Academic Press.

W. Härdle (1990). *Smoothing Technique with Implementation in S*, Springer-Verlag.

T.J. Hastie and R.J. Tibshirani (1990). *Generalized Additive Models*, Chapman & Hall/CRC.

G. Wahba (1990). *Spline Models for Observational Data*, Society for Industrial and Applied Mathematics.

W.A. Barnett, J. Powell, and G.E. Tauchen (Editors) (1991). *Nonparametric and Semiparametric Methods in Econometrics and Statistics: Proceedings of the 5th International Symposium in Economic Theory and Econometrics*, Cambridge University Press.

J.M. Chambers and T.J. Hastie (1991). *Statistical Models in S*, Chapman & Hall/CRC.

B.R. Ripley (1991). *Statistical Inference for Spatial Processes*, reprint edition, Cambridge University Press.

L.C. Hamilton (1992). *Regression with Graphics, A Second Course in Applied Statistics*. Duxbury Press.

W. Härdle (1992). *Applied Nonparametric Regression*, reprint edition, Cambridge University Press.

D.W. Scott (1992). *Multivariate Density Estimation*, Wiley-Interscience.

P. Spector (1993). *An Introduction to S and S-Plus*, Duxbury Press.

P.J. Green and B.W. Silverman (1994). *Nonparametric Regression and Generalized Linear Models, A Roughness Penalty Approach*, Chapman & Hall/CRC.

C.M. Bishop (1995). *Neural Networks for Pattern Recognition*, Oxford University Press.

M.P. Wand and M.C. Jones (1995). *Kernel Smoothing*, Chapman & Hall/CRC.

J. Fan and I. Gijbels (1996). *Local Polynomial Modelling and Its Applications*, Chapman & Hall/CRC.

W. Härdle and M.G. Schimek (Editors) (1996). *Statistical Theory and Computational Aspects of Smoothing*, Physica-Verlag.

B.D. Ripley (1996). *Pattern Recognition and Neural Network*, Cambridge University Press.

J.S. Simonoff (1996). *Smoothing Methods in Statistics*, Springer-Verlag.

A. Bowman and A. Azzalini (1997). *Applied Smoothing Techniques for Data Analysis: The Kernel Approach with S-Plus Illustrations*, Oxford University Press.

J.D. Hart (1997). *Nonparametric Smoothing and Lack-of-Fit Tests*, Springer-Verlag.

C.G. Looney (1997). *Pattern Recognition Using Neural Networks*, Oxford University Press.

J.O. Ramsay and B.W. Silverman (1997). *Functional Data Analysis*, Springer-Verlag.

T.P. Ryan (1997). *Modern Regression Methods*, Wiley-Interscience.

J.M. Chambers (1998). *Programming with Data: A Guide to the S Language*, Springer-Verlag.

J. Horowitz (1998). *Semiparametric Methods in Econometrics*, Springer-Verlag.

A.D.R. McQuarrie and C.L. Tsai (1998). *Regression and Time Series Model Selection*, World Scientific Publishing.

S. Efromovich (1999). *Nonparametric Curve Estimation: Methods, Theory, and Applications*, Springer-Verlag.

R.L. Eubank (1999). *Spline Smoothing and Nonparametric Regression*, 2nd edition, Marcel Dekker/CRC.

C. Loader (1999). *Local Regression and Likelihood*, Springer-Verlag.

A. Pagan and A. Ullah (1999). *Nonparametric Econometrics*, Cambridge University Press.

M.L. Stein (1999). *Interpolation of Spatial Data - Some Theory for Kriging*, Springer-Verlag.

R.O. Duda, P.E. Hart, and D.G. Stork (2000). *Pattern Classification*, 2nd edition, John Wiley & Sons.

J. Fox (2000). *Nonparametric Simple Tegression: Smoothing Scatterplots*, Sage Publications.

J. Fox (2000). *Multiple and Generalized Nonparametric Regression*, Sage Publications.

J. Gill (2000). *Generalized Linear Models: A Unified Approach*, Sage Publications.

R.H. Myers (2000). *Classical and Modern Regression with Applications*, 2nd edition, Duxbury Press.

W.N. Venables and B.D. Ripley (2000). *S Programming*, Springer-Verlag.

R. Christensen (2001). *Advanced Linear Modeling*, 2nd edition, Springer-Verlag.

C. De Boor (2001). *A Practical Guide to Splines*, revised edition, Springer-Verlag.

A.J. Dobson (2001). *An Introduction to Generalised Linear Models*, 2nd edition, Chapman & Hall/CRC.

B. Everitt (2001). *Handbook of Statistical Analyses Using S-Plus*, 2nd edition, Chapman & Hall/CRC.

L. Fahrmeir, and G. Tutz (2001). *Multivariate Statistical Modelling Based on Generalized Linear Models*, 2nd edition, Springer-Verlag.

T. Hastie, R. Tibshirani, and J. H. Friedman (2001). *The Elements of Statistical Learning: Data Mining, Inference, and Prediction*, Springer-Verlag.

L.C. Marsh and D.R. Cormier (2001). *Spline Regression Models (Quantitative Applications in the Social Sciences)*, Sage Publications.

R.H. Myers, D.C. Montgomery, and G.G. Vining (2001). *Generalized Linear Models: With Applications in Engineering and the Sciences*, John Wiley & Sons.

K.P. Burnham and D.R. Anderson (2002). *Model Selection and Multi-model Inference*, 2nd edition, Springer-Verlag.

J. Fox and G. Monette (2002). *An R and S-Plus Companion to Applied Regression*, Sage Publications.

C. Gu (2002). *Smoothing Spline ANOVA Models*, Springer-Verlag.

A. Krause and M. Olson (2002). *The Basics of S-Plus*, 3rd edition, Springer-Verlag.

J.O. Ramsay and B.W. Silverman (2002). *Applied Functional Data Analysis: Methods and Case Studies*, Springer-Verlag.

W.N. Venables and B.D. Ripley (2002). *Modern Applied Statistics with S*, 4th edition, Springer-Verlag.

J.P. Hoffmann (2003). *Generalized Linear Models: An Applied Approach*, Allyn & Bacon.

D. Ruppert, M.P. Wand, and R.J. Carroll (2003). *Semiparametric Regression*, Cambridge University Press.

A. Yatchew (2003). *Semiparametic Regression for the Applied Econometrician*, Cambridge University Press.

W. Härdle, M. Müller, S. Sperlich, and A. Werwatz (2004). *Nonparametric and Semiparametric Models*, Springer-Verlag.

B.D. Ripley (2004). *Spatial Statistics*, Wiley-Interscience.

INDEX

Introduction to Nonparametric Regression, Kunio Takezawa.
Copyright © 2006 John Wiley & Sons, Inc.

WILEY SERIES IN PROBABILITY AND STATISTICS

ESTABLISHED BY WALTER A. SHEWHART AND SAMUEL S. WILKS

Editors: *David J. Balding, Noel A. C. Cressie, Nicholas I. Fisher,*
Iain M. Johnstone, J. B. Kadane, Geert Molenberghs. Louise M. Ryan,
David W. Scott, Adrian F. M. Smith, Jozef L. Teugels
Editors Emeriti: *Vic Barnett, J. Stuart Hunter, David G. Kendall*

The *Wiley Series in Probability and Statistics* is well established and authoritative. It covers many topics of current research interest in both pure and applied statistics and probability theory. Written by leading statisticians and institutions, the titles span both state-of-the-art developments in the field and classical methods.

Reflecting the wide range of current research in statistics, the series encompasses applied, methodological and theoretical statistics, ranging from applications and new techniques made possible by advances in computerized practice to rigorous treatment of theoretical approaches.

This series provides essential and invaluable reading for all statisticians, whether in academia, industry, government, or research.

† ABRAHAM and LEDOLTER · Statistical Methods for Forecasting
AGRESTI · Analysis of Ordinal Categorical Data
AGRESTI · An Introduction to Categorical Data Analysis
AGRESTI · Categorical Data Analysis, *Second Edition*
ALTMAN, GILL, and McDONALD · Numerical Issues in Statistical Computing for the Social Scientist
AMARATUNGA and CABRERA · Exploration and Analysis of DNA Microarray and Protein Array Data
ANDĚL · Mathematics of Chance
ANDERSON · An Introduction to Multivariate Statistical Analysis, *Third Edition*
* ANDERSON · The Statistical Analysis of Time Series
ANDERSON, AUQUIER, HAUCK, OAKES, VANDAELE, and WEISBERG · Statistical Methods for Comparative Studies
ANDERSON and LOYNES · The Teaching of Practical Statistics
ARMITAGE and DAVID (editors) · Advances in Biometry
ARNOLD, BALAKRISHNAN, and NAGARAJA · Records
* ARTHANARI and DODGE · Mathematical Programming in Statistics
* BAILEY · The Elements of Stochastic Processes with Applications to the Natural Sciences
BALAKRISHNAN and KOUTRAS · Runs and Scans with Applications
BARNETT · Comparative Statistical Inference, *Third Edition*
BARNETT and LEWIS · Outliers in Statistical Data, *Third Edition*
BARTOSZYNSKI and NIEWIADOMSKA-BUGAJ · Probability and Statistical Inference
BASILEVSKY · Statistical Factor Analysis and Related Methods: Theory and Applications
BASU and RIGDON · Statistical Methods for the Reliability of Repairable Systems
BATES and WATTS · Nonlinear Regression Analysis and Its Applications
BECHHOFER, SANTNER, and GOLDSMAN · Design and Analysis of Experiments for Statistical Selection, Screening, and Multiple Comparisons
BELSLEY · Conditioning Diagnostics: Collinearity and Weak Data in Regression

*Now available in a lower priced paperback edition in the Wiley Classics Library.
†Now available in a lower priced paperback edition in the Wiley–Interscience Paperback Series.

† BELSLEY, KUH, and WELSCH · Regression Diagnostics: Identifying Influential Data and Sources of Collinearity

BENDAT and PIERSOL · Random Data: Analysis and Measurement Procedures, *Third Edition*

BERRY, CHALONER, and GEWEKE · Bayesian Analysis in Statistics and Econometrics: Essays in Honor of Arnold Zellner

BERNARDO and SMITH · Bayesian Theory

BHAT and MILLER · Elements of Applied Stochastic Processes, *Third Edition*

BHATTACHARYA and WAYMIRE · Stochastic Processes with Applications

† BIEMER, GROVES, LYBERG, MATHIOWETZ, and SUDMAN · Measurement Errors in Surveys

BILLINGSLEY · Convergence of Probability Measures, *Second Edition*

BILLINGSLEY · Probability and Measure, *Third Edition*

BIRKES and DODGE · Alternative Methods of Regression

BLISCHKE AND MURTHY (editors) · Case Studies in Reliability and Maintenance

BLISCHKE AND MURTHY · Reliability: Modeling, Prediction, and Optimization

BLOOMFIELD · Fourier Analysis of Time Series: An Introduction, *Second Edition*

BOLLEN · Structural Equations with Latent Variables

BOLLEN and CURRAN · Latent Curve Models: A Structural Equation Perspective

BOROVKOV · Ergodicity and Stability of Stochastic Processes

BOULEAU · Numerical Methods for Stochastic Processes

BOX · Bayesian Inference in Statistical Analysis

BOX · R. A. Fisher, the Life of a Scientist

BOX and DRAPER · Empirical Model-Building and Response Surfaces

* BOX and DRAPER · Evolutionary Operation: A Statistical Method for Process Improvement

BOX, HUNTER, and HUNTER · Statistics for Experimenters: Design, Innovation, and Discovery, *Second Editon*

BOX and LUCEÑO · Statistical Control by Monitoring and Feedback Adjustment

BRANDIMARTE · Numerical Methods in Finance: A MATLAB-Based Introduction

BROWN and HOLLANDER · Statistics: A Biomedical Introduction

BRUNNER, DOMHOF, and LANGER · Nonparametric Analysis of Longitudinal Data in Factorial Experiments

BUCKLEW · Large Deviation Techniques in Decision, Simulation, and Estimation

CAIROLI and DALANG · Sequential Stochastic Optimization

CASTILLO, HADI, BALAKRISHNAN, and SARABIA · Extreme Value and Related Models with Applications in Engineering and Science

CHAN · Time Series: Applications to Finance

CHARALAMBIDES · Combinatorial Methods in Discrete Distributions

CHATTERJEE and HADI · Sensitivity Analysis in Linear Regression

CHATTERJEE and PRICE · Regression Analysis by Example, *Third Edition*

CHERNICK · Bootstrap Methods: A Practitioner's Guide

CHERNICK and FRIIS · Introductory Biostatistics for the Health Sciences

CHILÈS and DELFINER · Geostatistics: Modeling Spatial Uncertainty

CHOW and LIU · Design and Analysis of Clinical Trials: Concepts and Methodologies, *Second Edition*

CLARKE and DISNEY · Probability and Random Processes: A First Course with Applications, *Second Edition*

* COCHRAN and COX · Experimental Designs, *Second Edition*

CONGDON · Applied Bayesian Modelling

CONGDON · Bayesian Statistical Modelling

CONOVER · Practical Nonparametric Statistics, *Third Edition*

COOK · Regression Graphics

*Now available in a lower priced paperback edition in the Wiley Classics Library.
†Now available in a lower priced paperback edition in the Wiley–Interscience Paperback Series.

---

*Now available in a lower priced paperback edition in the Wiley Classics Library.

†Now available in a lower priced paperback edition in the Wiley–Interscience Paperback Series.

*Now available in a lower priced paperback edition in the Wiley Classics Library.
†Now available in a lower priced paperback edition in the Wiley–Interscience Paperback Series.

---

*Now available in a lower priced paperback edition in the Wiley Classics Library.

†Now available in a lower priced paperback edition in the Wiley–Interscience Paperback Series.

---

*Now available in a lower priced paperback edition in the Wiley Classics Library.
†Now available in a lower priced paperback edition in the Wiley–Interscience Paperback Series.

---

*Now available in a lower priced paperback edition in the Wiley Classics Library.

†Now available in a lower priced paperback edition in the Wiley–Interscience Paperback Series.

SHAFER and VOVK · Probability and Finance: It's Only a Game!

SILVAPULLE and SEN · Constrained Statistical Inference: Inequality, Order, and Shape Restrictions

SMALL and McLEISH · Hilbert Space Methods in Probability and Statistical Inference

SRIVASTAVA · Methods of Multivariate Statistics

STAPLETON · Linear Statistical Models

STAUDTE and SHEATHER · Robust Estimation and Testing

STOYAN, KENDALL, and MECKE · Stochastic Geometry and Its Applications, *Second Edition*

STOYAN and STOYAN · Fractals, Random Shapes and Point Fields: Methods of Geometrical Statistics

STYAN · The Collected Papers of T. W. Anderson: 1943–1985

SUTTON, ABRAMS, JONES, SHELDON, and SONG · Methods for Meta-Analysis in Medical Research

TAKEZAWA · Introduction to Nonparametric Regression

TANAKA · Time Series Analysis: Nonstationary and Noninvertible Distribution Theory

THOMPSON · Empirical Model Building

THOMPSON · Sampling, *Second Edition*

THOMPSON · Simulation: A Modeler's Approach

THOMPSON and SEBER · Adaptive Sampling

THOMPSON, WILLIAMS, and FINDLAY · Models for Investors in Real World Markets

TIAO, BISGAARD, HILL, PEÑA, and STIGLER (editors) · Box on Quality and Discovery: with Design, Control, and Robustness

TIERNEY · LISP-STAT: An Object-Oriented Environment for Statistical Computing and Dynamic Graphics

TSAY · Analysis of Financial Time Series, *Second Edition*

UPTON and FINGLETON · Spatial Data Analysis by Example, Volume II: Categorical and Directional Data

VAN BELLE · Statistical Rules of Thumb

VAN BELLE, FISHER, HEAGERTY, and LUMLEY · Biostatistics: A Methodology for the Health Sciences, *Second Edition*

VESTRUP · The Theory of Measures and Integration

VIDAKOVIC · Statistical Modeling by Wavelets

VINOD and REAGLE · Preparing for the Worst: Incorporating Downside Risk in Stock Market Investments

WALLER and GOTWAY · Applied Spatial Statistics for Public Health Data

WEERAHANDI · Generalized Inference in Repeated Measures: Exact Methods in MANOVA and Mixed Models

WEISBERG · Applied Linear Regression, *Third Edition*

WELSH · Aspects of Statistical Inference

WESTFALL and YOUNG · Resampling-Based Multiple Testing: Examples and Methods for *p*-Value Adjustment

WHITTAKER · Graphical Models in Applied Multivariate Statistics

WINKER · Optimization Heuristics in Economics: Applications of Threshold Accepting

WONNACOTT and WONNACOTT · Econometrics, *Second Edition*

WOODING · Planning Pharmaceutical Clinical Trials: Basic Statistical Principles

WOODWORTH · Biostatistics: A Bayesian Introduction

WOOLSON and CLARKE · Statistical Methods for the Analysis of Biomedical Data, *Second Edition*

WU and HAMADA · Experiments: Planning, Analysis, and Parameter Design Optimization

YANG · The Construction Theory of Denumerable Markov Processes

* ZELLNER · An Introduction to Bayesian Inference in Econometrics

ZHOU, OBUCHOWSKI, and McCLISH · Statistical Methods in Diagnostic Medicine

---

*Now available in a lower priced paperback edition in the Wiley Classics Library.

†Now available in a lower priced paperback edition in the Wiley–Interscience Paperback Series.